HANDBOOK OF
PAINT RAW MATERIALS

HANDBOOK OF PAINT RAW MATERIALS

Second Edition

by

Ernest W. Flick

np NOYES PUBLICATIONS
Park Ridge, New Jersey, U.S.A.

Library of Congress Catalog Card Number: 89-30275
ISBN: 0-8155-1184-1
Printed in the United States

Published in the United States of America by
Noyes Publications
Mill Road, Park Ridge, New Jersey 07656

10 9 8 7 6 5 4 3 2 1

Library of Congress Cataloging-in-Publication Data

Flick, Ernest W.
Handbook of paint raw materials / by Ernest W. Flick. -- 2nd ed.
p. cm.
Includes index.
ISBN 0-8155-1184-1 :
1. Paint materials--Handbooks, manuals, etc. I. Title.
TP935.F617 1989
667'.6--dc19 89-30275
CIP

Preface

This handbook contains descriptions of nearly 4,000 raw materials which are currently available to the paint industry. It will be of value to technical and managerial personnel in paint manufacturing companies and companies which supply raw materials or services to these companies. This book will be useful to both those with extensive experience as well as those new in the field.

The data included represent selections from manufacturers' descriptions made at no cost to, nor influence from, the makers or distributors of the materials. Only the most recent information has been included. Lead-based chemicals and color pigments are excluded. It is believed that all of the products listed here are currently available, which will be of utmost interest to readers concerned with product discontinuances.

The raw materials have been divided into the following 13 sections:

1. Antifoams, Defoamers, Dispersants, Surfactants (258)
2. Driers and Antiskinning Agents (194)
3. Extenders, Fillers, Pigments (420)
4. Flame/Fire Retardants (33)
5. Flatting Agents (28)
6. Latex Emulsions (102)
7. Oils (92)
8. Preservatives, Bactericides, Fungicides (80)
9. Resins (2,064)
10. Rheological/Viscosity Control Agents (125)
11. Silicone Additives (51)
12. Titanium Dioxides (33)
13. Miscellaneous (469)

The numbers in parentheses indicate the number of products to be found in each section. When searching for a particular type of product, the reader is advised to consult each section which might be pertinent.

In addition to the sections listed above, a list of Suppliers' Addresses and a Trade Name Index are included.

The table of contents is organized in such a way as to serve as a subject index to the book. Companies are presented alphabetically in each section and appropriate products are listed for each company. The book lists the following product information, as available, in the manufacturer's own words:

(1) Company name and product category.

(2) Trade name and product numbers.

(3) Product description: a description of the product, its properties, and possible applications, as presented by the supplier.

My fullest appreciation is expressed to the companies and organizations who supplied the data included in this book.

Newburyport, Massachusetts
April 1989

Ernest W. Flick

NOTICE

To the best of our knowledge the information in this publication is accurate; however the Publisher does not assume any responsibility or liability for the accuracy or completeness of, or consequences arising from, such information. This industrial guide does not purport to contain detailed user instructions, and by its range and scope could not possibly do so. Mention of trade names or commercial products does not constitute endorsement or recommendation for use by the Author or Publisher.

In certain cases, paint raw materials could be toxic, and therefore due caution should always be exercised in the use of any potentially hazardous materials. Final determination of the suitability of any information or product for use contemplated by any user, and the manner of that use, is the sole responsibility of the user. We strongly recommend that users seek and adhere to a manufacturer's or supplier's current instructions for handling each material they use.

The Author and Publisher have used their best efforts to include only the most recent data available. The reader is cautioned to consult the supplier in case of questions regarding current availability.

Contents

SECTION 1
ANTIFOAMS, DEFOAMERS, DISPERSANTS, SURFACTANTS

SECTION 2
DRIERS AND ANTISKINNING AGENTS

SECTION 3
EXTENDERS, FILLERS, PIGMENTS

SECTION 4
FLAME/FIRE RETARDANTS

SECTION 5
FLATTING AGENTS

SECTION 6
LATEX EMULSIONS

SECTION 7
OILS

SECTION 8
PRESERVATIVES, BACTERICIDES, FUNGICIDES

SECTION 9
Resins

SECTION 10
RHEOLOGICAL/VISCOSITY CONTROL AGENTS

SECTION 11
Silicone Additives

SECTION 12
TITANIUM DIOXIDES

SECTION 13
MISCELLANEOUS

Section 1
Antifoams, Defoamers, Dispersants, Surfactants

ALLIED COLLOIDS: DISPEX Dispersing Agents:

DISPEX dispersants are widely and increasingly used in emulsion paint formulations in order to obtain higher production rates and long term storage stability.

DISPEX A40 and DISPEX N40 are respectively the ammonium and sodium salts of a polymeric carboxylic acid and can be regarded as the standard dispersants for matt emulsion paints, but they can also be used in certain semi-gloss and gloss formations.

DISPEX GA40 and DISPEX G40 are the ammonium and sodium salts respectively of a carboxylated copolymer and have been designed to give optimum efficiency in semi-gloss and gloss emulsion paints.

Typical Properties:

DISPEX N40:
Nature: Sodium Salt
Appearance: Pale Yellow
Active Content: 40+-1%
pH: 7.0 to 7.5
Specific Gravity: 1.30
Odour: Slight

DISPEX A40:
Nature: Ammonium Salt
Appearance: Pale Yellow
Active Content: 40+-1%
pH: 7.5 to 8.5
Specific Gravity: 1.16
Odour: Ammoniacal

DISPEX G40:
Nature: Sodium Salt
Appearance: Water White
Active Content: 40+-1%
pH: 7.0 to 8.0
Specific Gravity: 1.20
Odour: Slight

DISPEX GA40:
Nature: Ammonium Salt
Appearance: Water White
Active Content: 40+-1%
pH: 7.0 to 8.0
Specific Gravity: 1.13
Odour: Ammoniacal

COLLOIDS, INC.: Additives for Paints and Adhesives:

<u>Product:</u>

Defoamers:

COLLOID 60:
Outstanding defoamer for water-based systems employing resin latex binders. Quick foam knock-down. Excellent longevity. Effective in low concentrations.
pH: 5.8
Viscosity (cps.): 100
Specific Gravity (25C.): .90
Applications: Paints/Adhesives

COLLOID 581B:
Water-dispersible defoamer for use in paints and adhesives. Non-ionic base. Compatible and active in most systems.
pH: 5.5
Viscosity (cps.): 630
Specific Gravity (25C.): .80
Applications: Paints/Adhesives

COLLOID 640:
Hydrophobic silica-type defoamer. Stable, non-ionic. Good shelf life. Will not float to surface. Gives rapid bubble-break in brushing or roll-outs without causing fish-eye or orange peel problems.
pH: 8.2
Viscosity (cps.): 820
Specific Gravity (25C.): .90
Applications: Paints/Adhesives

COLLOID 643:
Light tan, opaque liquid defoamer especially formulated to give outstanding foam control when used in the manufacture of paints. Good stability. Quick bubble-break. High efficiency for maximum economy.
Viscosity (cps.): 780
Specific Gravity (25C.): .85
Applications: Paints

COLLOID 645:
Non-separating opaque liquid defoamer. Provides excellent longevity in both flat and semi-gloss paints.
Viscosity (cps.): 700
Specific Gravity (25C.): 0.88
Applications: Paints

COLLOIDS, INC.: Additives for Paints and Adhesives(Continued):

Product:

COLLOID 646:
Outstanding foam control agent for use in synthetic latices. Formulated for use in various aqueous coatings. Good product stability. Recommended for use in polyvinyl acetate, vinyl acrylic, and acrylic emulsion formulations.
pH: 8.4
Viscosity (cps.): 100
Specific Gravity (25C.): .91
Applications: Paints

COLLOID 647:
Hydrophobic, silica-type defoamer formulated for use in aqueous paints and water-borne industrial coatings. Rapid bubble break without causing any fish-eyes or orange peel effects. Non-ionic defoamer recommended for polyvinyl acetate, vinyl acrylic and acrylic emulsions.
Viscosity (cps.): 500
Specific Gravity (25C.): .90
Applications: Paints

COLLOID 675:
Offers superior foam control in latex paints and adhesive systems. Outstanding longevity when used in emulsion systems. Excellent foam knock-down.
pH: 6.5
Viscosity (cps.): 250
Specific Gravity (25C.): .88
Applications: Paints/Adhesives

COLLOID 677:
Superior defoamer for use in latex-based, neoprene and natural latex-based adhesives. No pre-mixing required. Effective even after long periods of aging.
pH: 6.0
Viscosity (cps.): 130
Specific Gravity (25C.): .90
Applications: Paints/Adhesives

COLLOID 679:
Defoamer recommended for latex-based paints, general latex compounding. Low viscosity liquid, permitting application with spray nozzles where desired. Excellent for color development. Does not cause fish-eyes or other interference in films. Free of silicone. Good shelf stability.
pH: 6.4
Viscosity (cps.): 300
Specific Gravity (25C.): .87
Applications: Paints

COLLOIDS, INC.: Additives for Paints and Adhesives (Continued):

Product:

COLLOID 680:
Outstanding defoamer for water-based coating systems employing resin latex binders and adhesives. Good leveling. Excellent longevity.
pH: 6.2
Viscosity (cps.): 120
Specific Gravity (25C.): .90
Applications: Paints/Adhesives

COLLOID 681F:
Highly efficient liquid defoamer for latex-based paints and adhesives. Low viscosity, permitting application with spray nozzles. Good drum stability. Effective performance in acid or alkaline media.
pH: 6.0
Viscosity (cps.): 340
Specific Gravity (25C.): .87
Applications: Paints/Adhesives

COLLOID 691:
Non-silica defoamer. Stable, uniform product. Exceptionally long-lasting. Highly efficient.
pH: 7.5
Viscosity (cps.): 880
Specific Gravity (25C.): .85
Applications: Paints

COLLOID 694:
Outstanding defoamer for latex paints and polyvinyl-acetate copolymer adhesives. Stable, uniform product. Produces fish-eye-free films.
pH: 7.3
Viscosity (cps.): 500
Specific Gravity (25C.): .85
Applications: Paints/Adhesives

COLLOID 711:
Clear, low viscosity, liquid defoamer for dextrine adhesives. Allows fish-eye-free films. No pre-dispersion required.
pH: 7.2
Viscosity (cps.): 15
Specific Gravity (25C.): .86
Applications: Adhesives

COLLOIDS, INC.: Additives for Paints and Adhesives (Continued):

Product:

COLLOID 770:
Effective, clear liquid defoamer recommended for polyvinyl alcohol, casein adhesives. Free from separation.
pH: 5.6
Viscosity (cps.): 60
Specific Gravity (25C.): .93
Applications: Adhesives

COLLOID 796:
Non-ionic defoamer for use in paints and adhesives. Homogeneous--has excellent product stability. Water-dispersible.
pH: 7.5
Viscosity (cps.): 880
Specific Gravity (25C.): .85
Applications: Paints/Adhesives

COLLOID 797:
An excellent general purpose defoamer for use in aqueous systems. Very effective at low concentrations. Has excellent product stability while readily dispersible in water systems.
pH: 6.4
Viscosity (cps.): 200
Specific Gravity (25C.): .87
Applications: Paints/Adhesives

COLLOID 999:
Foam control agent for water-based inks and adhesives. Exceptional compatibility provides smooth films and efficient use of colored pigments.
pH: 8.6
Viscosity (cps.): 300
Specific Gravity (25C.): .98
Applications: Adhesives

COLLOID 1010:
10% active, highly efficient silicone-based foam control agent designed for use in both latex emulsion and adhesive systems.
pH: 8.0
Viscosity (cps.): 220
Specific Gravity (25C.): 1.00
Applications: Paints/Adhesives

COLLOIDS, INC.: Additives for Paints and Adhesives(Continued):

Dry Defoamers:

Product:

COLLOID 513DD:
Highly efficient dry defoamer for polyvinyl acetate and polyvinyl alcohol adhesives. Ideally suited for dry mixing. water-dispersible and active in aqueous systems.
Bulk Density (Lbs./Cu.Ft.): 25
Applications: Adhesives

COLLOID 770DD:
Dry powder, ideally suited for blending with polyvinyl alcohol and starch adhesives in order to impart foam control during subsequent solution.
Bulk Density (Lbs./Cu.Ft.): 23
Applications: Adhesives

COLLOID 775DD:
Outstanding dry defoamer that effectively prevents air entrainment, froth and foam build-up. Ideally suited for blending with all components of dry-mix formulations.
Bulk Density (Lbs./Cu.Ft.): 38
Applications: Adhesives

COLLOID 778DD:
Highly efficient dry defoamer for use in various adhesive systems and casein glue solutions. Imparts excellent foam control in aqueous systems, including boiler water treatment systems.
Bulk Density (Lbs./Cu.Ft.): 35
Applications: Adhesives

Dispersants:

Product:

COLLOID 111:
Anionic dispersant for titanium dioxide and extender pigments used in latex paints. Excellent compatibility; permits efficient grinds with propylene glycol; provides maximum optical properties, maximum hiding and leveling.
pH: 10.0
Viscosity (cps.): 36.0
Specific Gravity (25C.): 1.10
Wt./Gal.: 9.2
Solids %: 25
Applications: Paints

COLLOIDS, INC.: Additives for Paints and Adhesives(Continued):

Dispersants(Continued):

Product:

COLLOID 225:
Anionic dispersant designed for aqueous coatings for both interior and exterior applications. Proven to be effective in formulations containing reactive pigments such as zinc oxide. Exhibits compatibility with glycols and with alkyd resins.
pH: 9.0
Viscosity (cps.): 250
Specific Gravity (25C.): 1.21
Wt./Gal.: 10.1
Solids %: 30
Applications: Paints

COLLOID 226/35:
Highly versatile, anionic dispersant offering exceptional viscosity control during storage. Especially suited for interior and exterior paints with or without reactive pigments. Imparts good color development and enhances gloss properties in semi-gloss and gloss paints.
pH: 7.5
Viscosity (cps.): 60.0
Specific Gravity (25C.): 1.20
Wt./Gal.: 10.0
Solids %: 35.
Applications: Paints

COLLOID 229:
Anionic dispersant for all pigments in interior and exterior wall paints. Good color development and compatibility; most economic in usage cost.
pH: 8.0
Viscosity (cps.): 1000
Specific Gravity (25C.): 1.28
Wt./Gal.: 10.7
Solids %: 40
Applications: Paints

COLLOID 261:
COLLOID 261 is a non-foaming anionic dispersant for enamel and semi-gloss paints. This unique acrylic copolymer provides the following features:
* High Gloss in a wide variety of formulations.
* High Solids for efficiency.
* Non-Foaming reduces defoamer additions.
* Pigment Dispersant compatible with all rheological additives

Non-Volatile Matter: 50.0
pH: 7.0
Viscosity: 100 cps.
Specific Gravity (25C.): 1.15

COLLOIDS, INC.: Additives for Paints and Adhesives(Continued):

Dispersants(Continued):

COLLOID 286:
COLLOID 286 is a non-foaming, general purpose anionic dispersant. This unique acrylic copolymer is recommended for the following uses:

* Pigment Dispersant with excellent performance at low use levels.
* Gloss Promoter for enamel and semi-gloss formulation.
* Gel Protection when using reactive pigments.
* Formulation Latitude by replacing two or more dispersants.

Non-Volatile Matter: 50.0
pH: 7.0
Viscosity: 100 cps.
Specific Gravity (25C): 1.17

Wetting Agent:

PENTEX 99:
Liquid form of dioctyl sodium sulfosuccinate. Chalk penetrant for water-based paints. Reduces surface tension at low concentrations.
pH: 6.5
Viscosity (cps.): 350
Specific Gravity (25C.): 1.10
Wt./Gal.: 9.2
Solids %: 75.
Applications: Paints

Thickener:

COLLOID 1560:
Emulsion form of anionic acrylic copolymer used to stabilize adhesives, textured paints and aqueous dispersions or emulsions. Use at pH 7-10.
pH: 2.5
Viscosity (cps.): 10
Specific Gravity (25C.): 1.03
Wt./Gal.: 8.6
Solids %: 28.
Applications: Paints Text

DANIEL PRODUCTS CO.: DAPRO Foam Suppressors:

Silicone-free and silicone modified additives to provide rapid air release and to prevent foam formation in aqueous coatings.

DAPRO Foam Suppressor DF1181

For Foam Prevention in Aqueous Coatings, Inks and Adhesives

DAPRO DF1181 has proven effective in many, but not all, clear and pigmented formulas based on a variety of both water soluble and latex vehicles. It resists emulsification and adsorption on pigment particles, therefore, it remains effective for long periods of time.

Composition: DAPRO DF 1181 is a synergistic combination of olefinic solids, silicone, and spreading agents in a parafinic liquid carrier.

DANIEL PRODUCTS CO.: DAPRO Silicone-Free Foam Suppressors

For Rapid Air Release from Aqueous Coatings, Inks, and Adhesives

Increase the rate of bubble-breaking, both in the container and in thin films.

Composition: DAPRO Foam Suppressors are dispersions of hydrophobic organic solids in parafinic liquid carriers.

Specifications:

Dispersions of Metallic Salt of Fatty Acid

DF 880:
Spreading Agent Level: Low
Wt./Gal.: 7.2 Lbs.
FDA Cleared: Yes
Outstanding Features: 1,2

DF 881S:
Spreading Agent Level: High
Wt./Gal.: 7.1 Lbs.
FDA Cleared: Yes
Outstanding Features: 4

DANIEL PRODUCTS CO.: DAPRO Silicone-Free Foam Suppressors (Continued):

Dispersions Of Olefinic Solids

DF 900:
Spreading Agent Level: Low
Wt./Gal. 7.0 Lbs.
FDA Cleared: No
Outstanding Features: 1,2,3

DF 911:
Spreading Agent Level: Med
Wt./Gal.: 7.4 Lbs.
FDA Cleared: No
Outstanding Features: 3,5

DF 944:
Spreading Agent Level: Med
Wt./Gal.: 7.4 Lbs.
FDA Cleared: Yes
Outstanding Features: 3,5

DF 975:
Spreading Agent Level: None
Wt./Gal.: 7.0 Lbs.
FDA Cleared: Yes
Outstanding Features: 1,2,3

1. Resistance to bubbling and pinholing.
2. Long-term stability.
3. Remains effective over wide pH range.
4. Ease of incorporation.
5. Resistance to crawling and fish-eyes.

DAPRO Defoamer NA 1621

For Use in Non-Aqueous Clear Urethane Wood Finishes

DAPRO Defoamer NA 1621 provides rapid bubble breaking in roller and brush applied solvent thinned urethane wood finishes. The general improvement in flow that occurs when using DAPRO NA 1621 can often reduce or eliminate cratering and crawling in finishes applied over hard-to-wet substrates. In low gloss coatings, DAPRO NA 1621 will often reduce silking and provide more uniform sheen.

DANIEL PRODUCTS CO.: DISPERSE-AYD:

Multi-Functional Pigment Dispersing Agents

DISPERSE-AYDs are blends of several surface active materials working synergistically to wet and deflocculate pigments and to stabilize the finished dispersions.

1. Rapid pigment wetting
2. Good flow at high pigment loading
3. Increased mill output
4. Maximum tinting strength development
5. Full chroma development

DISPERSE-AYDs provide:

6. Minimum floating, flooding and rub-up
7. Compatibility in a broad range of coatings
8. Long term viscosity stability
9. Optimum initial gloss with stability on aging
10. Freedom from hard settling

At recommended usage level, DISPERSE-AYD, will not adversely affect gloss, drying time, film hardness, water resistance or any other film property.

Compatibility:
DISPERSE-AYDs 1,6 and 8 are fully compatible with many solvent thinned vehicles.

DISPERSE-AYD 1 for use with most Organic and Inorganic Pigments
A proprietary reaction product of high molecular weight surface active materials with a long oil alkyd.

DISPERSE-AYD 6 and 8 for use with Carbon Black Pigments

Specifications:

DISPERSE-AYD 1:
Solids: 75.0%
Solvent: Mineral Spirits (<8% Aromatic)
Flash Point: 104F/40.0C
Viscosity (Gardner-Holdt): Q to T

DISPERSE-AYD 6:
Solids: 82.0%
Solvent: Xylol
Flash Point: 81F/27.2C
Viscosity (Gardner-Holdt): Q to S

DISPERSE-AYD 8:
Solids: 65.0%
Solvent: Mineral Spirits (<8% Aromatic)
Flash Point: 104F/40.0C
Viscosity (Gardner-Holdt): C to F

DANIEL PRODUCTS CO.: DISPERSE-AYD(Continued):

DISPERSE-AYD XL-1/80
For "Universal" Tinting Colors

DISPERSE-AYD XL-1/80 is a dispersing vehicle designed for use in manufacturing glycol-free colorants for in-plant and machine tinting.

A blend of surface active materials with the right lipophilic/hydrophilic balance for compatibility with water and solvent thinned systems.

Specifications:
- Solids: 75.0%
- Active Ingredients: 80.0%
- Solvent: Odorless Mineral Spirits
- Flash Point: 128F/53.3C
- Wt./Gal. (Sp.Gr.): 8.0 Lbs. (0.95)
- Color (Gardner): 8-10
- Viscosity (Gardner-Holdt): S-V

DISPERSE-AYD 15
For High Performance Solvent Thinned Coatings

DISPERSE-AYD 15 is a pigment dispersing vehicle that can be used without modification to prepare concentrated dispersions of most pigments for use in a wide range of high performance, solvent thinned coatings.

Modified thermoplastic acrylic in propylene glycol monomethyl ether acetate (PM Acetate).

Specifications:
- Solids: 65.0%
- Solvent: PM Acetate
- Flash Point: 108F/43.2C
- Wt./Gal. (Sp. Gr.): 8.7 Lbs. (1.04)
- Viscosity (Gardner-Holdt): V-Z
- Color (Gardner): 8 Max.

DISPERSE-AYD W-22 and W-28
For Water Thinned Coatings

Blends of anionic and non-ionic surface active materials.

Both DISPERSE-AYD W-22 and W-28 are effective with a wide range of pigments.

Specifications:	DISPERSE-AYD W-22	DISPERSE-AYD W-28
Active Ingredients	35.0%	55.0%
Viscosity (Gardner-Holdt)	R to V	G to J
Color (Gardner)	3 to 7	5 to 9
Wt./Gal. (Sp. Gr.)	8.8 Lbs. (1.06)	8.7 Lbs. (1.04)

DAYGLO COLOR CORP.: NALCO Antifoams:

Product Number:

2300:
Silica-silicone-organic
Flash Point, F (PMCC): 160
Weight % NVM: 73
Application Area: In-grind antifoam for latex paints, should be used with letdown defoamer.

2301:
Silicone-organic
Flash Point, F (PMCC): >200
Weight % NVM: 98
Application Area: Nonaqueous (solvent-based), high solids, and radiation cure coatings.

2302:
Silica-organic
Flash Point, F (PMCC): 160
Weight % NVM: 74
Application Area: Water-reducible coatings

2303:
Silica-organic
Flash Point, F (PMCC): 155
Weight % NVM: 75
Application Area: Water-based and water-reducible trade sales and industrial coatings

2305:
Silica-silicone-organic
Flash Point, F (PMCC): >200
Weight % NVM: 18
Application Area: Persistent antifoam, easy to mix in water-based and water-reducible coatings

2308:
Organic
Flash Point, F (PMCC): >200
Weight % NVM: 100
Application Area: High gloss latex and water-reducible coatings

2309:
Silica-organic
Flash Point, F (PMCC): 160
Weight % NVM: 64
Application Area: Water-reducible and cationic coatings

DAYGLO COLOR CORP.: NALCO Antifoams (Continued):

<u>Product Number:</u>

2311:
Silica-organic
Flash Point, F (PMCC): >200
Weight % NVM: 99
Application Area: Useful in low viscosity water based coatings and adhesives

2313:
Silica-organic
Flash Point, F (PMCC): >200
Weight % NVM: 100
Application Area: Water-reducible alkyd and acrylic based systems

2314:
Silica-silicone-organic
Flash Point, F (PMCC): 150
Weight % NVM: 60
Application Area: Persistent antifoam for water based and water-reducible coatings

2315:
Silica-organic
Flash Point, F (PMCC): >200
Weight % NVM: 100
Application Area: Excellent antifoam for acrylic based latex paints

2321:
Silica-organic
Flash Point, F (PMCC): >200
Weight % NVM: 100
Application Area: Excellent general purpose antifoam for latex paints

2322:
Organic
Flash Point, F (PMCC): >200
Weight % NVM: 100
Application Area: Semi-gloss and gloss water-based trade sales paints

2341:
Silica-organic
Flash Point, F (PMCC): >200
Weight % NVM: 100
Application Area: Water-based coatings and adhesives, broad pH range

DAYGLO COLOR CORP.: NALCO Antifoams/Dispersants:

Product Number:

2343:
Organic
Flash Point, F (PMCC): >200
Weight % NVM: 88
Application Area: Easily dispersed in low viscosity water-based coatings, inks and adhesives

OSS-271:
Silicone-organic
Flash Point, F (PMCC): 170
Weight % NVM: 55
Application Area: Water-reducible acrylic-based pigmented and clear coatings

Dispersants

2335:
Anionic dispersant
Flash Point F (PMCC): >200
Weight % NVM: 35
Application Area: Anionic dispersant for water-based coatings

2395:
Nonionic dispersant
Flash Point F (PMCC): >200
Weight % NVM: 90
Applications Area: Wide application nonionic dispersant for water-based coatings

G-12:
Cationic dispersant
Flash Point F (PMCC): >200
Weight % NVM: 95
Applications Area: Cationic dispersant for water-based coatings

DAYGLO COLOR CORP.: NALCO Coatings Additives:

Colloidal Silicas

Product Number:

1030:
Mar resistance aid
Flash Point F (PMCC): >200
Weight % NVM: 30
Application Area: Water-reducible coatings

1034-A:
Adhesion Promoter
Flash Point F (PMCC): >200
Weight % NVM: 34
Application Area: Latex paints and adhesives

1050:
Flatting agent
Flash Point F (PMCC): >200
Weight % NVM: 50
Application Area: Water-reducible coatings

1060:
Flatting agent
Flash Point F (PMCC): >200
Weight % NVM: 50
Application Area: Water-reducible coatings

1115:
Adhesion promoter
Flash Point F (PMCC): >200
Weight % NVM: 15
Application Area: Latex paints and adhesives

2327:
Mar resistance aid
Flash Point F (PMCC): >200
Weight % NVM: 40
Application Area: Water-reducible coatings

Coagulants and Flocculants

OSS-595:
Cationic coagulant
Weight % NVM: 45
Cationic coagulant for solids removal from latex systems

84SS-237:
Cationic coagulant
Weight % NVM: 40
Cationic coagulant for solids removal from latex systems

2332:
Anionic flocculant
Weight % NVM: 35
To improve clarification of solids removal systems

DREW INDUSTRIAL DIVISION: DREWFAX Specialty Surfactants:

DREWFAX 818:
Specialty Surfactant

DREWFAX 818 specialty surfactant effectively eliminates painted surface defects by improving wetting properties to the coating without adversely affecting gloss and color acceptance.

DREWFAX 818 is a formulated blend of silicone copolymers and glycol ether.

Typical Properties:
Appearance: Clear, light amber liquid
Specific Gravity: 0.95

DREWFAX 842:
Specialty Surfactant

DREWFAX 842 specialty surfactant acts as a flow and leveling agent in both aqueous and non-aqueous coating systems.

DREWFAX 842 is a formulated blend of silicone oils, surfactants and glycol ether.

Typical Properties:
Appearance: Amber, transparent liquid
Specific Gravity: 0.96

DREWFAX 412:
Specialty Surfactant

DREWFAX 412 specialty surfactant is formulated for use in various aqueous coatings.

DREWFAX 412 is a blend of silicone derivatives and glycol ethers.

Typical Properties:
Appearance @ 25C: Clear, light amber liquid
Specific Gravity: .94-.95

DREWFAX 420:
Specialty Surfactant

DREWFAX 420 specialty surfactant exhibits excellent slip and levelling properties in water-reducible and emulsion-based systems.

DREWFAX 420 is a blend of silicone derivatives and glycol ethers.

Typical Properties:
Appearance: Clear, light amber liquid
Specific Gravity: .96-.965

DREW INDUSTRIAL DIVISION: DREWFAX Specialty Surfactants (Continued):

DREWFAX 680:
Specialty Surfactant

DREWFAX 680 specialty surfactant is an effective air release agent for use in a variety of coatings including alkyd wood stains.

DREWFAX 680 is a blend of surfactants.

Typical Properties:
Specific Gravity: 1.03
Viscosity, Brookfield: 1200 cps

DREWFAX G-110:
Specialty Surfactant

DREWFAX G-110 specialty surfactant exhibits excellent color acceptance properties in tint-base formulations.

DREWFAX G-110 is a blend of formulated fatty acid derivatives.

Typical Properties:
Appearance: Amber liquid
Specific Gravity: 0.94

DREWGARD 432:
Corrosion Inhibitor

DREWGARD 432 corrosion inhibitor is specifically designed to prevent flash rusting water-based coatings are painted on metal surfaces.

DREWGARD 432 is a uniquely blended formulation of anti-rust agents and corrosion inhibitors.

Typical Properties:
Appearance @ 25C: White, free-flowing powder
Bulk Density: 55-65 lbs./cu.ft.

DREW INDUSTRIAL DIVISION: DREWPLUS Foam Control Agents:

DREWPLUS L-475:
Foam Control Agent

DREWPLUS L-475 foam control agent has demonstrated excellent defoaming ability in a broad range of paint systems.
DREWPLUS L-475 is a nonionic antifoam formulation comprised of a blend of mineral oils and silica derivatives.

Typical Properties:
Specific Gravity: 0.91
Viscosity (Brookfield): 1200 cps

DREWPLUS L-477:
Foam Control Agent

DREWPLUS L-477 foam control agent is an efficient all-purpose defoamer product.
DREWPLUS L-477 is a formulated blend of mineral oils and silica derivatives.

Typical Properties:
Specific Gravity: 0.90
Viscosity (Brookfield): 1100 cps

DREWPLUS L-493:
Foam Control Agent

DREWPLUS L-493 foam control agent is effective in synthetic latex paints and coatings as well as water reducible coatings for industrial applications.
DREWPLUS L-493 is formulated from a blend of mineral oils, silica derivatives and emulsifiers.

Typical Properties:
Specific Gravity: .90
Viscosity (Brookfield): 1400 cps

DREWPLUS Y-250:
Foam Control Agent

DREWPLUS Y-250 foam control agent is designed for aqueous foaming industrial systems that require quick foam knock-down and long lasting foam prevention
DREWPLUS Y-250 is a formulated blend of mineral oils, silica derivatives and esters.

Typical Properties:
Specific Gravity: 0.91
Viscosity (Brookfield): 1800 cps

DREW INDUSTRIAL DIVISION: DREWPLUS Foam Control Agents (Continued):

DREWPLUS L-483:
Foam Control Agent

DREWPLUS L-483 foam control agent has shown excellent defoaming and antifoaming properties in latex paints and coatings.

DREWPLUS L-483 is formulated from a blend of organic materials and hydrocarbons.

Typical Properties:
Appearance: Yellow, opaque liquid
Specific Gravity: 0.87

DREWPLUS L-484:
Foam Control Agent

DREWPLUS L-484 foam control agent has been specifically designed for use in paints containing rheology modifiers.

DREWPLUS L-484 is formulated from a blend of organic solids and mineral oils.

Typical Properties:
Specific Gravity: 0.87
Viscosity (Brookfield): 800 cps

DREWPLUS Y-381
Foam Control Agent

DREWPLUS Y-381 foam control agent has shown excellent defoaming and anti-foaming properties in synthetic latex paints.

DREWPLUS Y-381 is formulated from a blend of organic materials and hydrocarbons.

Typical Properties:
Appearance: Off-white liquid
Specific Gravity: 0.87

DREWPLUS Y-601:
Foam Control Agent

DREWPLUS Y-601 foam control agent has been especially developed to prevent and eliminate excessive foam in the manufacture of synthetic latices and in selected paint and coatings formulations.

DREWPLUS Y-601 is a proprietary formulation blend of mineral oils, organic solids and surface active compounds.

Typical Properties:
Appearance: Opaque, off-white colored liquid
Specific Gravity: 0.83
Viscosity (Brookfield): 400 cps

DREW INDUSTRIAL DIVISION: DREWPLUS Foam Control Agents (Continued):

DREWPLUS L-414:

DREWPLUS L-414 foam control agent is an economical silicone defoamer developed for aqueous paints and coatings where compatibility is a major concern.
DREWPLUS L-414 is a foam control agent formulated from a blend of silicone and silica derivatives.

Typical Properties:
Viscosity (Brookfield): 800 cps
Specific Gravity: 1.03

DREWPLUS L-418:
Foam Control Agent

DREWPLUS L-418 foam control agent is a highly compatible silicone defoamer recommended for use in trade sales paints and industrial coatings.
DREWPLUS L-418 is a foam control agent formulated from a blend of silica derivatives, hydrocarbons and silicones.

Typical Properties:
Specific Gravity: 0.98
Viscosity (Brookfield): 1250 cps

DREWPLUS L-419:
Foam Control Agent

DREWPLUS L-419 foam control agent is used as both an antifoam agent and defoamer in synthetic latex paints and water reducible coatings.
DREWPLUS L-419 is a foam control agent formulated from a blend of silica derivatives, hydrocarbons and silicones.

Typical Properties:
Appearance: Off white, opaque liquid
Specific Gravity: 0.96

DREWPLUS L-405:
Foam Control Agent

DREWPLUS L-405 foam control agent has shown excellent defoaming and antifoaming properties in synthetic latex paints and water-reducible coatings.
DREWPLUS L-405 is a foam control agent formulated from a blend of silicones and silica derivatives.
Typical Properties:
Appearance: Opaque, off-white
Specific Gravity: 0.98

DREW INDUSTRIAL DIVISION: DREWPLUS Foam Control Agents (Continued):

DREWPLUS L-464:
Foam Control Agent

DREWPLUS L-464 foam control agent is an economical general purpose defoamer product.
DREWPLUS L-464 is a proprietary blend of silica and organic solids.

Typical Properties:
Specific Gravity: 0.88
Viscosity (Brookfield): 1400 cps

DREWPLUS L-156:
Foam Control Agent

DREWPLUS L-156 foam control agent was specifically developed to prevent air entrainment in the flash tank and stripping column during latex monomer stripping.
DREWPLUS L-156 is a specially formulated proprietary dispersion of alcohols, fatty soaps and surfactants.

Typical Properties:
Specific Gravity: 0.97
Viscosity (Brookfield): 500 cps

Foam Control Agents for Industrial Coatings

DREWPLUS L-405:
DREWPLUS L-405 foam control agent has shown excellent defoaming and antifoaming properties in synthetic latex paints and water-reducible coatings.
DREWPLUS L-405 is a foam control agent formulated from a blend of silicones and silica derivatives.

Typical Properties:
Appearance: Opaque, off-white
Specific Gravity: 0.98

DREWPLUS L-414:
Foam Control Agent

DREWPLUS L-414 foam control agent is an economical silicone defoamer developed for aqueous paints and coatings where compatibility is a major concern.
DREWPLUS L-414 is a foam control agent formulated from a blend of silicone and silica derivatives.

Typical Properties:
Viscosity (Brookfield): 800 cps
Specific Gravity: 1.03

DREW INDUSTRIAL DIVISION: DREWPLUS Foam Control Agents (Continued):

DREWPLUS L-419:
Foam Control Agent

DREWPLUS L-419 foam control agent is used as both an antifoam agent and defoamer in synthetic latex paints and water reducible coatings.

DREWPLUS L-419 is a foam control agent formulated from a blend of silica derivatives, hydrocarbons and silicones.

Typical Properties:
- Appearance @ 25C: Off white, opaque liquid
- Specific Gravity: 0.96

DREWPLUS L-493:
Foam Control Agent

DREWPLUS L-493 foam control agent is effective in synthetic latex paints and coatings as well as water reducible coatings for industrial applications.

DREWPLUS L-493 is formulated from a blend of mineral oils, silica derivatives and emulsifiers.

Typical Properties:
- Specific Gravity: .90
- Viscosity (Brookfield): 1400 cps

DREWPLUS Y-250:
Foam Control Agent

DREWPLUS Y-250 foam control agent is designed for aqueous foaming industrial systems that require quick foam knock-down and long lasting foam prevention.

DREWPLUS Y-250 is a formulated blend of mineral oils, silica derivatives and esters.

Typical Properties:
- Specific Gravity: 0.91
- Viscosity (Brookfield): 1800 cps

DREWPLUS Y-281:
Foam Control Agent

DREWPLUS Y-281 foam control agent is recommended for the control of foam in the manufacturing application of trade sales paints, latex/rubber, industrial coatings, and adhesives.

DREWPLUS Y-281 is formulated from a blend of mineral oils, silica derivatives and surface active agents.

Typical Properties:
- Specific Gravity: 0.93
- Viscosity (Brookfield): 1000 cps

DUPONT CO.: ELVACITE AB Dispersants:

ELVACITE non-aqueous dispersants are a series of advanced technology pigment dispersants originally developed to meet the demanding requirements of the automotive finishes market.

ELVACITE AB dispersant series is now available to serve a broad range of coatings applications, including automotive, inks, colorants, and industrial coatings.

ELVACITE AB Dispersant Product:

Product:

AB 1010:
Weight % Solids: 59-61
Solvents: Toluene
Gardner-Holdt Viscosity: Z-Z3

AB 1015:
Weight % Solids: 46-50
Solvent: Toluene 70%
MEK 30%
Gardner-Holdt Viscosity: G-L

AB 1020:
Weight % Solids: 53-57
Solvents: Toluene
Gardner-Holdt Viscosity: X-Z

AB 1030:
Weight % Solids: 50-55
Solvents: Toluene
Gardner-Holdt Viscosity: Z1-Z3

"A" Segment Parameters

Product:

AB 1010:
AB 1020:
AB 1030:
Pigment Interaction:
- Acidic or amphoteric pigment surfaces
- Best with aromatic hydrocarbons

AB 1015:
Pigment Interaction:
- Basic or amphoteric pigment surfaces
- Tolerates medium polarity solvents, but aromatic hydrocarbons are best.

DUPONT CO.: Surfactants:

MERPOL 100 Surfactant

MERPOL 100 Surfactant is a nonionic surface active agent concentrate (100% active ingredient--octylphenoxypolyethoxy ethanol). It is useful in many applications requiring emulsification, wetting and/or detergency.

Uses:

MERPOL 100 is useful in any application requiring a nonionic surfactant, such as rapid wetting of substrates and surfaces. MERPOL 100 can also be formulated with other surfactants to promote and improve detergency. MERPOL 100 is compatible with most formulations. Other uses for MERPOL 100 include household detergents, industrial detergents, pesticide and herbicide formulations.

Typical Physical and Chemical Properties:

Ionic Type: Nonionic
Appearance: Clear liquid
Active Ingredient: 100
Color (Gardner): <4
Viscosity, 27C, cP: 222
Density, 25C
  g/mL: 1.063
  lb/gal: 8.9
pH (10% aqueous solution): 5-7

ZONYL FSO Fluorosurfactant

ZONYL FSO is a powerful surface tension depressant for use in water, acids, salts and bases.

Typical Properties:

Ionic Class: Nonionic
Composition: Fluoroalkyl Poly (Ethylene Oxide) Ethanol
Form: Turbid Liquid
Color: Brown
Active Ingredient, %: 50
Solvent: Ethylene Glycol/Water
Boiling Point, 760mm Hg: 100C (212F)
Melting Point: Below -28C (-18F)
Specific Gravity: 1.36
pH (100%): 7-8
Flash Point (PMCC): Above 93C (200F)

EXXON CHEMICALS: TPM-662:

TPM-662 from Exxon has shown itself to be an effective modifier of inorganic and organic particulate in both aqueous and oil based systems. This cationic surfactant attaches itself to clays, sands, ground slate, ground limestone, diatomaceous earth, perlite, mica, talcs and fiberous materials like cellulose and polyolefins. Particulate modification with TPM-662 alters the suspension, dispersion and wetting characteristics of the particulate within a given system by rendering the particles more hydrophobic (lipophilic).

Areas where TPM-662 finds application:

- Suspension of hydrophilic particles in oil bases (alkyd paints).
- Creation of oily films around hydrophilic particles for effective dewatering (kaolin slips in ceramic production).
- Screen/Mold release agent (ceramic slip molds).
- Modification of particulate surface to promote higher solids in water slurries.

Typical Physical Properties:

Color and form	light amber liquid
Pour point, typical	-6C
Weight per gallon, typical	7.8 lb/gal @ 60F
Flash point, typical	>200F (non-flammable)
Actives	98% min.
Solubility	Both oil and water dispersable

GENESEE POLYMERS CORP.: Antifoams:

DB-1 Defoamer:

A 100% active foam control agent based on a combination of organo-modified silicone polymers, silica, and waxes.

Typical Properties:
Appearance: Tan, opaque fluid
Density: 0.90
Wt./gallon: 7.5 lbs.
% Active: 100%
Flash Point (PMCC): >200F
Special Features:
Silicone defoaming properties
Compatible with soluble oils
Persistent foam control
Will not plate out during metalworking process

DB-12 Defoamer:

A 100% active foam control agent based on a combination of organo-modified silicone polymers, silica, and waxes.

Typical Properties:
Appearance: Tan, Opaque Fluid
Density: 0.90
Wt./Gallon: 7.5 lbs.
% Active: 100%
Flash Point (PMCC): >200F
Special Features:
Silicone defoaming properties
Compatible with soluble oils
Persistent foam control
Will not plate out during metalworking process

GP-210 Antifoam Emulsion:

A water dispersible industrial grade defoamer containing 10% active ingredient.

Typical Properties:
Silicone Content: 10%
Viscosity: 1500 cps
Density @ 25C: 1.0
Wt./Gallon: 8.3 lbs.
Emulsifier Type: Non-ionic
pH: 8.0

Special Features:
Has greater defoaming activity than conventional defoamers because it combines a silicone antifoam with an organic defoamer to make better use of the best defoaming properties of both materials.

GENESEE POLYMERS CORP.: Antifoams (Continued):

GP-300-1 Antifoam Compound:

A 100% active silicone defoamer.

Typical Properties:
- Silicone Content: 100%
- Viscosity & Appearance: Viscous white liquid
- Density, 25C: 1.0
- Wt./Gallon, 25C: 8.3 lbs.
- Flash Point (COC): >600F

Special Features:
- Excellent Foam Knock-down
- Good Alkaline Resistance
- Good for Hot or Cold Aqueous Systems

GP-310-1 Antifoam Emulsion:

A dilutable 10% active silicone anti-foam emulsion.

Typical Properties:
- Silicone Compound: 10%, wt.
- Viscosity: Thixotropic
- Density, 25C: 1.0
- Wt./Gallon: 8.3 lbs.
- Color and Appearance: White liquid
- Type Emulsifiers: Non-ionic
- Carrier: Water

Special Features:
- Excellent Foam Knock-down
- Good Alkaline Resistance
- Good for Hot or Cold Aqueous Systems

GP-330-1 Antifoam Emulsion:

A 30% active emulsion of high performance silicone antifoam.

Typical Properties:
- Silicone Compound: 30%, wt.
- Viscosity: Thixotropic
- Density, 25C: 1.0
- Weight/Gallon: 8.3 lbs.
- Color and Appearance: White liquid
- Type Emulsifiers: Non-ionic
- Carrier: Water

Special Features:
- Excellent Foam Knock-down
- Good Alkaline Resistance
- Good for Hot or Cold Aqueous Systems

GENESEE POLYMERS CORP.: Antifoams (Continued):

GP-700 Silicone Antifoam Compound:

A 100% active, high performance antifoam agent. Makes use of a unique silicone-silica combination.

Typical Properties:
- Appearance: Light tan, opaque fluid
- Density: 0.96
- Wt./Gallon: 8.0 lbs.
- % Active: 100%
- Flash Point (PMCC): >300F

Special Features:
- Silicone defoaming properties
- Greater compatibility with soluble oils
- Persistent foam control
- Resists plate-out in metalworking processes

DK-100 Non-Silicone Defoamer:

A high performance, non-silicone product designed to inhibit foam formation. Contains 100% active ingredient.

Typical Properties:
- Appearance: Opaque, gray-white liquid
- Density: .85
- Wt./Gallon: 7.4 lbs.
- Moisture Content: 0.2%
- Flash Point (COC): 200F

Special Features:
- Fast Foam Knock-down
- Persistence
- Alkaline Resistant
- Does not contain silicone

DK-230 Defoamer:

A 100% active, non-silicone defoamer.

FDA Status: DK-230 Defoamer contains ingredients which are FDA approved for use as defoaming agents in coatings.

Typical Properties:
- Appearance: Light Straw, Opaque Liquid
- Density: 0.82
- Wt./Gallon: 6.8 lbs.
- Composition: 100% Active Organic
- Flash Point (Closed Cup): 150F

Special Features:
- Fast foam knock-down
- Persistence
- Alkaline Resistance
- Does not contain silicone

GENESEE POLYMERS CORP.: Antifoams(Continued):

GP-1000-S Antifoam Compound:

A specially formulated product designed to control foam in aqueous systems.

Typical Properties:
- Silicone Content: None
- Color and Appearance: White opaque fluid
- Active Ingredient: 92.5%
- Specific Gravity: 0.76
- Wt./Gallon: 6.4 Pounds (77F)
- Flash Point (T.C.C.): 135F

Special Features:
- Fast foam knockdown
- Contains no silicone
- Contains no silica or abrasive material

W. R. GRACE & CO.: DAXAD High Activity Dispersants:

Polymethacrylates:

30:
Sodium salt
Solution 25%
Low foaming, low viscosity dispersant
Pigment dispersions for trade sales flat paints, polymerization, water treatment, agriculture, cosmetics, industrial cleaners.

30-30:
Sodium salt
Solution 30%
Higher solids DAXAD 30
Pigment dispersions for trade sales flat paints, polymerization, water treatment.

30S:
Sodium salt
Powder
Highly efficient; 90% active dispersing agent.
Water treatment, pigment dispersions for trade sales paints, dry mixes, agriculture.

32:
Ammonium salt
Solution 25%
Compatible with propylene glycol. Excellent dispersant for use in applications where sodium cannot be tolerated.
Ceramics, pigment dispersions.

34:
Acid
Solution 25%
For use where conventional alkaline dispersants cannot be used.
Ceramics, pigment dispersions, polymerization.

34A9:
Ammonium salt
Solution 25%
Daxad 34 buffered with ammonia to a pH of 9.
Polymerization, clay coating.

W. R. GRACE & CO.: DAXAD High Activity Dispersants(Continued):

Polyisobutylene-Maleic Anhydride Copolymer:

31:
Sodium salt
Solution 25%
Compatible with propylene glycol. Does reduce surface tension slightly.
Trade sales enamels, polymerization, leather tanning, water treatment.

Cationic Polyelectrolyte:

CP-2:
Solution 50%
Very effective in precipitating various anionics from aqueous streams
Water treatment, clay dispersions, anionic latexes.

HENKEL CORP.: Additives:

DYMSOL MS-40:
Appearance: Opaque, White Liquid
Density, Lbs./Gal.: 8.0
pH, 5% Emulsion: 7.0
Solids Content, %: 40
Emulsifier Type: Anionic/Nonionic, Protective Colloid
Type of Polyethylene: Non-Emulsifiable

FOAMASTER DS:
Defoamer for Water Base Coatings and Adhesives

FOAMASTER DS is a 100% active antifoam containing silicone and is designed to provide the foam-counteracting advantages of silicone while minimizing the disadvantages.

FOAMASTER DS demonstrates:
* High defoaming efficiency.
* Easy water dispesibility.
* Excellent defoaming persistence.
* Exceptional film compatibility.

Advantages:
* Provides high defoaming efficiency in alkaline or acidic systems.
* Imparts excellent antifoaming in water base coatings and adhesives.
* May be used in the letdown as well as in the grind.
* Minimizes film fisheye and crater formation tendencies.
* Is dispersible in water to an unusually high degree.
* Demonstrates excellent color development.
* Is waterfree.
* Contains FDA sanctioned ingredients.

Typical Properties:
Appearance: Opaque, yellow liquid
Activity, %: 100

FOAMASTER JMY

Antifoamer for Polyvinyl Acetate Emulsions

FOAMASTER JMY antifoamer is a 100% active liquid originally compounded especially for polyvinyl acetate emulsions.

Advantages:
* FOAMASTER JMY antifoamer protects the appearance and wearing qualities of paint by preventing the formation of foam.
* A liquid, FOAMASTER JMY antifoamer requires no preparation before being added to polyvinyl acatate emulsion.
* It is chemically compatible with the other ingredients
* It is effective in low pH systems.
* It does not affect color development.
* FOAMASTER JMY is totally silicone free.
* FOAMASTER JMY antifoamer is an economical product.

HENKEL CORP.: Additives(Continued):

FOAMASTER O
Defoamer for Latex Paints
Printing Inks and Adhesives

FOAMASTER O defoamer provides efficient foam and froth control in the manufacture of synthetic latex and subsequently compounded paints and adhesives. FOAMASTER O is also an extremely effective defoamer for aqueous printing inks. Liquid in nature and 100% active, this defoaming agent is recommended for SBR, PVA, acrylic and vinyl acrylic emulsions.

Typical Properties:
Appearance: Opaque, off-white liquid
Activity, %: 100

FOAMASTER R and FOAMASTER S
Defoamers for Synthetic Latex Systems

FOAMASTER R and FOAMASTER S provide efficient foam and froth control in the manufacture of paints and adhesives from synthetic latex.

Typical Properties:

FOAMASTER R:
Appearance: Opaque, off-white liquid
Density, lb/gal: 7.5
Solubility: Insoluble in water, dispersible in surfactant systems

FOAMASTER S:
Appearance: Opaque, light tan liquid
Density, lb/gal: 7.5
Solubility: Dispersible in water

FOAMASTER SA-3
Defoamer for Water Based Coatings

FOAMASTER SA-3 is a 100% active defoamer developed to provide efficient and economical defoaming in water based coatings. FOAMASTER SA-3 utilizes four different defoaming mechanisms, including a multiple hydrophobe system.

Advantages:
* High defoaming efficiency/economy.
* Rapid bubble break.
* Excellent defoaming persistence.
* Exceptional film compatibility.
* Minimal color development problems.
* Effective over broad pH range.
* Water free--results in reduced shipping costs and prevents internal microbial activity.

HENKEL CORP.: Additives(Continued):

FOAMASTER VL
Defoamer for Latex Tint Bases and Paints

FOAMASTER VL liquid is a 100% active antifoamer developed to meet the need for a foam-counteracting additive having:
* High defoaming efficiency.
* Easy water dispersibility.
* Complete compatibility with tints and colorants used in latex paint and adhesive tint bases.

Advantages:
* Provides high defoaming efficiency in alkaline or acidic systems.
* Is compatible and useful in all major latexes.
* Imparts excellent antifoaming in water based paints and adhesives.
* Is dispersible in water to an unusually high degree.
* Is water free.
* Does not detract from tinting color development.
* Produces films free of fish eyes and craters.

FOAMASTER 111
New Concept Defoamer

FOAMASTER 111 is a 100% active, liquid, broad spectrum silicone-free defoamer. It is specifically designed for the following applications:
* Water-based paints and coatings.
* Water-based printing inks.
* Latex adhesive systems.

Advantages:
* Broad spectrum--because FOAMASTER 111 employs multiple defoaming mechanisms. Effective in coatings, inks and adhesives.
* Totally silicone free. The non silicone nature of FOAMASTER 111 results in broad system compatability.
* Exceptional product stability
* No large particles or gels.

MODICOL S
Stabilizer for Natural and Synthetic Latex

MODICOL S is an unusual stabilizer that ensures the mechanical and chemical stability of both natural and syntehtic latexes. A highly sulfated fatty acid, MODICOL S is formulated to prevent premature coagulation of the latex under such conditions as high-speed agitation, acidification or pigmentation.

Typical Properties:
Ionic nature: Anionic
Moisture, %: 45.0

HENKEL CORP.: Additives(Continued):

NOPCO COLOR-SPERSE 188-A Surfactant

NOPCO COLOR-SPERSE 188-A is a 100% active nonionic surfactant which has proven to be particularly effective in aqueous and non-aqueous systems.

Applications:
NOPCO COLOR-SPERSE 188-A surfactant improves the color acceptance of latex paints when difficult-to-use predispersed colorants are used.

NOPCOSPERSE 44 AS
(NOPCOSPERSE 44, Ammonium Salt)

Typical Properties:
Appearance: Clear, light yellow liquid
% Solids: 35.0
pH (50% Solution): 7.8
Pounds per Gallon: 9.8
Solubility: Soluble in water

HERCULES INC.: HERCULES AR150 and AR160 Surfactants:

Low-Foaming, Nonionic Surface-Active Agents

HERCULES AR150 and AR160 surfactants are nonionic surface-active agents derived by ethoxylation of pale wood rosin. The first two digits of the product designation indicate the number of moles of ethylene oxide adducted. These polyethylene glycol esters of rosin are stable in alkaline and acidic media and are unaffected by hard water or the presence of metal ions. Their low-foaming properties and ability to control foaming make them well suited for use in industrial applications where a non-foaming detergent, emulsifier, or wetting agent is needed.

Typical Properties:

AR150:
- Physical form at 24C: liquid to soft wax
- Color, Gardner, at 40C: 13
- Viscosity at 38C, cps: 300
- pH, 1% solution: 9
- Cloudpoint, 2% solution, C: 60
- Surface tension, 0.1% in distilled water, dynes/cm: 39

AR160:
- Physical form at 24C: soft wax
- Color, Gardner, at 40C: 14
- Viscosity at 38C, cps: 250
- pH, 1% solution: 8
- Cloudpoint, 2% solution, C: 68
- Surface tension, 0.1% in distilled water, dynes/cm: 37.6

Outstanding Properties:

Effective at low concentrations; low-foaming; compatible with anionic and cationic systems; stable to acids and alkalies; retain efficiency in the presence of metal ions.

Soluble in all proportions in water, isopropanol, benzene, toluene, and ester solvents. Insoluble in hexane and kerosene.

Compatible with pine oil, quaternaries, and other germicidal agents; can be formulated with sulfate, alkaline, phosphate, silicate, and carbonate salts; can be used with most other nonionic surfactants and with anionic and cationic surfactants.

Typical Uses:

HERCULES AR150 and AR160 surfactants are useful wherever detergent, emulsifying, suspending, and/or dispersing action is needed. They offer a broad range of properties for use in industrial cleaners and detergents, and they are also used in suspending operations where no foaming is essential.

FDA Status:

HERCULES AR150 and AR160 surfactants are in compliance with the requirements of the U.S. Food and Drug Administration for uses where they may come in contact with food.

HI-TEK POLYMERS, INC.: PHOSFAC 5513 Pigment Dispersant:

PHOSFAC 5513, the potassium salt of a complex organic phosphate ester, was developed as a dispersant and surfactant for aqueous paints and coating formulations. It is effective for dispersing lead silicate, carbon black, zinc pigments and "reactive" pigments derived from chromium, strontium, etc. PHOSFAC 5513 is also highly effective in preventing the viscosity buildup in ZnO-based products containing certain preservatives. It is also suggested for use as a stabilizer for PVC.

Typical Properties:
Appearance: Clear to hazy, light amber gel
Odor: Mild, etherial
pH: 10% 50/50 IPA/water: 6.5-7.5
Activity: 65-70%
Melting Point: Approximately 100F

INTERSTAB CHEMICALS: Anti-Foaming Agents--Defoamers:

INTERSTAB defoamers have been designed to control foam during processing and application of coatings. The proper selection of INTERSTAB defoamers results in maximum persistency and color acceptance with minimal chance for film defects. While the actual amount of defoamer required depends on the specific formula, a good starting point is 0.2-0.5% on the weight of the coating. For difficult to defoam systems, use levels as high as 1.0% may be required.

ANTI-FOAMING AGENT:
ANTI-FOAMING AGENT is a 100% active silicone-free defoamer designed to reduce foam during preparation and application of water based systems.

HEVI-DUTY ANTI-FOAMING AGENT:
HEVI-DUTY ANTI-FOAMING AGENT is a 100% active silicone free defoamer which is effective in a wide variety of latex paints.

ANTI-FOAM AGENT 2079-87:
ANTI-FOAM AGENT 2079-87 is a 100% active liquid especially designed to eliminate foaming during the preparation and application of water based coatings. It is effective in a wide variety of paint formulations.

INTERFOAM X-1:
INTERFOAM X-1 is a 100% active defoamer, useful in a wide variety of coating systems. It is recommended for use in all emulsions.

INTERFOAM X-4:
INTERFOAM X-4 is a unique anti-foaming compound useful in a variety of systems.

INTERFOAM X-5:
INTERFOAM X-5 is a 100% active defoamer based on a hydrophobic particulate. It is effective in a wide range of systems.

INTERFOAM X-6:
INTERFOAM X-6 is a 100% active, efficient defoamer which is the most effective defoamer for many coatings types.

INTERFOAM X-7:
INTERFOAM X-7 is a 100% active defoamer which has a wide range of utility. Is particularly effective with polymeric associative type thickeners.

INTERSTAB CHEMICALS: Anti-Foaming Agents-Defoamers (Continued):

INTERFOAM X-100:
INTERFOAM X-100 is a high performance anti-foaming agent effective in many coatings systems.

INTERFOAM X-105:
INTERFOAM X-105 is a high performance anti-foaming compound effective in many coatings systems. It is useful in a variety of coatings and exhibits excellent performance initially and after aging in the paint.

INTERFOAM X-200:
INTERFOAM X-200 is an effective defoamer for use in a variety of coatings. It is particularly useful in tint bases since it does not adversely affect color acceptance with most colorants.

INTERFOAM X-210:
INTERFOAM X-210 is a versatile defoamer which is effective in many coatings types. It relies on several mechanisms to eliminate foam resulting in good across the board performance.

INTERFOAM S-500:
INTERFOAM S-500 is a high performance silicone containing defoamer, designed especially for coatings.

INTERFOAM W-600:
INTERFOAM W-600 is a new qeneration emulsion based defoamer. It has excellent stability and since it is water based, it has an excellent cost/performance profile.

Physical Properties:

INTERFOAM X-1:
Viscosity (G/H-Max.): Z-5
Form: Opaque Liquid

INTERFOAM X-2:
Viscosity (G/H-Max.): T
Form: Opaque Liquid

INTERFOAM X-3:
Viscosity (G/H-Max.): A-1
Form: Opaque Liquid

INTERFOAM X-4:
Viscosity (G/H-Max.): A
Form: Clear Liquid

INTERSTAB CHEMICALS: Anti-Foaming Agents--Defoamers(Continued):

Physical Properties(Continued):

INTERFOAM X-5:
Viscosity (G/H-Max.): Z-1
Form: Opaque Liquid

INTERFOAM X-6:
Viscosity (G/H-Max.): E
Form: Opaque Liquid

INTERFOAM X-7:
Viscosity (G/H-Max.): W
Form: Opaque Liquid

INTERFOAM X-100:
Viscosity (G/H-Max.): C
Form: Opaque Liquid

INTERFOAM X-105:
Viscosity (G/H-Max.): V
Form: Opaque Liquid

INTERFOAM X-200:
Viscosity (G/H-Max.): I
Form: Semi-Transparent Amber Liquid

INTERFOAM X-210:
Viscosity (G/H-Max.): F
Form: Cream Colored Opaque Liquid

INTERFOAM S-500:
Viscosity (G/H-Max.): M
Form: Opaque Liquid

ANTI-FOAM AGENT:
Viscosity (G/H-Max.): F
Form: Opaque Liquid

HEVI-DUTY ANTI-FOAM AGENT:
Viscosity (G/H-Max.): D
Form: Clear Liquid

INTERSTAB CHEMICALS: Wetting Agents and Dispersants:

INTERSPERSE:
INTERSPERSE is a dispersing agent effective with a wide variety of pigments. It is compatible with many resin systems and for this reason has gained widespread industry acceptance. Since INTERSPERSE is a film former it does not detract from coatings properties. Excellent performance can be obtained in alkyds, oils, oleoresinous vehicles, epoxy esters, epoxies, chlorinated rubber, plasticizers, and other solvent based coatings resins.

INTERSPERSE AQ:
INTERSPERSE AQ is an emulsifiable version of INTERSPERSE. It can be used in water reducible systems to improve pigment dispersion without detracting from film properties.

INTERWET 33:
INTERWET 33 is a 100% active, non-ionic wetting agent for use in all water based coatings. INTERWET 33 contributes to faster, more efficient pigment dispersion with minimal foam generation. INTERWET 33 also greatly improves the color acceptance of tint bases for many commercial universal colorant systems.

INTERWET 43:
INTERWET 43 is a 100% active, non-ionic wetting agent and emulsifier. It is effective in rendering many oil and alkyd systems water emulsifiable. It also finds use in many latex paint systems to improve pigment wetting and color acceptance (HLB=10.9).

INTERWET 212:
INTERWET 212 is a 100% active non-ionic wetting agent useful in reducing the viscosity of plastisols. It is effective in removal of entrapped air from plastisols as well as reducing the tendency for bubble formation during casting the molds. Unlike other surfactants INTERWET 212 also finds use in latex paints to improve pigment wetting and color acceptance (HLB=10.2).

INTERSPERSE:
Viscosity Max. (G/H-77F): Z
Form: Clear Liquid
INTERSPERSE AQ:
Viscosity Max. (G/H-77F): S
Form: Clear Liquid
INTERWET #33:
Viscosity Max. (G/H-77F): D
Form: Clear Liquid
INTERWET #43:
Viscosity Max. (G/H-77F): C
Form: Clear Liquid
INTERWET #212:
Viscosity Max. (G/H-77F): A
Form: Clear Liquid

C.J. PATTERSON CO.: PATCO Designed Chemicals:

PATCOTE 460

PATCOTE 460 is a 100% active liquid defoamer that does not contain silicone. It is designed for use in solvent based coatings and alleviates air entrapment and foam problems.

PATCOTE 500

PATCOTE 500 is a 100% active liquid defoamer containing silicone. It is designed for use a process defoamer for latex trade sales. It has good stability and does not cause surface imperfections generally attributable with these types of defoamers.

PATCOTE 501

PATCOTE 501 is a 100% active liquid defoamer that contains silicone. It is designed for use in alcohol fermentation.

PATCOTE 502

PATCOTE 502 defoamer is designed for use in potato processing, alcohol fermentation and latex coatings. It may be diluted with water up to 5:1 ratio with stirring.

PATCOTE 512

PATCOTE 512 is a 100% active liquid defoamer containing silicone. It is designed for use in urethane modified resins. It has shown excellent persistency in these systems.

PATCOTE 513

PATCOTE 513 is a 100% active liquid defoamer containing silicone. It is designed for use in water reducible acrylic coatings. It has demonstrated excellent persistence in these systems.

PATCOTE 519

PATCOTE 519 is a 100% active liquid defoamer containing silicone. It is designed for use in both trade sales and industrial acrylic lacquers. It has demonstrated excellent persistency in these systems.

PATCOTE 520

PATCOTE 520 is a 100% active liquid defoamer containing silicone. It is designed for use in water reducible alkyd resins. It has excellent persistence.

PATCOTE 525

PATCOTE 525 is a 100% active liquid defoamer containing silicone. It is designed for use in water reducible alkyd systems. Product will not separate.

C.J. PATTERSON CO.: PATCO Designed Chemicals(Continued):

PATCOTE 543

PATCOTE 543 is a 100% active liquid defoamer containing silicone. It is designed for use in PVA, vinyl-acrylic, or combination of PVA and acrylic based trade sales coatings. It has also found use in semi-gloss acrylic paints. Other characteristics include excellent color acceptance and persistence.

PATCOTE 550

PATCOTE 550 is a 100% active liquid defoamer containing silicone. It is designed for use in water reducible alkyds and acrylics as well as acrylic latices for both industrial and trade sales formulations. It has demonstrated excellent persistince.

PATCOTE 555

PATCOTE 555 is a 100% active liquid defoamer containing silicone. It is designed for use in acrylic latex systems for trade sales. It is generally used in conjunction with PATCOTE 834. It has excellent persistence and color acceptance.

PATCOTE 577

PATCOTE 577 is a 100% active liquid defoamer containing silicone. It is designed for use in PVA resins for trade sales and water reducible alkyds for industrial coatings. It has excellent persistency and color acceptance.

PATCOTE 801

PATCOTE 801 is a 100% active liquid defoamer that does not contain silicone. It is designed for use in PVA-acrylic copolymers and terpolymer emulsions for trade sales. It has excellent persistency and color acceptance in these systems.

PATCOTE 802

PATCOTE 802 is a 100% active liquid defoamer that does not contain silicone. It is designed for use in acrylic and terpolymer emulsions in trade sales. It has excellent persistency and color acceptance in these systems.

PATCOTE 803

PATCOTE 803 is a 100% active liquid defoamer that does not contain silicone. It is designed for use in acrylic and terpolymer emulsions for trade sales. It has excellent persistency and color acceptance in these systems.

PATCOTE 804

PATCOTE 804 is a 100% active liquid defoamer that does not contain silicone. It is designed for use in trade sales latex emulsions. It has excellent persistency and color acceptance in these systems.

C.J. PATTERSON CO.: PATCO Designed Chemicals(Continued):

PATCOTE 805

PATCOTE 805 is a 100% active liquid defoamer that does not contain silicone. It is designed for use in acrylic latexes for both trade sales and industrial applications. It has shown excellent persistence in these systems.

PATCOTE 806

PATCOTE 806 is a 100% active liquid defoamer that does not contain silicone. It is designed for use in acrylic and terpolymer emulsions in trade sales and some industrial applications.

PATCOTE 845

PATCOTE 845 is a 100% active liquid defoamer that does not contain silicone. It does not separate and has excellent persistence. Its primary use is in solvent, high solids and water base acrylic systems.

PATCOTE 847

PATCOTE 847 is a 100% active liquid defoamer that does not contain silicone. It is designed for use in both solvent and water based alkyds. It has excellent persistence and will not separate.

PATCOTE 883

PATCOTE 883 is a 100% active liquid defoamer thdt does not contain silicone. It is designed for use in trade sales acrylic emulsions. It has shown excellent persistence and color acceptance in these systems.

ROHM AND HAAS: TAMOL Dispersants:

Grade:

TAMOL 731:
Solids, %: 25
pH (As Packed): 9.5-10.5 (10% solution)
Viscosity, cps.: 65-238
Density lbs/U.S. gal: 9.2
Characteristics: General purpose; heat resistant

TAMOL 850:
Solids, %: 30
pH (As Packed): 9.5-10.5 (10% solution)
Viscosity, cps.: 125-325
Density lbs/U.S. gal: 9.9
Characteristics: Low foaming; Divalent ion stable

TAMOL 901:
Solids, %: 30
pH (As Packed): 8.0-9.5
Viscosity, cps.: 75-200
Density lbs/U.S. gal: 10.6
Characteristics: High solids pigment slurries

TAMOL 960:
Solids, %: 40
pH (As Packed): 8.0-9.0
Viscosity, cps.: 300-800
Density lbs/U.S. gal: 10.6
Characteristics: General purpose

TAMOL SG-1:
Solids, %: 35
pH (As Packed): 8.0-9.0
Viscosity, cps.: 120-400
Density lbs/U.S. gal: 9.6
Characteristics: Gloss paints; reactive pigment stable

TAMOL 963:
Solids, %: 35
pH (As Packed): 6.5
Viscosity, cps: 25
Density lbs/U.S. gal: 9.6
Characteristics: Good cost/performance; low foam

TAMOL 983:
Solids, %: 35
pH (As Packed): 8.2
Viscosity, cps.: 25
Density lbs/U.S. gal: 10.1
Characteristics: Good cost/performance; reactive pigment stable

ROHM AND HAAS: TRITON Wetting Agents:

Grade:

TRITON CF-10:
Active, %: 100
Viscosity, cps.: 250
Specific Gravity: 8.9
Surface Tension Dynes/Cm.: 31
Major Uses: Low foam; pigment dispersion

TRITON GR-7M:
Active, %: 64
Viscosity, cps.: 110
Specific Gravity: 8.4
Surface Tension Dynes/Cm.: 27
Major Uses: Prevent crawling in semi-gloss

TRITON N-57:
Active, %: 100
Viscosity, cps.: 240
Specific Gravity: 8.5
Surface Tension Dynes/Cm.: 29
Major Uses: Color acceptance

TRITON X-100:
Active, %: 100
Viscosity, cps.: 240
Specific Gravity: 8.9
Surface Tension Dynes/Cm.: 29
Major Uses: General purpose

TRITON X-102:
Active, %: 100
Viscosity, cps.: 330
Specific Gravity: 8.9
Surface Tension Dynes/Cm.: 32
Major Uses: Pigment dispersion and color acceptance

TRITON X-114:
Active, %: 100
Viscosity, cps.: 260
Specific Gravity: 8.9
Surface Tension Dynes/Cm.: 28
Major Uses: Color acceptance

ROHM AND HAAS: TRITON Wetting Agents(Continued):

Grade:

TRITON X-200:
Active, %: 28
Viscosity, cps.: 7000
Specific Gravity: 8.9
Surface Tension Dynes/Cm.: 30
Major Uses: Color acceptance

TRITON X-207:
Active, %: 100
Viscosity, cps.: 600
Specific Gravity: 8.2
Major Uses: Color acceptance

TRITON X-405:
Active, %: 70
Viscosity, cps.: 490
Specific Gravity: 9.2
Surface Tension Dynes/Cm.: 37
Major Uses: Pigment wetting in alkyd modified paint

TRITON CA:
Active, %: 100
Viscosity, cps: 385
Specific Gravity: 8.4
Surface Tension Dynes/Cm.: 29
Major Uses: Color acceptance

ROSS CHEM, INC.: Defoamers for Paints and Coatings:

Defoamers:

FOAM BURST 10, 20, 28 and 30:

100% active silicone antifoam for non-aqueous epoxies, urethanes and acrylate coatings. Effective at very low concentrations. Effectiveley de-airs and imparts internal slip for UV-curable coatings. Excellent defoamers for grind phase of pigment coatings.

FOAM BURST 301S:

Good general purpose defoamer for latex paints based on acrylic, butadiene-styrenes, polyvinyl acetate and ethylene vinyl acetate latexes.

FOAM BURST 320CT:

100% active organic defoamer for synthetic latex emulsions and paints. Highly effective with rapid foam knockdown.

FOAM BURST 350:

General purpose, longlasting non-silicone defoamer for paints and coatings based on synthetic latexes and water-reducible alkyds. Provides effective foam control and fish-eye free films.

FOAM BURST 373:

Hydrophobic silica type defoamer especially effective in gloss paints. Gives rapid bubble break and smooth fish-eye free films.

FOAM BURST 383:

100% active organic defoamer which gives rapid bubble break and excellent defoaming persistence. Exceptional "in can" aging efficiency. Minimal color development problems.

FOAM BURST 384:

Highly effective defoamer at relatively low cost for maximum economy. Gives rapid bubble break and smooth, fish-eye and crater free films when rolled out. Excellent persistence.

ROSS CHEM, INC.: Defoamers for Paints and Coatings(Continued):

Defoamers:

FOAM BURST 502:

Good general purpose organic defoamer for latex-based paints and coatings. Rapid bubble break and good film uniformity at a very economical price make it the defoamer of choice for many systems.

FOAM BURST 09-137-20:

Emulsified oil and organic blend for use in water based paint in the let down phase. 09-137-20 provides superior bubble break on rollout. Good "in can" aging persistence. Minimal effect on gloss.

FOAM BURST 1005:

Totally synthetic system, non-mineral oil, non-silicone designed for use in the let down phase of water based coatings. Excellent for use in clear or pigmented systems sensitive to oil or silicone.

FOAM BURST 327:

Highly effective defoamer for coatings based on polyvinyl alcohol, polyvinyl acetate, water-solubilized acrylics and acrylic copolymers. Offers good compatibility. Similar chemistry to FOAM BURST 1005 but with increased strength.

FOAM BURST 372:

Very effective defoamer for coatings based on synthetic latexes such as acrylics, polyvinyl acetates, and polyvinyl alcohol. Good compatibility in most systems. Similar chemically to FOAM BURST 1005 with added leveling ability.

TEGO CHEMIE SERVICE USA: Defoamers:

TEGO Foamex 1435:

Defoamer emulsion for water-thinnable printing inks.
Chemical Composition: W/O emulsion of a modified polysiloxane.

TEGO Foamex 7447:

Defoamer for water based emulsion paints and water thinnable systems
Chemical Composition: O/W emulsion of a hydrophobic polysiloxane-polyether-copolymer.

Physical data:

* Active matter content: 20%
* Density at 25C: approx. 1.0 g/cm3
* Non-volatile content: 24%+-1
* Volatile component: water
* Type of emulsifier: non-ionic/anionic
* Weight/gallon: 8.345 lbs

Properties:

TEGO Foamex 7447 inhibits the formation of macrofoam in the production and processing and water-based paint systems.

Areas of application:

* Building protection
  House paints, silk gloss and gloss paints based on pure acrylate and styrene-acrylate resins, polyvinyl acetate copolymers and terpolymers.
* Wood and Furniture Varnishes
  Water-thinnable systems for spray or pouring processes.
* Anti-corrosion general industrial paints
  Water-soluble systems based on alkyd, polyester and acrylic resins.
* Printing inks
  Water-thinnable gravure inks and flexographic printing inks.
* Other applications
  Production of polymer dispersions

TEGO CHEMIE SERVICE USA: Defoamers(Continued):

TEGO Foamex 1488:
Defoamer emulsion for water based emulsion paints and water thinnable systems.

Chemical Composition: O/W emulsion of a hydrophobic polysiloxane-polyether-copolymer.

Physical data:
* Active matter content: 20%
* Density (20C) approx. 1.0 g/cm3.
* Non-volatile content: 24%+-1
* Volatile Component: water
* Emulsifier type: non-ionic/anionic
* Weight/gallon: 8.345 lbs

Properties:
TEGO FOAMEX 1488 prevents macrofoam formation during the production and application of water based paint systems.

Areas of application:
Building protection
* exterior, satin gloss and gloss paints based on pure acrylic and styrene-acrylic resins, PVA-copolymers and terpolymers

Wood/furniture varnishes
* water thinnable systems which are applied by spraying

Anticorrosive coatings/general industrial paints
* water soluble systems based on alkyd, polyester and acrylic resins

Printing inks
* water thinnable gravure and flexographic inks

Miscellaneous
* production of polymer emulsions

TEGO FOAMEX 3062:
Defoamer for water based emulsion paints and water thinnable systems.

Chemical composition: Hydrophobic polysiloxane-polyether-copolymer.

Physical data:
* Active matter content: 100%
* Density (20C): approx. 1.0 g/cm3
* Colour index: max. 9 (Gardner)
* Flash Point: >212F (100C)
* Weight/gallon: 8.345 lbs

TEGO CHEMIE SERVICE USA: Defoamers(Continued):

TEGO FOAMEX 3062(Continued):

Properties:
TEGO FOAMEX 3062 prevents macrofoam formation during the production and application of water based systems. It is effective in the pH range 2-11

Areas of application:
Building protection
* exterior, satin gloss and gloss paints based on pure acrylic and styrene-acrylic resins, PVA-copolymers and terpolymers
Wood/furniture varnishes
* water thinnable systems which are applied by spraying
Printing inks
* water thinnable gravure and flexographic inks

CTFA adopted name: dimethicone-copolyol

TEGO FOAMEX N:

Defoamer concentrate for medium, high solid and solvent-free systems.

Chemical composition:
Dimethyl polysiloxane in combination with highly dispersed silicic acid.

Physical data:
* Active matter content: 100%
* Density (20C): 0.98 g/cm3
* Flash point: 392F (200C)
* Solvent: None
* Weight/gallon: 8.17810

Properties:
TEGO FOAMEX N prevents macrofoam during the production and application of medium, high solid and solvent-free paint systems.

Areas of application:
Building protection/anticorrosive coatings
* high viscosity paint systems based on unsaturated polyester (e.g. thick film systesm) and epoxy resins

CTFA adopted name: simethicone

TRI-STAR CHEMICAL CO.: TRI-STAR Antifoam 27:

Typical Properties:

Type	Synergistic blend of organic chemicals
Activity	100%
Appearance	Translucent to clear liquid
Odor	Mild
Specific Gravity@77F	0.84
Density, lbs./gal.	7.0
Flash Point	Over 240F
pH (1% dispersion)	4.8 typical
Water content	Nil
Dispersibilty	Readily disperses in water

Uses:

TRI-STAR Antifoam 27 is a 100% active, water dispersible, liquid antifoam. It is very fast acting and is outstanding for rapid knockout of existing foam and the prevention of foam formation in systems containing detergents, in latex emulsions, in industrial processes, paints, glues, paper coating formulations, water disposal systems, sewage plants, dye baths, textile printing, paper manufacturing, oil well drilling muds, refineries, and petrochemical plants. ANTIFOAM 27 should be considered in all industrial processes where foam is a problem.

TROY CHEMICAL CORP.: Defoamers:

TROYSOL AFL

Defoamer and Anti-Floating Agent for Solvent Systems

TROYSOL AFL is a silicone-free surfactant in liquid form to improve bubble release and reduce pigment floating in non-aqueous systems.

TROYSOL AFL is effective in solvent based and other non-aqueous systems where surface tension related problems such as foaming, bubble retention, bridging, and pigment floating are encountered. Typical applications include:

* improving bubble release in coatings applied by dip, curtain coater, flow-coater, roller, airless spray, etc.
* preventing foam buildup during manufacture or filling
* preventing bubble entrapment over wood, masonry, and other porous substrates
* preventing bridging over holes, grillwork, etc. in dip applications
* reducing pigment floating and flooding in multi-pigmented systems

Systems which have benefited from the incorporation of TROYSOL AFL include:

* roller applied alkyd, urethane, and chlorinated rubber coatings
* moisture curing urethanes
* nitrocellulose, acrylic, and vinyl lacquers
* two-component epoxy and urethane systems
* baking alkyd, polyester, and acrylic primers and topcoats
* polyester gel coats
* solvent gravure and silk screening inks
* high build thixotropic coatings

Advantages:

* high efficiency at low levels
* wide spectrum applicability
* easily incorporated with mild agitation
* silicone-free, minimal side effects
* eliminates film defects due to bubbling, floating, etc.

Typical Properties:

Composition: Polymeric surfactant solution
Appearance: Clear, amber liquid
Percent non-volatile: 28% minimum
Lbs/US gal. (25C): 7.25
Color (Gardner): 8 maximum
Flash Point: 104F. (40C) minimum TCC

TROY CHEMICAL CORP.: Defoamers(Continued):

TROYKYD D22
Effective General Purpose Defoamer for Latex Paints

TROYKYD D22 is a highly effective defoamer supplied as an off white opaque liquid. TROYKYD D22 is a quick dispersing hydrocarbon oil based blend containing unique surface active agents.

Use:

TROYKYD D22 is highly effective and compatible in latex paint formulations.

TROYKYD D22 provides excellent anti-foam action during pigment grinding and other latex paint processing operations.

TROYKYD D22 is effective at low levels in aqueous adhesive formulations based on PVA, acrylic and other commonly used latex polymers.

Advantages:

* rapid bubble breaking
* excellent persistency
* compatibile
* easily dispersed

Typical Properties:

Appearance: off white opaque liquid
Activity: 100%

TROYKYD D44

TROYKYD D44 is a highly effective general purpose defoamer. TROYKYD D44 is supplied as a particle free, low viscosity fluid.

Use:

TROYKYD D44 can be used in a wide variety of applications where solid particles are objectionable. TROYKYD D44 is especially effective in applications where longevity and compatibility are required. TROYKYD D44 is particularly suited for paints, coatings and inks. Typical applications include:

* latex production
* pigment slurries and dispersions
* latex paints
* aqueous adhesives
* flexographic and gravure inks
* joint compounds and sealants

Advantages:

* particle free--can be used with high gloss clear formulations
* broad utility
* high efficiency
* compatible
* excellent longevity
* economical

Typical Properties:

Appearance: Hazy amber
Non-Volatile: 99% Minimum

TROY CHEMICAL CORP.: Defoamers(Continued):

TROYKYD D55
High Potency Non-Silicone Defoamer for Aqueous Systems

TROYKYD D55 is a high potency, silicone-free liquid defoamer. A unique blend of surface-active ingredients, TROYKYD D55 is supplied as a low viscosity fluid.

Use:
TROYKYD D55 is especially effective in systems where the foam is unusually stable or difficult to eliminate. TROYKYD D55 can be used in a broad range of coatings and paints because of its high degree of compatibility and minimal effect on color development. Typical applications include:
* gloss and semi-gloss latex paints
* joint compounds and sealants
* pigment slurries and dispersions
* adhesives
* aqueous inks

Advantages:
* excellent persistency
* highly compatible
* silicone free
* high efficiency
* broad range of applications

Typical Properties:
Appearance: Off-white opaque liquid
Active Material, %: 100

TROYKYD D333
Single Phase, Silicone-Free Defoamer for Latex Paints and Aqueous Systems

Modified glycol-hydrocarbon blend
TROYKYD D333 is a clear, single phase, particle-free defoamer for latex paints and other aqueous systems. It is a stable, compatible, efficient defoamer which will not separate on standing
Use:
TROYKYD D333 is used in systems based on acrylic, PVA and SBR emulsion systems, emulsions of solvents and waxes.

Advantages:
* uniform and homogeneous--will not separate in storage
* cost efficient--100% active
* silicone free--minimum tendency to form fisheyes
* silica free--avoids grit formation
* easily added at any state of manufacture
* excellent compatibility
* contains FDA approved components

TROY CHEMICAL CO.: Defoamers(Continued):

TROYKYD D666
High Efficiency, General Purpose Defoamer for Nonaqueous Systems

Hydrocarbon based surfactant

TROYKYD D666 is a silicone free, high efficiency general purpose defoamer and anti-foam supplied as an off-white cloudy liquid.

Use:

TROYKYD D666 is designed for use with latex emulsions and water reducible resins.

TROYKYD D666 is highly effective and compatible with aqueous resin systems and provides excellent longevity and package stability. Typical applications include:

* latex paints
* water reducible coatings
* latex adhesives
* starch adhesives
* paper manufacture
* textile finishes
* warp sizes
* flexographic and gravure inks
* latex manufacture
* metal working fluids

Advantages:

* silicone free
* highly compatible
* excellent longevity
* does not cause fisheyes
* free flowing homogeneous liquid
* high efficiency

TROYKYD D999
High Efficiency, Silicone Free, Permanent Defoamer

Vegetable oil modified hydrocarbon

TROYKYD D999 is a silicone free, high efficiency defoamer supplied as an opaque yellow low viscosity liquid.

Use:

TROYKYD D999 is particularly effective for aqueous coatings and paints where excellent defoaming without adverse side effects is required. Specific applications include:

* gloss and semi-gloss latex paints
* water reducible industrial coatings
* emulsion based industrial coatings
* latex and starch adhesives
* water based inks
* latex and starch adhesives
* SBR paper coatings

Advantages:

* high efficiency
* excellent longevity
* superior compatibility
* minimal effect on color acceptance
* silicone free
* minimal effect on flow

TROY CHEMICAL CORP.: Dispersants/Surfactants:

TROYSOL AFP
Non-Ionic Wetting and Anti-Floating Agent for Solvent Systems

TROYSOL AFP is a micro-milled calcium carbonate powder treated with surface active agents and pigment dispersants to improve pigment wetting and reduce interfacial tension in systems to which it is added.

Use:

TROYSOL AFP is used effectively in solvent-based and other non-aqueous systems where flooding and/or floating are observed. Typical applications include:

* pigment systems based on mixtures of inorganic and fine particle-size organics
* vehicle systems based on alkyds, epoxies, chlorinated rubber, vinyls, styrenated alkyds, alkyd-amino baking enamels, nitrocellulose and acrylic lacquers, etc.
* systems based on high performance, poor wetting vehicles
* vehicle systems of two-component epoxies, urethanes, polyesters, thermoset acrylics, high solids systems, etc.
* tint bases to be subsequently tinted in the store, in the field, by paint contractors, etc.
* paint lines requiring color uniformity by spray/dip application

Advantages:

* positive control of flooding and floating
* extremely safe; minimal side effects
* wide spectrum activity
* particularly effective with poor wetting vehicles
* reduces color problems in spray/dip applications
* minimizes problems when paints are tinted on the job

Typical Properties:

Appearance: Fine white powder
Percent active: 100%

TROY CHEMICAL CORP.: Dispersants/Surfactants(Continued):

TROYSOL CD1
Pigment Dispersant for Solvent Based Systems

TROYSOL CD1 is a general purpose pigment dispersant and wetting agent for non-aqueous systems. A 70% active solution, TROYSOL CD1 offers efficient pigment grinding with improved gloss and color development.

Use:

TROYSOL CD1 allows the coating formulator to achieve higher pigment loadings and faster grind development in a non-aqueous coatings. TROYSOL CD1 will also improve color uniformity and development. The efficient dispersing action of TROYSOL CD1 provides higher initial gloss and improves gloss retention. Applications include:

* Short, medium and long oil alkyds
* polyesters
* urethanes
* epoxies
* pigment dispersions

Advantages:

* higher pigment loading
* reduced grinding time
* broad compatibility
* increased gloss
* improved gloss rentention
* maximum color development
* improved resin compatibility

TROYSOL CD2
Pigment Dispersant for Solvent Based Systems

Modified vegetable oil derivative

TROYSOL CD-2 is a a liquid pigment dispersant that has proven to be particularly effective in solvent-based industrial coatings. A 100% active liquid, TROYSOL CD2 provides efficient grinding at high pigment loadings. Dispersions made with TROYSOL CD2 exhibit enhanced color strength and stable viscosity.

Use:

TROYSOL CD2, because of its excellent compatibility and effectiveness at very low dosages, is particularly applicable to solvent-thinned industrial finishes. TROYSOL CD2 offers higher gloss, uniform color development and stable viscosity in most systems. Typical applications include:

* alkyd dip primers
* air dry alkyds
* acrylic lacquers
* epoxies
* alkyd amino baking enamels
* polyesters
* urethanes

Advantages:

* higher pigment loading
* stable viscosity
* reduced grinding time
* increased gloss
* enhanced color development
* FDA approval (175.300)

TROY CHEMICAL CORP.: Dispersants/Surfactants(Continued):

TROYSOL LAC
Flow Control Agent for Aqueous Systems

TROYSOL LAC is a high efficiency surface wetting agent supplied as a pale yellow liquid.

Use:

TROYSOL LAC is suggested for use in any aqueous system where surface contamination or film imperfections present problems. TROYSOL LAC is particularly effective in preventing crawling, craters, pinholes, fisheyes and other film discontinuities that develop from poor surface wetting. TROYSOL LAC should be used where oily, greasy or marginally treated substrates are encountered. TROYSOL LAC can also improve color acceptance in aqueous paints. Proven applications include:

* emulsion-based decorative enamels
* emulsion-based industrial coatings
* water-reducible alkyds
* water-borne metal finished
* water-based flexo inks
* latex based pressure sensitive adhesives

Advantages:

* effective at low levels
* compatible glycol-water solution
* improves color acceptance
* eliminates craters, crawling and fisheyes
* improves gloss in aqueous baking finishes
* provides better wetting of contaminated surfaces

TROYSOL LWD
Wetting Agent and Dispersant for Aqueous Systems

TROYSOL LWD is a premium grade lecithin-based dispersant and wetting agent for aqueous systems. TROYSOL LWD:

* prevents flocculation of dispersed pigments
* improves rate of pigment dispersion
* improves color uniformity and gloss in coating systems
* stabilizes viscosity
* improves freeze-thaw stability

Use:

TROYSOL LWD is used in:

* latex flat, semi-gloss and gloss paints based on PVA and acrylic resins
* water reducible industrial systems where wide compatibility and consistent performance are important
* water-diluted solvent paints based on oleo-resinous vehicles
* emulsion systems for adhesives, cleaning compounds, etc.

Advantages:

* light color for whites and light tints
* excellent color acceptance and uniformity
* cost effective

TROY CHEMICAL CORP.: Dispersants/Surfactants(Continued):

TROYSOL MS2
An Agent to Improve Mar Resistance
and Slip in Water-Based Systems

TROYSOL MS2 is a low viscosity liquid which mixes readily into water-thinned coatings and inks. TROYSOL MS2 produces the following improvements in film properties:
* increased surface slip
* mar resistance
* blocking resistance
* abrasion resistance
* surface flow

Use:
TROYSOL MS2 is used in air-drying and baking water-reducible systems where improvements in slip and flow are desired.
Typical applications include:
* air-drying and baking alkyd coatings
* air-drying and baking acrylic coatings
* baking polyester-melamine coatings
* aqueous flexographic inks
* clear water-reducible lacquers
* urethane emulsions
* gloss and semi-gloss emulsion coatings

Advantages:
* better recoatability versus competitive products
* effective at low levels
* compatible with broad range of aqueous systems
* improves slip and mar resistance
* improved abrasion and scratch resistance
* improved surface flow
* reduced "orange peel"
* no adverse effects on gloss or color
* no effect on water resistance

Typical Properties:
Composition: Solution of non-ionic oligomeric surfactants
Form: Clear pale liquid
Non-volatile: 28%
Viscosity (G-H): A-4 max.
Density: lbs/US gal: 7.92
Flashpoint: 43C (110F) TCC

TROY CHEMICAL CORP.: Dispersants/Surfactants(Continued):

TROYSOL S366
Flow Control Agent for Nonaqueous Systems

Nonionic silicone modified polymer

TROYSOL S366 is a specially formulated surface active compound supplied as a pale yellow liquid, that prevents craters, pinholes and fisheyes in solvent-thinned systems.

Use:

TROYSOL S366 is recommended for resin systems that generally exhibit inadequate flow properties, and for finishing applications where inadequate substrate preparations is encountered. TROYSOL S366 improves flow properties of poorly wetting vehicles and eliminates film imperfections caused by poorly cleaned substrates, or substrates contaminated by silicone or oil. Typical applications include:

* epoxies
* acrylics
* polyesters
* vinyls
* urethanes
* alkyd amino systems
* water-soluble alkyds/acrylics
* chlorinated rubber

Advantages:

* prevents craters, pinholes and fisheyes
* improves flow and leveling
* effective at low levels
* does not affect water resistance
* provides better substrate wetting
* broadly compatible
* occasionally defoams aqueous industrial systems

TROYSOL Q148
Flow Control Agent and Defoamer for Aqueous Systems

TROYSOL Q148 is a specially formulated flow control agent that may also function as a defoamer. Supplied as a 27% active solution, TROYSOL Q148 provides an optimum balance of defoaming, flow and substrate wetting in water reducible systems.

Use:

TROYSOL Q148 is effective in a wide variety of water reducible resins as well as emulsions and colloidal dispersions. In addition to coatings, TROYSOL Q148 has found application in inks and adhesives based on similar resins. These aqueous systems include:

* air-drying and baking alkyds
* maleinized oil primers
* polyesters
* thermoplastic acrylics
* air drying and baking epoxies
* urethanes
* thermosetting acrylics

Advantages:

* optimum defoaming, flow and leveling with one additive
* promotes spreading and wetting on difficult substrates
* eliminates craters and crawling
* does not interfere with intercoat adhesion
* effective at low levels
* broad compatibility with commonly used resins
* may eliminate floating and flooding

TROY CHEMICAL CORP.: Dispersants/Surfactants(Continued):

TROYSOL 98C
Dispersant and Wetting Agent for Non-Aqueous Systems

TROYSOL 98C is a unique amphoteric wetting and dispersing agent for non-aqueous systems.

Use:

TROYSOL 98C is used in a broad range of coatings, inks and other non-aqueous systems. It is compatible with aliphatic, aromatic and oxygenated solvents. Vehicle types in which TROYSOL 98C is effective include the following:

* long, medium and short oil alkyds
* drying oils
* amino resins
* vinyls
* acrylics
* oil modified urethanes
* epoxy esters
* two component epoxies
* saturated polyesters
* styrene and vinyltoluene modified alkyds

Advantages:

TROYSOL 98C provides these benefits:

* higher pigment loading
* faster pigment grinding
* improved dispersion
* better color development
* improved color acceptance
* reduced viscosity
* improved viscosity stability
* higher gloss
* better flow and leveling
* prevention of gelling and seeding
* reduced pigment settling

TROYSOL 307
Defoamer and Anti-Floating Agent for Nonaqueous Systems

Nonionic polymeric surfactant containing silicone copolymer

TROYSOL 307 is a specially formulated surface active liquid designed to improve bubble release and reduce pigment floating in nonaqueous systems.

Use:

TROYSOL 307 is a modification of TROYSOL AFL containing a flow control agent to minimize cratering or related flow defects. These systems include:

* alkyd-urea baking coatings
* polyester-melamine baking finishes
* moisture-curing polyurethanes
* two component systems--epoxies, urethanes, etc.

Advantages:

* high efficiency at low levels
* readily incorporated with mild agitation
* eliminates film defects resulting from bubble retention, pigment floating, etc.
* may eliminate cratering and other flow defects

ULTRA ADHESIVES, INC.: Antifoams/Defoamers:

DEE FO 97-2:

DEE FO 97-2 is a high efficiency foam control agent especially designed for aqueous paint systems. The product provides foam control at any point during processing and/or application. Excellent performance has been proven in a variety of emulsions, such as poly-vinyl acetate and acrylics. The initial defoaming activity is extremely high and the product retains its effectiveness on long term storage.

Specifications and Properties:

Activity	100%
Appearance	slightly hazy liquid
Density	7.33 lbs/gallon

DEE FO 278:

Antifoam/Defoamer for Adhesives Compounding

DEE FO 278 is a 100% active liquid defoaming and anti-foaming agent designed especially for emulsion polymers based on polyvinyl alcohol as a protective colloid and emulsifier.

It is particularly effective in adhesives compounding using PVAC homo-polymer and co-polymer emulsions and the newly developed ethylene-vinyl acetate (PVOH protected) emulsions.

DEE FO 278 can also be used in systems containing acrylic and styrene butadiene emulsions.

DEE FO 495:

Antifoam/Defoamer for Latex Flat, Semi-Gloss and Gloss Paints

DEE FO 495 is a 100% active liquid defoamer and anti-foaming compound.

DEE FO 495 has been designed specifically for highest efficiency in foam control for aqueous paint systems. Its well balanced composition provides for optimum compatibility with polyvinyl acetate emulsion based systems, vinyl acrylics, pure acrylics and butadiene/styrene based compositions.

The initial defoaming efficiency of DEE FO 495 is extremely high and the product retains its defoaming/anti-foam effectiveness on long term storage.

Specifications and Properties:

Activity	100%
Appearance	slightly hazy liquid
Weight per gallon, pounds	7.26

ULTRA ADHESIVES, INC.: Antifoams/Defoamers(Continued):

DEE FO 806-102:

DEE FO 806-102 is completely non-ionic and therefore highly functional over a broad pH range, ie. 2-14. The inertness of DEE FO 806-102 contributes and translates to highest defoaming efficiency over the entire liquid temperature range of water.

DEE FO 806-102 displays excellent defoaming and anti-foaming properties in such specific aqueous systems as enumerated below:

1. The newly introduced two package water thinned epoxy-acrylic coatings
2. Aqueous coatings applied electrophoretically or by spray, flow or dip coating technique
3. Acrylic emulsion based gloss and semi-gloss architectural and industrial finishes
4. Conventional emulsion polymer latex systems based on polyvinyl acetate, vinyl acrylic, vinyl acetate/ethylene, vinyl chloride/ethylene and SBR polymers.

Specifications and Properties:

Physical State	liquid
Color	light, opaque
Odor	bland
Activity	100%
Weight per gallon	7.45 pounds

DEE FO 1020-F:

Antifoam/Defoamer for Latex Flat, Semi-Gloss and Gloss Paints

DEE FO 1020-F represents a 100% active defoamer and antifoaming compund which was designed to function most efficiently in latex paints, flats, semi-gloss and gloss paint.

The well balanced composition of DEE FO 1020-F provides for optimum compatibility with various emulsion polymers (PVA, Acrylic and SBR) pigments, and especially dispersed colorants.

The initial defoaming efficiency of DEE FO 1020-F is extremely high and the defoamed total system retains its defoaming/antifoaming characteristics on long term storage.

Specifications and Properties:

Appearance	Opaque Liquid
Activity	100%
Weight Per Gallon	7.25-7.28 pounds

ULTRA ADHESIVES, INC.: Anti-Foams/Defoamers(Continued):

2001-1 Defoamer/Anti-Foam:

Maximum antifoam and defoaming efficiency with optimum economy. 2001-1 is finely balanced for optimum surface and interfacial surface tension properties; its development was directed towards immediate foam knock-down which is of extreme importance during the manufacture and processing of water-borne coatings and aqueous systems.

2001-1 is recommended for all major polymer systems such as 100% acrylic, PVAc, PVA-Acrylic, Ethylene-VA as well as water-borne formulations.

2001-1 is particularly effective in eliminating foam during application resulting in foam-free films. 2001-1 remains effective after long storage periods.

Typical Properties:
Appearance: opaque white liquid
Pounds per gallon: 7.75
Viscosity, cps (Brookfield): 1750+-500

Application:
Latex paints based on Acrylic, PVAc, PVA-Acrylic and Ethylene-Vinyl Acetate polymers.
Paper Coatings
Adhesives Manufacture
Amine neutralizable water reducible resin coatings and compounds.

DEE FO 2020A-Universal Defoamer:

Represents the latest non-silicone defoamer/anti-foamer technology. It was designed to give outstanding foam control when used in latex paints and aqueous coatings. Provides for excellent foam control during manufacture and processing of flat, semi-gloss and gloss paints. It will not adversely effect gloss retention or color development.

Special Features:
- 100% Defoaming activity
- Rapid foam break - upon mechanical agitation
  - upon application
- Cost effective
- Long term stability

Physical Properties:

Appearance:	Opaque, milky liquid
Weight/Gallon, pounds:	7.1
Viscosity (Brookfield):	3000-6000 cps

ULTRA ADHESIVES, INC.: Antifoamers/Defoamers(Continued):

DEE FO 2020E - Emulsifiable Universal Defoamer

DEE FO 2020E represents the latest non-silicone Defoamer/ Anti-foamer technology. It was designed to give outstanding foam control and provides ease of incorporation in many aqueous systems such as Inks, Adhesives, Textile, and Water Treatment Chemicals.

Special Features:
- 100% Defoaming Activity
- Rapid Foam Break - upon mechanical agitation
  - upon application
- Cost Effective
- Long Term Stability

Physical Properties:

Appearance	Opaque, off-white liquid
Weight/Gallon, Pounds	7.1
Viscosity (Brookfield)	2000-4000 cps.

Experimental DEE FO XRM-1158F:

Antifoam/Defoamer for Aqueous Inks and Clear Overprint Varnishes

A low viscosity liquid defoamer formulated specifically for aqueous inks and clear overprint varnishes. Defoamer design ensures long term product stability while allowing for rapid and complete dispersion in aqueous systems.

Specifications and Properties:

Appearance	Translucent, Amber Liquid
Weight/Gallon, Pounds @ 25C	8.4
Viscosity, Cps. @ 25C	500-600

FOAMTROL 101:

FOAMTROL 101 is one of a series of specially formulated foam control agents designed for application in water-base paints and coatings fields.

FOAMTROL 101 is recommended for use in the following systems:

Polyvinyl Acetate, Butadiene-Styrene Latex, Acrylics, Gums, Resins, Detergents, Wetting Agents and Fluorocarbons.

Specifications and Properties:

Physical State	Fluid (Thixotropic)
Active Ingredients	100%
Acid No.	5-6
pH	6-6.5
Weight/Gallon	6.74 Lbs.
Emulsifier System	Non-Ionic

ULTRA ADHESIVES, INC.: Antifoamers/Defoamers(Continued):

FOAMTROL 103:

FOAMTROL 103 is a 100% active product which contains no water or silicones. It is effective in both acid and alkaline media as well as some solvent systems.

Applications:
Polymerization Reactions: Polyvinyl Acetate, Polyvinyl Chloride, Acrylics, Butadiene-Styrene and Copolymers.
Paints and Coatings: All water-based paints and coatings containing the above polymers.
Paper and Textiles: Gelatins, Casein, Starch, Gums, Latex, Dextrin, Polyvinyl Alcohol, Fluorochemicals, Emulsifiers, Wetting Agents, Asphalt Emulsions and Resins.
Processing Industries: Cleaning Compounds, Waste Disposal, Detergents, Evaporation, Distillation and Fermentation.

Specifications and Properties:

Physical State:	Liquid Dispersion
Color	White
Active Ingredients	100%
Acid No.	9-10
pH	6.5-7
Weight/Gallon	6.7 Lbs.

FOAMTROL 160:

FOAMTROL 160 is a heavy duty defoamer designed specifically to meet the growing need for safe foam control without the side effects of "Fish Eyes".
FOAMTROL 160 is a 100% active product and is free of silicones and water.

Applications:
FOAMTROL 160 is recommended for application in the fields of Paints, Coatings, and Adhesives. It is effective in PVA, Acrylic and Styrene Latex Systems, and functions over a wide range of pH, temperature and viscosity conditions.

Specifications and Properties:

Physical State	Thixotropic Liquid Dispersion
Color	Straw
Active Ingredients	100%
Acid No.	7-8
pH	6.5-7
Emulsifier System	Non-Ionic
Weight/Gallon(Approx.)	7 Lbs.

ULTRA ADHESIVES, INC.: Products:

ULASPERSE 994-B:
Dispersant

ULASPERSE 994-B is a complex anionic phosphate ester. Its primary function is to readily disperse pigments used in aqueous systems.

ULASPERSE 994-B is very effective in deagglomerating such pigments as titanium dioxide, calcium carbonate, clay, magnesium silicate and other inorganic pigments. Recommended amounts of ULASPERSE necessary to effect complete dispersion of inorganic pigments range from 0.2-1.0 percent, based on solids.

The acceptability of organic color dispersions and universal colorants is greatly enhanced by the incorporation of ULASPERSE 994-B. ULASPERSE 994-B is a low molecular weight formulation; it should have little, if any, adverse effect on viscosity of the final product, when stored for long periods of time.

Specifications and Properties:

Type	Anionic
State	Liquid
Appearance	Transparent
Active Ingredients	25%
pH	6.2-6.8
Viscosity	8 cps
Specific Gravity	1.16

ZINPLEX 15:
Zinc Crosslinking Agent

ZINPLEX 15 can be used to crosslink aqueous carboxylated polymers to impart and improve the following physical properties:

1) Non-Blocking Characteristics
2) Chemical Resistance
3) Hardness
4) Water and Solvent Resistance
5) Detergent Resistance
6) Mar Resistance

Specifications and Properties:

Percent Solids (As Zinc Oxide)	15.0
Appearance	Colorless Liquid
pH	11.3-11.7
Weight per Gallon	10.1 Lbs.
Odor	Strongly Ammoniacal

Suggested Industrial Applications:

1) Floor Polishes
2) Paper and Paperboard Overprint Coatings
3) Clear and Pigmented Industrial Finishes
4) Printing Inks
5) Textile Finishes
6) Adhesives

R.T. VANDERBILT CO., INC.: Paint/Paper Products:

DARVAN No. 1
Pigment Dispersing Agent for Aqueous Systems

Chemical classification: Sodium salts of polymerized alkyl naphthalene sulfonic acids; anionic
Appearance: Buff granules
Active dispersing agent: 87% minimum

DARVAN No. 7
Minimum-Foaming Pigment Dispersing Agent for Aqueous Systems

Chemical classification: A polyelectrolyte dispersing agent in aqueous solution
Appearance: Clear to slightly opalescent colorless liquid
Total solids: 25+-1%
pH: 9.5 to 10.5
Density at 25C (77F): 1.16 Mg/m3
Viscosity at 25C (77F): 75 cps maximum

DARVAN No. 31
Organic and Inorganic Pigment Dispersing Agent for Aqueous Systems

Description and Properties:
Chemical classification: Sodium salt of a carboxylated polyelectrolyte
Physical state: Liquid
Color: Yellow
Total solids: 25% minimum
Density at 25C: 1.11+-.02 Mg/m3
Viscosity at 25C: 75 cps maximum
pH: 9 to 11
Remarks: Very effective in gloss as well as flat latex coatings

RHEOTOL
Organic Wetting and Dispersing Agent for Pigmented Coatings

RHEOTOL is a solution of polymerized alkyl phosphate. It aids pigment wetting and dispersion, improves flow, leveling, and gloss of oleoresinous coatings. It is a stable, nonreactive, film-forming organic composition, available as a pale heavy bodied liquid with a density of 0.97 Mg/m3. It consists of 73 to 75% non-volatile material and 25 to 27% mineral spirits. This product may be classed as a reagent or additive.

Uses:
RHEOTOL performs as a mixing and dispersing aid in many types of paint, particularly those based on poor wetting vehicles. In such cases, paste flow is increased, thus allowing a greater percentage of pigment to be mixed with a given amount of vehicle. Mixing and milling time is thereby reduced. RHEOTOL also improves leveling and gloss of many oleoresinous paints, such as those based on linseed oil and alkyd vehicles.

Section 2
Driers and Antiskinning Agents

COSAN CHEMICAL CO.: COATE Driers:

Metallic Soaps of Synthetic Acids.

Common Metal Salts and Their Function:

COATE Lead: Promotes hard, thorough drying throughout a film by acting as a polymerization catalyst.

COATE Zirconium: When lead free systems are desired, this is the auxiliary drier most often used. In comparison to a lead drier, it shows improved gloss, gloss retention, and color, but does not perform as well at low temperatures and high humidity.

COATE Cobalt: Primarily a strong oxidation catalyst, it promotes surface drying of films rapidly.

COATE Manganese: Also an active drier, but in addition to generating surface-dry characteristics, also promotes polymerization.

COATE Calcium: Can be used in one of three ways:
As a total replacement for lead;
In combination with zirconium to replace lead;
As an auxiliary drier in conjenction with lead.

COATE Zinc: An auxiliary drier that is used in conjunction with cobalt. It allows for optimum film hardness by keeping the film "open" and accessible to oxygen.

Typical Product Properties

Product:	Metal Content
Calcium 10%	10+-1%
Calcium 8%	8+-0.1%
Calcium 6%	6.0+-0.1%
Calcium 5%	5.0%+-0.1%
Calcium 4%	4.0%+-0.1%
Cobalt 12%	12.0+-0.1%
Cobalt 6%	6.0+-0.1%
Lead 36%	36.0+-0.2%
Lead 24%	24.0+-0.1%
Manganese 12%	12.0+-0.1%
Manganese 9%	9.0+-0.1%
Manganese 6%	6.0+-0.1%
Zinc 16%	16.0+-0.1%
Zinc 8%	8.0+-0.1%
Zirconium 24%	24.0+-0.1%
Zirconium 18%	18.0+-0.1%
Zirconium 12%	12.0+-0.1%
Zirconium 6%	6.0+-0.1%

DANIEL PRODUCTS CO.: DAPRO DP 5005:

Special Drier for Modified Polyurethanes and Alkyds

DAPRO DP 5005 is a chelating catalyst designed to replace cobalt drier where cobalt causes undesirable side effects. Unlike some metallic driers, it will not form soaps when used with high acid number vehicles and, therefore, will not interfere with viscosity or viscosity stability. DAPRO DP 5005 is particularly useful in clear and brightly pigmented oil or alkyd modified polyurethanes and in conventional alkyds which tend to discolor.

DAPRO DP 5005 can be used to:

* Minimize discoloration of clear polyurethane in the can
* Improve alkali resistance of polyurethane films
* Increase resistance to yellowing of air dry alkyd finishes exposed to prolonged heat or to alkaline fumes
* Prevent loss of drying on aging

Specifications:

Wt./Gal. (Sp.Gr.): 7.1 Lbs. (0.85)
Solvent: 1:1 Blend of Xylol and n-Butanol
Flash Point: 80F/26.7C

INTERSTAB CHEMICALS: Drier Catalysts:

ZIRCO:

ZIRCO, an oil-soluble polymeric zirconyl complex of a synthetic organic acid, is the best route to lead-free paints in compliance with legislative regulations.

Rare Earth:

Besides ZIRCO, Interstab offers two other important drier catalysts widely used in baking finishes. Comprised of Rare Earths, they provide the following advantages:

* Increased crosslinking in molecular structure which results in tougher films
* A wide range of catalytic activity at normal baking temperatures
* Excellent color retention in baked films
* Improved water and soap resistance of films

In addition, with heat cured alkyds (such as soya or dehydrated castor oil alkyds), the normal amount of cobalt metal can be replaced with as little as half the amount of Rare Earth metal.

INTERDRY CZ:

INTERDRY CZ is a special drier combination containing 6% cobalt metal and 9% zirconium metal.

ZIRCO:

Zirco 6%
Zirco 12%
Zirco 18%
Zirco 24%

INTERCAR ZIRCO:

Zirco 6%
Zirco 12%

Rare Earth Naphthenate:

Rare Earth 4%

Rare Earth Octoate:

Rare Earth 6%
Rare Earth 8%

INTERSTAB CHEMICALS: Driers:

Driers are metal soaps of organic acids, supplied as solutions in hydrocarbon solvent. Products are classified by metal content and acid type. The common metals include cobalt, manganese, calcium, zirconium, zinc, lead, rare earth, cerium and iron.

Naphthenate Driers:
These driers are metal soaps of highly refined and selected naphthenic acid dissolved in mineral spirits.

Hexogen Octoate Driers:
INTERSTAB Hexogen Driers are the pure metal soaps of 2-ethylhexoic acid dissolved in mineral spirits.

Noury-Dry Driers:
Noury-Dry driers are metal soaps of synthetic acids ranging from C-8 to C-12. Acids are carefully selected to give optimum performance for each individual metal.

Tallates:
INTERSTAB Tallate driers are made from tall oil acids.

INTERCAR Water-Dispersible Driers:
INTERCAR Driers are water dispersible, oil soluble driers, especially formulated for use in water-reducible alkyd and oil systems.

P.I. Driers:
INTERSTAB offers a complete line of paste driers, designed especially for the Printing Ink industry.

Noury-Dry Driers:
Calcium 4%
Calcium 5%
Calcium 6%
Calcium 8%
Calcium 10%
Cobalt 6%
Cobalt 12%
Iron 6%
Iron 9%
Iron 12%
Lead 24%
Lead 36%
Manganese 6%
Manganese 9%
Manganese 12%
Zinc 8%
Zinc 16%

INTERSTAB CHEMICALS: Driers(Continued):

Naphthenate Driers:
- Calcium 4%
- Calcium 5%
- Cerium 6%
- Cobalt 6%
- Iron 6%
- Lead 24%
- Manganese 6%
- Zinc 8%
- Zinc 10%

Hexogen Octoate Driers:
- Calcium 4%
- Calcium 5%
- Cobalt 6%
- Cobalt 12%
- Zinc 8%
- Zinc 18%
- Iron 6%
- Iron 9%
- Iron 12%
- Potassium 15% DEG
- Potassium 15% PEG
- Lead 24%
- Manganese 6%

P.I. Driers:
- Cobalt 4.4%
- Cobalt 5.6%
- Cobalt 6%
- Cobalt 12%
- Drier Blend CM
- Manganese 6%
- Manganese 9%
- Cobalt 6% Paste
- Manganese 6% Paste
- Drier Blend CM Paste

Tallate Driers:
- Calcium 4%
- Cobalt 6%
- Lead 24%

MOONEY CHEMICALS, INC.: MOONEY CEM-ALL Driers(Synthetics):

10% Calcium:
% NVM: 51
% Volatile: 49
Viscosity Gardner 25C: A-2

8% Calcium:
% NVM: 41
% Volatile: 59
Viscosity Gardner 25C: A-4

6% Calcium:
% NVM: 33
% Volatile: 67
Viscosity Gardner 25C: A-5

4% Calcium:
% NVM: 22
% Volatile: 78
Viscosity Gardner 25C: A-5

12% Cobalt:
% NVM: 66
% Volatile: 34
Viscosity Gardner 25C: F

6% Cobalt:
% NVM: 35
% Volatile: 35
Viscosity Gardner 25C: A-5

12% Copper:
% NVM: 71
% Volatile: 29
Viscosity Gardner 25C: R

36% Lead:
% NVM: 76
% Volatile: 24
Viscosity Gardner 25C: G

24% Lead:
% NVM: 50
% Volatile: 50
Viscosity Gardner 25C: A-5

12% Manganese:
% NVM: 64
% Volatile: 36
Viscosity Gardner 25C: C

MOONEY CHEMICALS, INC.: MOONEY CEM-ALL Driers(Synthetics) (Continued):

9% Manganese:
% NVM: 56
% Volatile: 44
Viscosity Gardner 25C: A-4

6% Manganese:
% NVM: 36
% Volatile: 64
Viscosity Gardner 25C: A-5

16% Zinc:
% NVM: 74
% Volatile: 26
Viscosity Gardner 25C: A

8% Zinc:
% NVM: 35
% Volatile: 65
Viscosity Gardner 25C: A-5

24% Zirconium:
% NVM: 86
% Volatile: 14
Viscosity Gardner 25C: Y

18% Zirconium:
% NVM: 64
% Volatile: 36
Viscosity Gardner 25C: A-3

12% Zirconium:
% NVM: 42
% Volatile: 58
Viscosity Gardner 25C: A-5

6% Zirconium:
% NVM: 21
% Volatile: 79
Viscosity Gardner 25C: A-5

MOONEY CHEMICALS, INC.: MOONEY HEX-CEM Driers(Octoates):

5% Calcium:
% NVM: 49
% Volatile: 51
Viscosity Gardner 25C: A

12% Cobalt:
% NVM: 64
% Volatile: 36
Viscosity Gardner 25C: A

6% Cobalt:
% NVM: 31
% Volatile: 69
Viscosity Gardner 25C: A-5

6% Manganese:
% NVM: 37
% Volatile: 63
Viscosity Gardner 25C: A-5

10% Nickel:
% NVM: 61
% Volatile: 39
Viscosity Gardner 25C: A-3

12% Rare Earth:
% NVM: 61
% Volatile: 39
Viscosity Gardner 25C: A-1

22% Zinc:
% NVM: 99
% Volatile: 1
Viscosity Gardner 25C: Y

18% Zinc:
% NVM: 80
% Volatile: 20
Viscosity Gardner 25C: A-5

8% Zinc:
% NVM: 35
% Volatile: 65
Viscosity Gardner 25C: A-5

18% Zirconium:
% NVM: 64
% Volatile: 36
Viscosity Gardner 25C: A-3

12% Zirconium:
% NVM: 42
% Volatile: 58
Viscosity Gardner 25C: A-5

6% Zirconium:
% NVM: 22
% Volatile: 78
Viscosity Gardner 25C: A-5

MOONEY CHEMICALS, INC.: MOONEY HYDRO-CURE and HYDRO-CEM Water-Dispersible Driers:

5% Calcium HYDRO-CEM:
% NVM: 60
% Volatile: 40
Viscosity Gardner 25C: H

Cobalt HYDRO-CURE:
% NVM: 52
% Volatile: 48
Viscosity Gardner 25C: A-2

Cobalt HYDRO-CURE II:
% NVM: 46
% Volatile: 54
Viscosity Gardner 25C: A

Manganese HYDRO-CURE:
% NVM: 45
% Volatile: 55
Viscosity Gardner 25C: A-4

Manganese HYDRO-CURE II:
% NVM: 64
% Volatile: 36
Viscosity Gardner 25C: K

12% Zirconium HYDRO-CEM:
% NVM: 47
% Volatile: 53
Viscosity Gardner 25C: A-5

DRI-RX:
% NVM: 30
% Volatile: 70
% Viscosity Gardner 25C: A-5

MOONEY CHEMICALS, INC.: MOONEY LIN-ALL Driers(Tallates):

6% Cobalt:
% NVM: 66
% Volatile: 34
Viscosity Gardner 25C: B

24% Lead:
% NVM: 67
% Volatile: 33
Viscosity Gardner 25C: A

6% Manganese:
% NVM: 62
% Volatile: 38
Viscosity Gardner 25C: A

8% Zinc:
% NVM: 66
% Volatile: 34
Viscosity Gardner 25C: F

MOONEY CHEMICALS, INC.: MOONEY NAP-ALL Driers(Naphthenates):

4% Calcium:
% NVM: 32
% Volatile: 68
Viscosity Gardner 25C: A-5

6% Cobalt:
% NVM: 55
% Volatile: 45
Viscosity Gardner 25C: A

6% Copper:
% NVM: 69
% Volatile: 31
Viscosity Gardner 25C: L

8% Copper:
% NVM: 75
% Volatile: 25
Viscosity Gardner 25C: X

12% Iron NEO-NAP (Modified Naphthenate):
% NVM: 72
% Volatile: 28
Viscosity Gardner 25C: G

6% Iron:
% NVM: 67
% Volatile: 33
Viscosity Gardner 25C: L

24% Lead:
% NVM: 60
% Volatile: 40
Viscosity Gardner 25C: A-2

6% Manganese:
% NVM: 56
% Volatile: 44
Viscosity Gardner 25C: A-1

8% Zinc:
% NVM: 60
% Volatile: 40
Viscosity Gardner 25C: F

10% Zinc:
% NVM: 67
% Volatile: 33
Viscosity Gardner 25C: H

MOONEY CHEMICALS, INC.: MOONEY TEN-CEM Driers(Neodecanoates):

5% Calcium:
% NVM: 42
% Volatile: 58
Viscosity Gardner 25C.: A-4

21% Cobalt Hydroxy TEN CEM:
% NVM: 83
% Volatile: 17
Viscosity Gardner 25C: 160

12% Cobalt:
% NVM: 69
% Volatile: 31
Viscosity Gardner 25C: A

6% Cobalt:
% NVM: 40
% Volatile: 60
Viscosity Gardner 25C: A-5

24% Lead:
% NVM: 60
% Volatile: 40
Viscosity Gardner 25C: A-3

2% Lithium (water carrier):
% NVM: 63
% Volatile: 37
Viscosity Gardner 25C: X

6% Manganese:
% NVM: 47
% Volatile: 53
Viscosity Gardner 25C: A-4

6% Rare Earth:
% NVM: 30
% Volatile: 70
Viscosity Gardner 25C: A-5

16% Zinc:
% NVM: 82
% Volatile: 18
Viscosity Gardner 25C: F

12% Zirconium:
% NVM: 43
% Volatile: 57
Viscosity Gardner 25C: A

6% Zirconium:
% NVM: 22
% Volatile: 78
Viscosity Gardner 25C: A-5

TROY CHEMICAL CORP.: Driers and Anti-skinning Agents:

TROYKYD PERMA DRY

Loss-of-Dry Inhibitor for Coatings and Printing Inks

TROYKYD PERMA DRY is a lead-free feeder drier designed to replace the driers which are preferentially adsorbed on pigment surfaces, or inactivated by reaction with other formulation ingredients. TROYKYD PERMA DRY maintains drying efficiency in paints and printing inks, and reduces gel time drift of polyester resins.

Use:

TROYKYD PERMA DRY is used in paints and inks containing high surface area pigments, such as carbon blacks, organic reds and colloidal silicas. TROYKYD PERMA DRY also provides excellent gel time stability in unsaturated polyester systems in which accelerators lose efficiency on aging. Typical applications include:

* linseed oil iron oxide stains
* pre-accelerated polyester resins
* long and medium oil alkyds paints pigmented with carbon black, organic red, chrome green, iron oxides and titanium dioxide
* offset printing inks

Advantages:

* prevents loss of drying on aging
* maintains gel time on aging
* may reduce cobalt drier requirements, thereby reducing wrinkling
* more cost effective than non-cobalt inhibitors
* minimal adverse effects on color and dry versus cobalt paste feeder driers
* easy to handle pourable liquid

Typical Properties:

Composition: Cobalt complex
Color: Blue
Form: Viscous, pourable liquid
Density: gm/cc: 1.07
lbs/U.S. gal.: 8.92
Non-volatiles (% by weight): 65.0 Min.

TROY CHEMICAL CORP.: Driers and Anti-skinning Agents(Continued):

TROYMAX Driers

Product:

Calcium 4%
Metal Content: 4.0+-0.1%
% Non-Vol: 45

Calcium 5%:
Metal Content: 5.0+-0.1%
% Non-Vol: 59

Calcium 6%:
Metal Content: 6.0+-0.1%
% Non-Vol: 73

Calcium 8%:
Metal Content: 8.0+-0.1%
% Non-Vol: 44

Calcium 10%:
Metal Content: 10.0+-0.1%
% Non-Vol: 55

Cobalt 6%:
Metal Content: 6.0+-0.1%
% Non-Vol: 37

Cobalt 12%:
Metal Content: 12.0+-0.1%
% Non-Vol: 73

Iron 9%:
Metal Content: 9.0+-0.1%
% Non-Vol: 63

Lead 24%:
Metal Content: 24.0+-0.1%
% Non-Vol: 54

Lead 36%:
Metal Content: 36.0+-0.2%
% Non-Vol: 78

Manganese 6%:
Metal Content: 6.0+-0.1%
% Non-Vol: 45

Manganese 9%:
Metal Content: 9.0+-0.1%
% Non-Vol: 68

TROY CHEMICAL CORP.: Driers and Anti-Skinning Agents(Continued):

TROYMAX Driers(Continued):

Product:

Manganese 12%:
Metal Content: 12.0+-0.1%
% Non-Vol: 78

Zinc 8%:
Metal Content: 8.0+-0.1%
% Non-Vol: 39

Zinc 16%:
Metal Content: 16.0+-0.1%
% Non-Vol: 70

Zirconium 6%:
Metal Content: 6.0+-0.1%
% Non-Vol: 27

Zirconium 12%:
Metal Content: 12.0+-0.1%
% Non-Vol: 49

Zirconium 18%:
Metal Content: 18.0+-0.1%
% Non-Vol: 68

Zirconium 24%:
Metal Content: 24.0+-0.1%
% Non-Vol: 91

TROYKYD W.D. Driers
(Water Dispersible)

Product:

Calcium 6%:
Metal Content: 6.0+-0.1%
% Non-Vol.: 73
Cobalt 6%:
Metal Content: 6.0+-0.1%
% Non-Vol: 66
Lead 24%:
Metal Content: 24.0+-0.1%
% Non-Vol: 62
Manganese 6%:
Metal Content: 6.0+-0.1%
% Non-Vol: 63
Zirconium 12%:
Metal Content: 12.0+-0.1%
% Non-Vol: 49

TROY CHEMICAL CORP.: Driers and Anti-Skinning Agents(Continued):

TROYKYD Driers
(Naphthenate)

Product:

Calcium 4%:
Metal Content: 4.0+-0.1%
% Non-Vol: 61

Cerium 6%:
Metal Content: 6.0+-0.1%
% Non-Vol: 55

Cobalt 6%:
Metal Content: 6.0+-0.1%
% Non-Vol: 66

Iron 6%:
Metal Content: 6.0+-0.1%
% Non-Vol: 77

Lead 24%:
Metal Content: 24.0+-0.1%
% Non-Vol: 62

Manganese 6%:
Metal Content: 6.0+-0.1%
% Non-Vol: 63

Rare Earth 4%:
Metal Content: 4.0+-0.1%
% Non-Vol: 34

Zinc 8%:
Metal Content: 8.0+-0.1%
% Non-Vol: 58

Zinc 10%:
Metal Content: 10.0+-0.2%
% Non-Vol: 73

TROY CHEMICAL CORP.: Driers and Anti-Skinning Agents(Continued):

TROYCHEM Driers
(Octoate)

<u>Product:</u>

Calcium 5%:
Metal Content: 5.0+-0.1%
% Non-Vol: 54

Cobalt 6%:
Metal Content: 6.0+-0.1%
% Non-Vol: 42

Cobalt 12%:
Metal Content: 12.0+-0.1%
% Non-Vol: 72

Iron 6%:
Metal Content: 6.0+-0.1%
% Non-Vol: 50

Lead 24%:
Metal Content: 24.0+-0.1%
% Non-Vol: 58

Manganese 6%:
Metal Content: 6.0+-0.1%
% Non-Vol: 46

Zinc 8%:
Metal Content: 8.0+-0.1%
% Non-Vol: 38

Zirconium 6%:
Metal Content: 6.0+-0.1%
% Non-Vol: 27

TROY CHEMICAL CORP.: Driers and Anti-Skinning Agents(Continued):

TROYKYD Anti-Skinning Agents
* Liquid or Powder
* Broad Spectrum
* Highly Efficient
* Varying Volatility

TROYKYD Product:

ANTI-SKIN:
Physical Form: Liquid
Type: Mixed Oximes
Most Universal
Applications:

Air Dry Varnishes	Air Dry Enamels
Air Dry Flats	Oil-Based House Paints
Oil-Based Caulks	Solvent-Based Adhesives
Pigment Dispersions	Letterpress & Offset Inks

ANTI-SKIN B:
Physical Form: Liquid
Type: Methylethyl Ketoxime
Best for air-drying alkyds
Applications:

Air Dry Varnishes	Air Dry Enamels
Air Dry Flats	Oil-Based Caulks
Solvent-Based Adhesives	Pigment Dispersions
Letterpress & Offset Inks	Oil-Based House Paints

ANTI-SKIN BTO:
Physical Form: Liquid
Type: Butyraldoxime
Best for air-drying oleoresinous
Applications:

Air Dry Varnishes	Air Dry Enamels
Air Dry Flats	Oil-Based House Paints
Dip Tank Coatings	Oil-Based Caulks
Solvent-Based Adhesives	Pigment Dispersions
Letterpress & Offset Inks	

ANTI-SKIN S:
Physical Form: Liquid
Type: Phenolic
Min. Discoloration in clear urethane
Applications:

Air Dry Varnishes	Air Dry Enamels
Air Dry Flats	Oil-Based House Paints
Dip Tank Coatings	Oil-Based Caulks
Solvent-Based Adhesives	Pigment Dispersions
Letterpress & Offset Inks	

TROY CHEMICAL CORP.: Driers and Anti-Skinning Agents (Continued):

TROYKYD Product:

ANTI-SKIN SM:
Physical Form: Liquid
Type: Phenolic
Good dip tank stability
Applications:

Air Dry Flats	Dip Tank Coatings
Oil-Based Caulks	Solvent-Based Adhesives
Pigment Dispersions	Letterpress & Offset Inks

ANTI-SKIN OP:
Physical Form: Powder
Type: Oxime
Odorless, minimal effect on drying
Applications:

Air Dry Varnishes	Air Dry Enamels
Air Dry Flats	Pigment Dispersions
Letterpress & Offset Inks	

ANTI-SKIN OL:
Physical Form: Liquid
Type: Phenolic
Odorless, high efficiency
Applications:

Air Dry Flats	Dip Tank Coatings
Solvent-Based Adhesives	Pigment Dispersions
Letterpress & Offset Inks	Hot Melt Compounds

ANTI-SKIN P:
Physical Form: Paste
Type: Phenolic
Odorless, FDA approved
Applications:

Air Dry Flats	Dip Tank Coatings
Solvent-Based Adhesives	Pigment Dispersions
Letterpress & Offset Inks	Hot Melt Compounds

R.T. VANDERBILT CO., INC.: Paint/Paper Products:

ACTIV-8 and ACTIV-8 in Hexylene Glycol
Drier Accelerators and Stabilizers

ACTIV-8 and ACTIV-8 in Hexylene glycol are different solution forms containing the same active ingredient, 1,10-phenanthroline. Both solutions are equally effective for drier acceleration and stabilization. ACTIV-8 in Hexylene Glycol is essentially odorless and has a high flash point whereas ACTIV-8 has the characteristic odor of n-butanol and is classified as a flammable liquid.

ACTIV-8:

1,10-Phenanthroline, %: 38
n-Butanol, %: 52
2-Ethylhexoic Acid, %: 10
Density, Mg/m3: 0.95
Weight/Gallon, lbs: 7.9
Flash Point, F/C: 97/36

ACTIV-8 in Hexylene Glycol:

1,10-Phenanthroline, %: 38
Hexylene Glycol, %: 62
Density, Mg/m3: 1.03
Weight/Gallon, lbs: 8.6
Flash Point, F/C: 211/99

Section 3
Extenders, Fillers, Pigments

AGRASHELL, INC.: Vegetable Shell Products:

Industrial Flour
Made from Vegetable Shell Raw Material

Agrashell's Industrial Flours are low specific gravity lignocellulose products manufactured from nut shells. The flours are light in color, non-toxic, free flowing, low in resin absoprtion, and easily dispersed.

Typical Applications:
Industrial Flour is an ideal filler/extender for many synthetic resins. Some other uses are listed below:

Rubber
Adhesives
Cosmetics
Texture Paints
Cast Polyesters
Chemical Carriers
Anti-Blocking Agents
Phenolic Molding Compounds
Cushioning Agent in Foundry Sands
Burn Out Medium in Grinding Wheels and Refractories

Properties:
Industrial Flour is available in two standard grades:

WF-5	-200 mesh
WF-7	-325 mesh

A minimum of ninety percent of each grade will pass through the designated U.S. Standard Sieve number. Both have the following typical properties:

Specific Gravity	1.35
Dry Packing Density (Lbs. per Cu Ft)	40
pH Value at 25C (In water)	5
Free Moisture (80C for 15 hrs.)	2-8%
Flash Point (Closed Cup)	380F
Oil Absorption	55-60%
Ash Content	0.5%

AGRASHELL, INC.: Vegetable Shell Products (Continued):

TEXSHEL
An Aggregate Made From Vegetable
Shell Raw Material for Texturing
Surface Coatings

TEXSHEL is designed for use in textured and non-skid coatings.

TEXSHEL is compatible with oil base paints, epoxy, acrylic and other synthetic resin coatings. Adhesion between TEXSHEL and the resin binder is excellent.

Typical Uses

"Sand Finish" Paints
Paddle Tennis Courts
Swimming Pool Decks
Marine Deck Coatings
Industrial Floor Coatings
Stairs, Catwalks or Any Area where Slipping Might Occur

Sizes:

TEXSHEL comes in six standard sizes from coarse to superfine.

AD-3C	(-12 + 20 mesh)	AD-7C	(-35 + 60 mesh)
AD-4C	(-14 + 30 mesh)	AD-9C	(-40 + 100 mesh)
AD-6C	(-18 + 40 mesh)	AD-10.5C	(-60 + 200 mesh)

Properties

TEXSHEL has the following characteristics:

Specific Gravity	1.35
Dry Packing Density (Lbs per Cu.Ft.)	42-47
Free Moisture (80C for 15 Hrs.)	3-10%
pH Value at 25C (In water)	4-6
Hardness-Vickers No.	25-30
Mohs Scale	3-4

AMERICAN COLLOID CO.: Clays:

HECTABRITE AW
Water-Washed Hectorite

A highly refined hectorite clay supplied as a free flowing flake.

Use: High viscosity, stabilizing and suspending agent for household and industrial specialties

Dry Brightness (G.E): 82-86

Purity: Hectorite content 85% minimum.

HECTABRITE LT
Cellulose Beneficated Hectorite

An economical blend of purified hectorite clay and hydroxyethylcellulose.

Use: Gellant for water-based coatings displaying superior pigment suspension and film reinforcement properties.

Dry Brightness (G.E.): 80-85

MAGNABRITE
Magnesium Aluminum Silicate (NF)

A highly refined, selected blend of white smectite clays supplied in flaked form to ease dispersion.

Use: Stabilizing and suspending agent for use in cosmetics and pharmaceuticals

Dry Brightness (G.E.): 83-87

Purity: Specially processed to control microbial content.

VOLCLAY HPM-20
Microfine Sodium Bentonite

High purity air-floated sodium bentonite consisting of micron-sized particles.

Use: Suspending agent, gellant, binder for use in household products, automotive products, aerosols, paints, enamels and other applications.

Purity: Hydrous aluminum silicate composed principally of the clay mineral montmorillonite.

BURGESS PIGMENT CO.: Aluminum Silicates:

OPTIWHITE

OPTIWHITE represents the most versatile pigment of the thermo-optic series. OPTIWHITE is not only capable of substantial TiO2 reduction, but acts as a flatting agent as well. Properly formulated OPTIWHITE approaches have resulted in significant properly improvements while saving substantially. Typical approaches result in equal or better hiding, 25% better scrub resistance with RMC savings.

OPTIWHITE P

OPTIWHITE P exhibits more opacity than OPTIWHITE. Film property advantages associated with OPTIWHITE are not as good but can still be demonstrated in many formulation approaches. OPTIWHITE P does contribute a buff-tone to coatings due to the 84 brightness. Angular sheen, depending on the formulation approach, may be equal to or slightly higher than OPTIWHITE. Overall, OPTIWHITE P offers more hiding and greater savings than OPTIWHITE without the exceptional advantages in film properties associated with optiwhite.

TISYN

TISYN exhibits the greatest degrees of opacity in this pigment series. Film properties and sheen properties more approach what is expected of conventional calcined clay. Again, coatings containing TISYN may exhibit a buff-tone, depending on the level of TISYN and other pigments in the formula.

Typical Specifications:

Thermo Optic:
OPTIWHITE:
- Particle Size Average Microns: 1.4
- G.E. Brightness %: 90-92
- Oil Absorption: 57
- U.S. Patents: 3,309,214

TISYN:
- Particle Size Average Microns: 0.8
- G.E. Brightness %: 84-85
- Oil Absorption: 73
- U.S. Patents: 3,021,195

OPTIWHITE P:
- Particle Size Average Microns: 1.3
- G.E. Brightness %: 84
- Oil Absorption: 65
- U.S. Patents: 3,021,195

BURGESS PIGMENT CO.: Aluminum Silicates(Continued):

Anhydrous:

ICECAP K:
Particle Size Average Microns: 1.0
G.E. Brightness %: 90-92
Oil Absorption: 55

ICEBERG:
Particle Size Average Microns: 1.4
G.E. Brightness %: 90-92
Oil Absorption: 55

Anhydrous Chemical Properties:
Ignition Loss %: 0-1.0
Silica %: 51.0-52.4
Alumina %: 42.1-44.3
Iron Oxide %: Trace
Titanium Dioxide %: 1.56-2.50
Manganese %: None

Hydrous:

THERMO GLACE H B:
Particle Size Average Microns: N/A
G.E. Brightness %: 88.0-90.0
Oil Absorption: 39

No. 10:
Particle Size Average Microns: 0.5
G.E. Brightness %: 87.0
Oil Absorption: 43.0

No. 17:
Particle Size Average Microns: 0.5
G.E Brightness %: 90-92
Oil Absorption: 43

No. 80:
Particle Size Average Microns: 0.5
G.E. Brightness %: 80.0-82.0
Oil Absorption: 42.0

No. 20:
Particle Size Average Microns: 0.75
G.E. Brightness %: 86.0
Oil Absorption: 35.0

No. 60:
Particle Size Average Microns: 0.75
G.E. Brightness: 86.0
Oil Absorption: 37.0

BURGESS PIGMENT CO.: Aluminum Silicates (Continued):

Hydrous Chemical Properties:

THERMO-GLACE HB--No. 10:
- Ignition Loss %: 13.7-14.1
- Silica %: 44.8-45.3
- Alumina %: 37.5-39.7
- Iron Oxide %: Trace
- Titanium Dioxide %: 1.35-2.27*
- Manganese %: None

*.8% for THERMO-GLACE H

No. 80:
- Ignition Loss %: 13.4-14.1
- Silica %: 43.5-44.5
- Alumina %: 38.0-40.0
- Iron Oxide %: 1.0-1.5
- Titanium Dioxide %: 1.4-2.0
- Manganese %: None

R. E. CARROLL, INC.: Fillers:

DOVER CLAY

Screen Size	99.5 minus 325 mesh screen
Particle Size Distribution	
Minus 2 Microns	80-85%
Plus 5 Microns	5-8%
pH	4.5-5.5
Brightness	74-76% G.E. Standard
Chemical Analysis	
Loss at 105C	.24
Loss on Ignition	14.13
Silicon Dioxide (SiO2)	45.95
Iron Oxide (Fe2O3)	1.33
Aluminum Oxide (Al2O3)	36.02

SWANEE CLAY

Chemical Analysis	
Loss on Ignition	13.55%
Silica SiO2	44.72%
Iron Oxide Fe2O3	0.36%
Alumina Al2O3	39.20%
Calcium Oxide	0.22%
Screen Size	99.5 minus 325 Mesh Screen
pH	5.0-5.4
Particle Size Distribution:	
Minus 2 microns	50-55%
Plus 6 microns	18-22%
Brightness	Approx. 78 G.E. Standard
Specific Gravity	2.58

YORK WHITE

Chemical Properties--Typical	
SiO2 plus insolubles	0.54
Fe2O3	0.03
Al2O3	0.27
CaCO3	95.80
MgCO3	3.45
Physical Properties:	
Oil Absorption	14 (rub-out method) 21.5 (Gardner-Coleman method)
Hardness (Mho Scale)	3
pH	9.4-9.7
Index of Refraction	1.59 (mean)
Specific Gravity	2.71
Sieve Analysis:	
Through U.S. No. 325 sieve	99.50% minimum

R. E. CARROLL, INC.: WHITE PIGMENT CORP.: VERFLAKE/VERWHITE:

VERFLAKE

VERFLAKE, one of a series of beneficiated, high calcium limestone products produced by White Pigment Corporation, has intermediate and closely sized particle distribution.

Low oil absorption, high brightness and good dispersion characteristics make it an excellent general purpose extender.

VERFLAKE is an ideal extender for plastics, rubber, coatings and caulks where a uniform product, free from oversize particles, is important. Polyester grade VERFLAKE is designed for use in sheet molding compounds (SMC), with finished parts exhibiting excellent physical and surface properties.

Typical Physical Properties:
- Particle size range (micrometers): Through 25
- Mean particle size (micrometers): 5.0
- Fineness of dispersion (Hegman): 5.0
- Mean Refractive Index: 1.59

Typical Chemical Composition:
- Calcium Carbonate ($CaCO_3$) %: 98.20
- Magnesium Carbonate ($MgCO_3$) %: 1.20

VERWHITE

VERWHITE, one of a series of beneficiated, high calcium limestone products produced by White Pigment Corporation, possess excellent fineness, color and dispersion characteristics.

Extensively employed in rubber and plastics, VERWHITE has proven effective as a bulk filler in plasticized PVC, especially in critical uses requiring the absence of grit. It also yields high volume resistivity for wire and cable applications and exhibits excellent heat stability.

VERWHITE's use in solvent and water-borne coatings provides consistent high fineness of grind readings and good gloss development. In putties, caulks and sealants, it is used at high filler levels because of its nominal oil absorption, low presence of fines and high brightness.

Superior brightness and retention make VERWHITE an economical filler for fine paper.

VERWHITE is available in both dry and slurry form.

Typical Physical Properties:
- Particle size range (micrometers): Through 12
- Mean particle size (micrometers): 3.0
- Fineness of dispersion (Hegman): 6.0
- Mean Refractive Index: 1.59

Typical Chemical Composition:
- Calcium Carbonate ($CaCO_3$) %: 98.20
- Magnesium Carbonate($MgCO_3$) %: 1.20

CYPRUS INDUSTRIAL MINERALS CO.: Extenders and Fillers:

BEAVERWHITE 200:
200 Mesh
Paint, Plastics, Rubber and Building Products
Physical Properties
Brightness: 85+-2%
Specific Gravity: 2.8
Hegman Fineness: 2
Median Particle Size: 10 um
Mineral Analysis:
Talc: 98%
Dolomite: <2%
Quartz: <1%

BEAVERWHITE 325:
325 Mesh
Paint, Plastics, Rubber and Building Products
Physical Properties:
Brightness: 86+-2%
Specific Gravity: 2.8
Hegman Fineness: 3.5-4.0
Median Particle Size: 6 um
Mineral Analysis:
Talc: 98%
Dolomite: <2%
Quartz: <1%

MISTRON MIST:
Moderate Fineness
Paint, Coatings and Rubber
Physical Properties:
Brightness: 90+-2%
Specific Gravity: 2.8
Hegman Fineness: 5.5
Median Particle Size: 3 um
Mineral Analysis:
Talc: 98%
Dolomite: <2%
Qaurtz: <1%

MISTRON MONOMIX:
Ultra Fine
Thixotrope
Plastics and Paints
Physical Properties:
Brightness: 89+-2%
Specific Gravity: 2.8
Hegman Fineness: 6
Median Particle Size: 1.5 um
Mineral Analysis:
Talc: 98%
Dolomite: <2%
Quartz: <1%

CYPRUS INDUSTRIAL MINERALS CO.: Extenders and Fillers (Continued):

MISTRON ULTRAMIX
Ultra Fine
High Oil Absorption
Paint
Physical Properties:
Brightness: 90+-2%
Specific Gravity: 2.8
Hegman Fineness: 6.5
Median Particle Size: 1.5 um
Mineral Analysis:
Talc: 98%
Dolomite: <2%
Quartz: <1%

MISTRON VAPOR:
Ultra Fine
Reinforcement
Rubber
Physical Properties
Brightness: 87+-2%
Specific Gravity: 2.8
Hegman Fineness: 5.5
Median Particle Size: 1.7 um
Mineral Analysis:
Talc: 98%
Dolomite: <2%
Quartz: <1%

MISTRON ZSC:
Ultra Fine - Hydrophobic
Zinc Stearate Treated
Paint, Plastic and Rubber
Physical Properties:
Brightness: 87+-2%
Specific Gravity: 2.8
Hegman Fineness: 6
Median Particle Size: 1.5 um
Mineral Analysis:
Talc: 98%
Dolomite: <2%
Quartz: <1%

CYPRUS INDUSTRIAL MINERALS CO.: Extenders and Fillers (Continued):

MISTRON 400:
Controlled Top Size
Paint, Rubber and Building Products
Physical Properties:
Brightness: 89+-2%
Specific Gravity: 2.8
Hegman Fineness: 4.0-4.5
Median Particle Size: 4.5 um
Mineral Analysis:
Talc: 98%
Dolomite: <2%
Quartz: <1%

MISTRON 600:
Fine Particle Size
High Loadings
Coatings Applications
Physical Properties
Brightness: 90+-2%
Specific Gravity: 2.8
Hegman Fineness: 6
Median Particle Size: 2.5 um
Mineral Analysis:
Talc: 98%
Dolomite: <2%
Quartz: <1%

DUPONT CO.: PULPLUS Polyethylene:

Asbestos Free
The Durable Reinforcing Filter Media

PULPLUS is a specially engineered fibrillated polyethylene pulp developed by DuPont to replace asbestos and other materials in many applications. Unlike other synthetics, PULPLUS is highly alkaline-resistant. It can serve as both a reinforcement and filter-assist medium in air-cured cement slurry system. When compared to asbestos cement products, PULPLUS is significantly tougher, while offering the flexural strength demanded in many end-use applications.

Benefits and Advantages of Using PULPLUS

Developed for use in a variety of applications such as building materials, coatings and paints, molded products, filters and paper products, PULPLUS has a wide range of distinct advantages over other available materials.

* Superior fines retention
* Excellent durability
* Hydrophilic surface
* Matrix compatibility
* Alkaline resistant
* Faster processing
* Good bonding and elasticity
* Economical and available
* Easy dispersion
* Excellent toughness

Product Properties:

Base Resin:
Type: Linear High Density Polyethylene
Density, gm/cc: 0.96
Melting Point, C: 134-138
Chemical Reactivity: Inert

Pulp:
Length, mm: 1 3
- Retained on 10 mesh, %: <15 4
- Through 10 mesh on 100 mesh, %: >60 4
- Through 100 mesh, %: <30 4

Diameter, Microns: 1-20
Surface Area, M 2/gm (BET, N2): 2-4

ECC AMERICA INC.: Calcium Carbonate:

ULTRAMITE/MICRO-WHITE 07:
An ultrafine wet-ground product available in slurry form only. Applications are restricted to water-based sytems, primarily coatings.
Particle Size Range (Micrometers): Up to 5
Mean Particle Size (Micrometers): 0.7

SUPERMITE/MICRO-WHITE 10:
An extra fine ground pigment to provide improved opacity and gloss properties in paints, coatings and paper applications. In plastics and elastomers, finer particle size aids development of superior physical properties and improved surface gloss.
Particle Size Range (Micrometers): Up to 8
Mean Particle Size (Micrometers): 1.0
Oil Absorption (Rub-out): 19-21

OPACIMITE:
A finely ground product with an engineered particle size distribution for use in paints and polymers. The unique particle size distribution and high brightness provide TiO_2 spacing for maximum opacity.
Particle Size Range (Micrometers): Up to 5
Mean Particle Size (Micrometers): 1.1
Oil Absorption (Rub-Out): 18 to 20

MACROMITE/MICRO-WHITE 15:
A very fine wet-ground product for the paint and polymer industries. Used where the ultra-fine products are not required but where a low top size is needed.
Particle Size Range (Micrometers): Up to 10
Mean Particle Size (Micrometers): 1.5
Oil Absorption (Rub-out): 17 to 19

ATOMITE/MICRO-WHITE 25:
A highly developed stir-in pigment that meets the most exacting requirements demanded by the paint, rubber, plastics, floor covering, and many other industries.
Particle Size Range (Micrometers): Up to 12
Mean Particle Size (Micrometers): 3.0
Oil Absorption (Rub-out): 14 to 16

SNOWFLAKE WHITE/MICRO-WHITE 50 (CODEX):
A general purpose stir-in pigment that has found wide use in practically all types of protective coatings, rubber, plastic, caulk, and glaze compounds, mastics and other end products.
Particle Size Range (Micrometers): up to 25
Mean Particle Size (Micrometers): 5.5
Oil Absorption (Rub-Out): 9-10

ECC AMERICA INC.: Calcium Carbonate(Continued):

SNOWFLAKE PE:

A medium ground pigment with the theoretical correct distribution of particles to ensure maximum loading in highly filled systems: e.g., polyester resins, SMC, BMC, TMC.

Particle Size Range (Micrometers): Up to 40
Mean Particle Size (Micrometers): 6.0
Oil Absorption (Rub-Out): 8 to 9

DRIKALITE/MICRO-WHITE 95:

A competitive grade of extender pigment for use in places where a maximum number of fine particles per unit cost is imperative.

Particle Size Range (Micrometers): Up to 44
Mean Particle Size (Micrometers): 6.5
Oil Absorption (Rub-Out): 8.5 to 9.5

DURAMITE:

A pigment with unique particle size distribution. Its easy dispersion characteristics have demonstrated its unusual merit as a mix-in type of pigment for a wide variety of applications such as carpet backing, roofing compounds, spackles, and coatings.

Particle Size Range (Micrometers): Up to 44
Mean Particle Size (Micrometers): 11.0
Oil Absorption (Rub-out): 5 to 8

No. 1 White:

A general purpose pigment for many uses where fineness is of secondary importance.

Particle Size Range (Micrometers): Up to 50
Mean Particle Size (Micrometers): 14.0
Oil Absorption (Rub-Out): 5 to 6

No. 3 White/MICRO-WHITE 100 (CODEX):

A white general purpose filler for low cost paints, rubber goods, mild abrasive compounds, and putties.

Particle Size Range (Micrometers): Up to 90
Mean Particle Size (Micrometers): 18.0

CP FILLER/CC-101:

A competitively priced filler for use in construction products where color is of secondary importance.

Particle Size Range (Micrometers): Up to 75
Mean Particle Size (Micrometers): 17.0

ECC AMERICA INC.: Calcium Carbonate(Continued):

MARBLE DUST:

A coarse ground natural calcium carbonate for use in putties, glazes and mild abrasive compounds. It is frequently blended with ATOMITE to achieve a wider distribution of particles which yields effective particle packing properties.

Particle Size Range (Micrometers): Up to 180
Mean Particle Size (Micrometers): 21.0

CC-103, CC-105 & SC-53:

A series of coarse ground calcium carbonate fillers used in polyolefins, carpet backing, caulks, sealants, and putties, and as mild abrasives in cleaners. These products offer a formulator a choice of fillers having low oil absorption, low resin demand, and relatively high dry brightness.

Mean particle Size (Micrometers): Up to 40
Oil Absoprtion (Rub-Out): Low

SUPERCOAT (Surface Treated):

A coated extra fine ground pigment. Easy incorporation and dispersion in plastics applications resulting in improved physical properties. Improved production throughput efficiency with low abrasiveness.

Particle Size Range (Micrometers): Up to 8
Mean Particle Size (Micrometers): 1.0
Oil Absorption (Rub-out): 15 to 17

KOTAMITE/MICRO-WHITE 25 SAM (Surface Treated):

A coated pigment designed for easy dispersion in plastic compounds e.g. polyolefins, rigid and flexible PVC. Hydrophobic for superior wire and cable insulation compounds and improved impact properties in polypropylene.

Particle Size Range (Micrometers): Up to 12
Mean Particle Size (Micrometers): 3.0
Oil Absorption (Rub-out): 12 to 14

OPACICOAT (Surface Treated):

A finely ground coated pigment with a unique particle size distribution. The clean top size and lack of ultrafines should permit higher loadings in polymer systems while still maintaining properties. The coating will aid in dispersion.

Particle Size Range (Micrometers): Up to 5
Mean Particle Size (Micrometers): 1.1
Oil Absorption (Rub-out): 14 to 16

ENGELHARD CORP.: Aluminum Silicate Pigments:

For Coatings, Inks, Adhesives, and Other Industrial Applications

ASP Aluminum Silicates:

Hydrous:

<u>Designation:</u>

072:
Physical Form: Spray Dried Beads
Special Modifications: Ultra-Flotation, Dispersant
Average Particle Size (um): 0.3
Brightness (G.E.): 90-92

100/200:
Physical Form: Highly Pulverized Powder
Special Modifications: None
Average Particle Size (um): 0.4
Brightness (G.E.): 86-88

101:
Physical Form: Highly Pulverized Powder
Special Modifications: 0.5% Stearate
Average Particle Size (um): 0.4
Brightness (G.E.): 86-88

102:
Physical Form: Spray Dried Beads
Special Modifications: Dispersant
Average Particle Size (um): 0.4
Brightness (G.E.): 86-88

170:
Physical Form: Highly Pulverized Powder
Special Modifications: Ultra-Flotation
Average Particle Size (um): 0.4
Brightness (G.E.): 90-92

172:
Physical Form: Spray Dried Beads
Special Modifications: Ultra-Flotation, Dispersant
Average Particle Size (um): 0.4
Brightness (G.E.): 90-92

600:
Physical Form: Highly Pulverized Powder
Special Modifications: None
Average Particle Size (um): 0.6
Brightness (G.E.): 85-87

ENGELHARD CORP.: Aluminum Silicate Pigments(Continued):

ASP Aluminum Silicates(Continued):

Hydrous:

Designation:

602:
Physical Form: Spray Dried Beads
Special Modifications: Dispersant
Average Particle Size (um): 0.6
Brightness (G.E.): 85-87

900:
Physical Form: Powder
Special Modifications: None
Average Particle Size (um): 1.5
Brightness (G.E.): 83-85

902:
Physical Form: Slurry Only
Special Modifications: Pre-dispersed
Average Particle Size (um): 1.5
Brightness (G.E.): 83-85

400:
Physical Form: Powder
Special Modifications: None
Average Particle Size (um): 4.8
Brightness (G.E.): 79-82

400P:
Physical Form: Highly Pulverized Powder
Special Modifications: None
Average Particle Size (um): 4.8
Brightness (G.E.): 79-82

Hydrous Delaminated:

NCF:
Physical Form: Highly Pulverized Powder
Special Modifications: Dispersant
Average Particle Size (um): 0.4
Brightness (G.E.): 87-89

NC/NC01:
Physical Form: Highly Pulverized Powder
Special Modifications: Dispersant
Average Particle Size (um): 0.7
Brightness (G.E.): 87-89

ENGELHARD CORP.: Aluminum Silicate Pigments(Continued):

Hydrous Delaminated:

Designation:

NCS:
Physical Form: Highly Pulverized Powder
Special Modifications: Dispersant
Average Particle Size (um): 0.8
Brightness (G.E.): 86-88

NC2:
Physical Form: Highly Pulverized Powder
Special Modifications: Dispersant
Average Particle Size (um): 1.3
Brightness (G.E.): 90-92

SATINTONE Aluminum Silicates:
Calcined:

5:
Physical Form: Highly Pulverized Powder
Special Modifications: Dehydroxylated by Controlled Heat Treatment
Average Particle Size (um): 0.8
Brightness (G.E.): 90-92

SPECIAL:
Physical Form: Highly Pulverized Powder
Special Modifications: Dehydroxylated by Controlled Heat Treatment
Average Particle Size (um): 1.2
Brightness (G.E.): 91-93

OP:
Physical Form: Highly Pulverized Powder
Special Modifications: Dehydroxylated by Controlled Heat Treatment
Average Particle Size (um): 1.3
Brightness (G.E.): 86-88

W:
Physical Form: Highly Pulverized Powder
Special Modifications: Dehydroxylated by Controlled Heat Treatment
Average Particle Size (um): 1.4
Brightness (G.E.): 90-92

PLUS:
Physical Form: Highly Pulverized Powder
Special Modifications: Dehydroxylated by Controlled Heat Treatment
Average Particle Size (um): 2.0
Brightness (G.E.): 93-95

EVANS CLAY CO.: Kaolin Clays & Extenders:

APEX:

APEX clay is an economical extender pigment that has fine particle size, high surface area, and high absorption properties in most vehicles.

APEX clay is particularly beneficial when compounded as an extender or filler in 1) rubber soles and heels, 2) rubber and vinyl flooring, 3) mechanical rubber goods, 4) rubber foam, 5) metal undercoatings and primers, 6) resin base mastics, 7) rug backing, and polyester "gunks".

Typical Physical Properties:	
Brightness, % Reflectance G.E.	79-81
Screen Residue + 325 Mesh	.7
Moisture	1.0
pH	5.5-7.0
Oil Absorption	40-45
Particle Size(Microns):	
Average	.8
% finer than 2	81-85
Typical Chemical Properties:	
Alumina (Al_2O_3)	39.5%
Silica (SiO_2)	44.5%
Loss On Ignition	14%

SNOBRITE:

SNOBRITE clay is a fine particle extender pigment that has consistently high brightness.

SNOBRITE clay is particularly beneficial when compounded as an extender or filler in 1) Adhesives, 2) wallpaper paste, 3) joint cements and calking compounds, 4) latex paints, 5) rubber and vinyl flooring, 6) resin based mastics, and 7) rubber compounding when color control is critical.

Typical Physical Properties:	
Brightness, % Reflectance G.E.	80-82
Screen Residue + 325 Mesh	.5
Moisture	1.0
pH	5.5-7.0
Oil Absorption	40-45
Particle Size(Microns):	
Average	.8
% finer than 2	81-85
Typical Chemical Properties	
Alumina (Al_2O_3)	39.5%
Silica (SiO_2)	44.5%
Loss on Ignition	14%

FRANKLIN LIMESTONE CO.: PEERLESS High Calcium Limestone "T" Products:

FRANKLIN T-11:

Hardness	3
pH	9.0-9.5
Specific Gravity	2.74
Hegman Grind	5-6
Mean Particle Size	4 microns
Maximum percent retained on US #325	.005%
Color	Neutral Beige

FRANKLIN T-13:

Hardness	3
pH	9.0-9.5
Specific Gravity	2.74
Hegman Grind	3-3 1/2
Mean Particle Size	6 microns
Maximum percent retained in US #325	.15%
Color	Neutral Beige

FRANKLIN T-14:

Hardness	3
pH	9-9.5
Specific Gravity	2.74
Hegman Grind	2 1/2 - 3
Mean Particle Size	7 microns
Maximum percent retained on US #325	.50%
Color	Neutral Beige

FRANKLIN T-325:

Hardness	3
pH	9.0-9.5
Specific Gravity	2.74
Hegman Grind	1 1/2 - 2 1/2
Mean Particle Size	10 microns
Maximum percent retained on US #200	.08%
Maximum percent retained on US #325	5.0%
Color	Neutral Beige

FRANKLIN MINERAL PRODUCTS: Water Ground Mica:

A Reinforcing Functional Pigment
for Use

As barrier in primers-sealants-protective paints-house paints
As mold release for tires and rubber products
As reinforcement in plastics
For improved slip in lubricants
For luster effects on wallpapers

Theoretical Chemical Formula	$H_2KAl_3(SiO_4)_3$
Specific Gravity	2.8-3
Hardness (Mohs Scale)	2.5
High Temperature Limit	Approximately 535C
Chemical Activity	Essentially inert except to H2F
Index of Refraction	1.58
Loss on Ignition	4-5%

Grade No. 12 325 mesh
Grade No. 4 160 mesh
Grade No. 88 100 mesh

As prepared by the Water Grinding Process

Bulking Value	.0424 Gals/lb.
Weight per Solid Gallon	23.57 lbs.
Nominal Particle Size	325, 160 or 100 Mesh respectively
Particle Shape	Very Thin & Platy
Apparent Density	10 lbs./cu.ft. approx
Moisture Loss @ 105-110C	Less than .25%
Oil Absorption	54 lbs./100 lbs mica
Color	White and Lustrous

Water Ground (Wet Ground) Mica is a clean, platy, high aspect ratio (area: thickness) pigment. Its unique particle shape acts to form as a barrier layer in paints and coatings, and provides reinforcing properties in plastics. The smooth, polished surfaces aid in lubrication and luster effects.

Franklin Mica has been carefully washed after milling to minimize it as a source of polyvalent metal ions and water soluble impurities.

GENSTAR STONE PRODUCTS CO.: CAMEL-CAL/CAMEL-CAL ST/CAMEL-CAL Slurry:

CAMEL-CAL is a fine-ground calcium carbonate available as a dry powder, a stearate-coated powder (CAMEL-CAL ST), and a stable, high solids (75%) water dispersion. In both the dry and slurry forms, the average particle size of 0.7 microns, promotes high scattering to develop dry hide, tinting strength, and TiO2 replacement values in coating formulations. In plastics compounding (both thermoset and thermoplastic), improved impact resistance, easier and lower energy cost compounding, and less barrel and die wear are promoted by the fine particle size of both regular CAMEL-CAL and CAMEL-CAL ST. Formulation and processing characteristics, formerly only obtained by use of precipitated carbonates, are now available at lower cost

Recommended Applications:
- Water Based Coatings, Inks
- Paper Filler, alkaline and neutral sized paper, and PVC pipe

Typical Properties:

Dry & ST Dry:
- Specific Gravity (solids): 2.70-2.71
- % Finer Than 2 Microns: 90
- pH: 9.5
- Dry Brightness (Hunter) (Min): 96

Slurry:
- Specific Gravity (Solids): 2.70-2.71
- Wt. per Gallon--Solids (Lbs): 22.57
- Wt. per Gallon--Slurry (Lbs.): 15.8-16.0
- % Finer Than 2 Microns: 90
- Solids, %W: 75+-0.5
- Solids, %V: 52.6
- pH: 9.5
- Brookfield Viscosity, cps: 90-150
- Dry Brightness (Hunter) (Min.): 96

GENSTAR STONE PRODUCTS CO.: CAMEL-CARB/CAMEL-TEX:

CAMEL-CARB:

CAMEL-CARB is a quality extender in the low-priced field. This product is manufactured from white Calcite, as are other Genstar carbonates. The quality control utilized in the production of CAMEL-CARB assures the user of uniform low vehicle demand, good color, and high brightness.

Recommended Applications:

Paint--Interior flat paint and exterior house paints

Rubber--Recommended for use as an inexpensive filler in all types of rubber compounds

Misc--Widely used in putty and caulk, ceramics, adhesives, linoleum, floor tile, and textile coatings

Typical Properties:

Specific Gravity: 2.70-2.71
Wt. per gallon solid (lbs.): 22.57
Average Particle Dia. (Microns, Sedigraph): 7.0
pH of Saturated Solution: 9.5
Solubility in water (%): 0.03
Dry Brightness (Hunter) (Min.): 93

CAMEL-TEX:

CAMEL-TEX is a fine ground general-purpose grade of calcium carbonate produced from extremely white Calcite. High whiteness, freedom from agglomerates, low vehicle demand and rapid dispersibility make CAMEL-TEX a highly valuable product in its field of application. CAMEL-TEX provides good color, low-angle sheen, scrubability and good tinting qualities.

Recommended Applications:

Paint--Interior flats, primers and sealers

Plastics--Polyester fiberglass premixes, preforms and hand lay-up gel coats

Rubber--Automotive products, household products, tubing, medical "goods," closures and other areas of manufacture

Misc--Used in putty, caulk, bath tub sealers, body deadeners and adhesives

Typical Properties:

Specific Gravity: 2.70-2.71
Wt. per gallon solid (lbs.): 22.57
Average Particle Dia. (Microns, Sedigraph): 5.0
pH of Saturated Solution: 9.5
Solubility in Water (%): 0.04
Dry Brightness (Hunter)(Min): 93

GENSTAR STONE PRODUCTS CO.: CAMEL-FIL:

CAMEL-FIL is a low cost, fine-ground calcite (calcium carbonate) specifically designed for high filler loading in glass-reinforced polyester (SMC, TMC, and BMC). The unique particle size distribution optimizes particle packing, leaving less void space to be filled by expensive resin. The result is lower viscosity and/or higher filler loading capability--levels of over 200 parts filler/100 parts resin are easily attainable in most formulations. Improved processing characteristics are also important--rapid dispersion and wet-out, excellent flow in the mold, and good surface smoothness of the molded parts are obtained. Low moisture content, product uniformity.

Recommended Applications:
Thermosets--Glass reinforced polyesters (SMC, BMC, TMC); Epoxies
Thermoplastics--PVC (rigid and flexible), polypropylene
Rubber--Automotive goods, floor tiles
Misc--Caulks, sealants, adhesives

Typical Properties:
Specific Gravity: 2.70-2.71
Wt. per gallon solid (lbs.): 22.57
Retained on 325 Mesh, % max.: .05
Average Particle Size, Microns: 5.5-6.0
pH of Saturated Solution: 9.5
Solubility in Water (%): 0.06
Dry Brightness (Hunter) (Min): 93

GENSTAR STONE PRODUCTS CO.: CAMEL-WITE/CAMEL-WITE ST/CAMEL-WITE Slurry:

The industry standard--CAMEL-WITE is an exceptionally white, fine particle size, wet-ground product, produced from high grade calcite limestone under the most exacting specifications. This attention to production and quality control means reproducible results, whether the application is a glass reinforced polyester, a latex paint, or a paper coating. Available as a dry powder, surface treated dry powder or a 72%w water slurry, its agglomerate-free characteristics promote rapid dispersion and minimum process time.

Recommended Applications:

Paint--Gloss, semi-gloss, flats, primers and sealers. Efficient pigment extender in water-based and solvent systems. pH control

Paper--Internal filler in alkaline and neutral sized paper.

Paper Coating--High Brightness, low binder demand. Coating colors of 70% with a viscosity of 2500 cps or lower.

PVC--Supported and unsupported vinyl film and sheeting; floor tile; wire and cable insulation; plastisols; extrusions. Rigid and flexible PVC.

Rubber--Automotive goods, footwear, medical supplies, etc.

Thermoplastics--Polypropylene, nylon, urethanes, high density and low density polyethylene, ABS, polystyrene.

Thermosets--Glass reinforced polyesters (SMC, BMC, TMC); epoxy compounds; alkaline phenolics.

Misc--Caulks, glazing compounds, ceramics, adhesives, food processing.

Typical Properties:

Dry & ST Dry:

Specific Gravity (Solids): 2.70-2.71
Wt. per Gallon (lbs) (25C): 22.57
Avg. Particle Dia. (Microns): 3.0
pH of Saturated Solution: 9.5
Solubility in Water: 0.08
Dry Brightness (Hunter) (Min): 95

Slurry:

Specific Gravity (Solids): 2.70-2.71
Wt. per Gallon (lbs) (25C): 15.3
Avg. Particle Dia. (Microns): 3.0
Solids (%w): 72.0+-0.5
Brookfield Viscosity: 100-200
pH of Saturated Solution: 9.5
Solubility in Water: 0.08
Dry Brightness (Hunter) (min.): 95

GEORGIA KAOLIN Co., INC.: Calcined Extender Pigments:

GLOMAX:

GLOMAX extender pigments offer important advantages in various paint systems:

* Contrast ratios are increased with the same prime white pigment volume
* Increasing the amount of GLOMAX extender pigments improves reflectance and decreases yellowness.
* Increasing the GLOMAX content reduces gloss.
* Imparts excellent salt spray resistance and fast grinding characteristics to a wide range of industrial primers.

GLOMAX extender pigments also provide significant advantages in rubber and plastics applications:

* GLOMAX imparts excellent physical and electrical properties to rubber and other polymer systems.
* Ease of dispersion improves processability in rubber and plastic compounds.

GLOMAX calcined aluminum silicate improves performance of a wide range of ceramic bodies:

* GLOMAX is pre-reacted to control shrinkage
* Improves thermal shock resistance

ALTOWHITE:

ALTOWHITE calcined aluminum silicate extender pigments exhibit improved optical properties in paint systems when compared with conventional calcined kaolins. In plastics, ALTOWHITE extender pigments are chemically stable over a wide range of pH, and easily dispersed in all systems.

ALTOWHITE offers four major advantages:

* Exhibits low gloss and sheen characteristics.
* Maintains excellent viscosity and stability.
* Useful as a TiO2 and pigment spacer in various coating and polymer systems.
* Exceptional hiding power over a wide range of PVC and TiO2 levels.

Typical Uses of GLOMAX/ALTOWHITE Extender Pigments:

Paint:
Oleoresinous and primer systems
Grade:
GLOMAX JDF/ALTOWHITE TE
Advantages:
Fast grinding,
least settling,
highest hiding,
controls flow.

Paint:
Eggshell finishes
Grade:
ALTOWHITE UF
Advantages:
Pigment spacer, gloss control

GEORGIA KAOLIN CO.: Calcined Extender Pigments(Continued):

Typical Uses of GLOMAX/ALTOWHITE Extender Pigments:

Paint:
Interior emulsions
Grade:
GLOMAX LL
ALTOWHITE TE
Advantages:
Excellent hiding and color, improves solvent release.

Paint:
Traffic paints
Grade:
GLOMAX LL
Advantages:
Abrasion resistance and bead retention.

PLASTICS AND RUBBER:

Electrical wire and cable compounds
Grade:
GLOMAX LL
Advantages:
Excellent mechanical and electrical characteristics.

Thermoplastics
Grade:
ALTOWHITE TE
Advantages:
TiO2 extender

Mechanical rubber compounds:
Grade:
GLOMAX LL
Advantages:
Improves abrasion resistance and processability.

CERAMICS:

Whitewares
Electrical porcelain
Refractories

Grade:
GLOMAX LL
Advantages:
Imparts dimensional stability. Improves fired properties.

GEORGIA KAOLIN CO.: Calcined Extender Pigments(Continued):

Typical Chemical Properties:
- Aluminum Oxide (combined): 44.40
- Silicon Dioxide (combined): 53.80
- Iron Oxide: 0.44
- Titanium Dioxide: 1.55
- Calcium Oxide: 0.05
- Magnesium Oxide: 0.23
- Sodium Oxide: 0.25
- Potassium Oxide: 0.09

Physical Constants:
- Refractive Index: 1.62
- Density: lbs./solid gallon: 21.91
- Bulking value; gallon/lb.: 0.046
- Specific Gravity: 2.62

Typical Physical Test Results:

GLOMAX JDF:
- Median Particle Size (Microns): 0.6
- Brightness (G.E. % of MgO): 90-92
- pH: 4.2-5.2
- Oil Absorption (%): 40-45

GLOMAX LL:
- Median Particle Size (Microns): 1.5
- Brightness (G.E. % of MgO): 90-92
- pH: 4.2-5.2
- Oil Absoprtion (%): 50-55

ALTOWHITE UF:
- Median Particle Size (Microns): 1.1
- Brightness (G.E. % of MgO): 92.5-93.5
- pH: 5.5-6.5
- Oil Absorption (%): 70-75

ALTOWHITE TE:
- Median Particle Size (Microns): 2.2
- Brightness (G.E. % of MgO): 92.5-93.5
- pH: 5.0-6.0
- Oil Absorption (%): 70-75

GEORGIA KAOLIN CO., INC.: Delaminated KAOPAQUE Aluminum Silicates:

For Paint and Coating Systems:

These large platey particles are readily dispersed in aqueous paint systems. They impart excellent whiteness, brightness and durability.

KAOPAQUE aluminum silicates offer the following specific advantages:

* outstanding hiding power
* excellent color and cleaner tints
* improved scrubbability and stain removal
* excellent holdout
* reduced "mud cracking"
* control of gloss and sheen

Typical Physical Properties:

KAOPAQUE 10:
Brightness (G.E. Meter), % of MgO: 89+-1
Average particle size, u (equivalent spherical diameter): 1
Moisture, % maximum: 1
Screen residue (325 mesh), % max.: 0.02
pH (20% solids slurry): 3.8-4.8

KAOPAQUE 10S:
Brightness (G.E. Meter), % of MgO: 89+-1
Average particle size, u (equivalent spherical diameter): 1
Moisture, % maximum: 1
Screen residue (325 mesh), % max.: 0.01
pH (20% solids slurry): 6.5-7.5

KAOPAQUE 10S Slurry:
Brightness (G.E. Meter), % of MgO: 89+-1
Average particle size, u (equivalent spherical diameter): 1
Moisture, % maximum: 32+-1
Screen residue (325 mesh), % max.: 0.01
pH (20% solids slurry): 6.5-7.5

KAOPAQUE 20:
Brightness (G.E. Meter), % of MgO: 87+-1
Average particle size, u (equivalent spherical diameter): 2
Moisture, % maximum: 1
Screen residue (325 mesh), % max.: 0.1
pH (20% solids slurry): 4.2-5.2

KAOPAQUE 20 Slurry:
Brightness (G.E. Meter), % of MgO: 87+-1
Average particle size, u (equivalent spherical diameter): 2
Moisture, % maximum: 32+-1
Screen residue (325 mesh), % max.: 0.01
pH (20% solids slurry): 6.5-7.5

GEORGIA KAOLIN CO.: HYDRITE Kaolinites:

Applications:
Paints--gloss and semi-gloss emulsions, primer systems
Adhesives--thixotropic aid
Caulks--thixotropic aid
Inks--letterpress
Engineering plastics
GK Grade: HYDRITE UF
Properties:
Finest particle size kaolinite commercially available, highest water and oil demand.
Advantages:
Highest gloss characteristics, best suspension.
Highest thixotropic characteristics.
Least abrasion.

Applications:
Paints--interior emulsions, primer systems, emulsion floor paints.
Plastics--bulk molding compounds, gel coats
Inks--letterpress
GK Grade: HYDRITE PX
Properties:
Good brightness, fine particle size, high water and oil demand.
Advantages:
Excellent gloss, color stain removal, hiding power and suspension.
Excellent rheological properties.
Excellent flow control

Applications:
Rubber--mechanical goods
GK Grade: HYDRITE PXS
Properties:
Predispersed form of HYDRITE PX.
Advantages:
High modulus characteristics

Applications:
Paints--interior emulsions, primer systems, universal tints
Plastics--bulk molding compounds
Adhesives--water base
Inks--letterpress
GK Grade: HYDRITE R
Properties:
Good brightness, free of coarse particles, medium oil demand.
Advantages:
Excellent color, stain removal, hiding power and suspension.
Rheological control.
Strike-in control.
Reduces cost with good finish.

GEORGIA KAOLIN CO.: HYDRITE Kaolinites(Continued):

Applications:
Rubber--mechanical and electrical goods
Plastics--bulk molding compounds, thermoplastic calendered goods
GK Grade: HYDRITE RS
Properties:
Predispersed form of HYDRITE R
Advantages:
Good physical, electrical and flow characteristics.

Applications:
Paints--interior emulsions, primer systems
Plastics--polyester premix
GK Grade: HYDRITE 121-S
Properties:
Intermediate particle size, intermediate water and oil demand. Predispersed.
Advantages:
Good stain removal, hiding power and suspension. Flow control.

APPLICATIONS:
Paints--interior emulsions, primer systems, exterior emulsions, exterior oleoresinous
Plastics--bulk molding compounds and preform epoxy molding phenolic molding
Rubber--mechanical goods
Adhesives-- epoxy type
water type
GK Grade: HYDRITE FLAT D
Properties:
Large particle size, lower water and oil demand.
Advantages:
Good enamel hold-out, suspension and hiding power
Low chalking rate.
Improves physical properties, non-reactive, permits high loadings.
Reduces shrinkage.
Very low modulus characteristics.
Non-reactive, thixotropic.
Controls penetration.

Chemical Analysis:
Aluminum Oxide: 38.38
Silicon Dioxide(combined): 45.30
Ignition Loss at 950C(combined water): 13.97
Iron Oxide: .30
Titanium Dioxide: 1.44
Calcium Oxide: .05
Magnesium Oxide: .25
Sodium Oxide: .27
Potassium Oxide: .04

GEORGIA KAOLIN CO.: HYDRITE Kaolinites(Continued):

Grade:

HYDRITE UF:
Median Particle Size(Microns): 0.20
Brightness (G.E. % of MgO): 82.0-85.0
pH: 4.2-5.2
Oil Adsorption %: 47

HYDRITE PX:
Median Particle Size(Microns): 0.68
Brightness (G.E. % of MgO): 87.0-89.0
pH: 4.2-5.2
Oil Adsorption %: 43

HYDRITE PXS:
Median Particle Size(Microns): 0.68
Brightness (G.E. % of MgO): 87.0-89.0
pH: 6.5-7.5
Oil Adsorption %: 42

HYDRITE R:
Median Particle Size(Microns): 0.77
Brightness (G.E. % of MgO): 85.0-86.5
pH: 4.2-5.2
Oil Adsorption %: 41

HYDRITE RS:
Median Particle Size(Microns): 0.77
Brightness (G.E. % of MgO): 85.0-86.5
pH: 6.5-7.5
Oil Adsorption %: 38

HYDRITE 121-S:
Median Particle Size(Microns): 1.5
Brightness (G.E. % of MgO): 83.0-84.5
pH: 6.5-7.5
Oil Adsorption %: 39

HYDRITE FLAT D:
Median Particle Size(Microns): 5.0
Brightness (G.E. % of MgO): 80.2-83.0
pH: 4.2-5.2
Oil Adsorption %: 34

GEORGIA MARBLE CO.: Calcium Carbonate Fillers:

Regular Coarse and Medium Ground Products:

No Color Specified

Non-color-critical color fillers are made from high brightness Alabama and Georgia sources, but color variation is not tightly controlled. Color is run-of-mine, typically above 85% brightness. These are good white products for use where the brightest fillers are not required.

Products:

350D:
% Passing 200 Wet Screen: 62+-4

RO-40:
% Passing 200 Wet Screen: 70+-5

Rock Dust:
% Passing 200 Wet Screen: 80+-4

D-70:
% Passing 325 Wet Screen: 67+-3

No. 9 NCS:
% Passing 325 Wet Screen: 91+-3

Color Controlled

These fillers are offered in a similar range of particle size distributions, but with controlled color for applications where consistently whiter fillers are needed.

No. 8:
% Passing 325 Wet Screen: 75+-5
Minimum Dry Brightness (%): 90

80-325:
% Passing 325 Wet Screen: 80+-5
Minimum Dry Brightness (%): 90

15M:
% Passing 325 Wet Screen: 88 Min.
Minimum Dry Brightness (%): 92

No. 9:
% Passing 325 Wet Screen: 91+-3
Minimum Dry Brightness (%): 92

No. 10:
% Passing 325 Wet Screen: 99.2 Min.
Minimum Dry Brightness (%): 92

GEORGIA MARBLE CO.: Calcium Carbonate Fillers(Continued):

Screen Controlled Products

Screened fillers have a more tightly controlled range of particle sizes.

Products:

XO:
- Maximum Retained (Coarse End): 1% on No. 16 Screen
- Minimum Passing (Fine End): 15%--No. 40 Screen

30-50:
- Maximum Retained (Coarse End): 1% on No. 30 Screen
- Minimum Passing (Fine End): 20%--No. 50 Screen

40-200:
- Maximum Retained (Coarse End): 1% on No. 40 Screen
- Minimum Passing (Fine End): 20%--No. 200 Screen

Finer Ground Products:

Georgia Marble finely ground fillers exhibit both tightly controlled particle size distribution and very high brightness for exacting applications.

General Use Fillers:

Products:

GAMA-SPERSE 80:
- Median Particle Size (microns): 2.3
- SSD: 1.2
- Retained on No. 325 Wet Screen: 0.005%
- Dry Brightness: 94
- Hegman Grind: 7

GAMA-SPERSE 6532
and
GAMACO:
- Median Particle Size (microns): 4.3
- SSD: 1.65
- Retained on No. 325 Wet Screen: 0.005%
- Dry Brightness: 94
- Hegman Grind: 6

WINGDALE WHITE:
- Median Particle Size (microns): 6.0
- SSD: 2.3
- Retained on No. 325 Wet Screen: 0.200%
- Dry Brightness: 94
- Hegman Grind: 3

GEORGIA MARBLE CO.: Calcium Carbonate Fillers(Continued):

General Use Fillers(Continued):

Products:

GAMA-SPERSE 6451
and
CALWHITE:
Median Particle Size (microns): 12.0
SSD: 2.3
Retained on No. 325 Wet Screen: 0.008%
Dry Brightness: 94
Hegman Grind: 5

GAMA-SPERSE 255:
Median Particle Size (microns): 12.0
SSD: 3.2
Retained on No. 325 Wet Screen: 0.020%
Dry Brightness: 94
Hegman Grind: 4

Fillers Tailored for Plastics:

Products:

GAMACO II:
Median Particle Size (microns): 3.9
SSD: 1.2
Retained on No. 325 Wet Screen: 0.005%
Dry Brightness: 94
Viscosity Range cps: 46,000

CALWHITE II:
Median Particle Size (microns): 7.0
SSD: 2.3
Retained on No. 325 Wet Screen: 0.008%
Dry Brightness: 94
Viscosity Range cps: 30,000

GAMA-PLAS:
Median Particle Size (microns): 7.0
SSD: 2.6
Retained on No. 325 Wet Screen: 0.200%
Dry Brightness: 94
Viscosity Range cps: 36,000

GEORGIA MARBLE CO.: Calcium Carbonate Fillers(Continued):

Stearate Surface Modified Fillers:

Product:

GAMA-SPERSE CS-II:
Median Particle Size (microns): 3.9
Retained on No. 325 Wet Screen: 0.005%
Stearate Content: 1.0%

Ultra Fine Products:
Very fine particle sizes, tightly controlled particle size distribution, and very high whiteness characterize the ultra fine product group.

Products:

GAMA-FIL 90:
Median Particle Size (microns): 0.8
% Less than 2 Microns: 92
% Less Than 1 Micron: 58
GE Brightness: 95
Viscosity Range cps: 300-500

GAMA-FIL 55:
Median Particle Size (microns): 1.8
% Less than 2 Microns: 55
% Less than 1 Micron: 28
GE Brightness: 95
Viscosity Range cps: 300-500

Typical Constants:

Chemical:
Total Carbonates (CaMg): 95% (Min.)
$MgCO_3$: 3.0% (Max.)
Acid Insolubles: 3.0% (Max.)
Physical:
Hardness (Moh's scale): 3
pH: 9.0-9.5
Specific Gravity (solids): 2.71

GREFCO, INC.: DIATOMITE Functional Fillers:

DICALITE Diatomite functional fillers are produced from diatomaceous earth, a versatile and valuable raw material.

Natural Diatomite Functional Fillers:

Color	Cream
Moisture, %	6.0
Ig. Loss (dry basis) %	5.0
pH (10% slurry)	6.5-8.5

104:
Brightness (G.E.): 76
Oil Absorption (GCOA), wt %: 160
Water Absorption, wt. %: 190
MPD (Micron): 2.4

CA-3:
Brightness (G.E.): 71
Oil Absorption (GCOA), wt %: 170
Water Absorption, wt. %: 205
MPD (Micron): 4.3

IG-33:
Brightness (G.E.): 70
Oil Absorption (GCOA), wt %: 170
Water Absorption, wt. %: 205
MPD (Micron): 4.6

143:
Brightness (G.E.): 72
Oil Absorption (GCOA), wt %: 175
Water Absorption, wt. %: 245
MPD (Micron): 4.8

SA-3:
Brightness (G.E.): 72
Oil Absorption (GCOA), wt %: 180
Water Absorption, wt. %: 250
MPD (Micron): 5.2

183:
Brightness (G.E.): 70
Oil Absorption (GCOA), wt %: 175
Water Absorption, wt. %: 200
MPD (Micron): 5.8

GREFCO, INC.: DIATOMITE Functional Fillers(Continued):

Processed Diatomite Functional Fillers:

WF:
Color: White
Brightness (G.E.): 90
Oil Absorption (GCOA), wt. %: 130
Water Absorption, wt. %: 200
MPD (Micron): 4.5

395:
Color: White
Brightness (G.E.): 90
Oil Absorption (GCOA), wt. %: 130
Water Absorption, wt. %: 200
MPD (Micron): 5.0

305:
Color: White
Brightness (G.E.): 90
Oil Absorption (GCOA), wt. %: 135
Water Absorption, wt. %: 190
MPD (Micron): 5.5

WB-5:
Color: White
Brightness (G.E.): 88
Oil Absorption (GCOA), wt. %: 140
Water Absorption, wt. %: 220
MPD (Micron): 7.5

L-5:
Color: White
Brightness (G.E.): 87
Oil Absorption (GCOA), wt. %: 150
Water Absorption, wt. %: 230
MPD (Micron): 9.0

L-10:
Color: White
Brightness (G.E.): 86
Oil Absorption (GCOA), wt. %: 160
Water Absorption, wt. %: 260
MPD (Micron): 11.0

SP-5:
Color: White
Brightness (G.E.): 87
Oil Absorption (GCOA), wt. %: 160
Water Absorption, wt. %: 250
MPD (Micron): 16.0

PS:
Color: Pink
Oil Absorption (GCOA), wt. %: 160
Water Absorption, wt. %: 180
MPD (Micron): 4.5

SF-5:
Color: Pink
Oil Absorption (GCOA), wt. %: 175
Water Absorption, wt. %: 230
MPD (Micron): 10.0

GREFCO, INC.: DICALITE Fillers and Extenders:

DICALITE WB-5:
Mineralogical: Diatomite
Type: Flux Calcined
Physical Form: Dry Powder
Particle Shape: Various Shapes of Diatom Structure
Color: White
Refractive Index: 1.46
Specific Gravity: 2.33
Median Particle Diameter, microns: 6.5-8.5

DICALITE L-5:
Mineralogical: Diatomite
Type: Flux Calcined
Physical Form: Dry Powder
Particle Shape: Various Shapes of Diatom Structure
Color: White
Refractive Index: 1.46
Specific Gravity: 2.33
Median Particle Diameter, microns: 8-10

GREFCO, INC.: DICAPERL Hollow Glass Microspheres:

DICAPERL is a family of lightweight, hollow glass bubble fillers. Available in a variety of particle size ranges, with or without surface modifications, DICAPERL is a very cost effective density reducing filler for any resin/binder system.

Silane-modified DICAPERL functional fillers substantially reduce weight, reduce shrinkage, improve impact, nailing, stapling, and sanding properties of molded parts. The finest grades are used in RP/Composites including glass fiber reinforced products formed by both spray-up and hand lay-up production processes, where surface detail is critical. The coarser grades are used in cast or molded parts not requiring ultrafine surface appearance.

Fine particle size, unmodified grades of DICAPERL are quite effective as thickening, and anti-sag additives in specialty coatings. The coarser grades are used to produce texturing and stipple effects in texture and acoustical coating mixes, block filler paints, etc.

Product Grade:

HP-110:
- Particle Shape: Bubble
- Average Particle Size (Microns): 310
- Effective Density: lb/ft 3: 11.0

HP-210:
- Particle Shape: Bubble
- Average Particle Size (Microns): 110
- Effective Density: lb/ft 3: 11.5

HP-510:
- Particle Shape: Bubble
- Average Particle Size (Microns): 70
- Effective Density: lb/ft 3: 14.0

HP-710:
- Particle Shape: Bubble
- Average Particle Size (Microns): 65
- Effective Density: lb/ft 3: 14.7

FP-1010:
- Particle Shape: Plate-like
- Average Particle Size (Microns): 25
- Effective Density: lb/ft 3: 74.9

All DICAPERL "Series 10" products are white, surface-modified materials with a softening point of approximately 1800F and a surface pH of 7. The surface modification is comparable to that obtained with conventional silanes containing organic functions such as amino, vinyl, cyclic epoxy, aliphatic epoxy, methacryloxy and mercapto groups.

Unmodified grades designated HP-100, HP-200 and HP-500 are also available. A "Series 12" product, HP-112, HP-212, HP-512, or HP-712 is recommended when superior stability is critical.

GREFCO, INC.: PERLITE Functional Fillers:

DICALITE Perlite functional fillers are a versatile family of mineral-based silicate products available throughout the world. These fillers are processed from an amorphous, naturally occurring glass found where there has been volcanic activity. When processed under the proper conditions the solid glass particles expand like popcorn up to 20 times their original volume. A specialized GREFCO process incorporating expansion, selective milling, and sizing determines the particle shape of each filler product.

A variety of flake or di-, tri- and tetrahedral particle shapes are obtained by this exclusive process. These unique shapes provide numerous functions including high surface area, permeability and reinforcement and have densities among the lowest of any of the mineral fillers. In addition to superior spatial loading this property affords the added bonus of density control in finished products. DICALITE Perlite fillers are off-white and do not interfere with the coloring of the products into which they are formulated.

DICALITE Perlite fillers are inert to all but strong acids or alkalis and have no residual surface reactivity. After processing they are free of any organic contamination.

Color	Off-White
Brightness (G.E.)	74
MOHS Hardness	5.5
Moisture, %	1.0
Ig. Loss (dry basis) %	1.5
pH (10% slurry)	5.5-8.5

FF416:

Oil Absorption (GCOA), wt. %:	210
Water Absorption, wt. %:	260
Effective Density g/cc:	2.3
MPD (Micron):	11

FF426:

Oil Absorption (GCOA), wt. %:	220
Water Absorption, wt. %:	275
Effective Density g/cc:	2.2
MPD (Micron):	13

FF436:

Oil Absorption (GCOA), wt. %:	240
Water Absorption, wt. %:	320
Effective Density g/cc:	2.2
MPD (Micron):	19

FF456:

Oil Absorption (GCOA), wt. %:	240
Water Absorption, wt. %:	330
Effective Density g/cc:	1.2
MPD (Micron):	25

FF476:

Oil Absorption (GCOA), wt. %:	240
Water Absorption, wt. %:	330
Effective Density g/cc:	1.2
MPD (Micron):	30

GREFCO, INC.: SOLKA-FLOC Functional Fillers:

SOLKA-FLOC is powdered cellulose obtained from mechanically disintegrated cellulose prepared by processing bleached or unbleached cellulose obtained as a pulp from such fibrous materials as wood or cotton. It occurs as a white or brown, odorless substance and consists of fibrous particles that may be compressed into self-binding tablets that disintegrate rapidly in water. It exists in various grades, exhibiting degrees of fineness ranging from a dense, free-flowing powder to a coarse, fluffy, nonflowing material. It is insoluble in water, in dilute acids, and in nearly all organic solvents. It is slightly soluble in sodium hydroxide TS. When bone dry, the white or bleached grades are at least 99.5% pure cellulose.

SOLKA-FLOC is an excellent filler for a wide variety of rubber and plastic products. SOLKA-FLOC improves dimensional stability, reduces green shrinkage, improves impact strength and improves drying rate of stable foams.

Applications include a wide variety of thermoset resins for injection molding and rubber compounds ranging from floor tiles to shoe soles.

SOLKA-FLOC is ashless and is used in the manufacturing of ceramic products as a burn out agent and to provide controlled porosity.

It is used as a plasticizer, bulking agent, arc intensifier, absorbent and lubricant.

SOLKA-FLOC can be used as a binder, texturizer and thickening agent in products such as latex paints, texture and adhesive compounds.

Grade:

AS-1016:
Average Fiber Length (microns): 210
Color: White

AS-1040:
Average Fiber Length (microns): 160
Color: White

KS-1016:
Average Fiber Length (microns): 290
Color: White

SW-40:
Average Fiber Length (microns): 120
Color: White

GREFCO, INC.: SOLKA-FLOC Functional Fillers(Continued):

Grade:

BW-20:
Average Fiber Length (microns): 100
Color: White

BW-40:
Average Fiber Length (microns): 60
Color: White

BW-60:
Average Fiber Length (microns): 50
Color: White

BW-100:
Average Fiber Length (microns): 40
Color: White

BW-200:
Average Fiber Length (microns): 35
Color: White

BW-300:
Average Fiber Length (microns): 22
Color: White

BW-2030:
Average Fiber Length (microns): 35
Color: White

Fine Granular:
Average Fiber Length (microns): Granules
Color: Cream

BNB-20:
Average Fiber Length (microns): 100
Color: Brown

BNB-40:
Average Fiber Length (microns): 70
Color: Brown

BNB-100:
Average Fiber Length (microns): 40
Color: Brown

HILL BROTHERS CHEMICAL CO.: HI-FIBE:

HI-FIBE is a specially formulated fiber filler that is used as an asbestos replacement. Because of its fibrillated structure, HI-FIBE functions to increase viscosity and improve crack and slump-resistance in oil or water based systems. It is chemically inert, with excellent resistance to acids, alkalis, bleaches, solvents, mildew and sunlight. HI-FIBE is an exceptionally uniform product, formulated for ease of dispersion, and will not separate on mixing. HI-FIBE offers the advantages of good dispersibility, wet and dry-slump resistance, very low shrinkage upon drying, and airless spraying.

Typical Properties:

HI-FIBE:

210X:
Bulk Density (lbs./cu.ft., loose): 49
Oil Absorption (lbs./100 lbs.): 55
Water Absorption (lbs./100 lbs.): 61
Specific Gravity: 2.9

230X:
Bulk Density (lbs./cu.ft., loose): 41
Oil Absorption (lbs./100 lbs.): 69
Water Absorption (lbs./100 lbs.): 78
Specific Gravity: 2.7

250X:
Bulk Density (lbs./cu.ft., loose): 27
Oil Absorption (lbs./100 lbs.): 93
Water Absorption (lbs./100 lbs.): 93
Specific Gravity: 2.6

250-R:
Bulk Density (lbs./cu.ft., loose): 41
Oil Absorption (lbs./100 lbs.): 68
Water Absorption (lbs./100 lbs.): 95
Specific Gravity: 2.9

254-SX:
Bulk Density (lbs./cu.ft., loose): 26
Oil Absorption (lbs./100 lbs.): 105
Water Absorption (lbs./100 lbs.): 168
Specific Gravity: 2.6

270X:
Bulk Density (lbs./cu.ft., loose): 23
Oil Absorption (lbs./100 lbs.): 113
Water Absorption (lbs./100 lbs.): 106
Specific Gravity: 2.6

HILL BROTHERS CHEMICAL CO.: HI-FIBE(Continued):

HI-FIBE:

290X:
Bulk Density (lbs./cu.ft., loose): 15
Oil Absorption (lbs./100 lbs.): 132
Water Absorption (lbs./100 lbs.): 114
Specific Gravity: 2.6

550:
Bulk Density (lbs./cu.ft., loose): 30
Oil Absorption (lbs./100 lbs.): 92
Water Absorption (lbs./100 lbs.): 116
Specific Gravity: 3.1

550-A:
Bulk Density (lbs./cu.ft., loose): 35
Oil Absorption (lbs./100 lbs.): 63
Water Absorption (lbs./100 lbs.): 63
Specific Gravity: 2.6

590-A:
Bulk Density (lbs./cu.ft., loose): 18
Oil Absorption (lbs./100 lbs.): 162
Water Absorption (lbs./100 lbs.): 207
Specific Gravity: 2.5

990:
Bulk Density (lbs./cu.ft., loose): 16
Oil Absorption (lbs./100 lbs.): 191
Water Absorption (lbs./100 lbs.): 222
Specific Gravity: 2.9

2000X:
Bulk Density (lbs./cu.ft., loose): 1.6
Oil Absorption (lbs./100 lbs.): 1760
Water Absorption (lbs./100 lbs.): 1823
Specific Gravity: .96

5000:
Bulk Density (lbs./cu.ft., loose): .97
Oil Absorption (lbs./100 lbs.): 2465
Water Absorption (lbs./100 lbs.): 3040
Specific Gravity: .90

J.M. HUBER CORP.: Synthetic Silica Products:

<u>Product/Grade:</u>

ZEOTHIX 265:
Average Particle Size: 1.7
Oil Absorption: 220
Structure: VHS
Suggested Application: Thickening/Thixotropy

ZEOTHIX 177:
Average Particle Size: 1.5
Oil Absoprtion: 215
Structure: VHS
Suggested Application: Thickening/Thixotropy

ZEOTHIX 95:
Average Particle Size: 2.5
Oil Absorption: 215
Structure: VHS
Suggested Application: Flatting

ZEOSYL 200:
Average Particle Size: 5.0
Oil Absorption: 200
Structure: HS
Suggested Application: Carrier/Rheology

HUBERSIL 162:
Average Particle Size: 12.0
Oil Absorption: 190
Structure: HS
Suggested Application: Reinforcing/Carrier

ZEOFREE 80:
Average Particle Size: 6.0
Oil Absorption: 190
Structure: HS
Suggested Application: Conditioning/Anti-caking

ZEOSYL 100:
Average Particle Size: 6.0
Oil Absorption: 190
Structure: HS
Suggested Application: Carrier/Reinforcing

ZEOSYL 110SD:
Oil Absorption: 190
Structure: HS
Suggested Application: Carrier/Reinforcing

J.M. HUBER CORP.: Synthetic Silica/Silicate Products(Continued):

Product/Grade:

ZEOMATT 155:
Average Particle Size: 4.0
Oil Absorption: 180
Structure: HS
Suggested Application: Flatting

ZEOFREE 153:
Average Particle Size: 5.0
Oil Absorption: 165
Structure: MS
Suggested Application: Carrier/Filler

ZEO 49:
Average Particle Size: 9.0
Oil Absorption: 90
Structure: LS
Suggested Application: Polishing Agent/Catalyst Support

ZEODENT 113:
Average Particle Size: 9.0
Oil Absorption: 85
Structure: LS
Suggested Application: Polishing Agent/Anti-Skid

HUBERSORB 600:
Average Particle Size: 3
Oil Absorption: 450
Structure: VHS
Suggested Application: Carrier/Conditioning

ZEOLEX 7A:
Average Particle Size: 6
Oil Absorption: 145
Structure: MS
Suggested Application: Carrier

HYDREX:
Average Particle Size: 4
Oil Absorption: 125
Structure: LS
Suggested Application: High Brightness Filler

HUBERFIL 96:
Average Particle Size: 4
Oil Absorption: 120
Structure: LS
Suggested Application: Paper Filler

J.M. HUBER CORP.: Synthetic Silicate Products(Continued):

Product/Grade:

ZEOLEX 80:
- Average Particle Size: 6
- Oil Absorption: 115
- Structure: LS
- Suggested Application: Titanium Dioxide Extender

ZEOLEX 23:
- Average Particle Size: 6
- Oil Absorption: 115
- Structure: LS
- Suggested Application: Reinforcing Filler

ZEOLEX 23A:
- Average Particle Size: 6
- Oil Absorption: 115
- Structure: LS
- Suggested Application: Conditioning/Anti-caking

ZEOLEX 7:
- Average Particle Size: 6
- Oil Absorption: 115
- Structure: LS
- Suggested Application: Conditioning/Anti-caking

HUBERFIL 93:
- Average Particle Size: 4
- Oil Absorption: 105
- Structure: VLS
- Suggested Application: Paper Filler

ZEOLEX 35P:
- Average Particle Size: 6
- Oil Absorption: 80
- Structure: VLS
- Suggested Application: Printing Ink

Structure--VHS(Very High Structure) products have OA values above 200; HS(High Structure) products have OA values between 175-200; MS(Medium Structure) have OA values between 125-175; LS(Low Structure) products have OA values between 75-125, and VLS(Very Low Structure) products OA values below 75.

ILLINOIS MINERALS CO.: Microcrystalline Silica:

Microcrystalline Silica - The Ideal, Inert Filler/Extender For a Broad Spectrum of Products

* Coatings
* Dielectric Products
* Adhesives and Ceramics
* Insulation
* Buffing and Polishing Compounds
* Broad Spectrum Usage

Chemical Analysis:
- Silica (SiO_2): 98.60%
- Aluminum Oxide (Al_2O_3): 0.70%
- Iron Oxide (Fe_2O_3): 0.05%
- Potassium Oxide (K_2O): 0.05%
- Titanium Oxide (TiO_2): 0.05%
- Calcium Oxide (CaO): 0.04%
- Magnesium Oxide (MgO): 0.04%
- Sodium Oxide (Na_2O): 0.03%
- Loss on Ignition (LOI): 0.44%

Physical Characteristics:
- Specific Gravity: 2.65
- Weight Per Solid Gallon, Lbs.: 22.07
- Bulking Value, Gallon per Pound: 0.04531
- pH Value: 6-7
- Refractive Index: 1.54-1.55
- Hardness (MOH's): 7.0
- Melting Point: 3100F (1722C)
- Specific Resistance (OHMS): 20,000

Benefits:
* inert
* low in soluble salts
* high purity
* high loading
* soft silica
* low moisture content
* low oil absorption
* good electrical resistance
* non-hygroscopic
* excellent suspension
* excellent dispersion
* soluble only in hydrofluoric acid
* economical

ILLINOIS MINERALS CO.: Microcrystalline Silica(Continued):

Chemical Analysis and Physical Data:

Micronized:

A-108:
Oil Absorption %: 27-31
Apparent Density Lbs/Cu Ft: 22-24
Avg. Particle Size Microns: 1.5
Mean Particle Size - Microns: 1.8

A-10:
Oil Absorption %: 27-31
Apparent Density Lbs/Cu Ft: 22-24
Avg. Particle Size Microns: 1.6
Mean Particle Size - Microns: 2.0

A-15:
Oil Absorption %: 27-31
Apparent Density Lbs/Cu Ft: 23-25
Avg. Particle Size Microns: 1.7
Mean Particle Size - Microns: 2.5

A-25:
Oil Absorption %: 27-31
Apparent Density Lbs/Cu Ft: 23-25
Avg. Particle Size Microns: 1.9
Mean Particle Size - Microns: 2.9

Air Floated:

O:
Oil Absorption %: 27-31
Apparent Density Lbs/Cu Ft: 27-29
Avg. Particle Size Microns: 2.1
Mean Particle Size - Microns: 4.4

1240:
Oil Absorption %: 27-31
Apparent Density Lbs/Cu Ft: 27-29
Avg. Particle Size Microns: 2.4
Mean Particle Size - Microns: 4.8

ILLINOIS MINERALS CO.: Microcrystalline Silica(Continued):

Chemical and Physical Data:

Air-Floated:

1160:
Oil Absorption %: 27-31
Apparent Density Lbs/Cu Ft: 27-29
Avg. Particle Size Microns: 2.5
Mean Particle Size - Microns: 5.2

54:
Oil Absorption %: 27-31
Apparent Density Lbs/Cu Ft: 28-30
Avg. Particle Size Microns: 2.5
Mean Particle Size - Microns: 5.6

250:
Oil Absorption %: 27-31
Apparent Density Lbs/Cu Ft: 29-31
Avg. Particle Size Microns: 2.5
Mean Particle Size - Microns: 6.4

200:
Oil Absorption %: 27-31
Apparent Density Lbs/Cu Ft: 29-31
Avg. Particle Size Microns: 2.5
Mean Particle Size - Microns: 6.6

MALVERN MINERALS CO.: Microcrystalline Arkansas Novaculite:

DAPER NOVACITE:

Particle Size:
Hegman Fineness 5

Chemical Analysis:
SiO_2: 99.490%
Fe_2O_3: .039
Al_2O_3: .102
TiO_2: .015
CaO: .014
MgO: .021

Typical Applications:
Silicone rubber extender
Silicone rubber dusting powder
Electrostatic coating
Molding compounds
Fluidized bed coatings
Casting and potting compunds
Most all coatings and paints

General Information:
DAPER NOVACITE lies between 1250 and L-207A in fineness. It is the coarsest of the finer grades. DAPER is a candidate as requirements become stricter with regard to fineness, gloss control and abrasivity. Even though it is quite fine, DAPER still has the uniqueness of low binder demand with superior rheological flow properties.

200 NOVACITE:

Average Particle Size (Range): 11-19u

Typical Applications:
Casting resins
Potting compounds
Coarser textured coatings
Chemicals
Fluid loss additives (Oil wells)
Abrasive medium (Wet blasting)
Ceramics

General Information:
200 NOVACITE is a product not considered to be very fine in particle size. Normally, 200 NOVACITE would be preferred where more grittiness, more abrasivity, and lower binder demand are desirable.

MALVERN MINERALS CO.: Microcrystalline Arkansas Novaculite (Continued):

325 NOVACITE:

Average Particle Size (Range): 8.9-14.9u

Typical Applications:
- Casting resins
- Potting compounds
- Molding compounds
- Abrasive medium (Wet blasting)
- Interior and Exterior latex paints

General Information:

325 NOVACITE is a typical 325 mesh product although its specifications vary rather widely. You will note the percentage passing 325 mesh can be as low as 88% and as high as 99%. A good product where lower binder demand in the 44u range is preferred.

550 EBONY NOVACITE:

Naturally occurring EBONY NOVACITE will produce truly black coatings regardless of surface, weather conditions or varying degrees of sun exposure.

550 Micron EBONY NOVACITE classification is to be considered very unique, black silicic pigment offering striking black pigmentation in various resinous mediums. Other properties such as opacity, low oil adsorption and good dispersion should be noted. While this product is typically (60% SiO_2) (novaculitic) the 3.5% carbon (C) is a manifest and seems to encapsulate the silica particles enhancing effective pigmentation.

EBONY NOVACITE is recommended for evaluation in most all coatings as well as in plastics and certain elastomers. A black pigment or extender of this nature should offer superior rheological behavior compared to black pigments with high surface area and high sorption properties.

550 EBONY NOVACITE:
Typical Physical Characteristics
Appearance:
Dry-Soft Black
Wet-Vivid Black

Apparent Bulk Density: Approx. 35#/cu.ft.
True Density: Approx. 21.5#/Gal.

Typical Chemical Composition:
- Silica: 63.15%
- Aluminum Oxide: 19.16%
- Carbon: 3.24%
- Ferric Oxide: 2.29%
- Titanium Dioxide: 1.80%

MALVERN MINERALS CO.: Microcrystalline Arkansas Novaculite (Continued):

20 Micron EBONY NOVACITE

Naturally occuring EBONY NOVACITE will produce truly black coatings regardless of surface, weather conditions or varying degrees of sun exposure.

20 Micron EBONY NOVACITE classification is to be considered very unique, black silicic pigment offering black pigmentation in various resinous mediums. Other properties such as opacity, low oil adsorption and good dispersion should be noted. While this product is typically (60% SiO_2) (novaculitic) the 3.5% carbon (C) is manifest and seems to encapsulate the silica particles enhancing effective pigmentation.

EBONY NOVACITE is recommended for evaluation in most all coatings as well as in plastics and certain elastomers. A black pigment or extender of this nature should offer superior rheological behavior compared to black pigments with high surface area and high sorption properties.

20 Micron EBONY NOVACITE
Typical Physical Characteristics
Appearance:
Dry-Soft Black
Wet-Black

Apparent Bulk Density: Approx. 35#/cu.ft.
True Density: Approx. 21.5#/Gal.

Typical Chemical Composition:
- Silica: 58.00%
- Aluminum Oxide: 21.06%
- Ferric Oxide: 2.29%
- Titanium Oxide: 1.40%
- Calcium Oxide: 6.88%

1250 NOVACITE:

Average Particle Size (Range): 7.3-12.9u

General Information:

1250 NOVACITE is a premium 325 mesh product. Normally, it is 100% finer than 44u. This outstanding product has been preferred for over 25 years in thermoset molding compounds. For almost any application in the polymer field in its range of fineness.

Typical Applications:
- Fluidized bed coatings
- Casting & potting resins
- Polyurethane grouts
- Molding compounds
- Electrostatic coatings
- Industrial coatings

MALVERN MINERALS CO.: Microcrystalline Arkansas Novaculite (Continued):

L-207A NOVACITE

Hegman Fineness: 6-7
Average Particle Size (Range) Fisher 3.5u to 4.9u

Chemical Analysis (Typical):
SiO_2: 99.49%

Typical Applications:
Silicone rubber extender
Silicone rubber dusting powder
Electrostatic coating
Molding compounds
Fluidized bed coatings
Casting and potting compounds
Most all coatings and paints
Light diffusing characteristic control

General Information:
L-207A NOVACITE is the "star" of the Line. Normally the product will disperse to a 7 Hegman Grind. All particles are individually platey with few clusters. 98% of the particles will fall between one and ten microns. L-207A as fine as it is has the uniqueness of low binder demand with superior flow properties.

S-325 NOVACITE

Hegman Fineness 2
Average Particle Size(Range): 7.4 - 16.3u

General Information:
S-325 NOVACITE is a choice selection as an extender pigment in most interior and exterior latex emulsion type trade sales coatings. The degree of fineness plus very low binder demand gives latex paints using S-325 excellent tint retention, good frost resistance, mar resistance and chalk resistance.

Typical Applications:
Exterior trade sales paint (emulsion)
Fluidized bed coatings
Interior latex paints
Corrosion resistant coatings
Casting and potting compounds
Abrasive medium (Wet blasting)

MANVILLE: CELITE Functional Fillers for the Coatings Industry:

CELITE Diatomite:
Diatomite is a soft powdery mineral with a unique particle stucture.

CELITE Grade Selection for Optimum Performance
Since CELITE extenders are essentially pure silica, they are inert and non-reactive in all commonly used solvent- and water-thinned organic coating vehicles. For the same reason, they do not interfere with the performance of other pigments.

CELITE Selector Chart:

Type of Finish:

Masonry Paints:
Required Hegman Fineness: 0
White Grade: 110
Natural Grade Off-White: 289
Useful Properties: Color and sheen uniformity on absorptive surfaces. High flatting efficiency.

Traffic Paints:
Required Hegman Fineness: 0-1
White Grade: 110-281
Natural Grade Off-White: 289

Required Hegman Fineness: 4
White Grade: 499
Useful Properties: Fast solvent release. Diffuse reflectivity, abrasion resistance, durability.

Primer Paints:
Required Hegman Fineness: 0-1
White Grade: 281
Natural Grade Off-White: 266, 289
Useful Properties: Good "tooth"-topcoat adhesion. Good sanding properties.

Flat Wall Paints Interior (Water-Emulsion or Solvent-Thinnable)
Required Hegman Fineness: 0-1
White Grade: 281
Natural Grade Off-White: 266
Useful Properties: High flatting efficiency. Gloss and sheen uniformity. Resistance to polishing.

MANVILLE: CELITE Functional Fillers for the Coatings Industry (Continued):

Celite Selector Chart(Continued):

Type of Finish:

House Paints, Exterior:
Required Hegman Fineness: 0-1
White Grade: 281
Natural Grade Off-White: 266

Required Hegman Fineness: 4
White Grade: 499
Useful Properties: High flatting efficiency at low PVC. Resistance to "flashing." Good application properties.

Maintenance Paints:
Required Hegman Fineness: 0-1
White Grade: 281
Natural Grade Off-White: 266
Useful Properties: Efficient flatting. Good application properties.

Satin-Luster (Eggshell Enamels):
Required Hegman Fineness: 4
White Grade: 499
Useful Properties: Efficient gloss and sheen control. Easy dispersibility.

Semi-Gloss Enamels:
Required Hegman Fineness: 4
White Grade: 499
Useful Properties: High flatting efficiency. Easy dispersibility.

Extra-Smooth Semi-Gloss Enamels:
Required Hegman Fineness: 5
White Grade: WHITE MIST
Useful Properties: Satin-smooth films. Readily dispersible with disc disperser.

Industrial Finishes:
Required Hegman Fineness: 5
White Grade: WHITE MIST
Useful Properties: Very high flatting efficiency permits low pigment to binder ratio.

Flat Varnishes and Lacquers (stir-in):
Required Hegman Fineness: 5
White Grade: WHITE MIST
Useful Properties: Efficient flatting with good film clarity. Readily dispersible with disc disperser.

MANVILLE: CELITE Functional Fillers for the Coatings Industry (Continued):

CELITE Selector Chart(Continued):

Type of Finish:

Adjustment of Finished Batches (stir-in):
Required Hegman Fineness: 4
White Grade: 499

Required Hegman Fineness: 5
White Grade: WHITE MIST
Useful Properties: Readily dispersible with disc disperser or high-speed mixer.

Coil Coatings:
Required Hegman Fineness: 5
White Grade: WHITE MIST
Useful Properties: High flatting efficiency.

Industrial Finishes:
Required Hegman Fineness: 5 1/2
White Grade: SUPER FINE
SUPER FLOSS
Useful Properties: Finest grade available.

Stains:
Required Hegman Fineness: 4-5
White Grade: 499
Useful Properties: Low gloss finisher.

CELITE Performance Chart:

Function:	Comment
Control of gloss and sheen	Excellent
Control of film permeability	Good
Control of film density (pigment packing)	Good
Tooth for topcoat adhesion	Good
Faster solvent release	Good
Sanding improvement	Good
Control mar resistance in flats	
Fine grades	Good
Coarse grades	Fair
Resistance to chipping	Good
Transparency in clears	Fair
Viscosity stability	Good
Effect on brightness	Good
Effect on tint retention	Good
"Touch-up" improvement	Excellent
Suspendability in pigmented coatings	Good
Ease of wetting	Excellent
Scrub resistance	Good
Effect on consistency	Good
Stain Removal	
Fine grades	Good
Coarse grades	Fair

MANVILLE: MICRO-CEL:

MICRO-CEL is the Manville trademark for synthetic calcium silicates made through the reaction of diatomite with calcium hydroxide. These unique pigments are processed in two grades for use in water emulsion paints: MICRO-CEL T-38 and MICRO-CEL T-70.

Because of their contribution to hiding, both are extensively used in latex flat wall paints to reduce the amount of titanium dioxide required, resulting in savings in raw material costs. In addition to excellent dry hiding, MICRO-CEL provides built-in gloss and sheen control, improvement in touch-up and uniformity of appearance.

MICRO-CEL also has good suspension properties which accounts for its use in traffic paints and other applications where non-settling is important.

Hiding Contribution:

MICRO-CEL T-38 and T-70 are classified as medium particle size extenders consisting of clusters of sub-micron particles.

Gloss and Sheen Control:

As a flatting agent, additions of 30 lbs of MICRO-CEL T-38 or 40 lbs. of MICRO-CEL T-70 per 100 gal. will result in very low 60 gloss and 85 sheen values.

Dispersion Techniques:

As with CELITE, MICRO-CEL should be the first pigment addition.

Typical Physical Properties:

MICRO-CEL T-38:

Color: White
Oil Absorption, % (Gardner-Colemen): 365
Bulking Value:
lbs./gal.: 18.80
gals./lb.: 0.532
% Retention 325 Mesh: 2.00
pH: 9.80
Refractive Index: 1.55

MICRO-CEL T-70:

Color: White
Oil Absorption, % (Gardner-Coleman): 220
Bulking Value:
lbs./gal.: 20.10
gals./lb.: 0.0498
% Retention 325 Mesh: 1.50
pH: 9.50
Refractive Index: 1.55

THE MILWHITE CO.,INC.: MICROBARITE:

Paints
Coatings
Glass
Rubber
Plastics

The High Quality Cost-Saving Alternative

Typical Physical Properties

	MB-1040:	MB-1065:
Specific gravity	4.3-4.4	4.3-4.4
pH	6-7	6-7
Oil absorption	9-10	11-13
Moisture (%), less than	0.3	0.3
Weight/solid gallon	35	37
Average particle size, microns	11-13	3-5
Hegman gauge	4+	6.5+
Dry brightness, Photovolt	88-92	90-92
Color	White	White

Typical Chemical Analysis

Barium Sulfate (BaSO4)	97-99%
Iron Oxide (Fe2O3)	<0.5%
Silica (SiO2)	<0.2%
Water Soluble Salts	<0.5%

NYCO: ATH Base Grades for ATH NYCOAT Surface Modified Product:

ATH products are available from NYCO in a broad range of sizes which cover the coarser ground or milled products as well as the finer precipitated products. They are also available in standard process products or a high brightness white product.

Typical Composition/Physical Data:
- Al_2O_3: 64.9%
- Total Na_2O: 0.28%
- Soluble Na_2O: 0.02%
- SiO_2: 0.01%
- Fe_2O_3: 0.01%
- Loss on Ignition (L.O.I.): 34.7%
- Absorbed (Surface) Moisture: 0.30%
- Specific Gravity: 2.42
- GE Brightness: 92-93
- Mohs Hardness: 2.5-3.5

Typical Size Distribution Data:

ATH 00:
- Micron Median: 6-8
- Bulk Loose lbs./cu.ft.: 28
- Density packed lbs./cu.ft.: 38

ATH 01:
- Micron Median: 7-9
- Bulk Loose lbs./cu.ft.: 30
- Density Packed lbs./cu.ft.: 42

ATH 02:
- Micron Median: 8-10
- Bulk Loose lbs./cu.ft.: 36
- Density Packed lbs./cu.ft.: 50

ATH 03:
- Micron Median: 12-14
- Bulk Loose lbs./cu.ft.: 36
- Density Packed lbs./cu.ft.: 51

ATH 04:
- Micron Median: 12-14
- Bulk Loose lbs./cu.ft.: 36
- Density Packed lbs./cu.ft.: 52

ATH 05:
- Micron Median: 13-15
- Bulk Loose lbs./cu.ft.: 38
- Density Packed lbs./cu.ft.: 56

NYCO: ATH Base Grades for ATH NYCOAT Surface Modified Products (Continued):

Typical Size Distribution Data (Continued):

ATH 06:
Micron Median: 14-16
Bulk Loose lbs./cu.ft.: 40
Density Packed lbs./cu.ft.: 58

ATH 07:
Micron Median: 16-18
Bulk Loose lbs./cu.ft.: 42
Density Packed lbs./cu.ft.: 60

ATH 08:
Micron Median: 2-2.5
Bulk Loose lbs./cu.ft.: 18
Density Packed lbs./cu.ft.: 26

ATH 09:
Micron Median: 3-4
Bulk Loose lbs./cu.ft.: 22
Density Packed lbs./cu.ft.: 34

ATH 10:
Micron Median: 6-7
Bulk Loose lbs./cu.ft.: 26
Density Packed lbs./cu.ft.: 38

ATH 11:
Micron Median: 0.8
Bulk Loose lbs./cu.ft.: 11
Density Packed lbs./cu.ft.: 22

NYCO: Developmental Products:

NYGLAS Surface Modified Crushed Glass:

NYGLAS is soda lime glass that has been crushed, screened and chemically modified to provide optimum dispersion and reinforcement in polymer composites.

Typical Chemical Properties:

Component:
- SiO2: 73.1%
- Na2O: 13.5
- CaO: 8.8
- MgO: 3.6
- Al2O3: 0.22
- K2O: 0.12
- Fe2O3: 0.37

Suggested Applications:

NYGLAS should be evaluated in:
- * RRIM systems versus glass flake
- * Casting systems
- * All thermoplastic resin systems
- * Most thermosetting resin

200 NYGLAS:

- +140 mesh: 1%
- +200 mesh: 3
- +200 mesh: 97
- -325 mesh: 78

325 NYGLAS:

- +140 mesh: 0%
- +200 mesh: 1
- +200 mesh: 99
- -325 mesh: 90

NYCO: NYAD Wollastonite:

NYAD G wollasonite High Aspect Ratio Grade:

Wollastonite is a naturally occurring calcium metasilicate. It is the only commercially availiable pure white mineral that is wholly acicular; typical length to diameter ratios range from 3:1 to 20:1. The average diameter of wollastonite is 3.5 microns.

NYAD G wollastonite is the basegrade for the G WOLLASTOKUP surface modified wollastonite product line.

Properties of NYAD wollastonite:

Appearance: Briiliant white
Shape: Acicular
Molecular weight: 116
Specific Gravity: 2.9
Refractive index: 1.63
pH (aqueous solution): 9.9
Water solubility (gms./100 cc): 0.0095
Density (lbs./solid gallons): 24.2
Bulking value (gal./lbs.): 0.0413
Moh's hardness: 4.5
Coefficient of expansion, (in./in./C): 6.5 x 10 -6
Melting point (C): 1540
Moisture--% (110C @ 24 hours): 0.5
Loss on ignition--% (volatiles @ 1000C): 0.8

Chemical Composition CaSiO3:

Component:	Typical
CaO	47.0%
SiO2	50.0
Fe2O3	1.0
Al2O3	1.0
MnO	0.1
MgO	0.3
TiO2	0.05

Typical Values of NYAD G wollastonite:

Brightness: 85
Oil Absorption: ASTM D-281: 40
Fineness of Grind: Hegman scale: 1
Bulk Density:
  Loose: lbs./cu.ft.: 25
  Tapped: lbs./cu.ft.: 40
Aspect Ratio: 20:1
Surface Area: B.E.T.: 0.8

NYCO: NYAD Wollastonite(Continued):

NYAD FP wollastonite Coarse Grade:

Typical Values:
- Color: 70
- Bulk Density
  - Loose: lbs./cu.ft.: 75
  - Tapped: lbs./cu.ft.: 98
- Surface Area: B.E.T.: 0.8

400 Mesh wollastonite Base Grade for WOLLASTOKUP:

The 400 mesh wollastonite is the base grade for 400 WOLLASTOKUP surface modified wollastonite product line.

Typical Values:
- Brightness: 90
- Oil Absorption: ASTM D-281: 22
- Fineness of Grind: Hegman scale: 4-5
- Bulk Density
  - Loose: lbs./cu.ft.: 40
  - Tapped: lbs./cu.ft.: 60
- Aspect Ratio: 5:1
- Surface Area: B.E.T.: 1.9
- Minus 400 mesh screen: %: 98
- Minus 325 mesh screen: %: 100

NYAD 1250 wollastonite Fine Particle Size Grade:

NYAD 1250 wollastonite is the base grade for the 10 WOLLASTOKUP surface modified wollastonite product line.

Typical Values:
- Brightness: 94
- Oil Absorption: ASTM D-281: 32
- Fineness of Grind: Hegman scale: 8
- Bulk Density:
  - Loose: lbs./cu.ft.: 30
  - Tapped: lbs./cu.ft.: 50
- Aspect Ratio: 3:1
- Surface Area: B.E.T.: 3.3
- Minus 30 micron: %: 100
- Minus 10 micron: %: 96

NYCO: NYAD Wollastonite(Continued):

NYAD 475 wollastonite Fine Particle Size Grade:

Typical Values:
- Brightness: 92
- Oil Absorption: ASTM D-281: 30
- Fineness of Grind: Hegman scale: 5-6
- Bulk Density:
  - Loose: lbs./cu.ft.: 50
  - Tapped: lbs./cu.ft.: 60
- Aspect Ratio: 3:1
- Surface Area: B.E.T.: 3.3
- Minus 30 Micron: %: 100
- Minus 10 Micron: %: 65

NYFLAKE Muscovite Mica:

MICACOAT:
- Muscovite mica, surface modified
- CAS Number: Mica 12001-26-2
- Muscovite 1318-94-1

NYSPHERES hollow glass microspheres:

NYSPHERES are tough, lightweight, hollow glass microspheres. They are inert, have no surface alkalinity and are non toxic.

Property:
- Form: Hollow glass microspheres
- Color: Off White
- Shell Thickness: Approx. 10% of diameter
- Melting Point: 1200C
- Hardness--Moh's: 5

Typical Values of NYSPHERES:

NYSPHERES 7-125-5:
- Specific Gravity: 0.70
- Bulk Density (lbs./cu. ft.): 25
- pH (water solution): 7.0

NYSPHERES 7-200-5:
- Specific Gravity: 0.70
- Bulk Density (lbs./cu.ft.): 25
- pH (water solution): 7.0
- Size (in microns): 10-200

NYSPHERES 7-350-5:
- Specific Gravity: 0.70
- Bulk Density (lbs./cu.ft.): 25
- pH (water solution): 7.0
- Size (in microns): 10-350

NYCO: NYCOR Barytes/Celestite:

NYCO is marketing two grades of celestite and two grades of barytes.

The celestite should be evaluated against bleached barytes, the barytes should be evaluated against unbleaced barytes. The NYCOR celestite products exhibit the same inherent properties of barytes such as low oil absorption, high density, and insolubility in water, acids and bases.

Typical Chemical Composition:

Barytes:
- Barium Sulphate (BaSO4): 94%
- Silica (SiO2): 2.75%
- Ferric oxide (Fe2O3): 1.75%
- Manganese (MnO): 0.33%
- Lead (PbSO4): 0.05%
- Loss on Ignition: 0.50%

Celestite:
- Strontium sulphate (SrSO4): 92-95%
- Barium sulphate (BaSO4): 2-3%
- Calcium Carbonate (CaCO3): 2%
- Alumina (Al2O3): 1% max.

Typical Physical Properties:

Barytes:
- Appearance: Off white
- Molecular wt. (theoretical): 201.4
- Specific Gravity: 4.25
- Refractive index: 1.64
- pH (5% water slurry): 7.0
- Moh's hardness: 3.0

Celestite:
- Appearance: White
- Molecular wt. (theoretical): 151.68
- Specific gravity: 3.95
- Moh's hardness: 3.5

NYCO: NYCOR Barytes/Celestite(Continued):

NYCOR 200 barytes:

Oil Absorption: ASTM D-281: 7-9
Fineness of Grind: Hegman Scale: 1-2
Bulk Density:
Loose: lbs/cu.ft.: 84
Tapped: lbs./cu.ft.: 125
Minus 200 mesh: %: 98

NYCOR 325 barytes:

Oil Absorption: ASTM D-281: 7-9
Fineness of Grind: Hegman Scale: 3-4
Bulk Density:
Loose: lbs./cu.ft.:86
Tapped: lbs./cu.ft.: 130
Minus 325 mesh: %: 95

NYCOR 200 celestite:

Oil Absorption: ASTM D-281: 6-8
Fineness of Grind: Hegman Scale: 2
Bulk Density:
Loose: lbs./cu.ft.:84
Tapped: lbs./cu.ft.:125
Minus 200 mesh: %: 95

NYCOR 325 celestite:

Oil Absorption: ASTM D-281: 6-8
Fineness of Grind: Hegman Scale: 4
Bulk Density:
Loose: lbs./cu.ft.: 83
Tapped: lbs./cu.ft.: 122
Minus 325 mesh: %: 95

OHIO LIME CO.: Pulverized Dolomitic Limestone:

STONELITE
Pulverized Dolomitic Limestone

STONELITE Pulverized Dolomitic Limestone is quarried, crushed, kiln dried and sized by the Ohio Lime Company. Its size, purity and flowability make it attractive for several industrial applications. Its low moisture content minimizes storage and handling problems during cold weather.

Filler: STONELITE Pulverized Dolomitic Limestone is used extensively as a filler in rubber, plastics, ceramics, caulking compounds, adhesives, sealants, roof coatings and latex or textured paint. Anywhere there is a need for an inexpensive and relatively inert filler where fineness is critical, STONELITE should be considered.

Typical Chemical Analysis:
- Calcium Carbonate: 54.50
- Magnesium Carbonate: 44.00

Typical Screen Analysis:
- Smallest Size: .21 Microns
- Mean Size: 10.00 Microns
- Top Size: 300.00 Microns

OHSO
Pulverized Dolomitic Limestone

Filler: OHSO Pulverized Dolomitic Limestone may be used as a filler in rubber, plastics, ceramics, caulking compounds, adhesives, sealants, roof coatings and textured paint. Anywhere there is a need for an inexpensive and relatively inert filler, OHSO should be considered.

Typical Screen Analysis:
- % Passing 30 Mesh (0.55 mm): 99.90
- % Passing 60 Mesh (0.25 mm): 85.00

PFIZER MINERALS, PIGMENTS & METALS DIVISION: MICROTALC - High Purity Montana Talcs:

PFIZER MICROTALC products are produced in Barretts, Montana, from an extensive deposit of exceptionally high quality talc ore. MICROTALC products are characterized by a platy grain shape, closely controlled particle size, low calcium carbonate content, and freedom from asbestiform fibers.

The PFIZER MICROTALC products include a series of seven products ranging from 10 microns top size to 30 microns top size.

Chemical Composition - (typical):

Silicon Dioxide	61.2%
Magnesium Oxide	31.7%
Calcium Oxide	0.2%
Aluminum Oxide	0.7%
Ferric Oxide	1.1%
Titanium Dioxide	0.05%
Sodium Oxide	0.06%
Potassium Oxide	0.03%
Loss On Ignition	5.1%

Physical Properties - (typical):

MP10-52:
Oil Absorption: 52
Hegman Fineness: 6.5+

MP12-50:
Oil Absorption: 50
Hegman Fineness: 6.0+

MP15-38:
Oil Absorption: 44
Hegman Fineness: 6.0

MP20-40:
Oil Absorption: 44
Hegman Fineness: 6-

MP25-38:
Oil Absorption: 43
Hegman Fineness: 5.5+

MP26-38:
Oil Absorption: 43
Hegman Fineness: 5.5

MP30-36:
Oil Absorption: 36
Hegman Fineness: 5.0

PFIZER MINERALS, PIGMENTS, & METALS DIVISION: TALCRON - High Purity Montana Talc:

Pfizer TALCRON products are produced in Barrets, Montana, from an extensive deposit of exceptionally high quality talc ore. TALCRON products are characterized by a platy grain shape, closely controlled particle size, low calcium carbonate content, and virtual freedom from asbestiform fibers.

The Pfizer TALCRON products include a series of four products.

Chemical Composition - (typical)
- Silicon Dioxide: 61.2
- Magnesium Oxide: 31.7
- Calcium Oxide: 0.19
- Aluminum Oxide: 0.7
- Ferric Oxide: 1.1
- Acid Solubles: 1.9
- Loss on Ignition: 5.0

Physical Properties:

MP40-27:
- Oil Absorption: 27
- Hegman Fineness: 3.5

MP44-26:
- Oil Absorption: 26
- Hegman Fineness: 3.25

MP45-26:
- Oil Absorption: 26

MP50-35:
- Oil Absorption: 31

PFIZER MINERALS, PIGMENTS & METALS DIVISION: ULTRAFINE Precipitated Calcium Carbonates:

ULTRAFINE precipitated calcium carbonates, especially developed as reinforcing agents in plastic and rubber compounds, are available in these grades:

MULTIFEX MM - Uncoated
MULTIFEX SC - Surface Treated
ULTRA-PFLEX - Surface Treated

All three grades are pure synthetic calcite with the following typical chemical composition before surface treatment:

$CaCO_3$: 98%
$MgCO_3$: <1%
Fe_2O_3: <0.05%
SiO_2: <0.1%
Al_2O_3: <0.02%
H_2O loss @110C: <0.5%

Typical Physical Properties:

MULTIFEX MM:
Specific Gravity: 2.71
Oil Absorption g/100 g: 55
Dry Brightness: 98
Average Particle Size: 0.07u

MULTIFEX SC:
Specific Gravity: 2.67
Oil Absorption g/100 g: 40
Dry Brightness: 96
Average Particle Size: 0.07u

ULTRA-PFLEX:
Specific Gravity: 2.65
Oil Absorption g/100 g: 35
Dry Brightness: 95
Average Particle Size: 0.07u

Applications:

Plastics and Rubber:
MULTIFEX SC and ULTRA-PFLEX are specifically surface treated to improve dispersion and physical properties in polymeric systems.

Coatings and Inks:
MULTIFEX MM disperses more easily in aqueous systems, while MULTIFEX SC and ULTRA-PFLEX disperse more easily in non-aqueous media.

Adhesives and Sealants:
The superior dispersion of the surface treated products, MULTIFEX SC and ULTRA-PFLEX make them the ideal choice.

PFIZER MINERALS, PIGMENTS & METALS DIVISION: ULTRAFINE Precipitated Calcium Carbonates(Continued):

MULTIFEX MM in Aqueous and Non-aqueous Coatings:

MULTIFEX MM is an Ultrafine Precipitated Calcium Carbonate developed specifically as an extender in surface coatings, as well as other applications. It is a high brightness, chemically pure, carefully controlled, high performance extender that offers many advantages to the paint formulator.

Typical Chemical Properties:
MULTIFEX MM is a pure synthetic calcite with the following chemical composition:

$CaCO_3$: 98%
$MgCO_3$: 1%
Fe_2O_3: 0.05%
SiO_2: 0.1%
Al_2O_3: 0.02%
H_2O loss @ 110C: 0.5%

Typical Physical Properties:

Average Particle Size (microns):	.07
Specific Gravity:	2.71
Oil Absorption g/100 g:	55
Dry Brightness:	98
Loose Bulk Density lbs./ft. 3:	15
Packed Bulk Density lbs./ft. 3	24

Advantages:
Gloss & Semi-Gloss Enamels
* Excellent film build
* Improved pigments suspension
* High brightness
* Elimination of flooding, floating, and streaking
* Non-sag properties
* Gloss retention
* Improved application characteristics

PFIZER MINERALS, PIGMENTS & METALS DIVISION: VICRON Ground Limestone (Eastern):

VICRON Ground Limestone is available in a full line of high brightness, high purity extender pigments of carefully controlled particle size distribution. They are manufactured from selected natural calcite which is processed at Adams, Massachusetts. The VICRON line has been engineered to contribute specific physical properties to paint, plastic, rubber, and other products.

Four VICRON pigment grades are currently available. Chemical composition is the same for all grades.

The uses for VICRON Ground Limestone include:

* Adhesives and glues
* Paint
* Plastics
* Putty and caulking

Chemical Composition - (typical)

Calcium Carbonate	96.0%
Magnesium Carbonate	1.5%
Silicon Dioxide	1.2%
Aluminum Oxide	0.3%
Ferric Oxide	0.08%
Moisture	0.25%

Physical Properties - (typical)

VICRON 15-15:
- Dry Brightness: 95
- Hegman Fineness: 6
- Oil Absorption: 19
- Surface Area m2/gm.: 2.4
- Bulk Density lb./ft. 3: 28

VICRON 25-11:
- Dry Brightness: 95
- Hegman Fineness: 5
- Oil Absorption: 16
- Surface Area m2/gm.: 1.8
- Bulk Density lb./ft. 3: 31

VICRON 31-6:
- Dry Brightness: 94
- Hegman Fineness: 3.5
- Oil Absorption: 16
- Surface Area m2/gm.: 1.3
- Bulk Density lb./ft. 3: 36

VICRON 41-8:
- Dry Brightness: 94
- Hegman Fineness: 3.0
- Oil Absorption: 16
- Surface Area m2/gm.: 1.1
- Bulk Density lb./ft 3: 38

SILBRICO CORP.: RYOLEX Industrial Perlites:

Typical Physical Properties:
Softening point: 1600-2000F
Fusion point: 2300-2450F
pH: 6.6 to 8.0
Specific heat: 0.20 Cal/G C
Specific Gravity: 2.2 to 2.4
Refractive index: 1.5
% Free moisture, max.: <0.5

Grade No. 1:
Mesh: +16: % Weight: 45
Mesh: -20+30: % Weight: 10
Mesh: -100: % Weight: 7

Grade No. 2-A:
Mesh: +16: % Weight: 12
Mesh: -20+30: % Weight: 23
Mesh: -100: % Weight: 10

Grade No. 39:
Mesh: +30: % Weight: 5
Mesh: -50+100: % Weight: 35
Mesh: -200: % Weight: 7

Grade No. 3-S:
Mesh: +16: % Weight: Trace
Mesh: -20+30: % Weight: 12
Mesh: -100: % Weight: 15

Grade No. 9:
Mesh: +16: % Volume: T
Mesh: -20+30: % Volume: 17
Mesh: -100: % Volume: 10

Grade No. 5:
Mesh: +8: % Volume: 50
Mesh: -8+16: % Volume: 35
Mesh: -16: % Volume: 15

SILBRICO CORP.: SIL-CELL Microcellular Filler:

SIL-CELL glass micro cellular fillers are inert, inorganic, non-toxic, hollow glass particles whose shapes vary to combine different geometries...spherical and irregular. These shapes provide advantages as a filler-reinforcement.

* Irregular Shape
* Multiple Cells
* Low Specific Gravity
* Strength

Typical Physical Properties of SIL-CELL

Chemical Analysis

Silicon Dioxide	73%
Aluminum Oxide	17%
Potassium Oxide	5%
Sodium Oxide	3%
Calcium Oxide	1%
Plus Trace Elements	

Grade SIL-32:
- Oil Absorption gms. oil per 100 cc: 30
- Average Particle Size, Microns: 75
- Particle Size Range, Microns: 1-300

Grade SIL-42:
- Oil Absorption gms. oil per 100 cc: 36
- Average Particle Size, Microns: 45
- Particle Size Range, Microns: 1-210

Grade SIL-43:
- Oil Absoprtion gms. oil per 100 cc: 40
- Average Particle Size, Microns: 35
- Particle Size Range, Microns: 1-150

Typical Particle Size Distribution:

SIL-32:
- +50 Mesh: 2
- -200: 50

SIL-42:
- +50 Mesh: Trace
- -200: 70

SIL-43:
- +50 Mesh: 0
- -200: 95

SIL-CELL is used in rotation molding; cultured marble; thermo-set castings; syntactic foam; SMC and BMC compounds; auto body putty; sealants and adhesives; patching compounds; FRP spray and hand lay up; tub and shower enclosures.

SMITH CHEMICAL & COLOR CO. INC.: Products Distributed:

CALWHITE

Typical Chemical Composition:	
Total Carbonates (Ca, Mg)	95.0% min.
$MgCO_3$	3.0% max.
Al_2O_3	0.1%-0.35%
Fe_2O_3	.02%-.16%
SiO_2	0.2%-1.4%
Mn	Trace
Typical Physical Composition:	
Hardness (Moh's Scale)	3
pH	9.0-9.5
Specific Gravity	2.71
Bulking Value	.0443 gal per pound
Specific Resistance	21.000 ohms
Particle Size Distribution	
SSD	2.1-2.4
CP	18
DP	300
Max. retained 325 wet screen	.008
Dry Brightness	96 % min.
Typical Oil Absorption	12-14

MICROFINE BARIUM SULFATE

Outstanding characteristics include high chemical purity, density, brightness, low oil absorption, excellent wettability and dispersion properties.

Typical Chemical Analysis:
- Barium Sulfate: 99.04%
- Silica: .31%
- Iron Oxide: .15%
- Calcium Carbonate: .40%
- Aluminum Oxide: .05%
- Ignition Loss (1200C): .05%

Typical Physical Properties:
- Color: Extra White
- Brightness (photo-volt analysis): 92-95
- Specific Gravity: 4.40
- Oil Absorption: 12

Standard Particle Size:
- Top Particle Size: 16.0 micron
- Mean Particle Size: 4.5 micron
- Low Particle Size: 3.0 micron
- Hegman Number: 5-6

Manufactured in France. Process technology includes ultra modern fluidized bed and alternate airstream micronization methods.

SMITH CHEMICAL & COLOR CO.,INC.: Products Distributed(Continued):

SNOW*TEX 45
Calcined Aluminum Silicate

Typical Physical Properties:

Color	White
G.E. Brightness	90-93
Oil Absorption	69-79
Moisture Content (maximum %)	0.4
pH Value	5.3-6.3
Specific Gravity	2.58
Bulking Value (lbs./gal.)	21.5
Refractive Index	1.62
Bulk Density lbs./cu.ft. (aerated)	18.8
Hegman Grind	5.0-6.0
Average Particle Size (microns)	.50
Residue +325 mesh (maximum %)	.02
Loss On Ignition	.23

Typical Chemical:

SiO_2:	56.91%
Al_2O_3:	39.68%
Fe_2O_3:	.93%
TiO_2:	.54%
CaO & MgO:	.32%
Na_2O & K_2O:	1.20%

SUPERFINE BESTONE TALC

Chemical:

Silicon Dioxide (SiO_2):	56.0
Magnesium Oxide (MgO):	30.7
Calcium Oxide (CaO):	7.0
Iron Oxide (Fe_2O_3):	0.1
Aluminum Oxide (Al_2O_3):	0.3
Moisture (H_2O):	0.6
Ignition Loss:	5.6

Typical Properties:

Specific Gravity:	2.85
Pounds per Gallon:	23.7
pH:	9.4
Plus 325 mesh, max. %:	0.4
Oil Absorption:	29
Consistency, KU:	74
Hegman Fineness:	4
Brightness (G.E.):	90
Median Particle Size:	5.5

SMITH CHEMICAL & COLOR CO., INC.: Products Distributed (Continued):

325 Mesh Water Ground Mica

Chemical Analysis:

Moisture 100%	0.10%
Loss On Ignition	4.30%
Silica (SiO_2)	47.90%
Iron Oxide (Fe_2O_3)	2.04%
Alumina (Al_2O_3)	33.13%
Calcium Oxide (CaO)	9.50%
Magnesia (MgO)	0.69%
Titanium Dioxide (TiO_2)	0.65%
Manganese (MnO_2)	0.01%
Sodium (Na_2O)	0.80%
Potash (K_2O)	9.80%
Phosphorus (P)	0.03%
Sulfur (S)	0.01%

Physical Properties:

Specific Gravity	2.82
Index of Refraction	1.58
Apparent Density Lbs./Cu.Ft.	6-18
Wet Bulking Value Gal./Lb.	.04257
Oil Absorption (Spatula Method)	40-47
Mesh Size	325
Hardness	2.5
pH	8.5

#611 White Silica

Chemical

Silica	99.90%
Iron Oxide	.02
Aluminum Oxide	.05
Titanium Dioxide	.015
Calcium Oxide	.01
Magnesium Oxide	.005

Mesh	99.4 thru 325 mesh
Weight per Solid Gallon	22.07
1 lb. Bulks in Gallons	.0453
Oil Absorption	26
pH Value	6.6-7.2
Moisture	Less than 1/2 of 1%
Loss on Ignition	.12
Particle Size Distribution (percent passing)	Microns
Finer than 40	97.6
Finer than 30	86.9
Finer than 20	68.9
Finer than 10	40.5
Finer than 5	22.0

TAMMSCO, INC.: GRANUSIL Industrial Sand: Ottawa, MN:

UNIMIN's Ottawa, MN GRANUSIL Industrial Sands are whole grain crystalline silica of highest quality. They exhibit hardness and fracture resistant characteristics superior to local river sands. GRANUSIL Industrial Sands can be used in applications such as the building and construction industry, and in plaster, stucco, mortar and paint formulations.

UNIMIN's Ottawa, MN GRANUSIL Industrial Sands have rounded grains which provide exceptional crush resistance. The grades are accurately sized and undergo a complete washing, drying and screening process.

Typical Sieve Analysis & Properties:

Sieve Analysis: Mesh(ASTM E-11):

		GRANUSIL				
	Mesh	10	16	18	20	40
Typical mean % Retained on Individual Sieves	16	16.0	0.1	-	-	-
	20	70.8	38.2	29.2	1.2	-
	30	13.0	58.5	27.2	28.5	0.2
	40	.2	2.7	15.2	52.8	32.0
	50	-	0.5	23.9	15.4	55.3
	70	-	-	3.9	2.1	8.9
	100	-	-	0.6	-	2.1
	pan	-	-	-	-	0.9

Property:
- Mineral: Quartz
- Color: White
- Hardness: 7
- Roundness: 0.8-0.9
- Moisture Content: <0.1
- Specific Gravity: 2.65
- Bulk Density -
  - Aerated: 92-95
  - Compacted: 98-100
- Melting Point: 3135/1724

Typical Chemical Analysis:
Mean Percent by Weight:

	10	16,18,20,40
Silicon dioxide	99.725	99.790
Iron oxide	0.025	0.017
Aluminum oxide	0.061	0.056
Calcium oxide	0.013	0.008
Titanium dioxide	0.004	0.004
Magnesium oxide	0.008	0.007
Potassium oxide	0.017	0.013
Sodium oxide	0.007	0.006
Loss on Ignition	0.016	0.099

TAMMSCO, INC.: NEOSIL A:

NEOSIL A Silica is a readily available, white, microcrystalline, extender pigment. NEOSIL A Silica has the advantages of chemical inertness, low oil or water demand, and does not reduce gloss as much as other extenders may. This pigment exhibits good enamel hold out and durability while it contributes desirable flow and leveling characteristics.

NEOSIL A Silica may be used in baking enamels, interior and exterior trade paints, maintenance finishes and floor coatings. NEOSIL A Silica may be used in melamine, epoxy, polyester, polyurethane, and phenolic resin systems. The pigment is well suited for use in water borne as well as solvent based systems.

Typical Values:

Specific Gravity: 2.65
pH: Approximately neutral
Moh Hardness: 6.5
Moisture, Maximum: <0.25
Refractive Index: 1.54
Hegman Grind: 6.5-7.0
Surface Area, cm2/g: 19,754
Oil Absorption, g/100 g: 27.9
Apparent Density, lbs./cu.ft.: 26.3
Average particle size, microns: 2.1
% Retained on 325 mesh sieve: 0
Dry Reflectance (green): 85.0

Chemical Analysis
Mean Percent by Weight

Silicon dioxide: 98.951
Iron oxide: 0.044
Aluminum oxide: 0.411
Calcium oxide: 0.027
Magnesium oxide: 0.015
Loss on Ignition: 0.463

TAMMSCO, INC.: NEOSIL XV:

NEOSIL XV Silica a is readily available, white, microcrystall-ine, extender pigment. In addition to chemical inertness, NEOSIL XV Silica also exhibits low oil or water demand. Desirable leveling and flow characteristics may be obtained with the use of NEOSIL XV Silica.

NEOSIL XV Silica has been used in exterior latex coatings where tint retention and durability are desirable qualities. Usage in industrial coatings includes coil coatings, baking enamels and air dry products. The controlled particle size will contribute to film uniformity.

Typical Values:

Specific Gravity: 2.65
pH: Approximately neutral
Moh Hardness: 6.5
Moisture, Maximum: 0.25
Refractive Index: 1.54
Hegman Grind: 5.5-6.5
Surface Area, cm 2/g: 15,391
Oil Absorption, g/100g: 29.8
Apparent Density, lbs./cu.ft.: 21.7
Average particle size, microns: 2.6
% Retained on 325 mesh sieve: 0
Dry Reflectance (green): 83.0

Chemical Analysis
Mean Percent by Weight

Silicon dioxide: 99.029
Iron oxide: 0.029
Aluminum oxide: 0.366
Calcium oxide: 0.018
Magnesium oxide: 0.016
Loss on Ignition: 0.462

TAMMSCO, INC.: NEOSIL 7:

NEOSIL 7 Silica is a white, microcrystalline, extender pigment. Close quality control assures uniform Hegman, particle size distribution and oil absorption characteristics. In addition to durability, NEOSIL 7 Silica contributes desirable flow qualities.

NEOSIL 7 Silica finds many uses in trade-sales paints, as well as uses in maintenance and floor coatings. The fineness of grind makes NEOSIL 7 Silica suitable for use in baking enamels as well as coil coatings, primers, and sealers. NEOSIL 7 may be used in melamine, epoxy, polyester, polyurethane, and phenolic resin systems. It is particularly well suited to use in high solids urethanes.

Typical Values:

Specific Gravity: 2.65
pH: Approximately neutral
Moh Hardness: 6.5
Moisture, Maximum: <0.25
Refractive Index: 1.54
Hegman Grind: 7+
Surface Area, cm 2/g: 16,516
Oil Absorption, g/100 g: 28.2
Apparent Density, lbs./cu.ft.: 22.1
Avarage particle size, microns: 2.6
% Retained on 325 mesh sieve: 0
Dry Reflectance (green): 85.0

Chemical Analysis
Mean Percent by Weight

Silicon dioxide:	98.885
Iron oxide:	0.038
Aluminum oxide:	0.474
Calcium oxide:	0.025
Magnesium oxide:	0.017
Loss on Ignition:	0.480

TAMMSCO, INC.: SILVER BOND B:

SILVER BOND B Silica is a ground crystalline, white, silica extender pigment. The toothy structure of SILVER BOND B Silica contributes to good adhesion to the substrate. The durability of silica makes coatings more abrasion resistant. The retention and stain resistance are also benefits of SILVER BOND Silica. Low resin demand makes this an efficient extender and contributes to the wide acceptance of SILVER BOND B Silica in paint formulations. SILVER BOND B Silica is easy to disperse and has a minimal effect on viscosity.

Typical Particle Analysis & Properties

Screen Analysis	Percent Passing
200 Mesh	100
270 Mesh	100
325 Mesh	99.0-99.8

Typical Values:

Specific Gravity: 2.65
pH: Approximately neutral
Moh Hardness: 6.5
Moisture, Maximum: >0.25
Refractive Index: 1.54
Hegman Grind: 3-4
Surface Area, cm2/g: 6,628
Oil Absorption, g/100g: 24.2
Apparent Density, lbs./cu.ft.: 32.7
Average particle size, microns: 10.0
% Retained on 325 mesh sieve: 0.2
Dry Reflectance (green): 80.0

Chemical Analysis
Mean Percent by Weight

Silicon dioxide: 99.404
Iron oxide: 0.036
Aluminum oxide: 0.273
Calcium oxide: 0.013
Magnesium oxide: 0.006
Loss on Ignition: 0.200

TAMMSCO, INC.: SILVER BOND 30:

SILVER BOND 30 Silica is a ground crystalline, white, extender pigment. The crystalline structure of SILVER BOND 30 Silica contributes to good adhesion to the substrate. The durability of silica makes coatings more abrasion resistant. Color acceptance and sheen control are also benefits of SILVER BOND 30 Silica. Low resin demand makes this an efficient filler and contributes to the wide acceptance of SILVER BOND 30 Silica in paint formulations.

Typical Values:

Specific Gravity: 2.65
pH: Approximately neutral
Moh Hardness: 6.5
Moisture, Maximum: <0.25
Refractive Index: 1.54
Hegman Grind: 4.5-5.5
Surface Area, cm2/g: 6,628
Oil Absorption, g/100g: 24.2
Apparent Density, lbs./cu.ft.: 34.3
Average particle size, microns: 8.5
% Retained on 325 mesh sieve: Trace
Dry Reflectance (green): 82.0

Chemical Analysis:
Mean Percent by Weight

Silicon dioxide:	99.404
Iron oxide:	0.036
Aluminum oxide:	0.273
Calcium oxide:	0.013
Magnesium oxide:	0.006
Loss on Ignition:	0.200

TAMMSCO, INC.: SILVER BOND 45:

SILVER BOND 45 Silica is a ground crystalline, white, silica extender pigment. The toothy structure of SILVER BOND B Silica contributes to good adhesion to the substrate. The durability of silica makes coatings more abrasion resistant. The retention and stain resistance are also benefits of SILVER BOND 45 Silica. Low resin demand makes this an efficient extender and contributes to the wide acceptance of SILVER BOND 45 Silica in paint formulations. SILVER BOND 45 Silica is easy to disperse and has a minimal effect on viscosity. Suggested uses include traffic paint.

Typical Values:

Specific Gravity: 2.65
pH: Approximately neutral
Moh Hardness: 6.5
Moisture, Maximum: <0.25
Refractive Index: 1.54
Hegman Grind: 0
Surface Area, cm2/g: 5,010
Oil Absorption, g/100 g: 25.3
Apparent Density, lbs./cu.ft.: 32.7
Average particle size, microns: 12.5
% Retained on 325 mesh sieve: 1.2
Dry Reflectance (green): 83.0

Chemical Analysis:
Mean Percent by Weight

Silicon dioxide:	99.592
Iron oxide:	0.030
Aluminum oxide:	0.213
Calcium oxide:	0.014
Potassium oxide:	0.008
Magnesium oxide:	0.007
Sodium oxide:	0.003
Loss on Ignition:	0.133

TAMMSCO, INC.: S MICRON:

S MICRON Silica is a stir-in, white, microcrystalline, extender pigment. Due to the fine particle size, when used as a replacement for coarser extender pigments, S MICRON Silica will impart a higher sheen.

Use of S MICRON Silica will not increase oil or water absorption from the resin system.

S MICRON Silica will contribute durability and high impact strength to the coating system. Stain removal, abrasion resistance, good scrubbability, tint retention, and long shelf life are important benefits derived from the use of S MICRON Silica. S MICRON Silica may be used in a variety of resin systems.

Typical Values:

Specific Gravity: 2.65
pH: approximately neutral
Moh Hardness: 6.5
Moisture, Maximum: <0.25
Refractive Index: 1.54
Hegman Grind: 4.75-5.5
Surface Area, cm2/g: 15,397
Oil Absorption, g/100 g: 29.8
Apparent Density, lbs./cu.ft.: 25.9
Average particle size, microns: 3.4
% Retained on 325 mesh sieve: 0
Dry Reflectance (green): 81.0

Chemical Analysis:
Mean percent by Weight

Silicon dioxide:	98.965
Iron oxide:	0.044
Aluminum oxide:	0.435
Calcium oxide:	0.019
Magnesium oxide:	0.013
Loss on Ignition:	0.444

TAMMSCO, INC.: SNOW WHITE 325 Mica/Spruce Pine, North Carolina:

UNIMIN's SNOW WHITE 325 Mesh Mica is a high quality muscovite mica. This product is produced from washed, beneficiated, graded feedstock. It is milled to yield a delaminated pigment particle which has a high diameter to thickness ratio.

SNOW WHITE 325 Mica exhibits an excellent whiteness characteristic, and is chemically inert. Its platey nature reduces cracking. It reinforces and strengthens the dried paint film and improves moisture vapor impermeability. SNOW WHITE 325 Mica is specially graded to meet all requirements of the industrial and trade sale paint manufacturing industries.

Typical Sieve Analysis & Properties

Screen Analysis	Percentage
+325 Mesh	12.0 maximum
-325 Mesh	88.0 minimum

Typical Values:

Specific Gravity: 2.87
pH: 5.75
Moh Hardness: 2.75
Refractive Index: 1.58
Chemical Activity: inert
Aspect Ratio: 15-35
Oil Absorption g/100g: 76.0
Apparent Density, lbs./cu.ft.: 13.5
Color: off white
Dry Reflectance (green): 62.0

Typical Chemical Analysis:
Mean Percent by Weight

Silicon dioxide	48.0
Iron oxide	5.0 max.
Aluminum oxide	31.5
Calcium oxide	0.5
Titanium dioxide	0.06
Potassium oxide	7.0
Sodium oxide	1.5
Loss on Ignition	4.8
Phosphorous	0.035 max.
Sulphur	0.015
Manganese dioxide	0.005

THIELE KAOLIN CO.: Clays:

RC-32

Typical Physical Properties:
- Specific Gravity: 2.63
- Moisture %: 0.5-1.5
- Screen Residue Retained
  - 325 Mesh %: 0.25
- Particle Size:
  - % Less Than 2 Microns: 89
- Surface Area M2/g B.E.T.: 20-26
- G.E. Brightness %: 79-81
- Oil Absorption: 36-40
- pH, Airfloated: 3.5-5.0*
- pH, Predispersed: 6.5-7.5*

* pH, dry clay at 20% Solids and slurry clay at % Solids shipped

Typical Chemical Properties:
- Ignition Loss %: 13.4-14.1
- Silica %: 43.5-44.5
- Alumina %: 38.0-40.5
- Iron Oxide %: 0.9-1.3
- Titanium Dioxide %: 1.4-3.5

RC-32 is a fine particle size hard clay designed primarily for the rubber and plastic industries. It has excellent rheology which permits high loading. RC-32 is easily dispersed in most elastomers.

T-77

T-77 is a fine particle size clay selected from some of the finest crude clay deposits in East Georgia. It is air refined to remove most of the quartz and coarse particles.

With its small particle size, T-77 will exhibit the same performance in thickening and strike-through as the fine particle size waterwashed clays at a much lower cost.

THIELE KAOLIN CO.: Clays(Continued):

B-80

THIELE B-80 is an air refined fine particle size kaolin produced under stringent quality controlled conditions. B-80 combines the benefits of high gloss, excellent "stir-in" hegman grinds, extremely low screen residues, low screen residues, low abrasive characteristics, and very low water soluble salt content with favorable economics.

B-80 offer the properties associated with the fine particle size waterwashed kaolins and has been successfully marketed in adhesives, ink, paint, and related industries. When used as a replacement for higher cost pigments, B-80 offers significant savings with distinct advantages in dispersion, non-settling properties, and gloss in particular systems.

PT-90

Typical Physical Properties:
- G.E. Brightness %: 87
- Oil Absorption: 43
- Mean Particle Size, Microns: .65

Available Forms:
- Predispersed-Slurry (69-71% solids) and bulk

KAOPLATE
(Delaminated)

Typical Physical Properties:
- G.E. Brightness %: 88-90
- Oil Absorption: 39

Available Forms:
- Predispersed - Slurry (67-68% Solids), bags and bulk

KAOPLATE is a delaminated clay composed of thin plate-like particles whose newly exposed surfaces yield a product of high whiteness, brightness, and better hiding power.

THOMPSON, WEINMAN & CO.: Calcium Carbonate - Barytes:

Calcium Carbonate

SUPERMITE:

An extra fine ground pigment to provide improved opacity and gloss properties in paints, coatings and paper applications. In plastics and elastomers, finer particle size aids development of superior physical properties and improved surface gloss.

Particle Size Range (Microns): Up to 8
Mean Particle Size (Microns): 1.0
Oil Absorption (Rub-out): 19 to 21

SUPERCOAT (Surface Treated):

A coated extra fine ground pigment. Easy incorporation and dispersion in plastics applications resulting in improved physical properties. Improved production throughput efficiency with low abrasiveness.

Particle Size Range (Microns): Up to 8
Mean Particle Size (Microns): 1.0
Oil Absorption (Rub-Out): 15 to 17

ATOMITE:

A highly developed stir-in pigment that meets the most exacting requirements demanded by the paint, rubber, plastics, floor covering, and many other industries.

Particle Size Range (Microns): Up to 12
Mean Particle Size (Microns): 3.0
Oil Absorption (Rub-out): 14 to 16

KOTAMITE (Surface Treated):

A coated pigment designed for easy dispersion in plastic compounds e.g. polyolefins, rigid and flexible PVC. Hydrophobic for superior wire and cable insulation compounds and improved impact properties in polypropylene.

Particle Size Range (Microns): Up to 12
Mean Particle Size (Microns): 3.0
Oil Absorption (Rub-out): 12 to 14

SNOWFLAKE WHITE:

A general purpose stir-in pigment that has found wide use in practically all types of protective coatings, rubber, plastic, caulk, and glaze compounds, mastics and other end products.

Particle Size Range (Microns): Up to 25
Mean Particle Size (Microns): 5.5
Oil Absorption (Rub-out): 9 to 10

SNOWFLAKE P.E.:

A medium ground pigment with the theoretically correct distribution of particles to ensure maximum loading in highly filled systems: e.g., polyester resins, SMC, BMC, TMC.

Particle Size Range (Microns): Up to 40
Mean Particle Size (Microns): 6.0
Oil Absorption (Rub-out): 8 to 9

THOMPSON, Weinman & Co.: Calcium Carbonate - Barytes(Continued):

Calcium Carbonate

DRIKALITE:

A competitive grade of extender pigment for use in places where a maximum number of fine particles per unit cost is imperative.

Particle Size Range (Microns): Up to 44
Mean Particle Size (Microns): 6.5
Oil Absorption (Rub-Out): 8.5 to 9.5

DURAMITE:

A pigment with unique particle size distribution. Its easy dispersion characteristics have demonstrated its unusual merit as a mix-in type of pigment for a wide variety of applications such as carpet backing, roofing compounds, spackles, and coatings.

Particle Size Range (Microns): Up to 44
Mean Particle Size (Microns): 11.0
Oil Absorption (Rub-out): 5 to 6

No. 1 White:

A general purpose pigment for many uses where fineness is of secondary importance.

Particle Size Range (Microns): Up to 50
Mean Particle Size (Microns): 14.0
Oil Absorption (Rub-out): 5 to 6

CP Filler:

A competitively priced filler for use in construction products where color is of secondary importance.

Particle Size Range (Microns): Up to 75
Mean Particle Size (Microns): 17.0

Marble Dust:

A coarse, ground natural calcium carbonate for use in putties, glazes, and mild abrasive compounds. It is frequently blended with ATOMITE to achieve a wider distribution of particles which yields effective particle packing properties.

Particle Size Range (Microns): Up to 180
Mean Particle Size (Microns): 21

MARBLEMITE:

A special grind from selected stone to give maximum loading levels with good color properties in highly filled systems such as cultured marble.

Particle Size Range (Microns): Up to 850
Mean Particle Size (Microns): 80.0

THOMPSON, WEINMAN & CO.: Calcium Carbonate - Barytes(Continued):

Barytes

BARIMITE XF:

A finely ground unbleached barytes with exceptionally good stir-in properties giving a typical Hegman Scale reading of about 6.

Particle Size Range (Microns): Up to 15
Mean Particle Size (Microns): 2.5
Oil Absorption (Rub-out): 9.0

BARIMITE:

An intermediate grade of finely ground unbleached barytes recommended for automotive primer use when a minimum grinding cycle is required. As a stir-in pigment, the Hegman Scale reading will be about 4.

Particle Size Range (Microns): Up to 25
Mean Particle Size (Microns): 4.0
Oil Absorption (Rub-out): 8.0

No.22 Barytes:

A coarser grade of unbleached barytes for use when fineness of grind is not critical.

Particle Size Range (Microns): Up to 60
99.5% thru 325 mesh
Mean Particle Size (Microns): 12.0
Oil Absorption (Rub-Out): 6.0

U.S. SILICA: MIN-U-SIL Ground Silica:

High Purity Filler and Extender

Architectural Paints:
MIN-U-SIL is an inert, low-cost, functional extender that can effectively space TiO2 for added opacity in both water-borne and solvent-borne systems.

Powder Coatings:
Replaces up to 50% TiO2 in formulations with no significant loss of hiding power.

Traffic Paint:
High durability, inertness, abrasion resistance, color acceptance, high loading and low cost make MIN-U-SIL an outstanding traffic paint filler.

Protective Coatings:
MIN-U-SIL offers superior resistance to corrosion, acid and heat. Also it's chemically inert.

Silicon Rubber:
MIN-U-SIL is an excellent semi-reinforcing filler for silicone rubber. Inert, dielectric.

Wire and Cable Coating Compounds:
MIN-U-SIL delivers excellent resistance to heat. It's chemically inert and low cost.

Adhesives and Sealants:
MIN-U-SIL gives fine elastomer extension, requiring less power consumption for mixing equipment. It's inert and costs less.

Plastics:
As a filler in thermoplastics, MIN-U-SIL supplies uniformity, heat resistance, inertness and low thermal expansion, all at a low cost.

Ceramic:
The high quality and consistent fineness of MIN-U-SIL can reduce or eliminate grinding time of ceramic glazes.

High Temperature Insulation:
MIN-U-SIL is naturally heat resistant, chemically inert and provides high loadings.

Epoxy Castings:
MIN-U-SIL provides the durability and inertness necessary to enhance the physical strength of large epoxy castings such as laboratory countertops and other chemical resistant laboratory components.

U.S. SILICA: MIN-U-SIL Ground Silica(Continued):

Name:

MIN-U-SIL 30:
Mine: Berkeley Springs, WV
Average Particle Size (Microns) Fisher: 4.19
Specific Surface Area (cm2/g): 5,400
Oil Absorption (lbs/100 lbs): 26.5
Hegman Grind: 5.5
Mine: Pacific, MO
Average Particle Size (Microns) Fisher: 3.5
Specific Surface Area (cm2/g): 6,659
Oil Absorption (lbs/100 lbs): 25.5
Hegman Grind: 5.25

MIN-U-SIL 15:
Mine: Berkeley Springs, WV
Average Particle Size (Microns) Fisher: 2.7
Specific Surface Area (cm2/g): 8,400
Oil Absorption (lbs/100 lbs): 31.7
Hegman Grind: 7.25
Mine: Pacific, MO
Average Particle Size (Microns) Fisher: 2.3
Specific Surface Area (cm2/g): 9,844
Oil Absorptuion (lbs/100 lbs): 33.0
Hegman Grind: 7.25

MIN-U-SIL 10:
Mine: Berkeley Springs, WV
Average Particle Size (Microns) Fisher: 2.0
Specific Surface Area (cm2/g): 11,320
Oil Absorption (lbs/100 lbs): 33.3
Hegman Grind: 7.25
Mine: Pacific, MO
Average Particle Size (Microns) Fisher: 1.7
Specific Surface Area (cm2/g): 13,320.6
Oil Absorption (lbs/100 lbs): 35.1
Hegman Grind: 7.5

MIN-U-SIL 5:
Mine: Mapleton Depot, PA
Average Particle Size (Microns) Fisher: 1.2
Specific Surface Area (cm2/g): 18,870
Oil Absorption (lbs/100 lbs): 44.0
Hegman Grind: 7.5
Mine: Mill Creek, OK
Average Particle Size (Microns) Fisher: 1.1
Specific Surface Area (cm2/g): 21,500
Oil Absorption (lbs/100 lbs): 42.0
Hegman Grind: 7.5
Mine: Berkeley Springs, WV
Average Particle Size (Microns) Fisher: 1.1
Specific Surface Area (cm2/g): 21,500
Oil Absorption (lbs/100 lbs): 44.0

U. S. SILICA: SIL-CO-SIL Ground Silica:

SIL-CO-SIL products are produced from high purity silica which is precision ground to microcrystalline sizes ranging from 9.5u to 3.7 um (average particle size expressed in microns). These ground sands are inherently inert, white, bright, have low moisture and are at least 99.5% SiO2.

<u>New</u> <u>Name:</u>

250:
Mine: Ottawa, IL
Avg. Particle Size (Microns): 9.5
Specific Surface Area, (cm2/g): 2380
Oil Absorption, (lbs/100 lbs): 20.0

150:
Mine: Pacific, MO
Avg. Particle Size (Microns): 8.0
Specific Surface Area, (cm2/g): 2830
Oil Absorption, (lbs/100 lbs): 18.0

125:
Mine: Pacific, MO
Avg. Particle Size (Microns): 7.5
Specific Surface Area, (cm2/g): 3020
Oil Absorption, (lbs/100 lbs): 18.5

Mine: Mill Creek, OK
Avg. Particle Size (Microns): 6.5
Specific Surface Area, (cm2/g): 3480
Oil Absorption, (lbs/100 lbs): 18.5

Mine: Millville, NJ
Avg. Particle Size (Microns): 7.5
Specific Surface Area, (cm2/g): 3020
Oil Absorption, (lbs/100 lbs): 19.0

Mine: Berkeley Springs, WV
Avg. Particle Size (Microns): 7.0
Specific Surface Area, (cm2/g): 3230
Oil Absorption, (lbs/100 lbs): 20.5

106:

Mine: Pacific, MO
Avg. Particle Size (Microns): 7.0
Specific Surface Area, (cm2/g): 3230
Oil Absorption, (lbs/100 lbs): 19.0

U. S. SILICA: SIL-CO-SIL Ground Silica(Continued):

<u>New Name:</u>

106:

Mine: Mapleton Depot, PA
Avg. Particle Size, (Microns): 7.2
Specific Surface Area, (cm2/g): 3140
Oil Absorption, (lbs/100 lbs): 19.0

Mine: Ottawa, IL
Avg. Particle Size, (Microns): 7.0
Specific Surface Area, (cm2/g): 3230
Oil Absorption, (lbs/100 lbs): 23.0

Mine: Mill Creek, OK
Avg. Particle Size, (Microns): 6.2
Specific Surface Area, (cm2/g): 3650
Oil Absorption, (lbs/100 lbs): 19.0

90:

Mine: Pacific, MO
Avg. Particle Size, (Microns): 6.5
Specific Surface Area, (cm2/g): 3480
Oil Absorption, (lbs/100 lbs): 20.0

Mine: Mapleton Depot, PA
Avg. Particle Size, (Microns): 7.0
Specific Surface Area, (cm2/g): 3230
Oil Absorption, (lbs/100 lbs): 21.0

Mine: Ottawa, IL
Avg. Particle Size, (Microns): 6.5
Specific Surface Area, (cm2/g): 3480
Oil Absorption, (lbs/100 lbs): 24.0

Mine: Berkeley Heights, WV
Avg. Particle Size, (Microns): 6.5
Specific Surface Area, (cm2/g): 3480
Oil Absorption, (lbs/100 lbs): 22.5

75:

Mine: Mapleton Depot, PA
Avg. Particle Size, (Microns): 6.7
Specific Surface Area, (cm2/g): 3380
Oil Absorption, (lbs/100 lbs): 23.0

Mine: Pacific, MO
Avg. Particle Size, (Microns): 6.0
Specific Surface Area, (cm2/g): 3770
Oil Absorption, (lbs/100 lbs): 20.5

U.S. SILICA: SIL-CO-SIL Ground Silica(Continued):

New Name:

75:

Mine: Columbia, SC
Avg. Particle Size, (Microns): 6.0
Specific Surface Area, (cm2/g): 3770
Oil Absorption, (lbs/100 lbs): 22.0

Mine: Mill Creek, OK
Avg. Particle Size, (Microns): 6.0
Specific Surface Area, (cm2/g): 3770
Oil Absorption, (lbs/100 lbs): 20.0

Mine: Berkeley Springs, WV
Avg. Particle Size, (Microns): 6.0
Specific Surface Area, (cm2/g): 3770
Oil Absorption, (lbs/100 lbs): 24.0

63:

Mine: Ottawa, IL
Avg. Particle Size, (Microns): 5.5
Specific Surface Area, (cm2/g): 4120
Oil Absorption, (lbs/100 lbs): 25.0

Mine: Mapleton Depot, PA
Avg. Particle Size, (Microns): 6.6
Specific Surface Area, (cm2/g): 3430
Oil Absorption, (lbs/100 lbs): 24.0

Mine: Millville, NJ
Avg. Particle Size, (Microns): 6.0
Specific Surface Area, (cm2/g): 3770
Oil Absorption, (lbs/100 lbs): 21.0

Mine: Berkeley Springs, WV
Avg. Particle Size, (Microns): 5.5
Specific Surface Area, (cm2/g): 4120
Oil Absorption, (lbs/100 lbs): 25.0

53:

Mine: Mill Creek, OK
Avg. Particle Size, (Microns): 5.5
Specific Surface Area, (cm2/g): 4120
Oil Absorption, (lbs/100 lbs): 21.0

U.S. SILICA: SIL-CO-SIL Ground Silica(Continued):

<u>New Name:</u>

53:

Mine: Mapleton Depot, PA
Avg. Particle Size, (Microns): 6.5
Specific Surface Area, (cm2/g): 3480
Oil Absorption, (lbs/100 lbs): 25.0

Mine: Ottawa, IL
Avg. Particle Size, (Microns): 4.9
Specific Surface Area, (cm2/g): 4620
Oil Absorption, (lbs/100 lbs): 26.0

52:

Mine: Berkeley Springs, WV
Avg. Particle Size, (Microns): 5.0
Specific Surface Area, (cm2/g): 4530
Oil Absorption, (lbs/100 lbs): 26.0

Mine: Columbia, SC
Avg. Particle Size, (Microns): 5.6
Specific Surface Area, (cm2/g): 4040
Oil Absorption, (lbs/100 lbs): 23.0

51:

Mine: Millville, NJ
Avg. Particle Size, (Microns): 4.5
Specific Surface Area, (cm2/g): 5030
Oil Absorption, (lbs/100 lbs): 22.0

Mine: Pacific, MO
Avg. Particle Size, (Microns): 5.0
Specific Surface Area, (cm2/g): 4530
Oil Absorption, (lbs/100 lbs): 21.0

49:

Mine: Columbia, SC
Avg. Particle Size (Microns): 5.5
Specific Surface Area, (cm2/g): 4120
Oil Absorption, (lbs/100 lbs): 24.0

Mine: Mill Creek, OK
Avg. Particle Size (Microns): 5.0
Specific Surface Area, (cm2/g): 4530
Oil Absorption, (lbs/100 lbs): 22.0

U.S. SILICA: SIL-CO-SIL Ground Silica(Continued):

New Name:

47:

Mine: Mapleton Depot, PA
Avg. Particle Size, (Microns): 6.0
Specific Surface Area, (cm2/g): 3770
Oil Absorption, (lbs/100 lbs): 26.0

45:

Mine: Ottawa, IL
Avg. Particle Size, (Microns): 4.0
Specific Surface Area, (cm2/g): 5660
Oil Absorption, (lbs/100 lbs): 27.0

40:

Mine: Pacific, MO
Avg. Particle Size, (Microns): 40
Specific Surface Area, (cm2/g): 5660
Oil Absorption, (lbs/100 lbs): 23.0

38:

Mine: Mapleton Depot, PA
Avg. Particle Size, (Microns): 5.8
Specific Surface Area, (cm2/g): 3900
Oil Absorption, (lbs/100 lbs): 33.0

Mine: Berkeley Springs, WV
Avg. Particle Size, (Microns): 4.5
Specific Surface Area, (cm2/g): 5030
Oil Absorption, (lbs/100 lbs): 26.5

37:

Mine: Ottawa, IL
Avg. Particle Size, (Microns): 3.7
Specific Surface Area, (cm2/g): 6120
Oil Absorption, (lbs/100 lbs): 27.5

U.S. SILICA: SNOW*TEX Anhydrous Aluminum Silicate:

SNOW*TEX is a thermally processed kaolin clay, calcined at 2000F under controlled conditions. Calcination removes organic impurities producing a white, high brightness, low moisture, low grit extender pigment which offers formulation advantages in many applications. Primary consuming industries are:

Paints & Coatings:

* SNOW*TEX reduces cost by extending TiO_2.
* SNOW*TEX adds hiding, through inherent opacity and by effectively spacing the highly refractive TiO_2 particles which tend to agglomerate.
* SNOW*TEX imparts excellent scrubbability, sheen control, plus water spotting and burnish resistance.
* SNOW*TEX is hydrophilic, easily dispersed, improves brush ease, flow/leveling, and viscosity.

Plastics & Rubbers:

SNOW*TEX is an inert, low moisture, white extender pigment recommended for use in PVC compounding, gel coats, color concentrates, elastomer, and mechanical rubber compounds.

* As an extender pigment, SNOW*TEX reduces resin costs.
* SNOW*TEX enhances electrical properties in polyvinyl chloride wire and cable coating.
* SNOW*TEX imparts excellent chemical and abrasion resistance, improves heat distortion, dimensional stability, tensile strength, flexural strength, and modulus.

Ceramics:

* SNOW*TEX improves fired properties and increases dimensional stability in white ware, electrical porcelain, and refractories.

Typical Physical Properties:

Color: White
G.E. Brightness: 90-93
Oil Absorption: 69-79
Moisture Content (maximum %): 0.4
pH Value: 5.3-6.3
Specific Gravity: 2.58
Bulking value lbs./gal. (aerated): 21.5
gal./lb. (packed): .0465

Typical Chemical:

SiO_2:	56.91%
Al_2O_3:	39.68%
Fe_2O_3:	.93%
TiO_2:	.54%
CaO & MgO:	.32%
Na_2O & K_2O:	1.20%

R.T. VANDERBILT CO., INC.: Paint/Paper Products:

BILT-PLATES 156
Clay

BILT-PLATES 156 is a hydrated aluminum silicate produced from South Carolina deposits of the mineral kaolinite. It is a grit-free fine particle size clay obtained by a dry process including air separation.

Uses:
Suggested uses for BILT-PLATES 156 include coatings such as primers, alkyd flats and latex paints. Because of its very fine particle size it is particularly applicable for titanium dioxide extension in coatings that do not require a high-brightness extender.

Typical Chemical Analysis:
(calculated as oxides)

Silicon dioxide (SiO_2):	44.4%
Aluminum oxide (Al_2O_3):	38.9
Iron oxide (Fe_2O_3):	0.5
Titanium dioxide (TiO_2):	1.9
Calcium oxide (CaO)	0.07
Loss on ignition	13.3

Typical Properties:
Density, Mg/m^3: 2.62
Pounds per gallon: 21.8
Brightness, G.E.: 76
Visual color: Cream
Plus 325 mesh, %: .04
Oil absorption, rub-out: 41
Hegman fineness, 3 lbs./gal. 4.0
Particle size:
Less than 1.0 um, %: 95
Less than 0.5 um, %: 75
Average diameter, um: 0.2

PEERLESS Clay

A series of PEERLESS Clays for use as extender pigments in the coatings field. They have found application in traffic paints, barn paints, floor coverings, metal primers, and in related materials such as crack fillers and caulking compounds. PEERLESS clays are available in many grades according to dry brightness and fineness. They are the familiar "China Clay" and are produced by dry-processing the mineral kaolinite as found in the deposits of South Carolina.

No. 1: Dry brightness: 73
No. 2: Dry brightness: 68
No. 3: Dry brightness: 65

R.T. VANDERBILT CO., INC.: Paint/Paper Products(Continued):

I.T. Industrial Talc

I.T. industrial talc is produced in several grades to provide a wide range of fineness and binder demand. The chemical inertness, easy "stir-in" dispersion, and excellent suspension qualities of I.T. talc combine to give the paint chemist a versatile extender pigment with which to formulate solvent-thinned and water-thinned coatings.

Types:

I.T. X has low binder demand with 1 to 2% retained on a No. 325 mesh sieve. As a relatively coarse type of inert, it is a useful extender in coatings where paint fineness and film smoothness are not of prime importance, but where some degree of film roughness or "tooth" is beneficial, such as primers.

I.T. 3X has medium binder demand with fineness similar to the I.T. X grade. I.T. 3X with higher binder demand increases paint consistency and improves pigment suspension where settling is likely to be a problem.

I.T. FT has binder demand similar to I.T. 3X but is a fine textured type of inert that readily stirs into paint to a 2 Hegman fineness when used at a level of three pounds per gallon.

I.T. 5X is similar in fineness to I.T. FT although slightly higher in binder demand to impart increased paint consistency.

I.T. 325 has medium binder demand and serves as a general purpose extender for all paints where fineness and film smoothness are of primary importance. This grade readily stirs into paint to a 4 Hegman fineness at the three-pound per gallon level.

Uses: Suggested uses for I.T. industrial talc include interior wall paints, exterior house paints, primers, maintenance paints and traffic paints.

Typical chemical analysis for all grades(calculated as oxides):

Silicon dioxide (SiO_2)	56.0%
Magnesium oxide (MgO)	30.7
Calcium oxide (CaO)	7.0
Iron Oxide (Fe_2O_3)	0.1
Aluminum oxide (Al_2O_3)	0.3
Moisture (H_2O)	0.6
Ignition Loss	5.6

R.T. VANDERBILT CO., INC.: Paint/Paper Products(Continued):

NYTAL Industrial Talc

NYTAL industrial talc is produced in three grades to provide a wide range of fineness and binder demand. The chemical inertness, easy "stir-in" dispersion, and excellent suspension qualities of NYTAL combine to give the paint chemist a versatile extender pigment with which to formulate solvent-thinned as well as water-thinned coatings.

Types:

NYTAL 200 has low binder demand with 1 to 2% retained on a No. 325 sieve. As a relatively coarse type of inert, it is a useful extender in coatings where paint and film smoothness are not of prime importance, but where some degree of film roughness or "tooth" is beneficial, such as primers.

NYTAL 300 has medium binder demand and serves as a general purpose extender for all paints where fineness and film smoothness are of primary importance. This grade readily stirs into paint to a 4 Hegman fineness when used at a level of three pounds per gallon. NYTAL 300, with its ease of dispersion, good dry brightness, and compatibility with other ingredients of the formulation, contributes to the working and suspension properties of the paint.

NYTAL 400 is a superfine talc pigment. It stirs in easily to a 5 Hegman fineness at the three-pound per gallon level and finds application in formulations where high paint fineness and film smoothness are required, such as semi-gloss enamels. The great ease of incorporation enables last minute batch adjustments by the formulator for desired consistency, sheen, or other special properties prior to the filling operation.

Uses: Suggested uses for NYTAL industrial talc include interior wall paints, exterior house paints, primers, maintenance paints and traffic paints.

Typical chemical analysis for all grades(calculated as oxides):

Silicon dioxide (SiO_2)	56.0%
Magnesium oxide (MgO)	30.7
Calcium Oxide (CaO)	7.0
Iron Oxide (Fe_2O_3)	0.1
Aluminum Oxide (Al_2O_3)	0.3
Moisture (H_2O)	0.6
Ignition loss	5.6

NYTAL 300H industrial talc is an addition to the NYTAL line. It exhibits the same desirable chemical inertness, easy "stir-in" dispersion and excellent suspension properties of other NYTAL grades. NYTAL 300H has a 4 Hegman fineness at a use level of three pounds per gallon and relatively high binder demand. These combined properties provide a coating with improved pigment suspension.

Suggested uses for NYTAL 300H in both solvent-thinned and water-thinned coatings include interior wall paints, exterior house paints, primers, maintenance paints and traffic paints.

R.T. VANDERBILT CO., INC.: Paint/Paper Products(Continued):

PYRAX B and PYRAX WA

PYRAX is a hydrated aluminum silicate produced from North Carolina deposits of the mineral pyrophyllite. PYRAX B and PYRAX WA are relatively coarse grades and are useful in coatings wherein pigment fineness is not a prerequisite. The platy particle shape of PYRAX imparts "mud-cracking" resistance to high build coatings such as texture paints and block fillers. For the same reason, it is useful also in crack fillers and joint cements.

Typical Pigment Properties:
Chemical Analysis, % (Calculated as oxides)

Aluminum oxide (Al_2O_3): 16.0
Silicon dioxide (SiO_2): 77.75
Iron Oxide (Fe_2O_3): 0.21
Titanium dioxide (TiO_2): 0.08
Calcium oxide (CaO): 0.06
Magnesium oxide (MgO): 0.06
Sodium oxide (Na_2O): 0.27
Potassium oxide (K_2O): 1.84
Ignition loss: 3.1

PYRAX B:

Brightness (G.E.): 80
Density (Mg/m^3) at 25C: 2.8
Oil Absorption (Gardner-Coleman): 36
Residue on 200 mesh, %: 1

PYRAX WA:

Brightness (G.E.): 78
Density (Mg/m^3) at 25C: 2.8
Oil Absorption (Gardner-Coleman): 36
Residue on 200 mesh, %: 3

R.T. VANDERBILT CO., INC.: Paint/Paper Products(Continued):

VANSIL W
Calcium Silicate-Wollastonite

VANSIL W is an acicular calcium silicate product available in three grades of fineness to meet paint industry requirements. Each grade exhibits lower consistency effect in paint than indicated by its oil absorption value. The low oil absorption characteristic of VANSIL W enables the formulator to reduce raw material cost by replacement of vehicle. The relatively high pH (9.8) of VANSIL W is particularly useful in maintaining the desired alkaline pH level of latex paints and for the prevention of early rusting in latex maintenance coatings.

VANSIL W is an excellent extender pigment for both solvent-thinned and latex paints.

Types:

VANSIL W-10 is a coarse grade with 2% retained on a 200-mesh sieve. It is useful as an extender in high-build coatings such as block fillers and texture paints where pigment fineness is not a prerequisite.

VANSIL W-20 is a relatively coarse grade with 2% retained on a 325-mesh sieve. It will disperse to a 0 to 1 Hegman fineness when used at a level of three pounds per gallon, thereby providing some degree of film roughness where desired.

VANSIL W-30 disperses to a 4 Hegman fineness when used at a level of three pounds per gallon. This grade provides smooth paint films and is a general purpose extender.

Typical Properties:

VANSIL W-10:
- Density (Mg/m3): 2.9
- Weight (lbs./solid gallon): 24.2
- pH (10% slurry): 9.8
- Screen Analysis
  - Plus 200 mesh, %: 2.7

VANSIL W-20:
- Density (Mg/m3): 2.9
- Weight (lbs./solid gallon): 24.2
- pH (10% slurry): 9.8
- Screen Analysis
  - Plus 325 mesh, %: 1.5

VANSIL W-30:
- Density (Mg/m3): 2.9
- Weight (lbs./solid gallon): 24.2
- pH (10% slurry): 9.8
- Screen Analysis
  - Plus 325 mesh, %: 0.03

R.T. VANDERBILT CO., INC.: Paint/Paper Products(Continued):

VANTALC 6H
Magnesium Silicate Pigment

VANTALC 6H is a high purity, platy, hydrous magnesium silicate pigment. It is a superfine grade exhibiting high brightness and high oil absorption. VANTALC 6H is effective for gloss control of coatings, particularly where film smoothness and pigment suspension are important factors.

Typical Chemical Analysis:
calculated as oxides
Magnesium Oxide (MgO): 31.3%
Silicon Dioxide (SiO2): 61.5
Calcium Oxide (CaO): 0.4
Aluminum Oxide (Al2O3): 0.1
Ferric Oxide (Fe2O3): 1.3
Moisture (H2O): 0.25
Loss on Ignition: 5.0

Typical Properties:
Density at 25C (Mg/m3): 2.75
Weight (lbs./solid gallon): 22.91
pH (ASTM D 1208): 9.5
Dry Brightness: 90
Plus 325 Mesh, %: Nil
Oil Absorption (ASTM D 281): 52
Hegman Fineness (3 lbs/gallon): 6.0

VANTALC 6H meets the requirements of ASTM Specification D 605 for magnesium silicate pigment.

R.T. VANDERBILT CO., INC.: Paint/Paper Products(Continued):

VEECOTE
Pyrophyllite

VEECOTE is a hydrated aluminum silicate produced from North Carolina deposits of the mineral pyrophyllite. It is a relatively coarse grade that is useful in coatings where fineness and film smoothness are not of prime importance. VEECOTE exhibits excellent properties as an extender pigment for exterior house paint. Its platy particle shape contributes resistance to checking and cracking. In addition, frosting of tinted latex house paints is minimized.

Typical Chemical Analysis:
(Calculated as oxides)

Aluminum oxide (Al_2O_3):	16.0
Silicon dioxide (SiO_2):	77.75
Iron Oxide (Fe_2O_3):	0.21
Titanium Oxide (TiO_2):	0.08
Calcium oxide (CaO):	0.06
Magnesium oxide (MgO):	0.06
Sodium Oxide (Na_2O):	0.27
Potassium oxide (K_2O):	1.84
Ignition loss:	3.1

Typical Properties:

Brightness, G.E.: 80
Density, Mg/m^3: 2.8
Pounds per gallon: 23.3
pH: 6.9
Oil Absoprtion, rub out: 31
Plus 325 mesh, %: 1.0
Hegman fineness: 0-1

Section 4
Flame/Fire Retardants

BUCKMAN LABORATORIES: BUSAN 11-M1:

For Fire Resistance
In Plastics, Paints, Textiles, Rubber, and Adhesives

BUSAN 11-M1, modified barium metaborate monohydrate, performs as a synergist in formulations containing phosphorous compounds and/or halogen donors. In many applications it can be used to replace antimony trioxide on an equal weight basis. In other applications a combination of BUSAN 11-M1 and antimony trioxide is preferred. Substitution of BUSAN 11-M1 for part or all of the antimony trioxide in fire-resistant products can provide the following benefits:

* Raw material cost savings
* Improved resistance to flameless combustion (afterglow suppression)
* Reduced toxicity of fumes emitted during combustion
* Reduced smoke emissions
* Greater transparency and hence brighter, cleaner colors
* Improved performance characteristics due to the multi-functionality of BUSAN 11-M1

Product Characteristics:

Active ingredient:	
Barium metaborate, calculated as BaB2O4-H2O (minimim)	90%
Inert ingredients	10%
Specific gravity	3.25-3.35
Weight per solid U.S. gallon	27.5 lb
Bulking	0.036 U.S. gal per lb
Refractive index	1.55-1.60
Oil absorption, spatula rub-out method	30

A Multifunctional Pigment for the Coatings Industry

BUSAN 11-M1 is a multifunctional pigment that is used in coatings as a package preservative, can corrosion inhibitor, mold inhibitor (fungicide), corrosion inhibitor, ultraviolet light stabilizer, tannin stain blocking agent, flame retardant, and white rust inhibitor (zinc rich paints).

DOVER CHEMICAL CORP.: Chlorinated Paraffins:

Product:

PAROIL 10:
Distinguishing Characteristics: Stable, lowest vis., high solubility
Chlorine Content % by Weight: 40
Viscosity: SUS @ 210F: 33

CHLORO-FLO 40:
Distinguishing Characteristics: Stable, low vis., excellent solubility
Chlorine Content % by Weight: 39
Viscosity: SUS @ 210F: 63

CHLORO-FLO 42:
Distinguishing Characteristics: Stable, low vis., excellent solubility
Chlorine Content % by Weight: 40
Viscosity: SUS @ 210F: 90

PAROIL 140:
Distinguishing Characteristics: Stable, low volatility, good solubility
Chlorine Content % by Weight: 42
Viscosity: SUS @ 210F: 150

PAROIL 45:
Distinguishing Characteristics: Good color, low vis. ex. stability
Chlorine Content % by Weight: 45
Viscosity: SUS @ 210F: 50

PAROIL 142-A:
Distinguishing Characteristics: Stable, med. vis., good solubility
Chlorine Content % by Weight: 45
Viscosity: SUS @ 210F: 200

PAROIL 145-A:
Distinguishing Characteristics: Stable, med. vis., good solubility
Chlorine Content % by Weight: 46
Viscosity: SUS @ 210F: 230

PAROIL 50:
Distinguishing Characteristics: Low vis., high chlorine
Chlorine Content % by Weight: 50
Viscosity: SUS @ 210F: 40

DOVER CHEMICAL CORP.: Chlorinated Paraffins(Continued):

PAROIL 152:
Distinguishing Characteristics: Very stable, very low volatility, good solubility
Chlorine Content: % by Weight: 51
Viscosity: SUS @ 210F: 70

PAROIL 150A:
Distinguishing Characteristics: Stable, very high vis., med solubility
Chlorine Content % by Weight: 50
Viscosity: SUS @ 210F: 450

PAROIL 150LVA:
Distinguishing Characteristics: Med. to high vis.
Chlorine Content % by Weight: 48
Viscosity: SUS @ 210F: 250

PAROIL 1057:
Distinguishing Characteristics: Low vis., ex. solubility
Chlorine Content % by Weight: 57
Viscosity: SUS @ 210F: 55

PAROIL 1061:
Distinguishing Characteristics: High chlorine, ex. stability
Chlorine Content % By Weight: 61 min
Viscosity: SUS @ 210F: 72

PAROIL 57-61:
Distinguishing Characteristics: Blend of 1057 and 1061
Chlorine Content % By Weight: 59
Viscosity: SUS @ 210F: 63

PAROIL 1650:
Distinguishing Characteristics: Med. vis., ex. stability, ex. solubility
Chlorine Content % by Weight: 62
Viscosity: SUS @ 210F: 90

PAROIL 170T:
Distinguishing Characteristics: High chlorine, lower vis. than LV.
Chlorine Content % by Weight: 70
Viscosity: SUS @ 210F: 75

PAROIL 170LV:
Distinguishing Characteristics: High vis., high chlorine
Chlorine Content % by Weight: 67
Viscosity: SUS @ 210F: 350

PAROIL 170HV:
Distinguishing Characteristics: Very high vis., high chlorine.
Chlorine Content % by Weight: 70
Viscosity: SUS @ 210F: 530

DOVER CHEMICAL CORP.: Chlorinated Paraffins (Continued):

CHLOREZ Resin Grades

CHLOREZ 700:
Physical Form: White Powder
Chlorine Content % By Weight: 70
Softening Point (Ball and Ring Method, C): 95-110

CHLOREZ 700-S:
Physical Form: White Powder
Chlorine Content % By Weight: 70
Softening Point (Ball and Ring Method, C): 95-110

CHLOREZ 725-S:
Physical Form: White Powder
Chlorine Content % By Weight: 71
Softening Point (Ball and Ring Method, C): 120-140

CHLOREZ 760:
Physical Form: White Powder
Chlorine Content % By Weight: 74
Softening Point (Ball and Ring Method, C): 160

REZ-0-SPERSE Grades

REZ-O-SPERSE A-1:
Physical Form: Water Dispersion of CHLOREZ 700
Solids, % by Weight: 65
Chlorine Content % by Weight: 65
Viscosity, Poise: 48

REZ-O-SPERSE 3:
Physical Form: Water Emulsion of PAROIL 170HV
Solids, % by Weight: 66.5
Chlorine Content % by Weight: 45
Viscosity Poise: 150-300

GREAT LAKES CHEMICAL CORP.: Flame Retardant Chemicals:

GREAT LAKES DE-83R

DE-83R, Decabromophenyl Oxide, is a high purity fully brominated aromatic based general purpose flame retardant with excellent thermal stability.

Applications:
DE-83R is recommended as an additive flame retardant for both thermoplastic, elastomeric and thermoset polymer systems. Its properties also make DE-83R an excellent flame retardant for coatings and adhesive systems.

Physical Properties:
Formula Weight: 959.2
Melting range: 300-310C
Organic Bromine Content (Theoretical): 83.3%

Bis(Tribromophenoxy)ethane
GREAT LAKES FF 680

Applications:
FF 680, a proprietary flame retardant, is a white, crystalline powder. High bromine content and excellent thermal stability make this material an outstanding flame retardant for many thermoplastic and thermoset systems. FF 680 is an efficient flame retardant for those applications where thermal stability at high processing temperatures is important. FF 680 has a good ultra-violet light stability making it an ideal candidate for light stable applications.

Physical Properties:
Bromine, %: 70.0
Melting range, C: 223-225
Density: 2.58

FIREMASTER 836
Halogenated Phosphate Ester

FIREMASTER 836 is a high efficiency liquid flame retardant additive that combines the flame retarding utility of bromine, chlorine, and phosphorus. The product is a low viscosity liquid making it easy to handle year round.

Applications:
FIREMASTER 836 is recommended as an additive flame retardant for applications where high efficiency is critical and a liquid product is desired.

Typical Properties:
Bromine Content, %: 35-37
Chlorine Content, %: 8-9.5
Phosphorus Content, %: 6-8
Viscosity @ 25C., cps: 200

HUMPHREY CHEMICAL CORP.: ZB-112*ZB-237*ZB-325 (Zinc Borates):

Fire Retardants*After-Glow Depressants*Biostats

Applications:

Humphrey Zinc Borates are used in fire retardant systems for:

- Elastomers
  - Natural Rubber
  - Butyl
  - Neoprene*
  - Hypalon*
  - Nitrile
  - SBR
- Textiles
  - Duck
  - Burlap
  - Osnaburg Cloth
- Window Shades
- Coatings
- Vinyl
  - Coated Fabrics
  - Calendered Sheet and Film
  - Organosols
  - Latices
  - Plastisols
- Polyamides
- Polyolefins
- Polyesters
- Phenolics
- ABS
- Backings for Carpets and Upholstery
- Paper

Performance at Combustion Temperatures

* Promotes degradation and decomposition of halogens to form HCl
* Loses water of hydration as steam (helps snuff out flames)
* Undergoes acid hydrolysis to form zinc chloride and free boric acid--both assist in fire retardation.
* Fuses and isolates substrate from atmospheric oxygen.

Typical Analysis:

ZB-112:
Zinc: % as ZnO: 45
Boron: % as BO: 35
Loss on Ignition %: 20

ZB-112 Special:
Zinc: % as ZnO: 45
Boron: % as BO: 35
Loss on Ignition: 20

ZB-112R:
Zinc: % as ZnO: 47
Boron: % as BO: 34
Loss on Ignition: 19

ZB-237:
Zinc: % as ZnO: 33
Boron: % as BO: 41
Loss on Ignition: 26

ZB-325:
Zinc: % as ZnO: 52
Boron: % as BO: 29
Loss On Ignition: 19

Section 5
Flatting Agents

DANIEL PRODUCTS CO.: FLAT-AYD:

Pre-Dispersed Flatting Bases for Gloss Control of Coatings

FLAT-AYD Bases are dispersions of selected flatting agents in a variety of vehicles. They provide simple stir-in incorporation, efficient gloss reduction and assured re-production of results. Those FLAT-AYD Bases containing olefin also provide a high degree of scuff and polish resistance. Coatings utilizing FLAT-AYD Bases produce films of excellent clarity and sheen uniformity.

FLAT-AYD Selector

FLAT-AYD:

For Use In Polyurethanes

Oil modified polyurethanes:

FA-NCO-6:
Flatting Agent: 22.0% Silica
Vehicle Solids: 22.5% Oil Modified Polyurethane
Solvent: 55.5% M.S.
Flash Point: 101F/38.3C
Wt./Gal. (Sp.Gr.): 8.0 Lbs. (0.96)

Two component polyurethanes:

FA-OL-2:
Flatting Agent: 25.0% Silica
Vehicle Solids: 28.0% Castor Polyol (OH-127)
Solvent: 47.0% Xylol
Flash Point: 80F/26.7C
Wt./Gal. (Sp.Gr.): 8.8 Lbs. (1.06)

For Use In Vehicles Compatible with Medium Oil Alkyd

FA-3B:
Flatting Agent: 18.0% Silica
Vehicle Solids: 27.0% Medium Oil Alkyd
Solvent: 55.0% M.S.
Flash Point: 101F/38.3C
Wt. Gal. (Sp.Gr.): 8.0 Lbs. (0.96)

FA-7A:
Flatting Agent: 25.0% Silica
Vehicle Solids: 20.0% Medium Oil Alkyd
Solvent: 55.0% M.S.
Flash Point: 101F/38.3C
Wt. Gal. (Sp.Gr.): 8.6 Lbs. (1.03)

DANIEL PRODUCTS CO.: FLAT-AYD(Continued):

For Use in Baking Finishes and Lacquers

FLAT-AYD:

FA-3B:
Flatting Agent: 18.0% Silica
Vehicle Solids: 27.0% Medium Oil Alkyd
Solvent: 55.0% M.S.
Flash Point: 101F/38.3C
Wt.Gal. (Sp.Gr.): 8.0 Lbs. (0.96)

FA-14C:
Flatting Agent: 23.0% Silica & Olefin
Vehicle Solids: 22.0% Short Oil Alkyd
Solvent: 55.0% Xylol
Flash Point: 80F/26.7C
Wt.Gal. (Sp.Gr.): 8.8 Lbs. (1.06)

FA-VNL-8:
Flatting Agent: 24.0% Silica & Olefin
Vehicle Solids: 20.0% Polyester Plasticizer
Solvent: 56.0% Xylol
Flash Point: 80F/26.7C
Wt./Gal.(Sp.Gr.): 8.3 Lbs. (1.00)

For Use in Vinyl Toluene Treated Vehicles

FA-VT-9:
Flatting Agent: 30.0% Silicate & Stearate
Vehicle Solids: 17.0% Vinyl Toluene Treated Alkyd
Solvent: 36.0% M.S.
17.0% V.M. & P.
Flash Point: 80F/26.7C
Wt. Gal. (Sp.Gr.): 8.1 Lbs. (0.97)

For Use in Aqueous Coatings

FA-W-34:
Flatting Agent: 31.5% Silica & Olefin
Vehicle Solids: 3.0% Surfactants
Solvent: 48.5% Water
17.0% Prop. Glycol
Flash Point: 225F/107.2C
Wt./Gal. (Sp.Gr.): 9.7 Lbs. (1.16)

DAVISON CHEMICAL DIVISION: SYLOID Silicas:

Flatting Agents for Coatings Industry

SYLOID Silica Characteristics: Purity
Advantages for the Coatings Formulator/Manufacturer:
- In-can stability
- Clear film clarity
- Exterior durability

SYLOID Silica Characteristics: Particle strength
Advantages for the Coatings Formulator/Manufacturer:
- Flatting consistency
- Overgrind resistence

SYLOID Silica Characteristics: Low particle density
Advantages for the Coatings Formulator/Manufacturer:
- High efficiency

SYLOID Silica Characterictics: Particle size range
Advantages for the Coatings Formulator/Manufacturer: Optimized flatting efficiency

SYLOID Silica Characteristics: Narrow particle size distribution
Advantages for the Coatings Formulator/Manufacturer:
- Added efficiency
- Ease of dispersion

SYLOID Silicas Typical Properties

SYLOID:

74:
- Loss on Ignition @ 1750F (%): 5.0
- Avg. Particle Size (microns): 6.0
- Oil Absorption (lb/100 lb): 200

83:
- Surface Treatment: X
- Loss on Ignition @ 1750F (%): 6.5
- Avg. Particle Size (microns): 4.0
- Oil Absorption (lb/100 lb): 275

161:
- Surface Treatment: X
- Loss on Ignition @ 1750F (%): 15.5
- Avg. Particle Size (microns): 6.0
- Oil Absorption (lb/100 lb): 155

162:
- Surface Treatment: X
- Loss on Ignition @ 1750F (%): 15.5
- Avg. Particle Size (microns): 7.0
- Oil Absorption (lb/100 lb): 155

DAVISON CHEMICAL: SYLOID Silicas(Continued):

SYLOID Silicas Typical Properties

<u>SYLOID:</u>

169:
  Surface Treatment: X
  Loss On Ignition @ 1750F (%): 15.0
  Avg. Particle Size (microns): 4.5
  Oil Absorption (lb/100 lb): 275
221:
  Loss On Ignition @ 1750F (%): 8.0
  Avg. Particle Size (microns): 6.5
  Oil Absorption (lb/100 lb): 275
234:
  Loss On Ignition @ 1750F (%): 8.0
  Avg. Particle Size (microns): 2.5
  Oil Absorption (lb/100 lb): 275
235:
  Loss on Ignition @ 1750F (%): 8.0
  Avg. Particle Size (microns): 4.0
  Oil Absorption (lb/100 lb): 275
244X1517:
  Loss on Ignition @ 1750F (%): 8.5
  Avg. Particle Size (microns): 3.0
  Oil Absorption (lb/100 lb): 220
245:
  Loss On Ignition @ 1750F (%): 8.0
  Avg. Particle Size (microns): 3.0
  Oil Absorption (lb/100 lb): 275
308:
  Surface Treatment: X
  Loss On Ignition @ 1750F (%): 5.0
  Avg. Particle Size (microns): 4.0
  Oil Absorption (lb/100 lb): 200
378:
  Surface Treatment: X
  Loss On Ignition @ 1750F (%): 5.0
  Avg. Particle Size (microns): 3.0
  Oil Absorption (lb/100 lb): 200
620:
  Loss on Ignition @ 1750F (%): 5.0
  Avg. Particle Size (microns): 15.0
  Oil Absorption (lb/100 lb): 180
AL-1:
  Loss on Ignition @ 1750F (%): 7.0
  Avg. Particle Size (microns): 9.0
  Oil Absorption (lb/100 lb): 60

DEGUSSA CORP.: Flatting Agent HK 188:

is a specially developed wet-process silica designed for low cost/high performance applications.

With a particle size range of 1-11 micrometers Flatting Agent HK 188 is almost universally applicable for a wide range of clear and pigmented coatings. Besides its general versatility, the low angular sheen produced by Flatting Agent HK 188 makes it especially suitable for coil coatings and similar industrial finishes which require this property.

Additional Advantages of Flatting Agent HK 188:

Easy adjustment of any desired sheen level by just altering the added quantity.

Incorporation by simple stirring, low or high shear.

Smooth finishes with excellent mar resistance will be obtained

Flatting Agent HK 188 produces little bodying of the paint at normal concentrations.

Technical Data:

SiO2-content: %: >99
Moisture: (2 h at 105C/220F): %: <6
Ignition loss: (2h at 1000C/1800F): %: 6
pH value: (in 4% aqueous suspension): approx 6.5
Compacted apparent density: g/l: 135
Density: g/cm3: 2.0
Refractive Index: 1.45
Oil absorption: g/100g: 155
Particle size of agglomerates:
Limiting value: um: 1-11
Average: um: 3

DEGUSSA CORP.: Flatting Agent OK 412/OK 500/HK 400:

Flatting Agent OK 412/OK 500:

are suited, for example, for lacquers, acid-catalyzed systems, polyurethane varnishes, unsaturated polyester lacquers, baking enamels, coil coatings, etc. They can be dispersed in every system within a short time by means of a high-speed stirrer. Grinding instruments with an intensive shearing effect should not be used.

Coating films which are delustered with Flatting Agent OK 412/OK 500 show an excellent surface development with a smooth, velvet-like appearance, and meet the most demanding requirements.

Flatting Agent OK 412:

is a finely divided, surface-treated silica produced according to a special precipitation process. The easy dispersibility of OK 412 makes efficient and economical production of flat lacquer possible through simple mixing and stirring. Depending on the amount added, every desired degree of flatting can be developed; this applies equally for both clear and pigmented systems.

Flatting Agent OK 412 is distinguished by the following proven characteristics:

1. It is versatile for application in almost all coating systems.
2. OK 412 offers simple development of every desired degree of flatting.
3. Flatting Agent OK 412 is superior over most flatting silicas in terms of fineness.
4. The product has only little influence on the rheological properties of coatings.
5. Coating films delustered with Flatting Agent OK 412 develop excellent surfaces with outstanding burnish and mar resistance.
6. The organic surface treatment of OK 412 results in very good suspension behavior.

Flatting Agent OK 500:

This product is a variant of Flatting Agent OK 412. It was developed especially for thick films. When applied in heavy coats, OK 412 may cause drying retardation due to its specific organic treatment. Flatting Agent OK 500 consists of the same finely divided particle structure as OK 412, and offers equal positive application characteristics.

Flatting Agent HK 400:

is an untreated flatting silica with the same fineness and complements the two organically treated grades OK 412 and OK 500. As a pure silica, HK 400 may tend to settle. With an addition of AEROSIL R 972 this disadvantage can easily be eliminated.

DEGUSSA CORP.: Flatting Agent TS 100:

Flatting Agent TS 100:

Flatting Agent TS 100 is a pure silica produced by a thermal process with an agglomerate particle size of 2-10 micrometers. It was developed for special applications and can be used in clear lacquers as well as in pigmented systems.

Flatting Agent TS 100 offers the following advantages:

1. It is particularly suitable for use in final coats on vinyl coated fabrics, leather, textiles, foils, photographic coatings and others.
2. Due to its outstanding flatting effect, TS 100 can solve the most difficult problems with very small additions.
3. TS 100 is easily dispersed by a normal stirrer. Due to its good grinding stability, a smoother surface can be achieved by more intensive dispersing methods.
4. TS 100 imparts a slight thixotropic effect to lacquers used in the above mentioned applications.
5. TS 100 imparts such excellent transparency to coatings, that even the deepest colour shades are hardly affected.

Due to its excellent flatting effect and very good transparency, TS 100 is especially suitable for final coatings on leathercloth, foil lacquers and polyurethane-textile coatings.

Furthermore, TS 100 is recommended for flatting lacquer systems which--due to their composition--are difficult to flatten, e.g. air-drying alkyd resin lacquers, lacquers based on urethane oil and reaction lacquers.

TS 100 imparts a certain thixotropic effect to the lacquer system which is an advantage for the above mentioned applications. The use of a thickening agent is rarely necessary.

Physical and Chemical Data:

- SiO2 content %: >99.8
- Drying loss %: <4
- Ignition loss %: <2
- pH-value: 7
- Bulk density g/l: 50
- Density g/cm3: 2,2
- Refractive index: 1,45
- Oil-absorption g/100 g: 390
- Agglomerate particle size
  - Limiting values: micrometers: 2-10
  - Average values: micrometers: 4

Section 6
Latex Emulsions

W. R. GRACE & CO.: EVERFLEX Paint/Coating Emulsions:

MA:

EVERFLEX MA is a unique polyvinyl acetate copolymer which combines very fine particle size with a high degree of chemical and mechanical stability. This combination of properties has made EVERFLEX MA suitable as a vehicle for specialty paints as well as the coating and impregnation of paper and textiles.

Typical Emulsion Properties:
- Total Solids: 50-52%
- Viscosity: 200-800 cps
- pH: 4.0-6.5
- Average Particle Size (approx.): 0.1 micron
- Weight per Gallon, Emulsion: 9.0 lbs.

GT:

EVERFLEX GT is a versatile emulsion polymer designed for high solids mastic compounds, caulks, and joint cements. Its outstanding chemical and mechanical stability enables the formulator to develop high solids, high viscosity compounds with exceptional flow and leveling characteristics.

Typical Emulsion Properties:
- Total Solids: 54-56%
- Viscosity: 1500-3000 cps
- pH: 4.0-6.5
- Average Particle Size: 1 micron

G:

EVERFLEX G is a high molecular weight vinyl acetate copolymer emulsion. It offers excellent flow, flexibility and machining properties. Consequently, it finds wide application in specialty coating applications.

Components of this emulsion comply with FDA Regulation 121.2520 for use in food packaging, adhesives and coatings.

Typical Emulsion Properties:
- Total Solids: 51-53%
- Viscosity: 300-700 cps
- pH: 4.0-6.5
- Average Particle Size (approx.): <2 microns

W. R. GRACE & CO.: EVERFLEX Paint/Coating Emulsions(Continued):

BG:

EVERFLEX BG is a premium quality, high molecular weight emulsion polymer designed for use in exterior paints. It has a long proven history of exceptional durability in paints designed for exterior surfaces, particularly wood, where flexibility, distensibility, low temperature consolidation, pigment binding capacity, tint retention, resistance to ultraviolet degradation, and good leveling are required.

Components of this emulsion comply with FDA regulation 121.2520 for use in food packaging, adhesives and coatings.

Typical Emulsion Properties:
- Total Solids: 50-52%
- Viscosity: 200-500 cps
- pH: 4.0-6.5
- Average Particle Size; <2 microns
- Weight per Gallon, Solids: 9.57 lbs.

515L:

A highly efficient pigment binder developed for use in such surface coatings as interior flat wall paints and primer sealers. Excellent pigment binding efficiency is obtained without sacrifice of pigment compatibility, allowing wide compounding latitude.

Typical Properties:
- Total Solids (%): 47-49
- pH: 7.0-8.5
- Viscosity (cps): 20-60
- Average Particle Size (microns): 0.2
- Weight/Gallon (lbs): wet: 8.4
  solids: 8.5

W. R. GRACE & CO.: VERSAFLEX Acrylic Coating Emulsions:

1:

A versatile acrylic prepolymer specifically designed to prevent blistering and peeling soon after application and before the film is leached of its water extractables. It is then that films are most susceptible to loss of adhesion due to extreme condensation-heavy dew or prolonged rain.

In addition, VERSAFLEX 1 provides excellent:

* Alkali Resitance
* Pigment Binding
* All-Weather Resistance
* Low Temperature Flexibility
* Tint Retention
* Scrub Resistance

VERSAFLEX 1 - a versatile emulsion for latex coatings.

Typical Properties:

Total Solids: 54-56%
Viscosity: 200-800 cps
pH: 5.0-6.0
Average Particle Size: 0.2 micron
Weight per Gallon, Emulsion: 8.8 lbs.

8:

VERSAFLEX 8 is a styrene acrylate copolymer emulsion, designed for use as a high gloss coating where excellent water resistance is required. On drying, it forms an exceptionally clear, non-blocking, water resistant film which will not re-whiten on exposure to water. It can also be used as a binder for clay or titanium dioxide where a good, scrub resistant surface is required.

VERSAFLEX 8 will form a film at 18C or above. Curing at 150C for 2 minutes will result in improved properties.

VERSAFLEX 8 is manufactured from materials approved for use under 21 CFR: 175:105.

Typical Properties:
Total Solids: 48-50%
Viscosity: <200 cps
pH: 2.5-3.5
Weight per Gallon: 8.8 lbs.
Particle Size: 0.1 to 1.2 microns

POLYVINYL CHEMICALS INC.: NEOCRYL Acrylic Emulsions for Water-Borne Coatings:

A-601:
Type: Acrylic
Solids: 32%
pH: 7.7
Viscosity (25C): 100 cps
Features/Applications: Outstanding exterior durability, wide adhesion latitude, and good pigment wetting characteristics. For use in general metal and plastic finishing, paper and wall board coatings.

A-602:
Type: Acrylic
Solids: 32%
pH: 8.0
Viscosity (25C): 100 cps
Features/Applications: Harder version of A-601 with improved bloock resistance and sandability. Recommended for clean and pigmented board coatings and plastics.

A-603A:
Type: Acrylic
Solids: 36.5%
pH: 7.8
Viscosity (25C): 180C
Features/Applications: Excellent brushing properties and sandability. For use as a non-yellowing brushing varnish.

A-604:
Type: Acrylic
Solids: 32%
pH: 8.1
Viscosity (25C): 250 cps
Features/Applications: Outsatnding mar resistance, sandability, and exterior durability. Recommended for use in wood and plastic furniture finishing.

A-612:
Type: Acrylic
Solids: 32%
pH: 7.7
Viscosity (25C): 150 cps
Features/Applications: Replacement product for A-602 with differenet cosolvent blend. Recommended for use in clear and pigmenetd board coatings.

POLYVINYL CHEMICALS INC.: NEOCRYL Acrylic Emulsions for Water-Borne Coatings(Continued):

A-614:
Type: Acrylic
Solids: 32%
pH: 8.0
Viscosity (25C): 200 cps
Features/Applications: Replacement product for A-604 with different cosolvent blend. Recommended for use in clear and pigmented lacquers for wood and plastics.

A-621:
Type: Acrylic Copolymer
Solids: 40%
pH: 8.3
Viscosity (25C): 300 cps
Features/Applications: High application solids and high pigmented gloss. Rapid tack-free time. Recommended for use on metals and plastics as a sole vehicle or blended with water reducible alkyds.

A-622:
Type: Acrylic Copolymer
Solids: 32%
pH: 8.0
Viscosity (25C): 100 cps
Features/Applications: Exceptional humidity, water and corrosion resistance. Wide adhesion latitude and broad compatibility with other film formers including water reducible alkyds and urethane dispersions. Recommended for use in high performance industrial primers and topcoats for automotive and farm implements.

A-623:
Type: Acrylic Copolymer
Solids: 35%
pH: 7.3
Viscosity (25C): 180 cps
Features/Applications: Fast hardness development, excellent gasoline resistance and solvent resistance. For use in high performance air dry industrial primers & topcoats where block resistance is critical.

A-625:
Type: Acrylic Copolymer
Solids: 45%
pH: 7.8
Viscosity (25C): 130 cps
Features/Applications: Good balance of properties for general use on metal and plastic substrates where economy is of primary concern. Can be modified with water reducible alkyds.

POLYVINYL CHEMICALS INC.: NEOCRYL Acrylic Emulsions for Water-Borne Coatings(Continued):

A-630:
Type: Acrylic Copolymer
Solids: 44.5%
pH: 7.7
Viscosity (25C): 150 cps
Features/Applications: Uncoasced version of A-621. Can be formulated into roll, flow or dip coatings that exhibit excellent gloss and hardness.

A-633:
Type: Acrylic Copolymer
Solids: 42%
pH: 7.5
Viscosity (25C): 40 cps
Features/Applications: Wide adhesion latitude with execllent corrosion and chemical resistance. For use in pigmented primers or topcoats on plastics requiring good chemical resistance.

A-634:
Type: Acrylic Copolymer
Solids: 35%
pH: 6.5
Viscosity (25C): 200 cps
Features/Applications: Excellent print resistance and clarity. Suitable for many types of wood furniture applications including fine furniture and kitchen cabinets.

A-636:
Type: Acrylic Copolymer
Solids: 35%
pH: 6.4
Viscosity (25C): 300 cps
Features/Applications: Rapid hardness development and short tack free time. For use on high speed board coating lines where block resistance under low temperature curing conditions is required.

A-639:
Type: Acrylic Copolymer
Solids: 45%
pH: 6.5
Viscosity (25C): 160 cps
Features/Applications: Suggested for board coatings, metal coatings, and paper coatings where block resistance and economy are important. Can be used as a modifier for water reducible alkyds.

POLYVINYL CHEMICALS INC.: NEOCRYL Acrylic Emulsions for Water-Borne Coatings(Continued):

A-640:
Type: Acrylic Copolymer
Solids: 40%
pH: 8.0
Viscosity (25C): 50 cps
Features/Applications: Uncoalesced version of A-622. Broad compatibility with water reducible alkyds and urethane dispersions. For use in primesr and topcoats with excellent water humidity and corrosion resistance.

A-644:
Type: Acrylic Copolymer
Solids: 40%
pH: 7.1
Viscosity (25C): 30 cps
Features/Applications: Outstanding mar resistance, sandability and block resistance. For use in clear and pigmented coatings including high build dip applications for many types of wood articles.

A-645:
Type: Acrylic Copolymer
Solids: 40%
pH: 7.5
Viscosity (25C): 50 cps
Features/Applications: Uncoalesced version of A-623. For use in clear and pigmented air dry coatings which require exceptioanl solvent resistance.

A-650:
Type: Acrylic Copolymer
Solids: 40%
pH: 6.3
Viscosity (25C): 50 cps
Features/Applications: Uncoalesced version of A-634. Can be formulated for use in wood furniture coatings which require exceptional clarity and alcohol resistance.

A-655:
Type: Acrylic Copolymer
Solids: 45%
pH: 7.8
Viscosity (25C): 250 cps
Features/Applications: Wide adhesion latitude including excellent adhesion to "oily" substrates and many plastics. For use primer or topcoat for engine components, business machines, and metal castings.

POLYVINYL CHEMICALS, INC.: NEOCRYL Alkali-Soluble Acrylic Emulsions for Water-Borne Coatings:

BT-24:

Type: Acrylic Copolymer
Solids: 45%
pH: 5.0
Glass Transition Temperature (Tg): 8C
Solubilized Viscosity (25% wt. solids pH 9.0, 25C): 300 cps
Features/Applications: High gloss, good pigment wetting with excellent adhesion to fiberglass-reinforced polyester(SMC). For use with water reducible monomeric melamines in thermoset systems. When amine solubilized, suggested as a pigment grinding vehicle for water-borne coatings.

BT-175:

Type: Acrylic Copolymer
Solids: 40%
pH: 3.0
Glass Transition Temperature (Tg): 26C
Solubilized Viscosity (25% wt. solids pH 9.0, 25C): 500 cps
Features/Applications: When solubilized, has low viscosity at high solids with stable viscosity over a wide alkaline pH range. Excellent pigment dispersant. Recommended for use as a pigment grinding vehicle for water-borne coatings.

BT-520:

Type: Acrylic Copolymer
Solids: 40%
pH: 7.0
Glass Transition Temperature (Tg): 26C
Solubilized Viscosity (25% wt. solids pH 9.0, 25C): 100 cps
Features/Applications: Hard films with excellent water and corrosion resistance. For use as a water-borne coating for bright metals such as brass and chrome.

POLYVINYL CHEMICALS, INC.: NEOCRYL Alkali-Soluble Acrylic Emulsions for Water-Borne Coatings(Continued):

S-1007:
Type: Acrylic
Solids: 36%
Viscosity (25C): 850 cps
Solvent System: Toluene/Methyl Ethyl Ketone
Features/Applications: High gloss with excellent exterior durability, humidity, and gasoline resistance. For use in clear or pigmented coatings such as auto and farm equipment refinishing. Can be applied in thick films without crazing or cracking at low temperatures.

LE-800:
Type: Acrylic
Solids: 60%
Viscosity (25C): 8000 cps
Solvent Systems: 2-butoxy ethanol
Features/Applications: Water reducible, low temperature cure thermoset. Can be crosslinked with water-borne methylated melamines and cured under 120C(250F) without external acid catalysts. Coatings have high gloss and hardness with excellent stain, chemical, corrosion and waterresistance. Fo ruse in coatings for small ppliances, lighting fixtures, metal shelving and general metal finishing.

S-1004:
Type: Acrylic
Solids: 50%
Viscosity (25C): 4,000 cps
Solvent System: Odorless mineral spirits
Features/Applications: Aliphatic-soluble solution acrylic. Excellent pigment dispersant for hard-to-wet pigments such as carbon blacks in solvent or oil solutions. Suggested for liquid photocopy toners, inks and wood floor polishes.

B-705:
Type: BMA Copolymer
Solids: 100%
Viscosity: 900 cps*
Features/Applications: Hard, durable, glossy films. Good compatibility with waxes and resins, excellent flow and leveling. Dry-bright solvent-based polishes for wood and parquet floors; clear coatings for plastics, paper and wood.
*40% wt. solids in toluen, 25C.

ROHM AND HAAS: RHOPLEX Acrylic Latex Vehicles:

<u>Grade:</u>

RHOPLEX AC-829:
Exterior vehicle with chalk adhesion and film build
Solids, %: 55
pH: 8.6-9.1
Density lbs/U.S. gal: 8.8

RHOPLEX AC-64:
Exterior vehicle with chalk adhesion
Solids, %: 60.5
pH: 9.0-9.7
Density lbs/U.S. gal: 8.8

RHOPLEX AC-388:
Exterior vehicle with film build
Solids, %: 50
pH: 9.4-9.7
Density lbs/U.S. gal: 8.8

RHOPLEX AC-417:
Interior semi-gloss vehicle with superior adhesion
Solids, %: 48
pH: 9.2-10.0
Density lbs/U.S. gal: 8.9

RHOPLEX AC-490:
Interior semi-gloss vehicle
Solids, %: 46
pH: 9.2-9.5
Density lbs/U.S. gal: 8.8

RHOPLEX AC-507:
Exterior semi-gloss vehicle
Solids, %: 46.5
pH: 9.5-10.0
Density lbs/U.S. gal: 8.8

RHOPLEX HG-74:
Interior-exterior high gloss vehicle
Solids, %: 42.5
pH: 7.7-8.3
Density lbs/U.S gal: 8.6

ROHM AND HAAS: RHOPLEX Acrylic Latex Vehicles(Continued):

Grade:

RHOPLEX AC-61:
Concrete or wood floor paint vehicle
Solids, %: 46
pH: 9.5-10.0
Density lbs/U.S. gal: 8.8

RHOPLEX MV-23:
Stain resistant primer vehicle
Solids, %: 43
pH: 9.4-10.0
Density lbs/U.S gal: 8.8

RHOPLEX AC-2000:
Vehicle component of System 2000
Solids, %: 53.5
pH: 9.0-10.0
Density lbs/U.S. gal: 8.7

SYSTEM 2000:

A unique approach to high performance semi-gloss at lower cost based on an optimized combination of vehicle--RHOPLEX AC-2000, opaque polymer--ROPAQUE OP-62, and rheology modifier-ACRYSOL RM-5.

ROPAQUE Opaque Polymers:

Grade:

ROPAQUE OP-42:
Solids, %: 40
pH: 9.0-10.0
Density lbs/U.S. gal: 8.6

ROPAQUE OP-62:
Solids, %: 37.5
pH: 9.0-10.0
Viscosity, cps.: 100
Density lbs/U.S. gal: 8.6

STANCHEM, INC.: Polymers and Polymeric Specialties: Paints and Building Products Polymers:

STANCHEM Product:

5011:

Vehicle base for interior and exterior paints, especially in high PVC range. Excellent compatibility with tinting systems.
Polymer Type: Polyvinyl acetate-acrylic copolymer
Viscosity (cps): 800-1400
Solids % (min.): 54.5
Average Particle Size, M: 0.2-0.4
Tg, C: 9
pH @ 25C: 2-4
Weight lb/gal: 9.0
Ionic Charge: Nonionic

6035:

Excellent scrubbability, freeze-thaw resistance, leveling properties. Outstanding exterior durability and long term tint retention.
Polymer Type: Acrylic Copolymer
Viscosity (cps): 500-1000
Solids % (min.): 46
Average Particle Size, M: 0.05-0.1
Tg, C (Theoretical): 8
pH @ 25C: 9-10
Weight lb/gal: 8.6
Ionic Charge: Nonionic

6064:

Emulsion for exterior paints with good adhesion to chalky surfaces.
Polymer Type: Acrylic copolymer
Viscosity (cps): 300-1000
Solids % (min.): 60
Average Particle Size, M: 0.2-1.0
Tg, C (Theoretical): 2
pH @ 25C: 9.0-9.5
Weight lb/gal: 8.9
Ionic Charge: Anionic

STANCHEM, INC.: Polymers and Polymeric Specialties:Paints and Building Products Polymers (Continued):

STANCHEM Product:

6088:

Excellent leveling and high film build properties, for use in formulating one coat systems. Resists dirt pick up.
Polymer Type: Acrylic copolymer
Viscosity (cps): 1500-2500
Solids % (min.): 49.5
Average Particle Size, M: 0.2-1.0
Tg, C (Theoretical): 8
pH @ 25C: 9-10
Weight lb/gal: 8.7
Ionic Charge: Anionic

7013:

Superior hardener with chemical resistance properties. No discoloration at high temperatures.
Polymer Type: Acrylic polymer solution
Viscosity (cps): 1200-2500
Solids % (min.): 39
Tg, C (Theoretical): 105
Weight lb/gal: 8.0
Ionic Charge: MEK system

STANCHEM, INC.: Polymers and Polymeric Specialties: Caulks and Sealants:

STANCHEM Product:

5009:

Wide compatibility with modifiers and high solvent tolerance for general purpose adhesive compounding. A higher viscosity version is available.

Polymer Type: Polyvinyl acetate homopolymer
Viscosity (cps): 850-1000
Solids % (min.): 54.5
Average Particle Size, M: 1-3
Tg, C (Theoretical): 38
pH @ 25C: 4.0-5.0
Weight lb/gal: 9.1
Ionic Charge: Nonionic

5019:

Good water resistance and high electrolyte tolerance for use in adhesives, joint cements and textile sizing.

Polymer Type: Polyvinyl acetate homopolymer
Viscosity (cps): 1000-1600
Solids % (min.): 54.5
Average Particle Size, M: 0.5-1.0
Tg, C (Theoretical): 36
pH @ 25C: 4.0-5.0
Weight lb/gal: 9.0
Ionic Charge: Nonionic

6050:

High pigment tolerance and excellent resistance to water and ultraviolet light.

Polymer Type: Acrylic copolymer
Viscosity (cps): <300
Solids % (min.): 54.5
Average Particle Size, M: 0.2-1.5
Tg, C (Theoretical): -4
pH @ 25C: 4.2-4.8
Weight lb/gal: 8.8
Ionic Charge: Anionic

THIBAUT AND WALKER: 100% Acrylic Emulsions:

PARCRYL:

Item:

200:
% NV: 46.5
Viscosity cps.: 50-500
pH: 9-10
Lbs./Gal. Wet: 8.80
Gloss: High
Description: Interior house paints and masonry paints.

250:
% NV: 46.5
Viscosity cps.: 50-500
pH: 9-10
Lbs./Gal. Wet: 8.80
Gloss: High
Description: Caulking compounds and masonry paints.

300:
% NV: 46.5
Viscosity cps: 50-500
pH: 9-10
Lbs./Gal. Wet: 8.80
Gloss: High
Description: Interior and exterior house paints and masonry paints. Suitable for roof and tennis court coatings.

311:
% NV: 46.5
Viscosity cps.: 50-500
pH: 9-10
Lbs./Gal. Wet: 8.80
Gloss: High
Description: Extremely flexible polymer for high pigment loading. Suitable for roof and tennis court coatings.

357:
% NV: 60
Viscosity cps.: 100-1000
pH: 8.5-9
Lbs./Gal. Wet: 8.80
Description: A fast drying vehicle for water-borne traffic paints with excellent weather resistance.

4001:
% NV: 50
Viscosity cps.: 750-1800
pH: 8.5-9.5
Lbs./Gal. Wet: 8.80
Description: Interior and exterior house paints, interior and exterior.

THIBAUT AND WALKER: 100% Acrylic Emulsions(Continued):

PARCRYL(Continued):

Item:

4501:
% NV: 46.5
Viscosity cps: 300-1800
pH: 8.5-9.5
Lbs./Gal. Wet: 8.80
Gloss: High
Description: A versatile polymer for use in quality interior and exterior gloss house paints.

5001:
% NV: 46.5
Viscosity cps: 300-1800
pH: 8.5-9.5
Lbs./Gal. Wet: 8.80
Gloss: High
Description: Exterior gloss paints where color retention is important.

870:
% NV: 55
Viscosity cps: 1000-2500
pH: 5.5-6.5
Lbs./Gal. Wet: 8.80
Description: Water-based caulking compounds.

900:
% NV: 43
Viscosity cps: 100-600
pH: 9-10
Lbs./Gal. Wet: 8.80
Description: Flash rust and rust-resistant metal primers and stain-resistant wood primers.

Vinyl Acrylic Copolymer Emulsion
PARCO:

58-C-55:
% NV: 55
Viscosity cps: 1000-3000
pH: 4-5
Lbs./Gal. Wet: 8.80
Gloss: High
Description: Interior wall paints, gloss and semi-gloss exterior house paints, and masonry paints.

THIBAUT AND WALKER: Vinyl Acrylic Copolymer Emulsions (Continued):

PARCO(Continued):

Item:

69-C-55:
% NV: 55
Viscosity cps: 500-1500
pH: 4-5
Lbs./Gals. Wet: 8.80
Gloss: High
Description: Excellent scrub and stain resistance and enamel hold-out in interior paints. For interior-exterior gloss and semi-gloss house paints and masonry paints.

71-C-60:
% NV: 60
Viscosity cps: 1500-4000
pH: 4-5
Lbs./Gal. Wet: 9.07
Description: High tack adhesion binder for construction adhesives.

72-C-60:
% NV: 60
Viscosity cps: 1500-4000
pH: 4-5
Lbs./Gal. Wet: 9.15
Description: Interior-exterior high gloss house paints with wide formulating latitude.

75-C-55:
% NV: 55
Viscosity cps: 1000-3000
pH: 4-5
Lbs./Gal. Wet: 8.80
Gloss: High
Description: An economy product for use in flat and semi-gloss house and masonry paints.

SWA-555:
% NV: 55
Viscosity cps: 200-2000
pH: 4-5
Lbs./Gal. Wet: 8.90
Gloss: High
Description: Interior-exterior high gloss paints with performance equal to acrylics at much reduced cost. The wet adhesion exceed those obtained with acrylics.

UNION CARBIDE CORP.: UCAR Acrylic:

UCAR Acrylic:

503:
  Solids, %: 58
  pH: 6
  Weight per Gallon, lb: 9.2
  Particle Size, Microns: 0.4
  Viscosity, cp: 1200
  Surfactant Type: Anionic/Nonionic
  Polymer Type: Modified Acrylic
  Applications: Trade Paints

505:
  Solids, %: 55
  pH: 6
  Weight per Gallon, lb: 9.1
  Particle Size, Microns: 0.6
  Viscosity, cP: 800
  Surfactant Type: Anionic/Nonionic
  Polymer Type: Modified Acrylic
  Applications: Trade Paints

515:
  Solids, %: 53
  pH: 8.7
  Weight per Gallon, lb: 9.1
  Particle Size, Microns: 0.6
  Viscosity, cP: 400
  Surfactant Type: Anionic/Nonionic
  Polymer Type: Modified Acrylic
  Applications: Trade Paints

516:
  Solids, %: 56
  pH: 8.7
  Weight per Gallon, lb: 9.1
  Particle Size, Microns: 0.6
  Viscosity, cP: 500
  Surfactant Type: Anionic/Nonionic
  Polymer Type: Modified Acrylic
  Applications: Trade Paints

518:
  Solids, %: 50
  pH: 8.7
  Weight per Gallon, lb: 9.0
  Particle Size, Microns: 0.6
  Viscosity, cP: 900
  Surfactant Type: Anionic/Nonionic
  Polymer Type: Modified Acrylic
  Applications: Trade Paints

522:
  Solids, %: 51 pH: 9.0
  Weight per Gallon, lb: 9.0 Particle Size, Microns: 0.6
  Viscosity, cP: 900 Surfactant Type: Anionic/Nonionic
  Polymer Type: Modified Acrylic
  Applications: Trade Paints

UNION CARBIDE CORP.: UCAR Latex:

UCAR Latex:

100:
- Solids, %: 62
- pH: 6
- Weight per Gallon, lb: 8.8
- Particle Size, Microns: 0.3
- Viscosity, cP: 750
- Polymer Type: Acrylic
- Applications:
  - Buildings Products:
    - Caulks and Sealants

123:
- Solids, %: 60
- pH: 6
- Weight per Gallon, lb: 8.9
- Particle Size, Microns: 0.5
- Viscosity, cP: 200
- Polymer Type: Acrylic
- Applications:
  - Trade Paints
  - Building Products:
    - Caulks and Sealants
    - Mastics
    - Other
  - Pressure Sensitive Adhesives

124:
- Solids, %: 60
- pH: 8
- Weight per Gallon, lb: 8.9
- Particle Size, Microns: 0.5
- Viscosity, cP: 300
- Polymer Type: Acrylic
- Applications:
  - Building Products:
    - Caulks and Sealants
    - Mastics

130:
- Solids, %: 58
- pH: 5
- Weight per Gallon, lb: 9.3
- Particle Size, Microns: 0.8
- Viscosity, cP: 2000
- Surfactant Type: Nonionic
- Polymer Type: PVA Homopolymer
- Applications:
  - Trade Paints
  - Industrial Finishes
  - Building Products:
    - Caulks and Sealants/Other

UNION CARBIDE CORP.: UCAR Latex(Continued):

UCAR Latex:

131:
Solids, %: 60
pH: 5
Weight per Gallon, lb: 9.2
Particle Size, Microns: 0.8
Viscosity, cP: 2800
Surfactant Type: Nonionic
Polymer Type: PVA Homopolymer
Applications:
Trade Paints
Building Products:
Caulks and Sealants
Mastics
Other

148:
Solids, %: 48
pH: 6
Weight per Gallon, lb: 8.9
Particle Size, Microns: 0.2
Viscosity, cp: 100
Surfactant Type: Anionic
Polymer Type: Styrene Acrylic
Applications: Building Products: Mastics

153:
Solids, %: 55
pH: 4
Weight per Gallon, lb: 8.9
Particle Size, Microns: 0.5
Viscosity, cp: 200
Surfactant Type: Anionic
Polymer Type: Acrylic
Applications:
Building Products: Caulks and Sealants
Mastics

154:
Solids, %: 60
pH: 4
Weight per Gallon, lb: 9.0
Particle Size, Microns: 0.4
Viscosity, cp: 300
Polymer Type: Acrylic
Applications:
Building Products:
Caulks and Sealants
Mastics

UNION CARBIDE CORP.: UCAR Latex(Continued):

UCAR Latex:

163:
- Solids, %: 58
- pH: 4
- Weight per Gallon, lb: 8.9
- Particle Size, Microns: 0.4
- Viscosity, cP: 200
- Polymer Type: Acrylic
- Applications:
  - Trade Paints
  - Building Products
    - Caulks & Sealants
    - Mastics

173:
- Solids, %: 50
- pH: 6
- Weight per Gallon, lb: 8.6
- Particle Size, Microns: 0.15
- Viscosity, cP: 400
- Surfactant Type: Anionic
- Polymer Type: Acrylic
- Applications:
  - Pressure Sensitive Adhesives

174:
- Solids, %: 50
- pH: 6
- Weight per Gallon, lb: 8.6
- Particle Size, Microns: 0.15
- Viscosity, cP: 600
- Surfactant Type: Anionic
- Polymer Type: Acrylic
- Applications:
  - Pressure Sensitive Adhesives

175:
- Solids, %: 49
- pH: 5
- Weight per Gallon, lb: 8.4
- Particle Size, Microns: 0.15
- Viscosity, cP: 1700
- Surfactant Type: Anionic
- Polymer Type: Acrylic
- Applications: Pressure Sensitive Adhesives

347:
- Solids, %: 55
- pH: 4.5
- Weight per Gallon, lb: 9.1
- Particle Size, Microns: 0.2
- Viscosity, cP: 600
- Surfactant Type: Anionic/Nonionic
- Polymer Type: Vinyl-Acrylic
- Applications: Trade Paints

UNION CARBIDE CORP.: UCAR Latex(Continued):

UCAR Latex:

351:
Solids, %: 65
pH: 5
Weight per Gallon, lb: 9.2
Particle Size, Microns: 0.2
Viscosity, cp: 2500
Surfactant Type: Nonionic
Polymer Type: Vinyl-Acrylic
Applications:
Trade Paints

367:
Solids, %: 55
pH: 5
Weight per Gallon, lb: 9.0
Particle Size, Microns: 0.3
Viscosity, cp: 600
Surfactant Type: Anionic/Nonionic
Polymer Type: Vinyl-Acrylic
Applications:
Trade Paints
Building Products:
Caulks and Sealants
Mastics

376:
Solids, %: 55
pH: 5
Weight per Gallon, lb: 9.0
Particle Size, Microns: 0.3
Viscosity, cp: 600
Surfactant Type: Nonionic
Polymer Type: Vinyl-Acrylic
Applications:
Trade Paints

407:
Solids, %: 45
pH: 4
Weight per Gallon, lb: 8.9
Particle Size, Microns: 0.3
Viscosity, cp: 80
Surfactant Type: Anionic
Polymer Type: Acrylic

UNION CARBIDE CORP.: UCAR Vehicle:

UCAR Vehicle:

438:
Solids, %: 46
pH: 8
Weight per Gallon, lb: 8.7
Particle Size, Microns: 0.2
Viscosity, cP: 150
Surfactant Type: Anionic
Polymer Type: Acrylic
Applications: Trade Paints
Industrial Finishes

441:
Solids, %: 45
pH: 8
Weight per Gallon, lb: 8.7
Particle Size, Microns: 0.2
Viscosity, cP: 65
Surfactant Type: Anionic
Polymer Type: Acrylic
Applications: Industrial Finishes

443:
Solids, %: 41
pH: 7
Weight Per Gallon, lb: 8.7
Particle Size, Microns: 0.15
Viscosity, cP: 600
Surfactant Type: Anionic
Polymer Type: Acrylic
Applications: Trade Paints
Industrial Finishes

447:
Solids, %: 59
pH: 6
Weight per Gallon, lb: 9.1
Particle Size, Microns: 0.3
Viscosity, cP: 1300
Surfactant Type: Nonionic
Polymer Type: Vinyl-Acrylic
Applications: Industrial Finishes

455:
Solids, %: 45
pH: 4
Weight per Gallon, lb: 8.9
Particle Size, Microns: 0.3
Viscosity, cP: 60
Surfactant Type: Anionic
Polymer Type: Acrylic
Applications: Industrial Finishes

UNION CARBIDE CORP.: UCAR Vehicle(Continued):

UCAR Vehicle:

462:
Solids, %: 45.5
pH: 8
Weight per Gallon, lb: 8.8
Particle Size, Microns: 0.3
Viscosity, cP: 400
Surfactant Type: Anionic
Polymer Type: Acrylic
Applications: Industrial Finishes

464:
Solids, %: 50
pH: 8
Weight per Gallon, lb: 8.8
Particle Size, Microns: 0.3
Viscosity, cP: 500
Surfactant Type: Anionic
Polymer Type: Acrylic
Applications: Industrial Finishes

4358:
Solids, %: 45
pH: 8.2
Weight per Gallon, lb: 8.7
Particle Size, Microns: 0.15
Viscosity, cP: 60
Surfactant Type: Anionic
Polymer Type: Acrylic
Applications: Trade Paints
Industrial Finishes

4510:
Solids, %: 42
pH: 8.5
Weight per Gallon, lb: 8.8
Particle Size, Microns: 0.2
Viscosity, cP: 150
Surfactant Type: Anionic
Polymer Type: Styrene-Acrylic
Applications: Industrial Finishes

4512:
Solids, %: 44
pH: 8
Weight per Gallon, lb: 8.7
Particle Size, Microns: 0.2
Viscosity, cP: 150
Surfactant Type: Anionic
Polymer Type: Acrylic
Applications: Industrial Finishes

Section 7
Oils

HERCULES INC.: YARMOR Pine Oil:

YARMOR 302 Pine Oil
Highest Quality, All-Purpose-Grade Pine Oil

YARMOR 302 pine oil is a clear, pale yellow to near water-white oily liquid with a distinct pine-like odor. Derived from terpene oils of pinewood origin, it is a blend of related compounds, mainly terpene alcohols. YARMOR 302 meets requirements of Federal Specification LLL-P-400a for Type I pine oil. It is especially indicated for manufacture of top-performance cleaners and disinfectants and for all other uses where a high-quality pine-oil of uniform, highest terpene alcohol content is required.

Product Specifications:
- Specific gravity at 15.6/15.6C: 0.938-0.946
- Total terpene alcohols, %: 85 min
- Moisture, %: 0.5 max
- Distillation range, C:
  - 5%: 209 min
  - 95%: 225 max

Outstanding Characteristics:

Clear, pale yellow; high terpene alcohol content; piney odor; high solvent activity; excellent wetting, penetrating, and dispersing properties; high bactericidal activity when properly formulated.

Miscible with most common organic solvents. Trace solubility in water.

YARMOR 320W Pine Oil
A General-Purpose-Grade Pine Oil

YARMOR 302W pine oil is a clear, pale-yellow to near-water-white oily liquid with a distinct pinelike odor. Derived from terpene oils of pinewood origin, it is a blend of related compounds, predominantly terpene alcohols with minor amounts of terpene hydrocarbons. It is suitable for all uses where a general-purpose grade of pine oil is required.

Product Specifications:
- Specific gravity at 15.6/15.6C: 0.918-0.928
- Total alcohols, %, min: 70
- Moisture, %, max: 0.6
- Distillation Range, C:
  - 5%: 190 min
  - 95%: 225 max

Outstanding Characteristics:

Clear, pale color, piney odor; high solvent activity; excellent wetting, penetrating, and dispersing properties; high bactericidal activity when properly formulated; uniform.

Miscible with most common organic solvents. Trace solubility in water.

NL CHEMICALS, INC.: LINAQUA:

Water-thinnable linseed oil for addition to latex paints, printing inks, in-plant tinting colors, plywood marking paints, rust preventive shop coats and artists' colors.

Percent Solids Weight: 85
Solvent: G4
Viscosity: G-H approx.: G-K

NL CHEMICALS, INC.: Linseed Oils:

Raw Linseed Oil:

Raw:
Paints, etc. Raw linseed oil is well settled and double filtered. It conforms to Federal Specification TT-L-215 and ASTM D 234-82.

Viscosity: G-H: A
Acid Value, Max.: 4
Iodine Value: 177-195

Boiled Linseed Oils:
Boiled: Dries in 12-16 hours. Conforms to Federal Specification TT-L-190 and ASTM Specification D260-80 Type I.

Double Boiled: Paints containing blacks and other dry-retardant pigments. Dries in 8 hours. Meets ASTM D260-80 Type II and AASHTD M233-79.

Boiled:
Viscosity: G-H: A
Acid Value:, Max.: 6
Iodine Value: 170-190
Double Boiled:
Viscosity: G-H: B
Acid Value, Max.: 7
Iodine Value: 165-180

Semi-Oxidized Linseed Varnish Oil:

Aged:
A slightly oxidized, heat break-free oil for litho varnishes, patent leather and linoleum. Fast kettle polymerization. (TS-Aged Linseed Oil)

Viscosity: G-H: D
Acid Value, Max.: 2-5
Iodine Value: 150-170

Oxidized Linseed Oils:

Diamond "K":
For mixing with vehicles for both interior and exterior paints. Diamond "K" dries in 15-20 hours. Improves wetting, flow and leveling.

Miscible Diamond "K":
House paints, vehicle blending. Mineral spirits tolerance of 100 to 1. Excellent wetting, flow and leveling.

NL CHEMICALS, INC.: Linseed Oils(Continued):

Linseed Varnish Oils(Continued):

DIAMOND "K":
Viscosity: G-H: Z2-Z3
Acid Value, Max.: 5-9
Iodine Value: 115-130

Miscible DIAMOND "K":
Viscosity: G-H: Z2-Z3
Acid Value, Max.: 4-11
Iodine Value: 120-135

Linseed Varnish Oils:

SUPERIOR, Neutral KVO:
For light-colored varnishes and alkyds, white enamels, etc. Both are alkali refined and offer excellent heat bleaching. Conforms to TT-L-1155.

SUPERIOR "C":
This is SUPERIOR plus catalyst. Polymerizes to Z5 viscosity in 65% of the time required for SUPERIOR.

K.V.O.:
A general purpose alkali-refined oil for alkyds, varnishes, enamels, exterior house paints and printing inks.

NON-BREAK:
A mechanically-refined, heat break-free oil for alkyds, varnishes, enamels, etc. when color is of secondary importance.

SUPERIOR:
Viscosity: G-H: A
Acid Value, Max.: 0.3
Iodine Value: 177-190

SUPERIOR "C":
Viscosity: G-H: A
Acid Value, Max.: 0.8
Iodine Value: 170-190

K.V.O.:
Viscosity: G-H: A
Acid Value, Max.: 2-3
Iodine Value: 170-190

Neutral K.V.O.:
Viscosity: G-H: A
Acid Value, Max.: 0.3
Iodine Value: 170-190

NON-BREAK:
Viscosity: G-H: A
Acid Value, Max.: 4
Iodine Value: 170-190

NL CHEMICALS, INC.: Linseed Oils(Continued):

Modified Linseed Oils:

KELLIN:
Chemically-treated linseed. Polymerizes rapidly, dries fast, excellent water resistance. For printing inks. Available also at S, Z viscosity.

ESSKOL:
Redwood finishes, varnishes and enamels. ESSKOL bodies faster than KELLIN.

CYKELIN & CYKELIN 70:
Varnishes, enamels, aluminum paints, trim and trellis, cold cut varnishes, reinforcement oil. Dicyclopentadiene copolymers of linseed oil that produce vehicles of quick hard dry and excellent water and alkali resistance. CYKELIN 70 at 70% solids in mineral spirits.

KELLIN U-V:
Viscosity: G-H: U-W
Acid Value, Max.: 8-14
Iodine Value: 120-145

ESSKOL Y-Z:
Viscosity: G-H: Y-Z
Acid Value, Max.: 12-17
Iodine Value: 135-155

CYKELIN:
Viscosity: G-H: Z3-Z4
Acid Value, Max.: 4
Iodine Value: 150-162

CYKELIN 70:
Viscosity: G-H: S-V
Acid Value, Max.: 1-4
Iodine Value: 100-115

Linseed Grinding Oils:

ACID REFINED:
For whites, light tints, etc. Oil is acid refined for excellent wetting properties.

A-R VARNISH and GRINDING:
For white paints, light-colored varnishes, this oil is alkali refined. Ideal for reactive pigments.

NL CHEMICALS, INC.: Linseed Oils(Continued):

Linseed Grinding Oils(Continued):

SUPERFLO Grinding:
For whites and light tints, oil is mechanically refined. Ideal for reactive pigments.

No.10, No.20 and No.30 Grinding:
For white and light tint house paints. Contains polymerized oil for excellent wetting flow, leveling. Both No. 10 and No. 20 Grinding meet linseed portion of vehicle composition of Federal Specification TT-P-103b, TT-P-104b.

ACID REFINED:
Viscosity: G-H: A_
Acid Value, Max.: 4-6
Iodine Value: 170-190

A-R Varnish & Grinding:
Viscosity: G-H: A_
Acid Value, Max.: 2-3
Iodine Value: 170-190

SUPERFLO Grinding:
Viscosity: G-H: A_
Acid Value, Max.: 1-4
Iodine Value: 170-190

No. 10 Grinding:
Viscosity: G-H
Acid Value, Max.: 4-6
Iodine Value: 150-170

No. 20 Grinding:
Viscosity: G-H
Acid Value, Max.: 4-6
Iodine Value: 150-170

No. 30 Grinding:
Viscosity: D-E
Acid Value, Max.: 3-5
Iodine Value: 150-170

Polymerized Linseed Oils:

S-70 OKO, M-2 1/2 OKO, M-7 1/2 OKO, M-25 OKO and M-37 OKO:
These are the highest quality polymerized oils available. They are produced to remove all traces of decomposed matter. Their low acid, light color, minimum after yellowing--suggest uses in paints, varnishes, enamels, flats, aluminum paints, printing inks, lithographic varnishes, and brake linings. OKO's are available to meet all requirements of Federal Specification TT-L-201, Type II.

NL CHEMICALS, INC.: Linseed Oils(Continued):

Polymerized Linseed Oils(Continued):

T-1215-X-Y, T-1215 Y-Z, T-1215 Z-Z1, T-1215 Z2-Z4 and T-1215 Z4-Z6:
Exterior house paints, varnishes, enamels, flats. T-1215's are available to meet all requirements of Federal Specification TT-L-201, Type I.

Special M-25:
Calcicoaters, varnishes, puffing agent for house paints. Adaptable to hot and cold liming.

S-70 OKO:
Viscosity: G-H: Z2-Z3
Acid Value, Max.: 3
Iodine Value: 115-130

M-2 1/2 OKO:
Viscosity: G-H: Z4-Z5
Acid Value, Max.: 3
Iodine Value: 115-130

M-7 1/2 OKO:
Viscosity: G-H: Z6+
Acid Value, Max.: 3
Iodine Value: 115-130

M-25 OKO:
Viscosity: G-H: Z8
Acid Value, Max.: 3
Iodine Value: 115-130

M-37 OKO:
Viscosity: G-H: Z9
Acid Value, Max.: 3
Iodine Value: 115-130

T-1215 X-Y:
Viscosity: G-H: X-Y
Acid Value, Max.: 6-9
Iodine Value: 122-132

T-1215 Y-Z:
Viscosity: G-H: Y-Z
Acid Value, Max.: 6-9
Iodine Value: 118-128

T-1215 Z-Z1:
Viscosity: G-H: Z-Z1
Acid Value, Max.: 6-9
Iodine Value: 116-126

NL CHEMICALS, INC.: Linseed Oils(Continued):

Polymerized Linseed Oils(Continued):

T-1215 Z2-Z4:
Viscosity: G-H: Z2-Z4
Acid Value, Max.: 6-9
Iodine Value: 115-125

T-1215 Z4-Z6:
Viscosity: G-H: Z4-Z6
Acid Value, Max.: 6-9
Iodine Value: 115-125

SPECIAL M-25:
Viscosity: Z8
Acid Value, Max.: 9-11
Iodine Value: 108-118

Miscellaneous Linseed Oil Products:

Anti-Spalling Compound CB-5855L:
A uniformly blended mixture of mineral spirits (ASTM D-235) and Boiled Linseed Oil (ASTM D-260). Effectively prevents pitting and scaling (spalling) of concrete surfaces. Meets AASHTD M 233-79.

Washed Recovered Linseed:
Oil recovered during process of refining linseed oil.

OKO Fatty Acids:
Acids recovered during process of linseed oil polymerization.

Linseed Fatty Acids:
Bleached twitchellized acids.

Anti-Spalling Compound:
Viscosity: G-H: A-5
Acid Value, Max.: 6
Iodine Value: 170-190
Washed Recovered Linseed:
Viscosity: G-H: B
Acid Value, Max.: 95-125
Iodine Value: 150-165
OKO Fatty Acids:
Viscosity: G-H: A
Acid Value, Max.: 140-170
Iodine Value: 135-150
Linseed Fatty Acids:
Viscosity: G-H: A_
Acid Value, Max.: 9
Iodine Value: 180-195

NL CHEMICALS, INC.: Marine Oils/Miscellaneous Modified Oils:

KELMER Polymerized Marine Oil:

KELMER Z2-Z3:
Extensively used in aluminum coatings, house paints, gloss and semigloss enamels, caulking compounds, mill whites, varnish, checking oil, nonsagging additives.

Percent Solids: 100
Viscosity: G-H: Z2-Z3
Acid Value, Max.: 10
Iodine Value: 100-105

KELLOX Oxidized Marine Oils:

Z3, Z6-Z7 and 375:
Maintenance finishes, nitrocellulose plasticizers, barn, fence and roof paints, freight car finishes, cold cut enamels, rustproofing liquids, antifouling paint, leveling additive.

Z3:
Percent Solids: 99
Viscosity: G-H: Z4+-
Acid Value, Max.: 12
Iodine Value: 85-115

Z6-Z7:
Percent Solids: 100
Viscosity: G-H: Z6+-
Acid Value, Max.: 12
Iodine Value: 80-100

375:
Percent Solids: 72-75
Solvent: M
Viscosity: G-H: Z4+
Acid Value, Max.: 13
Iodine Value: 65-90

Commercial Deck Oil:
Slashing oil to inhibit rust accumulation.

Viscosity: G-H: A1-A2
Acid Value, Max.: 12

Miscellaneous Modified Oils:

ADMEROL 75-M-70:
Flexible long oil oxidizing copolymer supplied at 70% solids for TT-V-81F and TT-V-121G.

ADMEROL 351-M-70:
Water, alkali and abrasion-resistant copolymer resin for interior primers and floor finishes. Low viscosity, fast bodying oil for use in house paints.

NL CHEMICALS, INC.: Soybean Oils/Soybean Lecithin:

Soybean Varnish Oils:

SUPERIOR:
Alkyds, varnishes, enamels for whites and light tints. Alkali refined. Excellent heat bleach and color. Meets TT-S-600. Type I and ASTM D 1462.

NON-BREAK:
Alkyds, varnishes, enamels. Mechanically refined. Heat and acid break-free.

DEGUMMED:
Economy oil for alkyds, varnishes and enamels. Crude soy with phosphatides removed. Meets Federal Specification TT-S-600. Type II and ASTM D 124.

SUPERIOR:
Viscosity: G-H: A-
Acid Value, Max.: 0.3 Max.
Iodine Value: 126-136

NON-BREAK:
Viscosity: G-H: A-
Acid Value, Max.: 0.5 Max.
Iodine Value: 126-136

Oxidized Soybean Oils:

AGED:
Lightly oxidized soybean oil, used as a plasticizer in caulks, putties and artists paints.

DIAMOND "K":
Paints, enamels, putty. Improves wetting, flow and leveling

SPECIAL T-BLOWN Z2-Z4, Z5-Z6, and Z7-Z8:
Caulking compounds, plasticizers. Contains no drier. Good package stability.

High Solubility Blown Q and Z1-Z2:
Paints, enamels, caulking compounds, lacquers. Mineral spirits tolerance 20 to 1, minimum.

AGED:
Viscosity: G-H: D-E
Acid Value, Max.: 5
Iodine Value: 119-125

NL CHEMICALS, INC.: Soybean Oils/Soybean Lecithin(Continued):

Oxidized Soybean Oils(Continued):

DIAMOND "K":
Viscosity: G-H: Z2-Z4
Acid Value, Max.: 4-9
Iodine Value: 75-95

SPECIAL T-BLOWN Z2-Z4:
Viscosity: G-H: Z2-Z4
Acid Value, Max.: 6-12
Iodine Value: 60-85

SPECIAL T-BLOWN Z5-Z6:
Viscosity: G-H: Z5-Z6
Acid Value, Max.: 7-14
Iodine Value: 60-80

SPECIAL T-BLOWN Z7-Z8:
Viscosity: G-H: Z8
Acid Value, Max.: 7-14
Iodine Value: 60-80

HIGH SOLUBILITY BLOWN Q:
Viscosity: G-H: P-R
Acid Value, Max.: 5

HIGH SOLUBILITY BLOWN Z1-Z2:
Viscosity: G-H: Z1-Z2
Acid Value, Max.: 3-8
Iodine Value: 70-100

Polymerized Soybean Oil:

Heat Bodied Soy Z2:
Paints and caulking compounds. High quality vacuum polymerized soybean oils.

Viscosity: G-H: Z1-Z3
Acid Value, Max.: 4-8
Iodine Value: 90-100

Modified Soybean Oils:

Low Acid Value DRISOY G-I, DRISOY Y-Z and Z2-Z3:

Varnishes, enamels, exterior house paints, printing inks, linoleum, chemically-treated soybean oils. Dry like linseed. Recommended for general replacement of linseed oil.

NL CHEMICALS, INC.: Soybean Oils/Soybean Lecithin(Continued):

Modified Soybean Oils(Continued):

Low Acid Value DRISOY G-1:
Viscosity: G-H: H-J
Acid Value, Max.: 6-8
Iodine Value: 110-130
DRISOY Y-Z:
Viscosity: G-H: Y-Z
Acid Value, Max.: 4-8
Iodine Value: 95-110
DRISOY Z2-Z3:
Viscosity: G-H: Z2-Z3
Acid Value, Max.: 4-8
Iodine Value: 90-105

Soybean Oil Plasticizers:

SOY SOLINOX T and SOY SOLINOX Z:
Lacquers, coated fabrics, artificial leather. Chemically-treated soybean oils. SOY SOLINOX T has hydroxyl value of 109-115. SOY SOLINOX Z, 109-115.

SOY SOLINOX T:
Viscosity, G-H: T-U
Acid Value, Max.: 8-12
Iodine Value: 75-90
SOY SOLINOX Z:
Viscosity, G-H: Z-Z1
Acid Value, Max.: 12-16
Iodine Value: 65-80

Soybean Lecithin:

KELECIN F, FB and FD:
Emulsifying and dispersing agents for protective coatings, mastics, animal feeds, automotive specialty chemicals, textiles, rubber processing, plastics, cosmetics and food products.

KELECIN F:
Viscosity: G-H: Z6
Acid Value, Max.: 32
KELECIN FB:
Viscosity: G-H: Z6
Acid Value, Max.: 32
KELECIN FD:
Viscosity: G-H: Z6
Acid Value, Max.: 32

KELECIN 1081:
Water-dispersible lecithin for water emulsion paints.
Viscosity: G-H: Z2
Acid Value, Max.: 22-31
Iodine Value: 95

SONNEBORN DIVISION: SONNEBORN White Oils:

Typical properties of standard Sonneborn white oils.

Product:

USP mineral oil

KAYDOL:
Viscosity (cst @ 40C): 64-70
(SSU @ 100F): 340-355
SG 25C 25C: 0.869-.885

ORZOL:
Viscosity (cst @ 40C): 61-64
(SSU @ 100F): 320-330
SG 25C 25C: 0.869-.885

BRITOL:
Viscosity (cst @ 40C): 57-60
(SSU @ 100F): 300-310
SG 25C 25C: 0.869-.885

GLORIA:
Viscosity (cst @ 40C): 39-42
(SSU @ 100F): 200-210
SG 25C 25C: 0.859-.880

PROTOL:
Viscosity (cst @ 40C): 35-37
(SSU @ 100F): 180-190
SG 25C 25C: 0.859-.875

Light mineral oil, NF

RUDOL:
Viscosity (cst @ 40C): 28-30
(SSU @ 100F): 145-155
SG 25C 25C: 0.852-.870

ERVOL:
Viscosity (cst @ 40C): 24-26
(SSU @ 100F): 125-135
SG 25C 25C: 0.849-.865

BENOL:
Viscosity (cst @ 40C): 18-20
(SSU @ 100F): 95-105
SG 25C 25C: 0.839-.855

BLANDOL:
Viscosity (cst @ 40C): 14-17
(SSU @ 100F): 80-90
SG 25C 25C: 0.839-.855

SONNEBORN DIVISION: SONNEBORN White Oils(Continued):

Light mineral oil, NF(Continued):

Product:

CARNATION:
Viscosity (cst @ 40C): 11-14
(SSU @ 100F): 65-75
S G 25C 25C: 0.837-.853

KLEAROL:
Viscosity (cst @ 40C): 7-10
(SSU @ 100F): 50-60
S G 25C 25C: 0.822-.833

Typical properties of standard SONNEBORN white oils, technical grade

SEMTOL 350:
Viscosity (cst @ 40C): 64-90
(SSU @ 100F): 340-425
SG 25C 25C: 0.850-.890

SEMTOL 100:
Viscosity (cst @ 40C): 18-20
(SSU @ 100F): 95-105
SG 25C 25C: 0.839-.855

SEMTOL 85:
Viscosity (cst @ 40C): 14-17
(SSU @ 100F): 80-90
SG 25C 25C: 0.839-.855

SEMTOL 70:
Viscosity (cst @ 40C): 11-14
(SSU @ 100F): 65-75
SG 25C 25C: 0.837-.853

SEMTOL 40:
Viscosity (cst @ 40C): 4-5
Viscosity (SSU @ 100F): 40-43
SG 25C 25C: 0.804-.820

THIBAUT AND WALKER: Processed Oils:

0-151:
NV: 100
Solvent: None
Viscosity (G-H): W-Y
Type Oil: Linseed

0-152:
NV: 100
Solvent: None
Viscosity (G-H): Q-R
Type Oil: Linseed

0-154:
NV: 100
Solvent: None
Viscosity (G-H): Z2-Z4
Type Oil: Linseed

0-159:
NV: 100
Solvent: None
Viscosity (G-H): G-J
Type Oil: Linseed

Description: Pale, kettle bodied linseed oils suitable for blending with other oils, bodied or unbodied and resins; for use in the manufacture of exterior house paints, varnishes, enamels, flats, printing inks and lithographic varnishes. T & W's polymerized linseed oils are available to meet all requirements of TT-L-201A, Type 1.

Section 8
Preservatives, Bactericides, Fungicides

ACETO CORP.: Biocides:

Tributyltin Fluoride
(n-C4H9)3SnF

Technical Data

Appearance	White, flowing powder
Tin Content	37.5%+-0.7
Melting Point	240-250C.
Weight per Liter	Approx. 0.6 kg.
Flash Point	Approx. 20C.

Compatability
Soluble in chlorinated hydrocarbons.
Partly soluble in ketones, ethers, esters, aliphatic & aromatic solvents.
Insoluble in water and alcohol.

Application
Highly effective biocide against micro-organisms, bacteria & fungi.
Toxicant for anti-fouling paints (underwater paints).

Tributyltin Oxide
(C4H9)3-Sn-O-Sn-(C4H9)3

Physical Properties:

Physical form	Clear, light yellowish liquid
Color (Gardner)	Max. 2
Specific Gravity d 20/4	1.17+-0.03
Refractive Index n 20/D	1.485+-0.01
Tin Content	Min. 37.7%
Volatile Matter	Max. 2.0%
Flash Point	>100C.

Compatibility:

Solubility	<20 ppm in water soluble in organic solvents

Application:
Highly effective Biocide against micro-organisms, bacteria and fungi. Toxicant for anti-fouling paints (underwater paints).

COSAN CHEMICAL CORP.: Preservatives:

COSAN PMA-30
E.P.A. Reg. No. 8489-2

COSAN PMA-30 is a solubilized phenylmercuric acetate formulation which has earned acceptance as the highly effective standard in the industry for protecting liquid emulsion systems and films. COSAN PMA-30 is water soluble and compatible with styrene-butadiene, acrylic, vinyl-acrylic, polyvinyl acetate, oil and alkyd modified systems. COSAN PMA-30 will not adversely affect freeze-thaw stability, whites and light tints.

Specifications:
Phenylmercuric acetate, minimum: 30%
Metal content, minimum: Mercury, 18.0%
Specific gravity: 1.30-1.32

COSAN PMA 100

COSAN PMA 100 is a 100% active phenylmercuric acetate based preservative in powder form. It is used to protect water-based coatings against spoilage during storage in the container and also to prevent mildew growth on the dried film. The product is supplied in water-soluble bags, available in a variety of sizes. These bags are added directly into the paint, thereby preventing worker contact with the chemical itself and also eliminating the need for drum disposal.

Typical Properties:
Active Ingredient:
Phenylmercuric Acetate: 100%
Typical Mercury Content: 59.5-60.0%
Appearance: Free-flowing white powder
Formula: C6H5HgO2CCH3
Bulking Value: 0.05 gal/lb.
Melting Point: 149C
Particle Size: 95% 75 microns

COSAN P
E.P.A. Reg. No. 476-2040-8489

Folpet N-(Trichloromethyl) Thiophthalimide
(U.S. Patent No. 2,553,770)

COSAN P is an off-white, organic compound of relatively high activity against a broad spectrum of fungi. COSAN P is recommended as a mildew inhibitor in oil based paints. COSAN's low order of low toxicity makes it well suited for use in food plants, dairies and breweries.

COSAN P is an excellent microbiocide for use in general vinyl applications such as wall coverings, upholstery, awnings and similar applications. It is heat stable under most conditions.

COSAN CHEMICAL CORP.: Preservatives(Continued):

COSAN 91
E.P.A. Reg. No. 8489-24

COSAN 91 is a liquid organic compound which is water soluble and recommended for use as an industrial preservative against bacterial deterioration in water-based systems such as resin emulsions, adhesives, dispersed colors, and ready-mixed joint cements.

Specifications:
Active Ingredient: 2-[(Hydroxymethyl)amino]ethanol: 100
Physical Form: Clear Amber Liquid
Specific Gravity (80F): 1.144-1.153

COSAN 145
E.P.A. Reg. No. 8489-26

COSAN 145 is a liquid organic preservative recommended for use in resin emulsions, latex paint, adhesives, dispersed colors, pigment slurries, and ready-mix joint cements.
COSAN 145 contains 50% active ingredient and, when used at recommended levels, will provide antimicrobial activity to a water-based system.

COSAN 145 offers the following advantages:
* non-yellowing
* is a mobile 50% aqueous solution
* does not discolor in conjunction with organic mildewcides
* is not temperature sensitive and can be added at any point during manufacture
* will not adversely affect physical properties of the final formulation in the wet or dry state
* has a high flash point and a low freezing point which makes storage and handling easy and trouble-free

Specifications:
Active Ingredient:
Methanol [[[2-(dihydro-5-methyl-3(2H)-oxazolyl)-1-methylethoxy]methoxy]methoxy]: 50%
Physical Appearance: clear liquid

COSAN 158
E.P.A. Reg. No. 8489-21
COSAN 158 is a yellow liquid of low viscosity which has been designed to insure preservation against in-can bacterial attack as well as providing the necessary antifungal ingredient for mildew resistance of the applied coating.
Specifications:
Active Content:
Tributyltin benzoate: 15%
Alkyd (derived from fatty acids of coconut oil) Amine hydrochlorides: 21.2%

INTERNATIONAL SOURCING INC.: Parabens:

Parabens, Esters of Parahydroxybenzoic Acid, have been widely used in the preservation of cosmetics, pharmaceuticals, personal care products, and foods for over fifty years.

Physical Properties

ISI Parabens at recommended use levels are:

1. Colorless, odorless, tasteless.
2. Non-toxic, chemically stable compounds.
3. Most effective at pH 3 to 8.
4. Readily soluble in most systems.
5. Non-irritating and non-sensitizing to the skin.
6. Economical and easy to employ.
7. Additive in effect when used in combination.
8. Non-hygroscopic, non-volatile, and safe in storage.

Chemical Properties

Methyl p-Hydroxybenzoate	(Methyl Paraben)
Ethyl p-Hydroxybenzoate	(Ethyl Paraben)
Propyl p-Hydroxybenzoate	(Propyl Paraben)
Butyl p-Hydroxybenzoate	(Butyl Paraben)

Specifications (Technical)

Methyl
Assay: 99.0% min.
Appearance: Ivory Powder
Melting Range (C): 124-128

Ethyl:
Assay: 99.0% min.
Appearance: Ivory Powder
Melting Range (C): 114-118

Propyl:
Assay: 99.0% min.
Appearance: Ivory Powder
Melting Range (C): 93-98

Butyl:
Assay: 99.0% min.
Appearance: Ivory Powder
Melting Range (C): 67-72

INTERSTAB CHEMICALS: Microbiostatic Agents: Mercurials:

Mercurial microbiocides are only allowed for use in water based paints and water based products where they are recognized as the most cost effective preservatives and fungicides. Interstab offers a complete line of mercurial microbiocides.

INTERCIDE 60:
INTERCIDE 60 is phenyl mercury acetate powder containing 60% mercury as metal, in water soluble, pre-weighed bags. These bags are convenient to use and offer the additional benefits of accuracy and economy.

INTERCIDE PMA-18%:
INTERCIDE PMA-18 is solubilized PMA 60 containing 18% mercury as metal (30% phenyl mercuric acetate). Like INTERCIDE 60, INTERCIDE PMA 18 is an effective in-can preservative and fungicide.

INTERCIDE PMO-11%:
INTERCIDE PMO 11% is a solution of phenyl mercuric oleate containing 11% mercury as metal.

INTERCIDE MDS:
INTERCIDE MDS is an effective in-can preservative and fungicide containing 10% mercury as di-(phenylmercuric)-dodecyl succinate. Paints containing INTERCIDE MDS are characterized by their excellent mildew resistance due to the low rate of leaching from dried films.

Physical Properties:

INTERCIDE 60:
Viscosity (G/H-77F.): N/A
% Metal: 59.5-60.0

INTERCIDE PMA-18%:
Viscosity (G/H-77F.): A
% Metal: 18.0-18.2

INTERCIDE PMO-11%:
Viscosity (G/H-77F.): A
% Metal: 11.0-11.2

INTERCIDE MDS:
Viscosity (G/H-77F.): A
% Metal: 10.0

INTERSTAB CHEMICALS: Microbiostatic Agents: Non Mercurials:

INTERCIDE 340-A:

INTERCIDE 340-A is a water-dispersible organotin biocide, effective against a wide range of microorganisms. It is designed for use as an in-can preservative and fungicide in water based coatings. It is also an effective ingredient in wood preservatives and wood preserving stains.

INTERCIDE 340-A will not adversely affect storage stability of the formulation. It will not discolor paint films or affect other coating properties and its use will not result in galvanic action when applied on metallic surfaces.

INTERCIDE T-O:

INTERCIDE T-O is a liquid organotin compound effective in the control of a broad variety of fungi and bacteria. Chemically Bis (tri-n-butyltin) oxide, it is formulated for use in non-aqueous coatings.

INTERCIDE N-628:

INTERCIDE N-628 is a complex trialkyltin compound. It is a semi-fluid paste which is effective in controlling fungi in vinyl sheet, pipe, tubing, and flooring.

INTERCIDE FC:

INTERCIDE FC is a solution of (tri-n-butyltin) oxide designed for use as a microbiocide in flexible polyvinylchloride, vinyl tubing and flooring.

INTERCIDE TJP:

INTERCIDE TJP is an organotin compound designed for controlling gram positive bacteria and fungi in tile grouts and wet mixed tape joint compounds. INTERCIDE TJP is a fine white powder which will not adversely affect the other properties of dry powdered compounds in the dry mix or wet state (e.g., does not cause staining over nail heads).

INTERCIDE Tributyltin Neodecanoate:

INTERCIDE Tributyltin Neodecanoate is an effective microbiostat for use in water cooling towers.

INTERCIDE TMP:

INTERCIDE TMP is a powdered non-metallic organic fungicide designed for use in non-aqueous coatings and caulks. INTERCIDE TMP is air milled to an average particle size of 10 microns (to aid dispersion). The active ingredient is N-trichloromethylthio phthalimide which is the most cost effective organic fungicide currently on the market. INTERCIDE TMP has a low order of toxicity and is highly effective against a wide variety of microorganisms. Water should not be included in formulations containing TMP to avoid possible hydrolysis.

INTERSTAB CHEMICALS: Microbiostatic Agents: Non Mercurials (Continued):

INTERCIDE Copper Naphthenate:
INTERCIDE Copper is the copper soap of refined naphthenic acid in mineral spirits. It is readily soluble in low cost petroleum solvents, oils, and alkyd resins. Copper Naphthenate is effective in preserving wood, textiles and other fibrous materials.

INTERCIDE Copper 10%:
INTERCIDE Copper 10% contains the copper soaps of naphthenic and 2-ethyl hexoic acids in O-dichlorobenzene. It is effective in preventing fungal attack of fibrous materials such as jute, rope, and cloth.

INTERCIDE Copper Octoate:
INTERCIDE Copper Octoate is the copper soap of 2-ethylhexoic acid in mineral spirits. Like the naphthenates, it is readily soluble in low cost petroleum solvents, oils, and alkyd resins. Copper octoates are the new generation of copper soaps effective in preserving wood, textiles, and other fibrous materials.

INTERCIDE QL:
INTERCIDE QL contains 1.8% copper as copper 8-quinolinolate. It is designed for use in solvent applications to protect fibrous substrates (e.g., textiles, cord, etc.) from fungal deterioration.

INTERCIDE Zinc Naphthenate 8%:
INTERCIDE Zinc Naphthenate 8% is the zinc soap of refined naphthenic acid dissolved in mineral spirits. It is readily soluble in low cost petroleum solvents, oils and alkyds and is effective in preserving wood, textiles, and other fibrous materials. Zinc Naphthenate is a transparent pale yellow liquid, and as such does not discolor the treated substrate.

INTERCIDE Zinc Octoate:
INTERCIDE Zinc Octoate is the zinc soap of 2-ethyl-hexoic acid in mineral spirits. INTERCIDE zinc octoates are readily soluble in low cost petroleum solvents, oils, and alkyd resins. Zinc octoates are the new generation of zinc soaps effective in preserving wood, textiles, and fibrous materials. Zinc octoate is even paler in color than the naphthenate, and as such does not discolor the treated substrate.

INTERSTAB CHEMICALS: Microbiostatic Agents: Non Mercurials (Continued):

Physical Properties:

Product:

INTERCIDE 340A:
Viscosity (G/H-77F): A
% Metal: N/A
INTERCIDE T-O:
Viscosity (G/H-77F): A
% Metal: 38.0% Min.
INTERCIDE FC:
Viscosity (G/H-77F): A
% Metal: N/A
INTERCIDE TJP:
Viscosity (G/H-77F): N/A
% Metal: N/A
INTERCIDE TMP:
Viscosity (G/H-77F): N/A
% Metal: N/A
Copper 10%:
Viscosity (G/H-77F): T
% Metal: 9.9-10.1
INTERCIDE QL:
Viscosity (G/H-77F): A
% Metal: 1.7-1.9
Zinc Naph. 8%:
Viscosity (G/H-77F): G
% Metal: 8.0 Min.
Copper Naph. 6%:
Viscosity (G/H-77F): E
% Metal: 6.0 Min.
Copper Naph. 8%:
Viscosity (G/H-77F): G
% Metal: 7.9-8.1
Copper Octoate 6%:
Viscosity (G/H-77F): A
% Metal: 5.9-6.1
Copper Octoate 8%:
Viscosity (G/H-77F): D
% Metal: 7.9-8.1
Copper Octoate 10%:
Viscosity (G/H-77F): K
% Metal: 9.9-10.1

M&T CHEMICALS, INC.: Additives:

BIOMET TBTF
(tributyltin fluoride)
Antifoulant for Marine Paints

BIOMET TBTF is an organotin compound for use in shipbottom paints to control the attachment of marine fouling organisns. BIOMET TBTF has been used successfully in a variety of coating formulations and in combinations with other additives to provide broad spectrum control of marine fouling organisms, thereby contributing to speed and operating economy of both commercial ships and pleasure craft.

The Advantages of BIOMET TBTF Are:

* Chemical compatibility (non reactive) with coating ingredients
* Allows development of formulations with a full range of colors
* Noncorrosive to aluminum and steel
* Low water solubility
* Broad spectrum control of marine life including Barnacles--Algae--Tube Worms--Hydroids

Typical Properties:
Physical form: White powder
Active ingredient: 95% min.
Tin content: 38.0%
Specific Gravity: 1.27
Solubility: Relatively insoluble in water and organic solvents

BIOMET TBTO
Brand of Bis (tri-n-butyltin) Oxide

BIOMET TBTO is a slightly yellow, clear liquid containing over 95% active ingredient. BIOMET TBTO is representative of a new class of compounds which exhibits control of most fungi, bacteria, and marine organisms.

Typical Properties:
Physical form: clear, straw colored liquid
Molecular weight: 596.0
% Assay as bis(tri-n-butyltin oxide): 96
% tin content: 38.8
Specific Gravity (25C/25C): 1.17
Lbs/Gal (25C): 9.76
Boiling Point, C (at 2mm): 180
Freezing Point, C: below -45
Flash Point, C (T.C.C.): above 100
Viscosity, centistokes (25C): 4.8
Surface tension, dynes/cm (25C): 30.6

M&T CHEMICALS, INC.: Additives (Continued):

BIOMET TBTO
Antifoulant
Bis(tributyltin) Oxide
For Shipbottom Paints

BIOMET TBTO Antifoulant is an organotin compound, which will provide antifouling properties in marine paint formulations. Originally used solely in pleasure craft systems, BIOMET TBTO is now receiving wide acceptance in heavy marine systems as an adjunct to other antifouling ingredients.

The advantages of BIOMET TBTO Antifoulant are:

* Miscible in a wide range of resin and solvent systems.
* Compatibility with coating ingredients.
* Formulations with full range of colors.
* Non-corrosive to aluminum and steel.
* Low water solubility.
* Broad spectrum control of marine life including barnacles, algae, tubeworms, and hydroids.

Typical Properties:

Physical form: clear, straw colored liquid
Molecular weight: 596.0
% Assay as bis(tri-n-butyltin oxide): 96
% Tin content: 30.0

BIOMET TBTO Antimicrobial
Bis (tri-n-butyltin) Oxide

For Use as a Wood Preservative

BIOMET TBTO (bis[tri-n-butyltin]oxide) is an organotin compound for use in manufacture of wood preservative solutions designed for brush, dip and pressure treatment of wood. BIOMET TBTO is effective against a wide spectrum of microorganisms and reduces attack of termites and wood boring beetles. It is an effective alternative to pentachlorophenol, creosote and inorganic arsenicals in the treatment of lumber and plywood for above ground use. TBTO is listed by the American Wood Preservers Association as an above ground wood treatment.

The advantages of BIOMET TBTO Antimicrobial are:

* Non-staining colorless liquid
* Will not cause swelling or distortion of wood
* Non-corrosive to aluminum and steel
* Can be applied from organic solvent or aqueous systems

Recommended Uses for BIOMET TBTO Wood Preservatives:

* Treatment of Construction Lumber and Plywood
* Treatment of Mill Work
* Treatment of Siding and Outdoor Wood Structures

M&T CHEMICALS, INC.: Additives(Continued):

BIOMET 300-Series
Polymeric Marine Antifoulants
BIOMET 300, 302, 303, and 304

BIOMET 300-Series Marine Antifoulants are film-forming polymers which have a biologically active tributyltin group chemically bound to the backbone of an acrylic binder. BIOMET 300-Series polymers are supplied at 50% solids in a high aromatic content organic solvent convenient for use in the manufacture of marine antifouling paints.

Long-term control of marine fouling is achieved through slow release of the biologically active organotin group from the antifouling coating via hydrolysis by seawater.

BIOMET 300-Series polymers are readily compounded into marine paints with pigments and other additives to meet a broad range of performance and erosion rate requirements for commercial ships and pleasure craft.

The Advantages of BIOMET 300-Series Polymers Are:

* Film-former and toxicant combined in one product.
* Controlled release of the organotin antifoulant.
* Allow formulation of slow, moderate and fast eroding coatings whose lifetine is directly proportional to film thickness.
* Chemically compatible (non-reactive) with coating ingredients.
* Allows formulations with full range of colors.
* Non-corrosive to aluminum and steel.
* Blends easily with pigments and other paint ingredients.
* Broad spectrum control of marine life including Barnacles--Algae--Tube Worms--Hydroids.

Composition:

Solution of Polyalkylmethacrylate polymer containing tributyltin methacrylate in a high aromtic content organic solvent of boiling range 310-360F.

BIOMET 430/45
Antifoulant for Marine Paints

BIOMET 430/45 is a stable dispersion of tri-n-butyltin fluoride in organic solvent which offers greater process flexibility and shorter milling times in paint manufacture than solid TBTF. This translates into lower manufacturing costs.

The Advantages of BIOMET 430/45 Are:

1. Easily dispersible with Cowles or other high-speed mixers.
2. Gives uniform fine paint grind.
3. Non-dusting, easy to handle.
4. Compatible in conventional and organotin polymer based paint systems.

NATIONAL MINERAL PRODUCTS: Dolomite Fillers & Extenders:

Dolomite Typical Chemical Analysis:

- Moisture: .1%
- Loss on Ignition: 47.10%
- SiO_2: .47%
- Fe_2O_3: .12%
- Al_2O_3: .08%
- CaO: 32.05%
- MgO: 20.37%
- $CaCO_3$: 57.20%
- $MgCO_3$: 42.57%

Dolomite Typical Sieve Analysis:

Sieve Analysis(% Passing)

DSF-AF:

U.S.Std. Sieve	
50	99.1
70	97
100	88
140	76
200	60
325	36

#325:
D-303:

U.S. Std. Sieve	
50	100
70	100
100	99.7
140	99.3
200	99
325	97

D-220:

U.S. Std. Sieve	
50	99.9
70	99.8
100	99.2
140	95
200	80
325	48

Select Fillers:
#100:

U.S Std. Sieve	
50	99.3
70	97
100	90
140	76
200	57
325	40

D-205:

U.S. Std. Sieve	
50	100
70	100
100	100
140	99.5
200	98
325	73

Select Fillers:
#200:

50	100
70	99.9
100	99.8
140	98
200	86
325	58

ROHM AND HAAS: Fungicide/Preservative:

SKANE Fungicide:

Grade:

SKANE M-8:
- Active, %: 45
- Viscosity, cps.: 40
- Density lbs/U.S. gal: 8.7

KATHON Preservative:

Grade:

KATHON LX:
- Solids, %: 14
- pH: 2-4
- Density lbs/U.S. gal: 10.4

KATHON LX, 1.5%:
- Solids, %: 1.5
- pH: 3-5
- Density lbs/U.S. gal: 8.4

TROY CHEMICAL CORP.: Biocides-Mercurial:

Troysan PMA-100
E.P.A. Registration No. 5383-8
Organo-metallic Biocide
Packaged in Water Soluble Bags

TROYSAN PMA-100 is an organo metallic powdered bactericide/ fungicide specially packaged in water soluble bags designed for safety in handling.

TROYSAN PMA-100 is a broad spectrum biocide. It can be used for in-can preservation. It can also be used as a mildewcide in exterior formulations.

Advantages:

* protects against bacterial and fungal organisms
* water soluble bags provide ease of handling
* long lasting protection
* over 20 years of successful field experience
* standard of the coatings industry
* broad spectrum biocide
* provides in-can preservation and mildew resistance
* cost effective
* easily dispersible powder

Typical Properties:
Active Ingredient:
Phenyl Mercury Acetate: 100%
Mercury Metal Content: 59.5-60%
Appearance: Free flowing white powder
Melting Point: 151C
Particle Size: 95% < 75 microns

TROYSAN PMA-30
E.P.A. Registration No. 5383-4
Organo Metallic Liquid Biocide

TROYSAN PMA-30 is an organo-metallic water soluble liquid biocide.

TROYSAN PMA-30 can be used as an in-can preservative as well as a mildewcide for exterior formulations.

Advantages:

* water soluble liquid
* broad spectrum efficacy
* over 20 years successful field experience
* provides both in-can preservation, and mildew resistance

Typical Properties:
Active Ingredient:
Phenylmercury Acetate: Minimum 30%
Mercury Metal Content: 18.0% Min.
Specific Gravity: 1.275-1.320
pH: 7.5-8.0

TROY CHEMICAL CORP.: Biocides-Mercurial(Continued):

TROYSAN CMP
(TROYSAN CMP Acetate)
E.P.A. Registration No. 5383-10
Lowest Cost Organo-Metallic Biocide

TROYSAN CMP is an organo-metallic water soluble liquid. It is especially useful in aqueous systems where viscosity drift from spoilage is a problem.

TROYSAN CMP is the most cost effective mercurial in-can preservative.

Advantages:
* water soluble liquid
* most cost effective mercurial biocide
* long lasting protection
* 20 years of successful field experience
* lends viscosity stability to contaminated systems

Typical Properties:
Active Ingredient:
Chlormethoxypropyl mercuric acetate: 20%
Mercury Metal Content: 10% Min.
Appearance: Clear, mobile liquid
Color (Gardner): 3 Max.
pH: 10.0+-0.5
Specific Gravity: 1.120-1.146

TROYSAN PMDS-10
E.P.A. Registration No. 5383-62
Organo-metallic Liquid Biocide for Aqueous and Oil-modified Latex Formulations

TROYSAN PMDS-10 is an organo-metallic liquid biocide. It has broad spectrum efficacy for both in-can preservation and mildew resistance applications.

TROYSAN PMDS-10 may be used in a variety of aqueous and oil modified latex formulations. It is an excellent all-purpose preservative that is effective in a variety of systems.

Advantages:
* less pigment shock in comparison to other phenylmercury compounds
* long lasting
* broad spectrum activity
* twenty years of proven performance

Typical Properties:
Active Ingredients:
Di(phenylmercury) dodecenyl succinate: 21%
Mercury Metal Content, minimum: 10.0%
Specific Gravity: 0.925-0.947
Lbs./Gallon: 7.7-7.9

TROY CHEMICAL CORP.: Biocides--Non-Mercurial:

TROYSAN 142
E.P.A. Registration No. 5383-7
Non-metallic organic powdered biocide

TROYSAN 142 is a non-metallic organic powder. It has broad spectrum efficacy in a variety of aqueous systems.

TROYSAN 142 is used in a variety of aqueous systems, including adhesives, pigment dispersions, resin emulsions, powder joint cements, and protein colloids.

Advantages:
* highly effective biocide
* broad spectrum (bactericide/fungicide) activity
* long lasting
* 15 years of successful field experience
* cost effective protection
* F.D.A. Clearances

Typical Properties:
Active Ingredient:
3,5-dimethyltetrahydro 1,3,5,2H-Thiadiazine-2-Thione: 95%
Specific Gravity: 1.39
Melting Point: 99.5C

TROYSAN 174
E.P.A. Registration No. 5383-11
Non-metallic, Low Cost, Water-soluble Biocide

TROYSAN 174 is a non-metallic, organic biocide designed for use in systems which are subject to bacterial deterioration. Examples of such systems are: latex paints, resin emulsions, adhesives, pigment dispersions, joint cements, cutting oils, drilling additives and any other system where water is a primary component. TROYSAN 174 is a broad spectrum biocide.

Advantages:
* lowest cost biocide
* broad spectrum biocide
* ease of incorporation
* excellent freeze-thaw stability as well as low freezing point
* over 10 years of successful field performance
* long lasting efficacy

Typical Properties:
Active Ingredient:
2[(hydroxymethyl)amino]ethanol: 100%
Physical Form: Clear amber liquid
Specific Gravity: 1.14-1.16
pH: 11.0+-0.5
Viscosity (Gardner): L Max.
Color (Gardner): 7 Max.

TROY CHEMICAL CO.: Biocides-Non-Mercurial(Continued):

TROYSAN 192
E.P.A. Registration No. 5383-17
Non-metallic, Low Cost, Water-Soluble Biocide

TROYSAN 192 is a non-metallic, organic biocide designed for use in systems which are subject to bacterial deterioration. Examples of such systems are: latex paints, resin emulsions, adhesives, pigment dispersions, joint cements, cutting oils, drilling additives and any other system where water is a primary component. It is supplied as a clear water soluble liquid.

TROYSAN 192 is a broad spectrum biocide.

Advantages:

* cost effective biocide
* broad spectrum biocide
* ease of incorporation
* excellent freeze-thaw stability as well as low freezing point
* one of the first non-metallic biocides
* over 10 years of successful field performance
* long lasting efficacy

Typical Properties:

Active ingredient:
2[(hydroxymethyl)amino-2-methyl-propanol: 100%
Physical Form: Clear amber liquid
Specific Gravity: 0.96-0.99
Lbs/Gallon: 8.0-8.5
pH: 11.2+-0.5
Viscosity (Gardner): A3-A5
Color (Gardner): 3 Max.
Solubility: Completely soluble in water

TROYSAN 364
E.P.A. Registration No. 5383-47
Liquid Water Dispersible
Broad Spectrum Biocide

TROYSAN S364 is used in a variety of applications.

TROYSAN S364 is designed primarily for interior applications except for certain caulks, sealants and adhesives.

Advantages:

* single component bactericide-fungicide
* cost effective
* broad spectrum
* liquid
* long lasting
* ten years successful field application

Typical Properties

Composition: TROYSAN 174 21%/TROYSAN TBTO 29%
Inert: 50%
Specific Gravity: 1.01-1.03

TROY CHEMICAL CORP.: Biocides-Non-Mercurial(Continued):

TROYSAN POLYPHASE AF1
EPA Registration 5383-63

TROYSAN POLYPHASE AF1 is a broad spectrum, liquid, non-metallic fungicide which is effective against a wide variety of fungal organisms. It may be used to protect both interior or exterior coatings from the growth of fungal organisms on the film surface. It may also be used as a fungicide in cutting oils, textiles, paper coatings, inks, plastics, adhesives, and canvas and cordage. It is also recommended as a wood preservative.

Uses:
- Paints and Stains
- Plastics
- Cutting Oils
- Wood
- Textiles
- Paper Coatings
- Canvas and Cordage
- Inks
- Adhesives

Typical Properties:
- Active Ingredient: 40% min. 3-iodo-2-propynyl butyl carbamate
- Appearance: Amber mobile liquid
- Color (Gardner): 9 maximum
- Viscosity (Gardner Bubble) 25C: C maximum

TROYSAN POLYPHASE P100
EPA Registration No. 5383-50

TROYSAN POLYPHASE P100 is a broad spectrum, solid, non-metallic fungicide which is effective against a wide variety of fungal organisms. It may be used to protect both interior or exterior coatings from the growth of fungal organisms on the film surface, as a fungicide in cutting oils, textiles, paper coatings, inks, plastics, adhesives, and canvas and cordage. It is also used as a wood preservative.

Uses:
- Paints and Stains
- Plastics
- Cutting Oils
- Wood
- Textiles
- Paper Coatings
- Inks
- Adhesives

Typical Properties:
- Active Ingredient: 97% min. 3-iodo-2-propynyl butyl carbamate
- Appearance: Off white granular powder
- Melting Point: 64-66C

R.T. VANDERBILT CO., INC.: Paint/Paper Products:

VANCIDE 89
Industrial Preservative

Vinyl:
To protect vinyl products as e.g. PVC from unsightly mildew growth and stiffening due to plasticizer depletion caused by bacteria and fungi, add 0.5 to 1 part of VANCIDE 89 based on 100 parts, by weight of the resin.

Polyethylene:
To protect polyethylene from the growth of fungi and bacteria or to make bacteriostatic-fungistatic polyethylene, mill in 0.75% VANCIDE 89 by weight of the polyethylene.

Paint:
Protect paint from damage by mold by milling in 0.25 to 1.0% VANCIDE 89 based on the weight of the paint.

Lacquer:
To protect lacquer used in coating electrical equipment from the growth of mildew, add 1% to 3% VANCIDE 89 based on the solids.

Soap:
To make sanitizing deodorant powdered hand soap, add 1.0% VANCIDE 89 by weight of the anhydrous soap.

Wallpaper flour paste:
For preservation of wallpaper flour paste against growth of mold, use 0.25% VANCIDE 89 based on the weight of the paste.

Rubber and rubber coated articles:
To protect bacteriostatic latex for making cast or coated products, add 0.4% VANCIDE 89 based on the total dry weight of the latex compound.

Chemical composition:
N-trichloromethylthio-4-cyclohexene-1,2-dicarboximide

Assay:
92% minimum

R.T. VANDERBILT CO., INC.: Paint/Paper Products(Continued):

VANCIDE 51Z
Mold Inhibitor for Latex Paint
U.S. Patents 3,725,327 and 3,737,551

General Description:
VANCIDE 51Z is a very effective mold inhibitor in latex paints. It will not become discolored during exposure in sulfide atmospheres. VANCIDE 51Z is a dry powder and can be dispersed in a paint in the same manner as the addition of a pigment.

Chemical Composition:
Zinc dimethyldithiocarbamate: 87.0%
Zinc 2-mercaptobenzothiazole: 7.5%
(Total zinc 19.8%)

Color and State: White to cream powder

Density, Mg/m3: 1.72+-0.03

VANCIDE 51Z
Mold Inhibitor of Caulking Compounds

VANCIDE 51Z has been shown to inhibit mold growth in many commercial caulking compounds. It is readily compatible with formulations of either butyl rubber or acrylic copolymers containing petroleum or alcohol solvents without affecting performance or color.

VANCIDE MZ-96
Industrial Preservative
U.S. Patent No. 3,725,327 and 3,737,551

VANCIDE MZ-96 is a free-flowing, white, water-dispersible powder used as an industrial preservative.
VANCIDE MZ-96 acts as a preservative in starch and adhesive formulations.
VANCIDE MZ-96 is a very effective mold or mildew inhibitor in latex or water-based coatings.

Chemical Composition: Zinc dimethyldithiocarbamate

Molecular Formula: C6H12N2S4Zn

Molecular Weight: 305.82

Assay: 98% Zinc dimethyldithiocarbamate (ZIRAM)

Total Zinc as Metallic: 20.6% minimum

R.T. VANDERBILT CO., INC.: Paint/Paper Products(Continued):

VANCIDE TH
Preservative for Latex Paints
U.S. Patent 3,755,590

VANCIDE TH is an effective preservative for latex paints. It prevents attack by microorganisms while the paint is in storage. This nonmetallic liquid, which is completely soluble in water, can be added at any point during paint preparation.

This active ingredient has been cleared for use as an antimicrobial agent in manufacture of paper and paperboard. (21 CFR 176.170)

Physical and Chemical Properties:
- Chemical composition: Hexahydro-1,3,5-triethyl-s-triazine
- Physical form: Colorless to light yellow liquid

VEEGUM T in Latex Paints

VEEGUM T is a complex colloidal magnesium aluminum silicate. It is derived from a naturally-occurring base by a refining process. Its aqueous dispersions are high viscosity thixotropic gels at low solids. This indicates its use for imparting thixotropy to latex paints.

Average Properties of VEEGUM T:
- Appearance: Small flakes
- Color: Near white
- Odor: None
- Density: 2.8 Mg/m3
- Moisture: Less than 8% (at time of shipment)
- 4% Water gel pH: 9 to 9.5

VEEGUM T in latex paints develops thixotropy which allows formulation of desirable properties such as: no separation, "dripless" application and leveling with sag resistance.

VAN GEL B
Thixotrope for Aqueous Systems

VAN GEL B is derived from a naturally-occurring colloidal magnesium aluminum silicate that has been highly refined to produce uniform thixotropic aqueous dispersions at low concentrations.

Application:
VAN GEL B is used in latex paint for the development of smooth thixotropic consistency. It prevents syneresis and settling, provides "dripless" application, controls leveling with sag resistance and, in addition, eliminates or minimizes the need to stir paint prior to its use.

VININGS INDUSTRIES: Antimicrobial Agents:

AMA-420
(Antimicrobial Agent)

AMA-420 is an effective, low cost, liquid antimicrobial agent for use as a preservative in slurries, adhesives, coatings, and high viscosity suspensions.

Composition:

Active Ingredient:	
tetrahydro-3,5-dimethyl-2H-1,3,5-thiadiazine-2-thione	20%
Inert Ingredients:	80%
Total	100%

Typical Physical Properties:
Physical Form: Creamy liquid
Specific Gravity: 1.06 @ 25C

AMA-424
(Antimicrobial Agent)

AMA-424 is an effective, low cost, liquid antimicrobial agent for use as a preservative in slurries, adhesives, coatings, and high viscosity suspensions.

Composition:

Active Ingredient:	
tetrahydro-3,5-dimethyl-2H-1,3,5-thiadiazine-2-thione	24%
Inert Ingredients:	76%
Total	100%

Typical Physical Properties:
Physical Form: Liquid
Specific Gravity: 1.161 @ 25C

AMA-35D-P
(Antimicrobial Agent for Use in Aqueous Suspensions & Oilfield Operations)

Vinings' AMA-35D-P is a nearly odorless, dustless, free flowing crystalline solid. It is registered for use as a preservative of industrially prepared water based suspensions (pigment slurries, adhesives, polymers and starches), and in oilfield operations.

Composition:

Active Ingredient:	
Tetrahydro-3,5-dimethyl-2H-1,3,5-thiadiazine-2-thione	99%
Inert ingredient:	1%
Total:	100%

Typical Physical Properties:
Physical form: Crystalline solid
Melting Point: 106C

Section 9
Resins

AO POLYMER CORP.: AOTHANE Urethanes:

AOTHANE Moisture Curing Urethanes

AOTHANE Moisture Curing Urethanes are oil free, one package moisture curing polyurethane resins. Outstanding features include fast dry, high gloss, excellent abrasion resistance.

Properties:

2008:
- Solids, %: 40+-1
- Solvent: Rule 66 Xylol-CELLOSOLVE Acetate
- Viscosity, Brookfield, cps: 120
- NCO Content, %; 3.0+-0.2

2008H:
- Solids, %: 40+-1
- Solvent: Rule 66 Xylol-CELLOSOLVE Acetate
- Viscosity, Brookfield, cps: 500
- NCO Content, %: 3.0+-0.2

2009:
- Solids, %: 60+-1
- Solvent: Rule 66 Xylol-CELLOSOLVE Acetate
- Viscosity, Brookfield, cps: 950
- NCO Content, %: 3.5+-0.2

Applications:

AOTHANE Moisture Curing Urethane Resins were developed specifically for use in formulating fast drying protective coatings for numerous materials including concrete, wood, metals and fabric, etc. Suggested uses include floors in schools, stores, gymnasiums, as well as tough finishes for furniture, toys, sports equipment, etc.

AOTHANE 3001 & 3002

AOTHANE 3001 is a one-package, non-aromatic, oil-free, moisture-curing polyurethane resin formulated especially for outstanding color and gloss retention. Properties include exceptional resiliency, toughness, abrasion, and mar resistance. It is ready to use as received. AOTHANE 3002 is the same polymer but at a higher solids.

Typical Properties:

3001:
- Solids, %: 40+-1%
- Solvent: Rule 66 Xylene-CELLOSOLVE Acetate
- Viscosity @ 25C, cps: 100-125
- Available NCO, %: 3+-0.2

3002:
- Solids, %: 60+-1%
- Solvent: Rule 66 Xylene-CELLOSOLVE Acetate
- Viscosity @ 25C, cps: 3200-3900 cps
- Available NCO, %: 4.1+-2

AO POLYMER CORP.: AOTHANE Urethanes (Continued):

AOTHANE 3001 & 3002(Continued)

Applications:

AOTHANE 3001 is suggested for numerous clear non-yellowing coatings on metals, masonry, wood, paper, polyesters, epoxies and leather. It has excellent exterior and interior durability as well as good chemical and solvent resistance. Suggested uses include furniture, floors, walls, boats, wood paneling, etc.

AOTHANE 3001 can be pigmented with various castor oil dispersions prior to use to obtain pigmented glossy finishes. A two-component epoxy/polyamide primer is recommended for metal and concrete applications.

AOTHANE 1999

AOTHANE 1999 is a low molecular weight polyether prepolymer developed for use where high tensile strength must be achieved under rapid curing conditions.

Properties:
- Viscosity @ 25C, cps: 1500 to 3000
- Available NCO, %: 25+-1
- Wt/Gal , lbs.: 9.1
- Solids, %: 99+-1
- Color, Gardner: 1-2

AOTHANE 2000

AOTHANE 2000 is a low molecular weight prepolymer developed for use in cold molding and foam applications. It is recommended for use in combination with various polyether, polyester, polylactone, and castor oil polyols to yield elastomers at room temperature for cold molding, encapsulation, coatings, sealants, castings, etc.

Typical Properties:
- Viscosity @ 25C, cps: 4000 average
- Available NCO, %: 17+-1
- Specific Gravity @ 25/25C: 1.07
- Weight/gallon, lbs.: 8.9
- Solids, %: 100
- Color, Gardner: 1-2

AO POLYMER CORP.: AOTHANE Urethanes (Continued):

AOTHANE 2001

AOTHANE 2001 is a polyester prepolymer developed specifically for adhesive, sealant and encapsulating applications. The formulated product offers a wide range of hardness (50-80 Shore A), flexibility and resiliency.

Properties:
- Viscosity @ 25C, cps: 10,000-15,000
- Available NCO, %: 7.0+-0.2
- Wt/Gal., lbs: 8.7
- Solids %: 99.5
- Color, Gardner: 1

AOTHANE 2002

AOTHANE 2002 is a moderately crosslinked prepolymer developed for sealant and adhesive applications. When properly formulated product offers a wide range of hardness, flexibility, and resiliency.

Typical Properties:
- Viscosity @ 25C, cps: 7,000-12,000
- Available NCO, %: 4.0+-0.1
- Specific Gravity @ 25C: 1.04
- Wt./Gallon, lbs.: 8.7
- Solids, %: 100%
- Color, Gardner: 1

Suggested Uses:
AOTHANE 2002 is used in conjunction with AOCURE H to formulate two-part self-leveling, low durometer joint-sealants.

AOTHANE 2003

AOTHANE 2003 is a medium molecular weight polyether prepolymer developed for use in sealant and coating applications.

Properties:
- Viscosity @ 25C, cps: 2000-3500
- Available NCO, %: 2.8-3.0
- Wt/gal. lbs.: 8.5
- Solids, %: 99
- Color, Gardner: 1

AO POLYMER CORP.: AOTHANE Urethanes (Continued):

AOTHANE 2004

AOTHANE 2004 is a moderately crosslinked prepolymer developed for sealant and adhesive applications. When properly formulated, the product offers a wide range of hardness, flexibility, and resiliency.

Typical Properties:

Viscosity @ 25C, cps: 6000-10000
Available NCO: 3.0+-0.2
Specific Gravity @ 25C: 1.03
Wt./Gallon, lbs.: 8.6
Solids, %: 99.5
Color, Gardner: 1

Suggested Uses:

AOTHANE 2004 is used in combination with AOCURE H to formulate two-part self-leveling, low durometer joint sealants.

AOTHANE 2005

AOTHANE 2005 is a urethane prepolymer which when properly formulated offers a wide range of hardness, flexibility and resiliency for one and two-part seamless elastomeric waterproofing systems. Addition of AOTHANE 2006 will provide a wider range of lower hardness and increase elongation for both sealant and membrane applications.

Both hand and mechanical mix urethane/coal tar, expansion joint sealants for highway as well as Federal Specification SS-S-00200d for jet fuel resistance can be compounded using a combination of AOTHANE 2005 and 2006.

Typical Properties:

Viscosity, cps: 6,000-8,500
Available NCO, %: 3.0+-0.2
Specific Gravity 25C: 1.03
Wt./Gallon, lbs.: 8.6
Solids content, %: 99
Color, Gardner: 1

AO POLYMER CORP.: AOTHANE Urethanes(Continued):

AOTHANE 2006

AOTHANE 2006 is a very high molecular weight urethane prepolymer with low crosslinking density. It is recommended for use in combination with AOTHANE 2005 for one or two-part seamless elastomeric water-proofing systems. Both hand and machine mix urethane/coal tar expansion joint sealants for highway and Federal Specification SS-S-00200d for jet fuel resistance can be compounded using a combination of AOTHANE 2005 and 2006.

Typical Properties:

Viscosity, cps: 8,000-16,000
Available NCO, %: 1.8+-0.2
Specific Gravity @ 25C: 1.04
Wt./Gallon, lbs.: 8.7
Solids %: 99.5
Color, Gardner: 1
Appearance: Clear

AOTHANE 2007

AOTHANE 2007 is a prepolymer with moderate crosslinks developed specifically for adhesive and elastomeric coatings giving exceptionally high abrasion resistance. It is usually used with AOCURE MC or AOCURE MCF, alone or in combination with other curing agents.

Typical Properties:

Viscosity cps: 12,000-18,000
Available NCO %: 5.0+-0.1
Specific Gravity: 1.06
Solids, %: 90
Color, Gardner: 2
Wt. per gal., lbs.: 8.85

Applications:

Suggested uses include high abrasion and wear resistant coatings (Shore A 50 to 80) for parking decks, ramps, balcony areas, pool decks, equipment rooms. Also suggested for concrete floor areas that must maintain absolute integrity against water infiltration to areas below.

AO POLYMER CORP.: AOTHANE Urethanes(Continued):

AOTHANE 2056

AOTHANE 2056 is a urethane prepolymer which when properly formulated offers a wide range of hardness, flexibility and resiliency for one and two-part seamless elastomeric waterproofing systems. Addition of AOTHANE 2006 where desired will provide a lower hardness and increase elongation for both sealant and membrane applications.

Both hand and mechanical mix urethane/coal tar expansion joint sealants for highway as well as Federal Specification SS-S-00300d for jet fuel resistance can be compounded using AOTHANE 2056.

Typical Properties:

Viscosity @ 25C., cps: 5,000-8,000
Available NCO, %: 2.5+-0.2
Specific Gravity @ 25C: 1.03
Weight/gallon, lbs.: 8.6
Solids %: 99
Color, Gardner: 1

AOTHANE 2065

AOTHANE 2065 is a softer version of the AOTHANE 2056 urethane prepolymer which, when properly formulated, offers a wide range of hardness, flexibility and resiliency for one and two-part seamless elastomeric waterproofing systems.

Both hand and mechanical mix urethane/coal tar expansion joint sealants for highway as well as Federal Specification SS-S-00300D for jet fuel resistance can be compounded using AOTHANE 2065.

Typical Properties:

Viscosity @ 25C., cps: 4,000-8,000
Available NCO, %: 2.3+-0.2
Specific Gravity @ 25C: 1.03
Wt/gal., lbs.: 8.6
Solids, %: 99
Color, Gardner: 1

AO POLYMER CORP.: AOTHANE Urethanes(Continued):

AOTHANE 3002-5

AOTHANE 3002-5 is a moderately crosslinked MDI based prepolymer developed for sealant and adhesive applications. When properly formulated product offers a wide range of hardness, flexibility, and resiliency.

Typical Properties:

Viscosity, cps: 30,000-55,000
Available NCO, %: 5.0+-0.2%
Specific Gravity @ 25C: 1.04
Wt./Gal. lbs: 8.7
Solids, %: 100
Color, Gardner: 1

Suggested Uses:

AOTHANE 3002-5 is used in combination with AOCURE H to formulate two-part self-leveling, low durometer joint-sealants.

AOTHANE 3007

AOTHANE 3007 is an MDI based prepolymer with moderate crosslinks developed specifically for adhesive and elastomeric coatings giving exceptionally high abrasion resistance. It is usually used with AOCURE MC or MCF alone or in combination with other curing agents.

Typical Properties:

Viscosity, cps: 30,000-45,000
Available NCO, %: 5.0+-0.2
Specific Gravity: 1.06
Solids, %: 90
Color, Gardner: 2
Wt. per gallon. lbs. : 8.85

Applications:

Suggested uses include high abrasion on wear resistant coating (Shore A 50 to 85) for parking decks, ramps, balcony areas, pool decks, equipment rooms. Also suggested for concrete floor areas that must maintain absolute integrity against water infiltration to areas below.

AO POLYMER CORP.: AOTHANE Urethanes(Continued):

AOTHANE 3075

AOTHANE 3075 is an MDI based urethane prepolymer which when properly formulated offers a wide range of hardness, flexibility and resiliency for one and two-part seamless elastomeric waterproofing systems. Addition of AOTHANE 2006 where desired will provide a lower hardness and increase elongation for both sealant and membrane applications.

Both hand and mechanical mix urethane/coal tar expansion joint sealants for highway as well as Federal Specification SS-S-00300d for jet fuel resistance can be compounded using AOTHANE 3075.

Typical Properties:

- Viscosity, cps: 20,000-35,000
- Available NCO, %: 2.5+-0.2
- Specific Gravity @ 25C: 1.03
- Weight/gallon, lbs.: 8.6
- Color, Gardner: 1
- Solids %: 99

AO POLYMER CORP.: Products:

MOCRYL W-150, MOCRYL W-151 and MOCRYL W-160
Water Soluble Acrylic Polymers

MOCRYL W-150 and MOCRYL W-160 are thermosetting acrylic polymers developed specifically for use in water based coatings. Being hydroxycarboxy functional they usually require addition of a crosslinking agent such as melamine-formaldehyde or urea-formaldehyde resins.

MOCRYL resins exhibit excellent mechanical and chemical durability and outstanding adhesion to a variety of substrates including aluminum, steels, copper, PVC, wood, paper, etc.

MOCRYL W-150:
Appearance: clear liquid
N.V.: 65+-1
Volatiles: IPA/BU CELLOSOLVE
Gardner Viscosity: Z5-Z6
Acid No.: 39-42

MOCRYL W-151:
Appearance: clear liquid
N.V.: 65+-1
Volatiles: IPA
Gardner Viscosity: Z6-Z7
Acid No.: 39-42

MOCRYL W-160:
Appearance: clear liquid
N.V.: 70+-1
Volatiles: BU CELLOSOLVE
Gardner Viscosity: Z3-Z4
Acid No.: 56-59

Applications:

MOCRYL W-150 is suggested for use where hard, solvent resistant, mar resistant, durable coatings are necessary. Decorative and functional coatings for appliances, corrosive environments, railroad cars, metal siding, and automotive finishes are a few applications. It can also be used as a hard thermo-plastic coating (without crosslinker) for flake board, panel board, wood grained polystyrene furniture, paper, etc.

MOCRYL W-160 is suggested for use where excellent UV and weather resistance is required with great film flexibility. Coil and roller coating applications for post-formed metal siding as well as venetian blind slats find use for MOCRYL W-160. As a flexibilizing polymer, MOCRYL W-160 may be added from 10-50% to MOCRYL W-150.

Coatings prepared form MOCRYL resins may be applied by conventional methods such as: spray, dip, roller coating and curtain coating.

AO POLYMER CORP.: Products(Continued):

KETONE RESINS

AO Polymer thermoplastic Ketone resins are used to improve the characteristics of other resins by promoting adhesion, improving gloss, lowering viscosity and increasing solvent release. Ketone resins do not form films themselves and must be used with other resins or vehicles. They are compatible with nitrocellulose, acrylics, phenolics, alkyds, vinyls, epoxies and urethanes.

Physical Properties:

Alicyclic Series
MR-85
Solids: 100%
color (50% MEK-Gardner Max.): 2
softening point (C): 80-90
acid value: neutral
specific gravity at 25C: 1.17-1.2

Aliphatic Series
MR-97
solids: 100%
color (50% MEK-Gardner Max.): 8
softening point (C): 90-110
acid value: neutral
specific gravity: 1.16-1.19

AO KETOMUL 200
(Ketone Emulsion)

Physical Properties of the Emulsion:
Solids: 48+-1%
Viscosity: 4000-8000 cps (Brookfield)
pH: 6-8
Weight per Gallon @ 20C: 9.45#
Surfactant System: Anionic
Color: Off-White

Physical Properties of the Resin Solids:
Type: Thermoplastic Alicyclic Ketone Resin
Softening Point (C): 80-90
Color (50% MEK): 2 Gardner (Max.)
Acid Value: Neutral

Remarks:

AO Ketone Emulsion 200 is an anionic emulsion of a thermoplastic alicyclic ketone resin. Being dispersed in a water system, this product allows the formulator to blend with various latexes, such as acrylics, ethylene-vinyl acetate, polyvinyl acetate, and others to improve such properties as adhesion, gloss, wet and dry tack development, etc. Compatibility with any specific polymer emulsion should be checked by the formulator. Some 1,1,1 trichlorethane is also emulsified in the product.

ARAKAWA CHEMICAL (USA) INC.: List of Materials for Paints:
Alkyd Resins:

Products:

Tall Oil Modified Alkyds:

ARAKYD 1A-120-60:
Solid %: 60+-1
Solvent: Xylene
Colour (Gardner): 6 max.
Acid Number: 9 Max.
Viscosity at 25C: Z1-Z5
Phthalic Anhydride %: 42
Oil Length: 39

ARAKYD 1772:
Solid %: 50+-1
Solvent: Xylene
Colour (Gardner): 5 Max.
Acid Number: 13 Max.
Viscosity at 25C: U-Y
Phthalic Anhydride %: 11
Oil Length: 36

ARAKYD 1781-60:
Solid %: 60+-1
Solvent: Toluene
Colour (Gardner): 7 Max.
Acid Number: 8 Max.
Viscosity: U-Z2
Phthalic Anhydride %: 11
Oil Length: 37

ARAKYD 1782-60:
Solid %: 60+-1
Solvent: Xylene
Colour (Gardner): 7 Max.
Acid Number: 8 Max.
Viscosity: Z3-Z6
Phthalic Anhydride %: 42
Oil Length: 37

ARAKYD 2302-60:
Solid %: 60+-1
Solvent: Xylene
Colour (Gardner): 7 Max.
Acid Number: 13 Max.
Viscosity: Z2-Z5
Phthalic Anhydride %: 44
Oil Length: 33
Characteristics: Heat resistance, High gloss, Alkali resistance, Hard film when baking at low temperature, Water resistance, High gloss, Hard film, Alkali resistance.
Uses: Baking paint mixing with melamine resin, Lacquer.

ARAKAWA CHEMICAL (USA) INC.: List of Materials for Paints:
Alkyd Resins(Continued):

Tall Oil Modified Alkyds(Continued):

Products:

ARAKYD 8042-80:
Solid %: 80+-1
Solvent: Xylene
Colour(Gardner): 5 Max.
Acid Number: 10 Max.
Viscosity at 25C: Z-Z4
Phthalic Anhydride %: 42
Oil Length: 37

ARAKYD 1778-80:
Solid %: 80+-1
Solvent: Alcohols
Colour(Gardner): 8 Max.
Acid Number: 21 Max.
Viscosity at 25C: Z3-Z6
Phthalic Anhydride %: 41
Oil Length: 34

Characteristics: High gloss, Hard film, Quick drying, Hard film
Uses: Baking paint by mixing with melamine resin, Paint for wood

ARAKYD 8012:
Solid %: 50+-1
Solvent: Xylene
Colour(Gardner): 10 Max.
Acid Number: 6-10
Viscosity at 25C: U-X
Phthalic Anhydride %: 35
Oil Length: 50
Characteristics: Quick drying, Hard film, High gloss
Uses: Phthalic resin based paint

ARAKYD 1465-60:
Solid %: 60+-1
Solvent: Mineral spirit
Colour(Gardner): 7 Max.
Acid Number: 6 Max.
Viscosity at 25C: U-X
Phthalic Anhydride %: 24
Oil Length: 66
Characteristics: Good water resistance, High gloss, Hard film
Uses: Ready mixed paint of synthetic resin

ARAKAWA CHEMICAL (USA) INC.: List of Materials for Paints:
Alkyd Resins(Continued):

Tall Oil Modified Alkyds(Continued):

Products:

ARAKYD 8055-60:
Solid %: 60+-1
Solvent: Mineral spirit
Colour (Gardner): 9 Max.
Acid Number: 7 Max.
Viscosity at 25C: X-Z2
Phthalic Anhydride %: 24
Oil Length %: 63
Characteristics: Good gloss retention
Uses: Ready mixed paint of synthetic resin

Drying Oil Modified Alkyds:

ARAKYD 3301-60:
Solid %: 60+-1
Solvent: Toluene
Colour(Gardner): 6 Max.
Acid Number: 9 Max.
Viscosity at 25C: W-Z
Phthalic Anhydride %: 42
Oil Length %: 42
Characteristics: High gloss, Hard film, Good colour retention

ARAKYD 4401X-60:
Solid %: 60+-1
Solvent: Xylene
Colour(Gardner): 6 Max.
Acid Number: 6 Max.
Viscosity at 25C: Z3-Z5
Phthalic Anhydride %: 41
Oil Length %: 37
Characteristics: Heat resistance

ARAKYD 4450X-60:
Solid %: 60+-1
Solvent: Xylene
Colour(Gardner): 6 Max.
Acid Number: 13 Max.
Viscosity at 25C: Z2-Z5
Phthalic Anhydride %: 45
Oil Length %: 28

ARAKAWA CHEMICAL (USA) INC.: Drying Oil Modified Alkyds:
Alkyd Resins(Continued):

Drying Oil Modified Alkyds(Continued):

ARAKYD 4451X-60:
Solid %: 60+-1
Solvent: Xylene
Colour(Gardner): 6 Max.
Acid Number: 15 Max.
Viscosity at 25C: W-Z1
Phthalic Anhydride %: 35
Oil Length: 27

Characteristics: Hard film when baking at low temperature. High gloss, Alkali resistance
Uses: Baking paint by mixing with melamine resin, Lacquer

ARAKYD 1185-50:
Solid %: 50+-1
Solvent: Mineral spirit
Colour(Gardner): 7 Max.
Acid Number: 6 Max.
Viscosity at 25C: Y-Z3
Phthalic Anhydride %: 35
Oil Length: 52
Characteristics: Quick drying, High gloss, Hard film, Good water resistance
Uses: Baking enamel, Air setting enamel

ARAKYD 3135:
Solid %: 50+-1
Solvent: Mineral spirit
Colour(Gardner): 7 Max.
Acid Number: 7 Max.
Viscosity at 25C: X-Z
Phthalic Anhydride %: 28
Oil Length: 52
Characteristics: Good weather resistance
Uses: Phthalic resin paint

ARAKYD 5301:
Solid %: 50+-1
Solvent: Xylene
Colour(Gardner): 6 Max.
Acid Number: 6 Max.
Viscosity at 25C: U-X
Phthalic Anhydride %: 30
Oil Length %: 40
Characteristics: Quick drying, Hard film
Uses: Ready mixed paint of synthetic resin

ARAKAWA CHEMICAL (USA) INC.: List of Materials For Paints:
Alkyd Resins(Continued):

Drying Oil Modified Alkyds(Continued):

ARAKYD 3165-60:
Solid %: 60+-1
Solvent: Mineral spirit
Color(Gardner): 7 Max.
Acid Number: 4-8
Viscosity at 25C: V-Y
Phthalic Anhydride %: 23
Oil Length %: 66
Characteristics: Quick drying, Good weather resistance
Uses: Ready mixed paint of synthetic resin

ARAKYD 1675-80:
Solid %: 80+-1
Solvent: Mineral spirit
Colour(Gardner): 7 Max.
Acid Number: 1 Max.
Viscosity at 25C: T-V
Phthalic Anhydride %: 20
Oil Length %: 67
Characteristics: Good reactivity
Uses: Combination with urethane resin

ARAKYD 400T:
Solid %: 54+-1
Solvent: Toluene
Colour(Gardner): 6 Max.
Acid Number: 12 Max.
Viscosity at 25C: X-Z
Phthalic Anhydride %: 31
Oil Length %: 35
Characteristics: Quick drying, Good weather resistance
Uses: Traffic paint

ARAKYD 221:
Solid %: 100
Colour(Gardner): 11 Max.
Acid Number: 8 Max.
Viscosity at 25C: V-X
Phthalic Anhydride %: 13
Oil Length %: 78

ARAKYD 251:
Solid %: 100
Colour(Gardner): 8 Max.
Acid Number: 13 Max.
Viscosity at 25C: T-W
Phthalic Anhydride %: 9
Oil Length %: 77
Characteristics: Good weather resistance, Good plasticizing effect
Uses: Plasticizer for traffic paint

ARAKAWA CHEMICAL (USA) INC.: List of Materials for Paints: Alkyd Resins(Continued):

Phenol Modified Alkyds:

ARAKYD 1232-60:
Solid %: 60+-1
Solvent: Xylene
Colour(Gardner): 7 Max.
Acid Number: 8 Max.
Viscosity at 25C: V-Y
Phthalic Anhydride %: 27
Oil Length: 28
Characteristics: Good weather and alkali resistance
Uses: Undercoat

ARAKYD 7100:
Solid %: 50+-1
Solvent: Xylene
Colour(Gardner): 10 Max.
Acid Number: 8 Max.
Viscosity at 25C: P-U
Characteristics: Hard film, Good corrosion resistance
Uses: Undercoat

ARAKYD 3422:
Solid %: 50+-1
Solvent: Xylene
Colour(Gardner): 6 Max.
Acid Number: 7 Max.
Viscosity at 25C: U-X
Characteristics: Quick drying, High gloss
Uses; Phthalic resin paint

Vinyl Modified Alkyds:

ARAKYD S-502:
Solid %: 50+-1
Solvent: Xylene
Colour(Gardner): 3 Max.
Acid Number: 7 Max.
Viscosity at 25C: V-Z
Characteristics: Quick drying, High gloss, good weather resistance
Uses: Quick drying paint, Hammer-tone paint

ARAKYD M-302:
Solid %: 50+-1
Solvent: Xylene
Colour(Gardner): 1 Max.
Acid Number: 5 Max.
Viscosity at 25C: R-U
Characteristics: Quick drying, High gloss, Hard film, Good chemical proof
Uses: Lacquer, Quick drying paint

ARAKAWA CHEMICAL (USA) INC.: List of Materials for Paints: Alkyd Resins(Continued):

Vinyl Modified Alkyds(Continued):

ARAKYD M-902:
Solid %: 50+-1
Solvent: Xylene
Colour(Gardner): 6 Max.
Acid Number: 3 Max.
Viscosity at 25C: J-M
Characteristics: Hard film, High gloss, Good recoating
Uses: Quick drying paint

ARAKYD M-7300X-55:
Solid %: 55+-1
Solvent: Xylene
Colour(Gardner): 2 Max.
Viscosity at 25C: W-Z
Characteristics: Quick drying, Good weather resistance
Uses: Quick drying paint

Urethane Modified Alkyds:

ARAKYD 7500X:
Solid %: 50+-1
Solvent: Xylene
Colour(Gardner): 8 Max.
Acid Number: 8 Max.
Viscosity at 25C: W-Z
Characteristics: Quick drying, Hard film
Uses: Quick drying paint

ARAKYD 7501X:
Solid %: 50+-1
Solvent: Xylene
Colour(Gardner): 8 Max.
Acid Number: 8 Max.
Viscosity at 25C: V-Y
Characteristics: Quick drying, Good recoating
Uses: Quick drying paint

ARAKAWA CHEMICAL (USA) INC.: List of Materials for Paints: Polyester:

Saturated Polyester:

KA1051:
Solid (%): 35+-1
Solvent: Special solvent
Colour(Gardner): 2 Max.
Acid Number: 2 Max.
Viscosity at 25C: W-Z1
Characteristics: Excellent adhesion
Uses: Precoating for metal

ARAKAWA CHEMICAL (USA) INC.: List of Materials for Paints: Hard Resins:

Products:

Rosin Ester:

Ester Gum AA-L:
Colour (USDA): WW Min.
Acid Number: 7 Max.
Softening Point (C. R&B): 82 Min.
Characteristics: Neutral resin. Very light colour

Ester Gum 105:
Colour (USDA): WG Min.
Acid Number: 20 Max.
Softening Point (C. R&B): 100-110
Characteristics: Neutral resin, Good weather resistance

Uses: Oil varnish, Enamel, Lacquer, Emulsion paint

Rosin Modified Maleics:

MALKYD No. 1:
Color (USDA): WG Min.
Acid Number: 25 Max.
Softening Point (C. R&B): 120-130

MALKYD No. 5:
Color (USDA): WG Min.
Acid Number: 25 Max.
Softening Point (C. R&B): 140-150

Characteristics: Partial compatibility with nitrocellulose. Very good weather resistance & colour retention
Uses: Oil varnish, Enamel & High gloss varnish

MALKYD No. 2:
Colour (USDA): WG Min.
Acid Number: 39 Max.
Softening Point (C. R&B): 120-130

MALKYD No. 6:
Colour (USDA): WG Min.
Acid Number: 39 Max.
Softening Point (C. R&B): 140-150

MALKYD No. 8:
Colour (USDA): WG Min.
Acid Number: 39 Max.
Softening Point (C. R&B): 130-140

Characteristics: Very light colour, Very good compatibility, weather resistance & colour retention
Uses: Clear lacquer, Enamel, Oil varnish & High gloss varnish

ARAKAWA CHEMICAL (USA) INC.: List of Materials for Paints: Hard Resins(Continued):

Products:

Rosin Modified Phenolics:

TAMANOL 135:
Colour (USDA): WG Min.
Acid Number: 18 Max.
Softening Point (C. R&B): 130-140

TAMANOL 145:
Colour (USDA): WG Min.
Acid Number: 18 Max.
Softening Point (C. R&B): 140-155

Characteristics: Very good oil solubility. Excellent characters for oil varnish
Uses: Oil varnish, Enamel, Baking paint, Electric insulation varnish

100% Phenolics:

TAMANOL AS:
Appearance: Reddish brown solution
Characteristics: Thermo-setting resin, Excellent electric insulation
Uses: Baking enamel, Electric insulation varnish

TAMANOL PA:
Appearance: Pale yellowish lump
Softening Point (C. R&B): 90-100
Characteristics: Novolac type resin
Uses: Shellac substitution

TAMANOL 100S:
Appearance: Pale yellowish lump
Softening Point (C. R&B): 100-130
Characteristics: Novolac type resin, Good durability & alkali resistance
Uses: Oil varnish, Electric insulation varnish

TAMANOL 100R:
Appearance: Pale yellowish lump
Softening Point (C. R&B): 75-95
Characteristics: Thermo-setting resin, Very good water and chemical resistance, Good durability, Excellent electric insulation
Uses: Oil varnish, Electric insulation varnish

TAMANOL 1010R:
Appearance: Pale yellowish lump
Softening Point: 85-110
Characteristics: Excellent thermo-setting resin, Very good water and chemical resistance and electric insulation
Uses: Curing varnish

ARAKAWA CHEMICAL (USA) INC.: List of Materials for Paints:
Polyols for Urethane:

Products

ARAKYD 4103:
Description: Drying oil modified alkyd
Solid (%): 50+-1
Solvent: Toluene
Colour (Gardner): 5 Max.
Acid Number: 10 Max.
OH Value: 50-70
Viscosity at 25C: R-U
Characteristics: Good anti-yellowing. Excellent adhesion & abrasion resistance
Uses: Paint for wood

ARAKYD 9000:
Description: Polyester
Solid (%): 100
Colour (Gardner): 1 Max.
Acid Number: 4 Max.
OH Value: 270-310
Viscosity: 500-2000 cps at 70C
Characteristics: Excellent adhesion & abrasion resistance. High gloss

ARAKYD 9001:
Description: Polyester
Solid (%): 70+-1
Solvent: Butyl acetate
Colour (Gardner): 3 Max.
Acid Number: 10 Max.
OH Value: 70-100
Viscosity at 25C: W-Z1
Characteristics: Excellent adhesion & abrasion resistance. Quick drying
Uses: Paint for metal

ARAKYD 9106:
Description: Acrylic
Solid: 50+-1 Solvent: Butyl acetate
Colour(Gardner): 1 Max.
Acid Number: 7 Max.
OH Value: 47-53 Viscosity at 25C: R-V
Characteristics: Hard film. Excellent adhesion & abrasion resistance, Good weather resistance
Uses: Paint for wood

ARAKYD 9101:
Description: Acrylic
Solid (%): 50+-1
Solvent: Butyl acetate
Colour (Gardner): 1 Max.
Acid Number: 7 Max.
OH Value: 47-53
Viscosity at 25C: Y-Z3
Characteristics: Quick drying, Hard film, Good weather resistance, Good chemical proof

ARAKAWA CHEMICAL (USA) INC: List of Materials for Paints: Water-Dispersion Resins:

Products:

ARAKYD WE-100:
Description: Styrene-acrylic
Solid (%): 37.5+-1
Viscosity: 100-400 (cps)
pH: 8.5-8.3
Characteristics: Quick drying. Good water resistance
Uses: Paint for wood & interior

ARAKYD WS-200:
Description: Epoxy ester
Solid (%): 50+-1
Colour (Gardner): 10-12
Viscosity: Z4-Z6
pH: 8.0-10.0
Characteristics: Good water resistance, Good corrosion resistance
Uses: Anti-corrosive paint

ARAKYD WS-300:
Description: Alkyd
Solid (%): 60+-1
Colour (Gardner): 5 Max.
Viscosity: X-Z1
pH: 8.0-9.0

ARAKYD WS-301:
Description: Alkyd
Solid (%): 80+-1
Colour (Gardner): Z7-Z8
Viscosity: Z7-Z8

Characteristics: Hard film, High gloss, Good salt resistance
Uses: Baking paint by mixing with melamine resin

CARGILL: Acrylics:

Solid Thermoplastic Acrylic Resins:

Prod. No.:

16 0115:
DIANAL BR-115:*
Viscosity in toluol: D-G @ 40% nvm
Acid Value(Max.): 0
MW(ave): 55,000
Description: Offers good compatibility with long and medium oil alkyds.

16 0116:
DIANAL BR-116:*
Viscosity in toluol: L-O @ 40% nvm
Acid Value(Max.): 8.5
MW(ave): 45,000
Description: 16 0106 type with improved pigment dispersability for paints, inks and adhesives.

16 0117:
DIANAL BR-117:*
Viscosity in toluol: T-W @ 40% nvm
Acid Value(Max.): 0
MW(ave): 140,000
Description: Moderately hard with good flexibility for adhesives and inks.

16 0118:
DIANAL BR-118:*
Viscosity in toluol: U-Y @ 40% nvm
Acid Value(Max): 0
MW(ave): 190,000
Description: Properties between 16 0101 and 16 0112 for inks, adhesives and temporary binders for ceramics.

16 0120:
DIANAL BR-120:*
Viscosity in toluol: E-H @ 40% nvm
Acid Value(Max.): 1.0
MW(ave): 47,000
Description: Harder, more durable than 16 0115 for modification of long and medium oil alkyds.

16 0121:
DIANAL BR-121:*
Viscosity in toluol: W-Z @ 40% nvm
Acid Value(Max.): 4.0
MW(ave): 87,000
Description: Excellent hardness, adhesion and weatherability on untreated metal.

*DIANAL is a registered trademark of Mitsubishi Rayon Co., Ltd.

CARGILL: Acrylics(Continued):

Thermoplastic Acrylic Lacquer Resins:

Prod. No.:

16 7073:
% N.V.(By Wt.): 40
Volatile: MEK
Viscosity(Gardner @ 77F): U-X
Base Resin: 16 0073

16 8073:
% N.V.(By Wt.): 30
Volatile: P.M. Acetate
Viscosity(Gardner @ 77F): V-Y
Base Resin: 16 0073

16 1079:
% N.V.(By Wt.): 51
Volatile: Toluol
Viscosity(Gardner @ 77F): Z4-Z7
Base Resin: 16 0079

16 1093:
% N.V.(By Wt.): 50
Volatile: Toluol
Viscosity(Gardner @ 77F): Z2-Z5
Base Resin: 16 0093

16 1106:
% N.V.(By Wt.): 51
Volatile: Toluol
Viscosity(Gardner @ 77F): Y-Z3
Base Resin: 16 0106

16 2106:
% N.V.(By Wt.): 50
Volatile: Toluol
Viscosity(Gardner @ 77F): X-Z3
Base Resin: 16 0106

16 3106:
% N.V.(By Wt.): 51
Volatile: Xylol
Viscosity(Gardner @ 77F): Y-Z3
Base Resin: 16 0106

16 4106:
% N.V.(By Wt.): 50
Volatile: Xylol
Viscosity(Gardner @ 77F): X-Z3
Base Resin: 16 0106

CARGILL: Acrylics(Continued):

Thermoplastic Acrylic Lacquer Resins(Continued):

<u>Prod. No.:</u>

16 9111:
% N.V.(By Wt.): 40
Volatile: Mineral Thinner Solvent
Viscosity(Gardner @ 77F): X-Z1
Base Resin: 16 0111

16 5115:
% N.V.(By Wt.): 45
Volatile: VM & P Naphtha
Viscosity(Gardner @ 77F): V-Y
Base Resin: 16 0115

16 6115:
% N.V.(By Wt.): 51
Volatile: VM & P Naphtha
Viscosity(Gardner @ 77F): Z2-Z5
Base Resin: 16 0115

16 8115:
% N.V.(By Wt.): 45
Volatile: Mineral Spirits
Viscosity(Gardner @ 77F): V-Y
Base Resin: 16 0115

16 1121:
% N.V.(By Wt.): 45
Volatile: Toluol
Viscosity(Gardner @ 77F): Z3-Z6
Base Resin: 16 0121

CARGILL: Alkyd Copolymers:

Styrenated Copolymers:

Prod. No.:

53 5364:
% N.V. By Wt.: 60
Solvent: Xylene
Viscosity: Z1-Z3
Type Oil: Oxidizing Oil
Description: Highly styrenated alkyd for very fast dry. Suitable for air dry, force dry and baking finishes.

53 5365:
% N.V. By Wt.: 60
Solvent: Xylene
Viscosity: Z4-Z5
Type Oil: Oxidizing Oil
Description: Specifically designed for hammer finishes. No additives needed for hammer effect. Air dry, force dry or baking finishes.

53 5367:
% N.V. By Wt.: 50
Solvent: Xylene, MS
Viscosity: Z1-Z3
Type Oil: Oxidizing Oil
Description: Flexible styrenated alkyd.

53 5390:
% N.V. By Wt.: 60
Solvent: Xylene
Viscosity: Z-Z2
Type Oil: Mixed Oxidizing Oils
Description: A styrenated alkyd suitable for fast drying industrial and hammer finishes.

Vinyl Toluenated Copolymers:

Prod. No.:

53 5303:
% N.V. By Wt.: 50
Solvent: VM&P
Viscosity: Y-Z1
Type Oil: Mixed Oxidizing Oils
Description: A vinyl toluenated alkyd with extremely fast air dry.

CARGILL: Alkyd Copolymers(Continued):

Vinyl Toluenated Copolymers(Continued):

Prod. No.:

53 5307:
% N.V. By Wt.: 50
Solvent: VM&P/M.S.
Viscosity: Z1-Z3
Type Oil: Mixed Oxidizing Oils
Description: A rapid drying VT alkyd copolymer with excellent leveling properties.

53 5322:
% N.V. By Wt.: 50
Solvent: VM&P
Viscosity: Z3-Z5
Type Oil: Mixed Oxidizing Oils
Description: A high drink VT alkyd copolymer which features rapid dry and excellent durability.

53 5348:
% N.V. By Wt.: 60
Solvent: VM&P
Viscosity: Z-Z2
Type Oil: Oxidizing
Description: Economical copolymer offering fast dry and film hardness development. Suitable for aerosol applications.

53 5356:
% N.V. By Wt.: 60
Solvent: VM&P
Viscosity: Z1-Z3
Type Oil: Linseed
Description: A fast drying VT copolymer with good aerosol compatibility.

53 5370:
% N.V. By Wt.: 60
Solvent: VM&P
Viscosity: Z1-Z4
Type Oil: Mixed Oxidizing Oils
Description: Forms a flexible coating with a fast dry. Used in aerosols, industrial enamels, wood coatings, and zinc stearate sanding sealers. Available at 50% solids as 53 5371.

CARGILL: Alkyd Copolymers (Continued):

Styrene/Vinyl Toluene Alkyd Copolymers:

Prod. No.:

53 5323:
% N.V. By Wt.: 60
Solvent: VM&P
Viscosity: Z1-Z4
Type Oil: Soya
Description: A styrenated VT alkyd copolymer with good flexibility and adhesion along with rapid air dry. (Available at 50% nvm as 53 5324.)

Acrylic Alkyd Copolymers:

53 5345:
% N.V. By Wt.: 60
Solvent: VM&P
Viscosity: Z-Z2
Type Oil: Soya
Description: A VT/acrylic alkyd copolymer which dries extremely fast for a wide range of industrial coatings and finishes.

53 5359:
% N.V. By Wt.: 50
Solvent: Xylene
Viscosity: S-V
Type Oil: DCO
Description: Acrylic modified alkyd copolymer resin designed to give outstanding exterior durability and fast dry with excellent adhesion to all metal surfaces.

53 5373:
% N.V. By Wt.: 50
Solvent: Xylene
Viscosity: X-Z
Type Oil: LSO
Description: A coreaction product of an alkyd, Styrene and acrylonitrile. Designed for early chemical resistance to lube oils, aliphatic solvents and gasoline.

53 5374:
% N.V. By Wt.: 50
Solvent: 20% Xylene
80% Ethyl Benzene
Viscosity: X-Z
Type Oil: LSO
Description: A coreaction product of an alkyd, Styrene and acrylonitrile. Designed for early chemical resistance to lube oils, aliphatic solvents and gasoline.

CARGILL: Butylated Urea Resins:

Prod. No.:

20 2000:
% N.V. By Wt.: 60
Solvent: 60% n-Butanol
40% Xylene
Viscosity: M-P
Description: Excellent general purpose urea especially adaptable to single coat enamels of high solids content. Outstanding properties include leveling, gloss, flexibility, compatability and stability.

20 2001:
% N.V. By Wt.: 60
Solvent: n-Butanol
Viscosity: G-J
Description: Exempt solvent version of 20 2000.

20 2051:
% N.V. By Wt.: 50
Solvent: Isobutanol
Viscosity: W-Y
Description: 20 2053 @ 50% NVM

20 2052:
% N.V. By Wt.: 50
Solvent: 100% Isobutanol
Viscosity: W-Z
Description: Exempt solvent version of 20 2053

20 2053:
% N.V. By Wt.: 60
Solvent: 60% Isobutanol, 40% Xylene
Viscosity: Y-Z1
Description: Designed for rapid curing application at ordinary baking temperatures or for use at lower temperatures such as those encountered with infra-red equipment. Recommended for alkyd-urea conversion varnishes.

20 2058:
% N.V. By Wt.: 70
Solvent: 82% Isobutanol, 18% VM&P
Viscosity: Z4-Z6
Description: Similar to 20 2052 with higher solids and higher viscosity.

High Solids Butylated Urea Resins:

21 2101:
% N.V. By Wt.: 75
Solvent: n-Butanol
Viscosity: Z3-Z4+
Description: High solids butylated urea offering excellent adhesion and recoatability.

CARGILL: High Solids Melamine Resins:

Prod. No.:

RESIMENE 717:*
CARGILL 23 2317:
% N.V. By Wt.: 84
Solvent: n-Butanol
Viscosity: Y-Z3
Description: A high solid polymeric methylated melamine designed to exhibit low temp cure. Uncatalyzed cure responses as low as 250F & catalyzed cure responses as low as 200F are possible. Exhibits wide range compatibility with Polyester, Alkyds, Acrylic, Vinyls, Epoxies and Cellulosics. Alkylation alcohol is methanol. Used in high solids coatings and water-borne coatings (with Co-Solvents).

RESIMENE 735:*
CARGILL 23 2335:
% N.V. By Wt.: 80
Solvent: Isopropanol
Viscosity: T-W
Description: A high solid methylated polymeric melamine, solvent and water soluble offering fast cure response and uncatalyzed enamel stability. Able to achieve flexible impact resistant films.

RESIMENE 741:*
CARGILL 23 2341:
% N.V. By Wt.: 89
Solvent: Isopropanol
Viscosity: W-Z3
Description: A high solid methylated polymeric melamine that makes this a very versatile product. Used in both aqueous and solvent borne coatings due to excellent balance between compatability and cure response. Excellent cure response is achieved with minimal catalyst (less than 0-2% PTSA solids on melamine solids).

RESIMENE 745:*
CARGILL 23 2345:
% N.V. By Wt.: 98 min.
Viscosity: Y-Z3
Description: Liquid hexamethoxymethyl melamine. Its monomeric composition offers excellent catalyzed cure response between 180F--over 600F.

CARGILL: High Solids Melamine Resins(Continued):

Prod. No.:

RESIMENE 797:*
CARGILL 2397:
% N.V. By Wt.: 92
Solvent: Ethanol
Viscosity: W-Z
Description: A modified methylated melamine, recommended for use in conventional solvent and high solids coatings. With high levels of acid catalyst, very low cure schedules are possible, as low as 150F.

23 2390:
% N.V. By Wt.: 80
Solvent: n-Butanol
Viscosity: Z3-Z5
Description: A high solids butylated melamine with wide compatability with resins, including baking alkyds, silicone polyesters, oil-free polyesters, nitrocellulose, cellulose acetate butyrates and thermosetting acrylics.

* RESIMENE is a registered trademark of Monsanto Company

CARGILL: High Solids Resins:

Resin Code #:

21-2101:
Description: Butylated Urea
Theroetical NV: Wt. %: 75
Vol %: 67.4
Wt/Gal: 8.76+-.1
G-H Visc: Z 3 1/2 - Z 4 1/2
Solvent Composition: n-BuOH
Cure Conditions: Bake
Features: Stability, retained measured solids, and intercoat adhesion following overbake.

57-1250:
Description: Epoxy Ester
Theoretical NV: Wt. %: 80
Vol %: 77.8
Wt/Gal: 8.78+-.1
Theo OH# (Solids): 133+-10
G-H Visc: Z2-Z4
Solvent Composition: EEP/Xylene 95:5
Cure Conditions: Bake
Features: Primers and one coat systems. Excellent adhesion with melamines; flexibility with ureas.

57-4368:
Description: Oil Modified Urethane
Theoretical NV: Wt. %: 70
Vol %: 67.0
Wt/Gal: 8.20+-.1
Theo OH# (Solids): 64+-5
G-H Visc: Z1-Z3
Solvent Composition: BuA/Xylene 95:5
Cure Conditions: Air/Force
Features: Flexibility, fast dry, exterior durability and humidity resistance. Ideal alkyd modifier.

57-4550:
Description: Blocked Aromatic Isocyanate
Theoretical NV: Wt. %: 75
Vol %: 66.0
Wt/Gal: 9.16+-.1
Theo OH # (Solids): 393+-10*
G-H Visc: Z5-Z 6 1/2
Solvent Composition: MIAK
Cure Conditions: Bake
Features: Similar to 57-5797 but intended for interior application.

* NCO Equivalent Wt.

CARGILL: High Solids Resins(Continued):

Resin Code #:

57-5702:
Description: Air Dry Polymer
Theoretical NV: Wt. %: 95
Vol %: 93.9
Wt/Gal: 8.70+-.1
G-H Visc: X-Z
Solvent Composition: Xylene
Cure Conditions: Air/Bake
Features: Cures at 200F without amino resins. Hardness, low viscosity/VOC.

57-5705:
Description: Short TOFA Alkyd
Theoretical NV: Wt. %: 90
Vol %: 88.0
Theo OH # (Solids): 140+-10
G-H Visc: Z5-Z6 1/2
Solvent Composition: EEP
Cure Conditions: Bake
Features: Economical general industrial baking enamels. Good flow.

57-5707:
Description: Modified TOFA Alkyd
Theoretical NV: Wt. %: 80
Vol %: 76.0
Theo OH# (Solids): 65+-5
G-H Visc: Z4-Z6
Solvent Composition: BuA
Cure Conditions: Air/Bake
Features: Highest gloss, pigment dispersion, mar and abrasion resistance.

57-5711:
Description: Short TOFA Alkyd
Theoretical NV: Wt. %: 90
Vol %: 88.9
Theo OH# (Solids): 100+-4
G-H Visc: Z3-Z5
Solvent Composition: PMA
Cure Conditions: Bake

CARGILL: High Solids Resins(Continued):

Resin Code #:

57-5717:
Description: Coconut Alkyd
Theoretical NV: Wt. %: 85
Vol %: 83.0
Wt/Gal: 8.90+-.1
Theo OH# (Solids): 158+-10
G-H Visc: Z3-Z5
Solvent Composition: EGMBE
Cure Conditions: Bake
Features: Color retention, non-yellowing when overbaked.

57-5720:
Description: Short TOFA Alkyd
Theoretical NV: Wt. %: 75
Vol %: 67.0
Wt/Gal: 8.65+-.1
Theo OH# (Solids): 66+-5
GH Visc: Y-Z2
Solvent Composition: MPK/BuA/Xylene 83:9:8
Cure Conditions: Air/Force
Features: Fast dry, gloss, implement and machinery coatings.

57-5722:
Description: Short SOYA Alkyd
Theoretical NV: Wt. %: 85
Vol %: 82.0
Wt/Gal: 8.83+-.1
Theo OH# (Solids): 140+-10
G-H Visc: Z1-Z3
Solvent Composition: EGMBE
Cure Conditions: Bake
Features: Flexibility, viscosity/VOC, gloss. Near polyester performance.

57-5731:
Description: Short SOYA Alkyd
Theoretical NV: Wt. %: 75
Vol %: 70.0
Wt/Gal: 8.80+-.1
Theo OH# (Solids): 72+-5
G-H Visc: Z1-Z3
Solvent Composition: BuA/MPK/Xylene 65:27:8
Cure Conditions: Air/Force
Features: Fast dry, gloss and gloss retention. Ideal resin for low VOC.

CARGILL: High Solids Resins(Continued):

Resin Code #:

57-5737:
Description: Phenolic Modified Alkyd
Theoretical NV: Wt. %: 75
Vol %: 67.8
Wt/Gal: 8.72+-.1
G-H Visc: Z2-Z4
Solvent: MPK/BuA/Xylene 83:10:7
Cure Conditions: Air/Force
Features: Primer vehicle. Salt spray performance and very short recoat intervals.

57-5742:
Description: Short TOFA Alkyd
Theoretical NV: Wt. %: 90
Vol %: 87.0
Wt/Gal: 9.33+-.1
Theo OH# (Solids): 152+-10
G-H Visc: Z 5 1/2 - Z 6 1/2
Solvent Composition: Xylene
Cure Conditions: Bake
Features: Economical general industrial baking enamels.

57-5743:
Description: VT Alkyd
Theoretical NV: Wt. %: 75
Vol %: 71.2
Wt/Gal: 8.45+-.1
G-H Visc: Y-Z2
Solvent Composition: BuA/Xylene 91:9
Cure Conditions: Air
Features: Fastest dry, gloss, and viscosity/VOC.

57-5747:
Description: Silicone Alkyd
Theoretical NV: Wt. %: 80
Vol %: 75.0
Wt/Gal: 8.70+-.1
Theo OH# (Solids): 22+-5
G-H Visc: Y-Z1
Solvent Composition: MAK/BuA 75:25
Cure Conditions: Air/Force
Features: Durability, flexibility. Ideal modifier to improve performance of air dry alkyds.

CARGILL: High Solids Resins(Continued):

Resin Code #:

57-5749:
Description: Acrylic Modified Alkyd
Theoretical NV: Wt. %: 70
Vol %: 67.4
Wt/Gal: 8.30+-.1
Theo OH# (Solids): 26+-5
G-H Visc: Z-Z2
Solvent Composition: BuA/MPK/TOL/MAK 59:31:8:2
Cure Conditions: Air/Force
Features: Fast dry, adhesion, gloss, gloss retention.

57-5750:
Description: Short TOFA Alkyd
Theoretical NV: Wt. %: 90
Vol %: 89.0
Wt/Gal: 8.90+-.1
Theo OH# (Solids): 143+-10
G-H Visc: Z3-Z4
Solvent Composition: EEP
Cure Conditions: Bake
Features: Economical general industrial baking enamels.

57-5754:
Description: Phenolic Modified Alkyd
Theoretical NV: Wt. %: 75
Vol %: 67.8
Wt/Gal: 8.72+-.1
G-H Visc: Z-Z2
Solvent Composition: MPK/BuA 90:10
Cure Conditions: Air/Force
Features: Similar in performance to 57-5737. Lower viscosity/ VOC is achievable.

57-5761:
Description: Oil Free Polyester
Theoretical NV: Wt. %: 90
Vol %: 86.8
Wt/Gal: 9.75+-.1
Theo OH# (Solids): 144+-10
G-H Visc: Z3-Z5
Solvent Composition: Xylol/PMA 80:20
Cure Conditions: Bake
Features: Economical, gloss, color retention on overbake, viscosity/VOC.

CARGILL: High Solids Resins(Continued):

Resin Code #:

57-5763:
Description: Oil Free Polyester
Theoretical NV: Wt. %: 100
Vol %: 100
Wt/Gal: 9.70+-.1
Theo OH# (Solids): 224+-10
G-H Visc: Z5-Z6
Solvent Composition: None
Cure Conditions: Bake
Features: Very low viscosity/VOC, flexibility, water extendable.

57-5766:
Description: Long Oil Alkyd
Theoretical NV: Wt. %: 90
Vol %: 87.7
Wt/Gal: 8.20+-.1
G-H Visc: Z1-Z3
Solvent Composition: MS/Xylene 70:30
Cure Conditions: Air
Features: Lowest VOC air dry architectural coating.

57-5767:
Description: Long Oil Alkyd
Theoretical NV: Wt. %: 90
Vol %: 87.7
Wt/Gal: 8.20+-.1
Solvent Composition: MS/Xylene 90:10
Cure Conditions: Air
Features: Thixotropic version of 57-5766. Sag resistance.

57-5776:
Description: Oil Free Polyester
Theoretical NV: Wt. %: 85
Vol %: 82.0
Wt/Gal: 9.31+-.1
Theo OH# (Solids): 175+-10
G-H Visc: Z4-Z6
Solvent Composition: PMA
Cure Conditions: Bake
Features: High performance polyester. Excellent overall properties.

CARGILL: High Solids Resins(Continued):

Resin Code #:

57-5778:
Description: Oil Free Polyester
Theoretical NV: Wt. %: 90
Vol %: 89.0
Wt/Gal: 9.10+-.1
Theo OH# (Solids): 215+-10
G-H Visc: Z3-Z5
Solvent Composition: EGMBE
Cure Conditions: Bake
Features: Gloss, hardness, application viscosity, mar resistance, water extendable.

57-5782:
Description: Oil Free Polyester
Theoretical NV: Wt. %: 95
Vol %: 94.0
Wt/Gal: 8.65+-.1
Theo OH# (Solids): 244+-15
G-H Visc: Z3-Z5
Solvent Composition: MIBK
Cure Conditions: Bake
Features: High megohm resistance. Ideal polyester for electrostatic applications.

57-5786:
Description: Silicone Modified Polyester
Theoretical NV: Wt. %: 80
Vol %: 74.6
Wt/Gal: 9.20+-.1
Theo OH# (Solids): 83+-5
G-H Visc: Y-Z1
Solvent Composition: MAK/EEP 60:40
Cure Conditions: Bake
Features: Exterior durability, flexibility, gloss, and hardness.

57-5789:
Description: Oil Free Polyester
Theoretical NV: Wt. %: 85
Vol %: 82.7
Wt/Gal: 9.30+-.1
Theo OH# (Solids): 153+-10
G-H Visc: Z3-Z5
Solvent Composition: EEP
Cure Conditions: Bake
Features: Most flexible high performance polyester.

CARGILL: High Solids Resins(Continued):

Resin Code #:

57-5794:
Description: Oil Modified Polyurethane
Theoretical NV: Wt. %: 80
Vol %: 75.0
Wt/Gal: 8.04+-.1
G-H Visc: Z1-Z2
Solvent Composition: MS/Toluene 90:10
Cure Conditions: Air
Features: Interior architectural; flexibility, abrasion resistance.

57-5797:
Description: Blocked Aliphatic Isocyanate
Theoretical NV: Wt. %: 70
Vol %: 62.5
Wt/Gal: 8.44+-.1
Theo OH# (Solids): 355+-10*
G-H Visc: X-Z
Solvent Composition: MIAK
Cure Conditions: Bake
Features: Stable one component urethane systems. Humidity and water stain resistance. Exterior application.

57-5811:
Description: Medium SOYA Alkyd
Theoretical NV: Wt. %: 80
Vol %: 74.0
Wt/Gal: 8.68+-.1
Theo OH# (Solids): 65+-5
G-H Visc: Z-Z3
Solvent Composition: MIAK/Xylene 89:11
Cure Conditions: Air/Force
Features: Designed for 1,1,1-trichloroethane VOC solvent exempt reduction.

57-5820:
Description: Short TOFA Alkyd
Theoretical NV: Wt. %: 70
Vol %: 66.0
Wt/Gal: 8.55+-.1
Theo OH# (Solids): 85+-5
G-H Visc: Z4-Z6
Solvent Composition: MIAK/Xylene 94:6
Cure Conditions: Air/Bake
Features: Designed for 1,1,1-trichloroethane VOC solvent exempt reduction.

* NCO Equivalent Wt.

CARGILL: High Solids Resins(Continued):

Resin Code #:

57-5836:
Description: Short TOFA Alkyd
Theoretical NV: Wt. %: 70
Vol %: 60.0
Wt/Gal: 9.13+-.1
Theo OH# (Solids): 108+-5
G-H Visc: Z4-Z6
Solvent Composition: EGMEA/Xylene 94:6
Cure Conditions: Bake
Features: Designed for 1,1,1-trichloroethane VOC solvent exempt reduction.

57-5885:
Description: Oil Free Polyester
Theoretical NV: Wt. %: 85
Vol %: 81.5
Wt/Gal: 8.95+-.1
Theo OH# (Solids): 220+-10
G-H Visc: Z1-Z2
Solvent Composition: HeAc/MAK 72:28
Cure Conditions: Bake
Features: Hardness. A polyester with acrylic properties.

57-5890:
Description: Oil Free Polyester
Theoretical NV: Wt. %: 85
Vol %: 82.2
Wt/Gal: 9.40+-.1
Theo OH# (Solids): 152+-10
G-H Visc: Z4-Z-5 1/2
Solvent Composition: EEP
Cure Conditions: Bake
Features: Best performing polyester for telegraphing and corrosion resistance and adhesion to galvanized steel.

XP1111-18:
Description: Oil Modified Urethane
Theoretical NV: Wt. %: 75
Vol %: 71.6
Wt/Gal: 8.35+-.1
Theo OH# (Solids): 32+-5
G-H Visc: Z-Z2
Solvent Composition: BuA/MAK 95:5
Cure Conditions: Air/Force
Features: Flexibility, gloss, gloss retention, and humidity resistance.

CARGILL: Long Oil Alkyds:

General Purpose Trade Sales and Industrial:

Prod. No.:

50 5000:
% N.V. By Wt.: 60
Solvent: MS
Viscosity: Y-Z1
Type Oil: Soya
Description: High viscosity long oil alkyd, improved dry properties. Versatile resin, coatings can be formulated from satin to high gloss finishes.

50 5004:
% N.V. By Wt.: 40
Solvent: OMS
Viscosity: M-Q
Type Oil: Soya
Description: High quality long oil alkyd in an odorless solvent. Available at 45% nvm.

50 5048:
% N.V. By Wt.: 60
Solvent: MS
Viscosity: U-W
Type Oil: TOFA
Description: 60% N.V. version of 50 5091 for easier bulk handling.

50 5054:
% N.V. By Wt.: 70
Solvent: MS
Viscosity: Z-Z3
Type Oil: Linseed
Description: A long oil linseed alkyd with good drying properties. Used in enamels and general maintenance coatings. TT-R-266-d, type II, class A.

50 5060:
% N.V. By Wt.: 60
Solvent: MS
Viscosity: U-W
Type Oil: Soya
Description: 60% N.V. version of 50 5070 for easier bulk handling. Meets the requirements of TT-R-266-d, Type 1, Class B.

CARGILL: Long Oil Alkyds(Continued):

General Purpose Trade Sales and Industrial(Continued):

Prod. No.:

50 5065:
% N.V. By Wt.: 60
Solvent: OMS
Viscosity: Z-Z2
Type Oil: Soya
Description: Basically a 50 5070 resin in an odorless solvent. Meets the requirements of TT-P-30E.

50 5070:
% N.V. By Wt.: 70
Solvent: MS
Viscosity: Z-Z4
Type Oil: Soya
Description: Meets the requirements of TT-R-266-d, Type 1, Class A. A work horse alkyd for high quality exterior or interior enamels.

50 5071:
% N.V. By Wt.: 70
Solvent: MS
Viscosity: Z4-Z6
Type Oil: Soya
Description: High drink version of 50 5070.

50 5074:
% N.V. By Wt.: 70
Solvent: MS
Viscosity: R-T
Type Oil: Soya
Description: A soya isophthalic alkyd which features excellent exterior durability.

50 5076:
% N.V. By Wt.: 70
Solvent: OMS
Viscosity: Z1-Z2
Type Oil: Soya
Description: A versatile resin in an odorless solvent at 70% N.V.

50 5081:
% N.V. By Wt.: 60
Solvent: MS
Viscosity: U-W
Type Oil: Sun
Description: A high quality sunflower alkyd used in a wide variety of trade sales and general maintenance coatings.

CARGILL: Long Oil Alkyds(Continued):

General Purpose Trade Sales and Industrial(Continued):

Prod. No.:

50 5091:
% N.V. By Wt.: 70
Solvent: MS
Viscosity: Z3-Z5
Type Oil: TOFA
Description: TOFA version of 50 5070 that meets the requirements of TT-R-266-d, Type V, Class A.

50 5092:
% N.V. By Wt.: 60
Solvent: MS
Viscosity: W-Y
Type Oil: TOFA
Description: TOFA version of 50 5060 with a high reduced viscosity.

50 5097:
% N.V. By Wt.: 80
Solvent: Mild Aromatic TS-28R
Viscosity: X-Z1
Type Oil: TOFA
Description: Versatile grinding vehicle. Compatible with most resins & polymers.

50 5080:
% N.V. By Wt.: 70
Solvent: Xylene
Viscosity: O-S
Type Oil: Soya
Description: Vehicle to meet the requirements of TT-R-266-d, Type I, Class C.

CARGILL: Long Oil Alkyds(Continued):

Latex Modifiers:

Prod. No.:

55 5501:
% N.V. By Wt.: 98.5 Min.
Viscosity: Z1-Z3
Type Oil: Soya
Type Acid: PA
Description: Versatile very long alkyd at 98% solids. Used as a modifier in latex and oil and oil based stains and paints.

55 5503:
% N.V. By Wt.: 98 Min.
Viscosity: X-Z3
Type Oil: TOFA
Type Acid: IPA
Description: Very long oil alkyd for use as a modifier for latex and oil based paints and stains.

48 5521:
% N.V. By Wt.: 99
Viscosity: Y-Z1
Type Oil: Linseed
Type Acid: IPA
Description: #1 Lithographic ink resin.

CARGILL: Medium Oil Alkyds:

General Purpose Industrial or Maintenance:

Prod. No.:

51 5103:
% N.V. By Wt.: 50
Solvent: MS
Viscosity: Z1-Z3
Oil: Linseed
Description: Versatile linseed alkyd with rapid hard and through dry. It is also an excellent grinding vehicle.

51 5114:
% N.V. By Wt.: 50
Solvent: MS
Viscosity: V-Y
Oil: Soya/Linseed
Description: Used in air dry primers and enamels. Good polyurethane compatibility.

51 5127:
% N.V. By Wt.: 50
Solvent: MS
Viscosity: X-Z
Oil: Soya
Description: Medium oil alkyd with good flattening efficiency

51 5135:
% N.V. By Wt.: 50
Solvent: VM&P
Viscosity: Z2-Z4
Oil: Soya
Description: A very durable, fast drying medium alkyd used in implement and structural coatings.

51 5141:
% N.V. By Wt.: 55
Solvent: MS
Viscosity: Z2-Z4
Oil: Soya/Linseed
Description: Used in air dry brushable industrial maintenance coatings. Can be reduced to 50% nvm to meet the requirements of TT-R-226-d, Type III.

CARGILL: Medium Oil Alkyds(Continued):

General Purpose Industrial or Maintenance(Continued):

Prod. No.:

51 5156:
% N.V. By Wt.: 55
Solvent: MS
Viscosity: Z2-Z4
Oil: Soya/Linseed
Description: Used in air dry brushable industrial maintenance coatings. Can be reduced to 50% nvm to meet the requirements of TT-R-266-d, Type III.

51 5178:
% N.V. By Wt.: 50
Solvent: MS
Viscosity: Y-Z2
Oil: TOFA
Description: TOFA version of 51 5184.

51 5184:
% N.V. By Wt.: 50
Solvent: MS
Viscosity: Z-Z2
Oil: Soya
Description: A general purpose medium alkyd with excellent drying and working properties. Possesses excellent gloss and color retention.

51 5186:
% N.V. By Wt.: 50
Solvent: MS
Viscosity: Z4-Z6+
Oil: TOFA
Description: TOFA version of 51 5187.

51 5187:
% N.V. By Wt.: 50
Solvent: MS
Viscosity: Z4-Z6+
Oil: Soya
Description: A high drink medium alkyd.

51 5196:
% N.V. By Wt.: 50
Solvent: MS
Viscosity: V-Y
Oil: Soya
Description: Isophthalic medium-long resin with excellent exterior durability. Used in a wide variety of interior and exterior trade sales and industrial applications.

CARGILL: Medium Oil Alkyds(Continued):

General Purpose Industrial or Maintenance(Continued):

Prod. No.:

51 7602:
% N.V. By Wt.: 60
Solvent: Xylene
Viscosity: Z-Z2
Oil: Linseed
Description: A medium alkyd with fast wetting and fast drying properties. It makes an excellent metal primer or enamel.

51 7619:
% N.V. By Wt.: 60
Solvent: MS
VM&P
Iso-butanol
Viscosity: Z1-Z4
Oil: Soya
Description: Features good flexibility and adhesion and a positive air dry. Used in air dry primers and enamels.

51 7633:
% N.V. By Wt.: 50
Solvent: MS
Viscosity: U-W
Oil: TOFA
Description: A general purpose medium oil alkyd with excellent drying properties.

51 7645:
% N.V. By Wt.: 50
Solvent: MS
Viscosity: Z2-Z4
Oil: Soya/Tung
Description: Modified with Tung Oil to speed up dry, ideal for fast air dry enamels for interior and exterior applications.

51 7647:
% N.V. By Wt.: 60
Solvent: Xylene
Viscosity: S-U
Oil: TOFA
Description: A shorter oil length non-drying alkyd resin recommended for use in high quality nitrocellulose lacquers.

CARGILL: Medium Oil Alkyds(Continued):

Chain-Stopped:

Prod. No.:

51 5110:
% N.V. By Wt.: 50
Solvent: MS/EEP
Viscosity: Z4-Z6
Oil: Soya
Description: A chain stopped medium alkyd with excellent pigment wetting properties. Well suited for high gloss, durable exterior coatings. Available @ 40% nonvolatile as 51 7631.

51 5119:
% N.V. By Wt.: 50
Solvent: VM&P
Viscosity: Z2-Z4
Oil: Soya
Description: Forms a durable coating with a fast dry. Used for industrial product coatings and implement coatings.

Traffic Alkyds:

54 5465:
% N.V. By Wt.: 61
Solvent: VM&P
Viscosity: Z4-Z7
Oil: Soya
Description: General purpose traffic alkyd that meets Texas traffic alkyd requirements.
TTA

54 5472:
% N.V. By Wt.: 65
Solvent: VM&P/Toluene
Viscosity: Z2-Z4
Oil: Soya
Description: Traffic alkyd offering fast drying characteristics and good gloss.

54 5475:
% N.V. By Wt.: 50
Solvent: VM&P
Viscosity: Z1-Z3
Oil: Soya
Description: Meets Colorado traffic requirements. Has excellent speed of dry, abrasion resistance and color retention properties. Recommended as a good general purpose traffic alkyd or where fast solvent release is required in wood or metal finishes.
CTA

CARGILL: Medium Oil Alkyds(Continued):

Traffic Alkyds(Continued):

Prod. No.:

54 5482:
% N.V. By Wt.: 61
Solvent: Toluene
Viscosity: G-I
Oil: Soya
Description: Traffic alkyd offering fast dry and good gloss for road marking paint.

Flat Wall Alkyd:

56 5603:
% N.V. By Wt.: 30
Solvent: OMS
Viscosity: W-Z
Oil: TOFA
Description: Flat wall alkyd. Provides good dry, excellent holdout and good soil removal. In odorless solvent.

Thixotropic Vehicles:

56 5640:
% N.V. By Wt.: 40
Solvent: Mineral Spirits
Brookfield Viscosity @ 77F: 1,200,000-1,600,000 cps
Type Oil: Soybean/TOFA
Description: 56 5640 is a 40% nonvolatile oxidizing medium oil thixotropic alkyd designed to be used as a 20-25% modification of the vehicle in flat, satin and semi-gloss interior or exterior enamels. The gel structure is hard, which imparts good thixotropy and ease of application.

56 5650:
% N.V. By Wt.: 50
Solvent: Mineral Spirits
Brookfield Viscosity @77F: 1,200,000-1,600,000 cps
Type Oil: Soybean/TOFA
Description: 56 5650 is a 50% nonvolatile version of 56 5640.

CARGILL: Medium Oil Alkyds(Continued):

Thixotropic Vehicles(Continued):

Prod. No.:

56 5662:
% N.V. By Wt.: 60
Solvent: Mineral Spirits
Type Oil: Mixed
Description: 56 5662 is a thixotropic alkyd which is used as a modifier in other alkyds, to impart a controlled consistency to clear and pigmented coatings or in stains.

56 5663:
% N.V. By. Wt.: 45
Solvent: Mineral Spirits
Type Oil: Mixed
Description: 56 5663 is a thixotropic alkyd used as a modifier in other alkyds to improve consistency to clear and pigmented coatings.

56 5670:
% N.V. By Wt.: 70
Solvent: Mineral Spirits
Brookfield Viscosity: 650,000-850,000 cps
Type Oil: Soybean
Description: 56 5670 is a 70% nonvolatile oxidizing medium to long oil thixotropic alkyd designed to be used as the sole vehicle or 50% of the vehicle in satin, semi-gloss and gloss interior or exterior enamels. The gel structure is soft, which allows for easy handling, yet imparts good rheological, non-settling and brushing properties.

56 5691:
% N.V. By Wt.: 70
Solvent: Mineral Spirits
Brookfield Viscosity @ 77F: 25,000-35,000 cps
Type Oil: TOFA/Soybean
Description: 56 5691 is a 70% nonvolatile oxidizing thixotropic alkyd, designed to be used as the sole vehicle or as a modifier in interior or exterior enamels.

57 5767:
% N.V. By Wt.: 90
Solvent: Mineral Spirits/Xylol
Viscosity 70% in M.S.: H-K
Type Oil: Oxidizing
Description: 57 5767 is a thixotropic long oil alkyd at 90% solids designed for both interior and exterior architectural coatings which can be formulated to meet voc levels of <250 g/l.

CARGILL: Medium Oil Alkyds(Continued):

Water Extendable Alkyds:

Prod. No.:

51 7680:
% N.V. By Wt.: 50
Solvent: MS
Viscosity: Y-Z2
Oil: Mixed
Description: Water extendable alkyd.

Rosin and/or Phenolic Modified:

51 5144:
% N.V. By Wt.: 60
Solvent: VM&P
Toluol
n-Butyl Alcohol
Viscosity: Z3-Z4
Type Oil: Linseed
Description: Very fast dry rosin and phenolic modified alkyd.

CARGILL: Melamine Resins:

Prod. No.:

22 2218:
% N.V. By Wt.: 50
Solvent: 60% Isobutanol/40% Xylene
Viscosity: M-Q
Description: A good general purpose melamine formaldehyde resin having rapid cure rate, outstanding gloss retention and heavy duty resistance to weathering, detergents, acids, alkalies and household stains.

22 2219:
% N.V. By Wt.: 50
Solvent: Isobutanol
Viscosity: M-Q
Description: Exempt solvent version of 22 2218.

22 2229:
% N.V. By Wt.: 55
Solvent: 100% n-Butanol
Viscosity: C-F
Description: Exempt solvent version of 22 2228

22 2228:
% N.V. By Wt.: 55
Solvent: 55% n-Butanol
45% Xylene
Viscosity: D-F
Description: A specially designed low viscosity melamine formaldehyde coating resin.

22 2240:
% N.V. By Wt.: 50
Solvent: 75% Isobutanol
25% Xylene
Viscosity: P-S
Description: An iso-butylated melamine designed for rapid cure applications.

22 2260:
% N.V. By Wt.: 60
Solvent: 55% n-Butanol
45% Xylene
Viscosity: X-Z
Description: 60% solids version of 22 2228.

CARGILL: Oil Free Polyesters:

For Reaction with Melamines:

Prod. No.:

66 6611:
Solvent: P.M. Acetate/Ethyl Acetate 66/34
% N.V. By Wt.: 55
Viscosity: B-E
Equivalent Wt. (NVM): 435
Description: Lacquer plasticizer. For use with Nitrocellulose lacquer finishes. Yields low viscosity.

66 6612:
Solvent: Solvent G
% N.V. By Wt.: 70
Viscosity: Z-Z2
Equivalent Wt. (NVM): 488
Description: Oil free polyester designed for coil coatings. Gives excellent flexibility, color retention and adhesion.

66 6613:
Solvent: Xylene
% N.V. By Wt.: 60
Viscosity: V-X
Equivalent Wt. (NVM): 660
Description: Economical polyester for coil coatings and general purpose spray applications.

66 6616:
Solvent: P.M. Acetate/Xylol 86/14
% N.V. By Wt.: 65
Viscosity: Z3-Z6
Equivalent Wt. (NVM): 212
Description: Designed for use in two component systems for aircraft and military type coatings. With resistance to lubricating oils.

Polyol Type for 2 Component Systems:

66 6638:
% N.V. By Wt.: 60
Solvent: P.M. Acetate/Methyl Isoamyl Ketone/Methyl Ethyl Ketone 44/19/37
Viscosity: I-L
Equivalent Wt. (NVM): 350
Description: This polyester is designed as a coreactant vehicle for use with polyurethane prepolymers. When combined with CARGILL 45 4506 tough, durable non-yellowing, abrasion resistant films are achieved. This resin may also be used in combination with melamine or urea formaldehyde resins in baking systems where top of the line results are desired. Yields excellent flexibility and impact resistance.

CARGILL: Oil Modified Polyurethanes:

Aromatic:

Prod. No.:

43 4301:
% N.V. By Wt.: 50
Solvent: MS
Viscosity: Z1-Z3
Type Oil: Linseed
Description: This polyurethane is noted for its toughness and excellent gloss retention. It is recommended for maintenance primers and enamels.

43 4307:
% N.V. By Wt.: 60
Solvent: MS
Viscosity: Y-Z2
Type Oil: Soya
Description: Soya version of 43 4310.

43 4309:
% N.V. by Wt.: 50
Solvent: MS
Viscosity: Z1-Z3
Type Oil: Linseed
Description: Designed for Satin Varnishes.

43 4310:
% N.V. by Wt.: 60
Solvent: MS
Viscosity: Y-Z3
Type Oil: Linseed
Description: This polyurethane resin features an excellent combination of hardness, toughness, and abrasion resistance along with a fast through dry.

43 4311:
% N.V. by Wt.: 60
Solvent: Xylene
Viscosity: Z1-Z3
Type Oil: Linseed
Description: Xylene version of 43 4310

43 4320:
% N.V. by Wt.: 55
Solvent: MS
Viscosity: U-W
Type Oil: Soya
Description: This polyurethane resin features excellent hardness and abrasion resistance. Provides good intercoat adhesion along with a fast dry.

CARGILL: Oil Modified Polyurethanes(Continued):

Prod. No.:

43 4328:
% N.V. By Wt.: 50
Solvent: MS
Viscosity: X-Z
Type Oil: Soya
Description: Good flattening efficiency oil modified urethane.

43 4344:
% N.V. By Wt.: 61
Solvent: MS
Viscosity: Y-Z2
Type Oil: Linseed
Description: This polyurethane resin features toughness and fast dry and is widely compatible with other resins. It is recommended for use in heavy duty floor finishes and other clear finishes.

43 4355:
% N.V. By Wt.: 55
Solvent: MS
Viscosity: W-Y
Type Oil: Soya
Description: This polyurethane resin combines toughness and abrasion resistance with outstanding color retention. It is recommended for general purpose urethane varnishes and enamels.

43 4358:
% N.V. By Wt.: 50
Solvent: MS
Viscosity: Y-Z1
Type Oil: Soya
Description: This polyurethane resin is very fast drying and provides excellent hardness and abrasion resistance.

43 4359:
% N.V. By Wt.: 50
Solvent: MS
Viscosity: Z1-Z3
Type Oil: Linseed
Description: This polyurethane resin is very fast drying and provides excellent hardness and abrasion resistance.

CARGILL: Oil Modified Polyurethanes(Continued):

Aromatic(Continued):

Prod. No.:

43 4388:
% N.V. By Wt.: 50
Solvent: MS
Viscosity: Y-Z1
Type Oil: Sun/Soya
Description: A fast drying polyurethane with excellent color retention in addition to typical urethane properties.

53 5335:
% N.V. By Wt.: 50
Solvent: Xylene
Viscosity: Z-Z2
Type Oil: Soya
Description: A fast drying copolymer with good gasoline resistance and scuff resistance. Can be used in Energy Saver Formulas.

53 5338:
% N.V. By Wt.: 50
Solvent: Xylene
Viscosity: W-Y
Type Oil: Soya Fatty Acid
Description: A fast drying copolymer used in industrial wood coatings and touch up enamels.

53 5347:
% N.V. By Wt.: 50
Solvent: VM&P
n-butyl acetate
Viscosity: V-Y
Type Oil: Soya Fatty Acid
Description: Exempt solvent version of 53 5338

57 5794:
% N.V. By Wt.: 80
Solvent: MS
Viscosity: Z1-Z2
Type Oil: Soya
Description: A high solids oil modified urethane designed for interior trade sales applications.

CARGILL: Oil Modified Polyurethanes(Continued):

Aliphatic:

Prod. No.:

43 4367:
% N.V. By Wt.: 61
Solvent: VM&P
Viscosity: Z4-Z6
Type Oil: DCO
Description: Designed for industrial applications requiring gloss retention and flexibility.

57 4368:
% N.V. By Wt.: 70
Solvent: N-Butyl acetate
Viscosity: Z1-Z3
Type Oil: DCO
Description: Higher solids version of 43 4367.

43 4364:
% N.V. By Wt.: 60
Solvent: MS
Viscosity: Z5-Z7
Type Oil: DCO
Description: M.S. version of 43 4367.

URETHANE PREPOLYMERS:

Prod. No.:

45 4506:
% N.V. By Wt.: 60
Solvent: Mixed
Viscosity: A1-D
Type Isocyanate: Aliphatic

45 4518:
% N.V. By Wt.: 57.5
Solvent: Mixed
Viscosity: A2-C
Type Isocyanate: Aliphatic

Description: Aliphatic isocyanate prepolymers designed primarily for reaction with 6600 series polyesters. May be reacted with other commercially available polyesters, polyols, epoxy resins, hydroxyl functional acrylic copolymer resins or amide systems. Gives extremely tough, durable glossy films with excellent color retention and abrasion resistance.

CARGILL: Short Oil Alkyds:

Non-Oxidizing--For Baking Applications or Plasticizers for Nitrocellulose:

Prod. No.:

52 5205:
% N.V. By Wt.: 60
Solvent: Xylene
Viscosity: Z3-Z5
Type Oil: Coconut
Description: A non-drying alkyd recommended for high quality baking finishes and as a lacquer plasticizing resin or in non-yellowing acid catalyzed alkyd/urea conversion varnishes.

52 5206:
% N.V. By Wt.: 60
Solvent: VM&P
Isobutyl Alcohol
Viscosity: Z3-Z5
Type Oil: Coconut
Description: Exempt solvent version of 52 5205.

52 7566:
% N.V. By Wt.: 60
Solvent: Toluene
Viscosity: X-Z
Type Oil: Coconut
Description: Non-drying alkyd designed for improved adhesion. Used extensively for wood lacquer finishes, it also can be used in alkyd-amino thermoset coatings.
Koppers 99-4.

52 7589:
% N.V. By Wt.: 60
Solvent: Xylene
Viscosity: Z2-Z4
Type Oil: Non-oxidizing Fatty Acid
Description: Non-drying alkyd designed for improved adhesion. Can be crosslinked with amino resins used for wood lacquer finishes.
Koppers 99-5.

52 7594:
% N.V. By Wt.: 60
Solvent: Xylene
Viscosity: Z2-Z3
Type Oil: Non oxidizing Fatty Acid
Description: Non-drying alkyd designed for maximum chalk retention and chemical resistance. Used extensively in both lacquer and alkyd-amino thermoset coatings.
Koppers 94-5-60.

CARGILL: Short Oil Alkyds(Continued):

Oxidizing--Baking or Air Dry Applications:

Prod. No.:

52 5228:
% N.V. By Wt.: 50
Solvent: VM&P
Isobutyl Alcohol
Viscosity: Z-Z3
Type Oil: Soya
Description: Soybean Oil version of 52 5234.

52 5230:
% N.V. By Wt.: 50
Solvent: Xylene, VM&P
Viscosity: Z3-Z5
Type Oil: TOFA
Description: A short TOFA alkyd recommended for use in general purpose industrial baking finishes.

52 5235:
% N.V. By Wt.: 50
Solvent: Xylene, VM&P
Viscosity: Z3-Z5
Type Oil: Soya
Description: Soybean version of 52 5230.

52 5234:
% N.V. By Wt.: 50
Solvent: VM&P
Isobutyl Alcohol
Viscosity: Z-Z3
Type Oil: TOFA
Description: A general purpose alkyd recommended for use with urea and melamine resins for bake coatings.

52 5236:
% N.V. By Wt.: 55
Solvent: Xylene
Viscosity: Z7 Max
Type Oil: TOFA
Description: Recommended for use in general purpose industrial baking finishes. Can be used in Energy Saver Formulas.

CARGILL: Short Oil Alkyds(Continued):

Oxidizing--Baking or Air Dry Applications(Continued):

Prod. No.:

52 5251:
% N.V. By Wt.: 50
Solvent: Xylene
Viscosity: Z2-Z4
Type Oil: TOFA
Description: Possesses very fast bake speed and excellent hardness. Has a slow viscosity reduction curve and excellent overbake properties.

52 5265:
% N.V. By Wt.: 50
Solvent: VM&P
Isobutyl Alcohol
Viscosity: Z1-Z4
Type Oil: TOFA
Description: A good general purpose short oil alkyd recommended for use with urea and melamine resins for bake coatings.

52 5284:
% N.V. By Wt.: 55
Solvent: VM&P
Isobutyl Alcohol
Viscosity: Z7 max.
Type Oil: TOFA
Description: Exempt solvent version of 52 5236.

52 7554:
% N.V. By Wt.: 55
Solvent: 150 Solvent
n-Butanol
Viscosity: X-Z
Type Oil: TOFA
Description: A short alkyd designed for roller-coating applications.

52-7570:
% N.V. By Wt.: 60
Solvent: Xylene
Viscosity: Z2-Z5
Type Oil: Soya
Description: General purpose air dry and baking resin. Exhibits a short to medium oil length dry.

CARGILL: Short Oil Alkyds(Continued):

Oxidizing--Baking or Air Dry Applications(Continued):

Prod. No.:

52 7585:
% N.V. By Wt.: 40
Solvent: Xylene
Viscosity: Y-Z1
Type Oil: Linseed
Description: High phthalic content alkyd designed with a high acid number for use as a resinous catalyst.

52 7595:
% N.V. By Wt.: 50
Solvent: SC-150, n-Butanol
Viscosity: Z1-Z3
Type Oil: TOFA/Castor Oil
Description: High phthalic content alkyd designed for rollercoater applications.

Oxidizing/Chain Stopped--For Rapid Air Dry or Force Cure Applications:

52 5254:
% N.V. By Wt.: 50
Solvent: VM&P
Isobutyl Alcohol
Viscosity: Y-Z2
Type Oil: TOFA
Description: A short, TOFA alkyd designed for fast air dry or baking finishes. Forms a flexible but durable coating.

52 5257:
% N.V. By Wt.: 50
Solvent: VM&P
Isobutyl Alcohol
Viscosity: Y-Z2
Type Oil: Soya
Desscription: Soybean Oil version of 52 5254.

52 5270:
% N.V. By Wt.: 50
Solvent: Xylene
Viscosity: W-Z
Type Oil: TOFA
Description: Imparts extremely fast tack free time, with good corrosion, humidity and early water resistance.

CARGILL: Short Oil Alkyds(Continued):

Oxidizing/Chain Stopped--For Rapid Air Dry or Force Cure Applications(Continued):

Prod. No.:

52 7716:
% N.V. By Wt.: 50
Solvent: VM&P/Toluene/Isobutanol
Viscosity: Z5-Z7
Type Oil: TOFA
Description: 52 5251 exempt version.

52 7733:
% N.V. By Wt.: 50
Solvent: Xylene
Viscosity: Z3-Z5
Type Oil: TOFA
Description: Imparts good fast through dry and good gloss retention.

Rosin and/or Phenolic Modified:

Prod. No.:

52 5262:
% N.V. By Wt.: 50
Solvent: Xylene
Viscosity: W-Y
Type Oil: Linseed
Description: A phenolic modified short alkyd having excellent water and solvent resistance. Features a rapid air dry.

52 5203:
% N.V. By Wt.: 50
Solvent: Xylene
Viscosity: X-Z
Type Oil: Linseed
Description: A rosin and phenolic modified short alkyd, exhibits very fast dry properties, good adhesion and corrosion resistant coatings. Systems can be formulated to meet TT-P-664-C, MiL-P-22332B.

52 5266:
% N.V. By Wt.: 50
Solvent: Xylene
Viscosity: U-W
Type Oil: TOFA
Description: Economical phenolic vehicle designed for primers.

CARGILL: Short Oil Alkyds(Continued):

Rosin and/Or Phenolic Modified(Continued):

Prod. No.:

52 5268:
% N.V. By Wt.: 50
Solvent: Xylene
Viscosity: W-Y
Type Oil: TOFA
Description: Economical version of 52 5203 type resin.

52 5269:
% N.V. By Wt.: 50
Solvent: Xylene
Viscosity: W-Y
Type Oil: Soya
Description: Primers can be formulated to meet TTE-485d, Type IV.

52 7701:
% N.V. By Wt.: 50
Solvent: PM Acetate
n-Butanol
VM&P
Type Oil: Linseed
Description: Exempt solvent version of 52 5203. Systems can be formulated to meet TT-P-664-C, Mil-E-52929, MIl-P-52995.

CARGILL: Silicone Modified:

Oil Free Polyester--For Baking Applications:

Prod. No.:

57 5786:
% N.V. By Wt.: 80
Solvent: Methyl Amyl Ketone/Ethyl-3-Ethoxy Propionate 60/40
Viscosity: Y-Z1
Type Oil: Oil Free Polyester/Silicone
Description: For use in industrial baking enamels. Coatings based on 57 5786 offer excellent gloss and hardness as well as good flexibility.

Oxidizing Alkyds--For Air Dry and/or Bake Applications:

57 5747:
% N.V. By Wt.: 80
Solvent: Methyl Amyl-Ketone/n-butyl Acetate 75/25
Viscosity: Y-Z1
Type Oil: Oxidizing Fatty Acid
Description: Designed for use in industrial air dry systems, exterior maintenance coatings. Coatings exhibit excellent flexibility and exterior durability.

62 6260:
% N.V. By Wt.: 60
Solvent: MS
Viscosity: W-Y
Type Oil: Soya
Description: An air drying silicone alkyd which meets the requirements of Federal Specification TT-E-490. Recommended for exterior high gloss and semi-gloss enamels where color and gloss retention are particularly required.

62 6283:
% N.V. By Wt.: 50
Solvent: MS
Viscosity: J-K
Type Oil: Soya
Description: Will not meet TT-P-28 requirements. Designed for quick drying gloss enamels with excellent exterior durability. Similar to 62 6285 but has better color retention. Faster drying than 62 6260.

62 6285:
% N.V. By Wt.: 50
Solvent: MS
Viscosity: Q-T
Type Oil: Soya/Linseed
Description: Similar to 62 6283. Can be formulated to pass the TT-P-28 high heat requirements.

CARGILL: Silicone Modified(Continued):

Non-Oxidizing Alkyds--For Baking Applications:

Prod. No.:

62 6256:
% N.V. By Wt.: 55
Solvent: Xylol
Viscosity: U-W
Type Oil: Non-Oxidizing
Description: Designed for high heat application. Has excellent gloss, color retention and adhesion to metal in one coat applications. Excellent durability and good resistance to chemical and solvent attack. Non-drying, must be baked with melamine, urea, triazine, etc.

Copolymers--For Air Dry and/or Bake Applications:

Prod. No.:

62 6247:
% N.V. By Wt.: 50
Solvent: Xylol
Viscosity: U-X
Type Oil: Soya
Description: A silicone-epoxy copolymer for air dry or baking applications. Gives excellent chemical resistance and adhesion to difficult surfaces.

62 6281:
% N.V. By Wt.: 60
Solvent: Xylol
Viscosity: A2-D
Type Oil: None
Description: Fluxing Vehicle as a binder for aluminum pigmented finishes.
Not recommended as sole vehicle for air dry or bake applications.

62 6291:
% N.V. By Wt.: 61
Solvent: MS/n-butyl acetate
Viscosity: M-T
Type Oil: None
Description: Mineral spirits cut of 6281 solids. For use where xylol emissions are restricted.
Not recommended as sole vehicle for air dry or bake applications.

CARGILL: Water Reducible Resins:

Resin Code #:

72-7203:
Description: Oil Free Polyester
Theoretical NV: Wt. %: 75
Vol %: 68.0
Acid # (Solids): 50-60
Wt/Gal: 8.90+-.1
G-H Visc: Z3-Z5
Solvent Composition: EGMBE/s-BuOH 75:25
Cure Conditions: Bake
Features: Economics, ease of formulation and low VOC.

72-7289:
Description: Oil Free Polyester
Theoretical NV: Wt. %: 75
Vol %: 68.0
Acid # (Solids): 50-60
Wt/Gal: 9.20+-.1
G-H Visc: Z5 1/2 - Z 6 1/2
Solvent Composition: PGMPE
Cure Conditions: Bake
Features: Flexibility, corrosion resistance and exterior durability. A high performance polyester.

73-7352:
Description: Modified Linseed Oil
Theoretical NV: Wt. %: 80
Vol %: 77.7
Acid # (Solids): 68-78
Wt/Gal: 8.35+-.1
G-H Visc: Z2-Z5
Solvent Composition: EGMBE
Cure Conditions: Air
Features: Wood penetration. Use to formulate semi-transparent and opaque stains.

74-7412:
Description: Proprietary
Theoretical NV: Wt. %: 98
Vol %: 98.0
Acid # (Solids): 0-2
Wt/Gal: 9.45+-.1
G-H Visc: X-Z1
Solvent Composition: None
Cure Conditions: Force/Bake
Features: Acid catalyzed low temperature cure for paper and wood substrates. Low VOC/viscosity.

CARGILL: Water Reducible Resins(Continued):

Resin Code #:

74-7415:
Description: Medium Oil Alkyd
Theoretical NV: Wt. %: 80
Vol %: 76.1
Acid # (Solids): 53-58
Wt/Gal: 8.65+-.1
G-H Visc: Z 5 1/2 - Z 6 1/2
Solvent Composition: PGTBE
Cure Conditions: Air
Features: High gloss, low VOC/viscosity, brushability, stability. 4 hour enamels.

74-7416:
Description: Long Oil Alkyd
Theoretical NV: Wt. %: 80
Vol %: 76.6
Acid # (Solids): 53-58
Wt/Gal: 8.50+-.1
G-H Visc: Z4-Z6
Solvent Composition: PGTBE
Cure Conditions: Air
Features: Low VOC/viscosity, stability, flexibility, brushability, high gloss.

74-7426:
Description: Short Oil Alkyd
Theoretical NV: Wt. %: 75
Vol %: 69.1
Acid # (Solids): 40-46
Wt/Gal: 8.85+-.1
G-H Visc: Z 5 1/2 - Z 6 1/2
Solvent: PGMPE/s-BuOH 50:50
Cure Conditions: Air/Bake
Features: High gloss, hardness, economical, stable with anti-corrosive pigments in dip tanks.

74-7432:
Description: Acrylic Mod. Alkyd
Theoretical NV: Wt. %: 70
Vol %: 64.5
Acid # (Solids): 45-49
Wt/Gal: 8.50+-.1
G-H Visc: Z 5 1/2 - Z7
Solvent Composition: PGMPE/s-BuOH 67:33
Cure Conditions: Air/Bake
Features: Fast dry, stability, adhesion to many plastics, low temperature cure with melamine.

CARGILL: Water Reducible Resins(Continued):

Resin Code #:

74-7435:
Description: Silicone Mod. Alkyd
Theoretical NV: Wt. %: 75
Vol %: 70
Acid # (Solids): 43-48
Wt/Gal: 9.00+-.1
G-H Visc: Z5 - Z 6 1/2
Solvent Composition: EGMBE
Cure Conditions: Air
Features: High gloss, exterior durability, 30% silicone modification.

74-7445:
Description: Chain Stopped Alkyd
Theoretical NV: Wt. %: 75
Vol %: 69.8
Acid # (Solids): 36-42
Wt/Gal: 8.80+-.1
G-H Visc: Z4-Z6
Solvent: EGMBE/i-BuOH 75:25
Cure Conditions: Air/Bake
Features: Exterior durability, corrosion resistance, early water resistance.

74-7450:
Description: Short SOYA Alkyd
Theoretical NV: Wt. %: 70
Vol %: 62.5
Acid # (Solids): 47-54
Wt/Gal: 8.50+-.1
G-H Visc: Z3-Z5
Solvent Composition: n-BuOH
Cure Conditions: Bake
Features: High gloss, economy, ease of formulation, adhesion to metal

74-7451:
Description: Short TOFA Alkyd
Theoretical NV: Wt. %: 70
Vol %: 64.8
Acid # (Solids): 47-54
Wt/Gal: 8.80+-.1
G-H Visc: Z4-Z6
Solvent Composition: EGMBE
Cure Conditions: Bake
Features: Dip tank and flowcoat stability, adhesion to steel, hardness, ease of formulation.

CARGILL: Water Reducible Resins(Continued):

Resin Code #:

74-7452:
Description: Short TOFA Alkyd
Theoretical NV: Wt. %: 75
Vol %: 70.2
Acid # (Solids): 36-44
Wt/Gal: 8.95+-.1
G-H Visc: Z 3 1/2 - Z 5 1/2
Solvent Composition: EGMBE
Cure Conditions: Bake
Features: Economics, hardness, high gloss, ease of formulation.

74-7455:
Description: Coconut Alkyd
Theoretical NV: Wt. %: 70
Vol %: 64.0
Acid # (Solids): 40-45
Wt/Gal: 9.00+-.1
G-H Visc.: Z4-Z6
Solvent Composition: EGMBE
Cure Conditions: Bake
Features: Color and color retention on overbake, hardness, flexibility. Will direct rollercoat.

74-7461:
Description: Short TOFA Alkyd
Theoretical NV: Wt. %: 70
Vol %: 61.0
Acid # (Solids): 47-56
Wt/Gal: 8.80+-.1
G-H Visc: Z4-Z6
Solvent Composition: s-BuOH/EGMBE 82:18
Cure Conditions: Bake
Features: Dip tank and flowcoat stability, adhesion to steel, hardness when baked, ease of formulation.

74-7462:
Description: Acrylic Alkyd Copolymer
Theoretical NV: Wt. %: 70
Vol %: 66.2
Acid # (Solids): 40-46
Wt/Gal: 8.00+-.1
G-H Visc: Z2-Z5
Cure Conditions: Air/Force

CARGILL: Water Reducible Resins(Continued):

Resin Code #:

74-7470:
Description: Chain Stopped Alkyd
Theoretical NV: Wt. %: 75
Vol %: 69.0
Acid # (Solids): 36-41
Wt/Gal: 8.70+-.1
G-H Visc: Z 5 1/2 - Z 6 1/2
Solvent Composition: EGMBE/s-BuOH 50:50
Cure Conditions: Air/Force
Features: Early water resistance, salt spray and humidity resistance, high gloss, stability.

74-7472:
Description: Fatty Acid Alkyd
Theoretical NV: Wt. %: 75
Vol %: 69.4
Acid # (Solids): 40-45
Wt/Gal: 8.60+-.1
G-H Visc: Z6-Z6 1/2
Solvent Composition: PGMPE/s-BuOH 50:50
Cure Conditions: Air/Force
Features: Very early water resistance, fast dry, high gloss, humidity resistance.

74-7474:
Description: Chain Stopped Alkyd
Theoretical NV: Wt. %: 75
Vol %: 69.0
Acid # (Solids): 37-43
Wt/Gal: 8.80+-.1
G-H Visc: Z 6 - Z 6 1/2
Solvent Composition: EGMBE/s-BuOH 50:50
Cure Conditions: Air/Force
Features: Best overall water reducible. Stability with anti-corrosives, high gloss, corrosion resistance, gloss retention.

74-7476:
Description: Fatty Acid Alkyd
Theoretical NV: Wt. %: 75
Vol %: 69.0
Acid # (Solids): 39-44
Wt/Gal: 8.60+-.1
G-H Visc: Z6-Z 6 1/2
Solvent Composition: EGMBE/s-BuOH 50:50
Cure Conditions: Air/Force
Features: Salt spray and humidity resistance, flexibility, high gloss, stability.

CARGILL: Water Reducible Resins(Continued):

Resin Code #:

74-7478:
Description: Phenolic Modified Alkyd
Theoretical NV: Wt. %: 75
Vol %: 70.0
Acid # (Solids): 39-44
Wt/Gal: 8.65+-.1
G-H Visc: Z5 1/2 - Z 6+
Solvent Composition: EGMBE/s-BuOH 50:50
Cure Conditions: Air
Features: Early recoatability. Best corrosion resistance in the market.

XP10515-56:
Description: Styrenated Alkyd Copolymer
Theoretical NV: Wt. %: 70
Vol %: 64.6
Acid # (Solids): 59-65
Wt/Gal: 8.40+-.1
G-H Visc: Z 5 1/2 - Z 6 1/2
Solvent Composition: PGMBE/s-BuOH 67:33
Cure Conditions: Air
Features: Very fast dry, economical, adhesion to many plastics.

XP10515-57:
Description: Styrenated Alkyd Copolymer
Theoretical NV: Wt. %: 70
Vol %: 65.4
Acid # (Solids): 67-73
Wt/Gal: 8.65+-.1
G-H Visc: Z 4 1/2 - Z 6
Solvent Composition: EGMBE
Cure Conditions: Air
Features: Lowest cost water reducible, adhesion to many plastics.

SOLVENT ABBREVIATIONS:

n-BuOH	Normal butyl alcohol
i-BuOH	Isobutyl alcohol
s-BuOH	Secondary butyl alcohol
EGMBE	Ethylene glycol mono-butyl ether
PGME	Propylene glycol mono-butyl ether
PGMPE	Propylene glycol monopropyl ether
PGTBE	Propylene glycol tertiary butyl ether

COOK PAINT AND VARNISH CO.: COROC Resins and Additives:
Acrylics:

A-55-M
Hydroxyl Functional Acrylic

A-55-M is an automotive quality hydroxyl functional thermosetting acrylic resin. It is particularly reactive with butylated melamines and if catalyzed properly it will react with the methylated type. Enamels based on A-55-M exhibit hardness, high gloss and gloss retention with good exterior durability. A-55-M is supplied at 60 nonvolatile in a blend of high flash VM&P naphtha and butanol.

Properties:
- Outstanding exterior durability and weatherability
- Tough, mar resistant finishes
- High gloss
- Gloss retention
- Good U.V. resistance
- Excellent color retention

A-69-M
Thermosetting Acrylic Copolymer

A-69-M is a hydroxyl functional acrylic copolymer which is designed to be blended with melamine resins and cured at elevated temperatures. It is supplied in a high boiling aromatic and glycol ether, a solvent combination which is particularly well suited to roller and coil coating applications. Enamels based on A-69-M, both in gloss and flats, have unusually good flexibility and adhesion combined with excellent exterior durability.

A-69-M has good compatibility with many kinds of melamine and urea resins which gives the formulator considerable freedom to arrive at the properties needed for a particular application. The utility of A-69-M is enhanced by its curing characteristics since it may be baked from 1 minute at 500F. to 30 minutes at 250F.

Properties:
- Outstanding exterior durability
- Very good adhesion to metal substrates
- Excellent flexibility
- Excellent impact resistance
- Excellent color retention
- Good fabrication and forming

Suggested Uses:

Aluminum siding enamels	Drapery hardware
Awnings and venetian blinds	Metal furniture
Coating whites	Lighting fixtures

COOK PAINT AND VARNISH CO.: COROC Resins and Additives:
Acrylics(Continued):

A-399-A2
Thermosetting Acrylic Copolymer

A-399-A2 is a hydroxy functional acrylic copolymer whose composition was arrived at through years of exterior exposure history. It offers exceptional exterior gloss retention when designed into high gloss automotive quality enamels and exhibits excellent color stability in high heat enamels. When modified with conventional butylated melamines, it may be cured at bake schedules as low as 250F; and with methylated melamines, 275F.

Physical Characteristics:
- Nonvolatile: 50+-1
- Solvent:
  - xylene: 100%
- Viscosity: U-W

Properties:
- Automotive quality.
- Gloss retention.
- Overbake color retention in high heat enamels.
- Excellent melamine compatibility.
- Develops good hardness and solvent resistance at baking schedules as low as 250F with conventional melamines.
- Excellent salt spray and humidity resistance over BONDERITE 1000 or META BOND 36 treated cold rolled steel.
- Heat resistant finishes withstanding temperatures up to 400F.

Suggested Uses:
- Automotive quality finishes
- Outdoor lawn furniture
- Exterior exposed appliances (ice machines, etc.)
- Computer hardware
- Metal cabinets
- Metal shelving
- Light fixtures

COOK PAINT AND VARNISH CO.: COROC Resins and Additives:
Acrylics(Continued):

A-879-M1
Hydroxyl Functional Acrylic Copolymer

A-879-M1 is an automotive quality hydroxyl functional acrylic copolymer designed to be crosslinked with melamine resins and cured at conventional bake schedules. A-879-M1 is supplied at 50 nonvolatile in an exempt solvent blend which blend has evaporation and solvency characteristics of xylene.

Enamels based on A-879-M1 may be used for general purpose industrial applications where high gloss, hardness, and exterior durability are specified. The properties may also be of interest in appliance coatings.

Properties:
Exempt solvent
Superior exterior durability
Very high gloss
Exceptional hardness and mar resistance

Suggested Uses:
General industrial enamels
Appliance enamels
Metal furniture and shelving
Lighting fixtures
Automotive accessories
Vending machines
Exterior signs
Electrical equipment

A-2115-M
Thermosetting Acrylic Resin

COROC Resin A-2115-M is a carboxy functional acrylic resin which has been designed for modification with melamine or urea resins, particularly the methylated types. In addition, it can be modified with epoxy resins, and/or polyester resins, to produce quality films. The formulation has a wide latitude for varying both type and level of modifier to optimize the most exacting properties suited to the desired application.

Properties:
Excellent salt spray and humidity resistance
Excellent exposure
Excellent detergent resistance
Good hardness and abrasion resistance
Good flexibity and adhesion
Excellent stain resistance

Suggested Uses: Highest quality appliance enamels.

COOK PAINT AND VARNISH CO.: COROC Resins and Additives:
Acrylic(Continued):

A-3284-E3
Carboxy Functional Thermosetting Acrylic

A-3284-E3 is a carboxy functional thermosetting acrylic resin that is designed to be crosslinked with amine resins and baked at elevated temperatures. It is supplied at 60 nonvolatile in propylene glycol monomethyl ether acetate*, a solvent suited to roller and coil application. A-3284-E3 may also be blended with other resin systems; particularly polyester, to optimize the most exacting properties suited to the desired application.

* DOWANOL PM Acetate(Dow Chemical)

A-3621-E5
Thermosetting Acrylic Resin

A-3621-E5 is a high solids, low viscosity hydroxy functional acrylic resin which has been designed to be modified with melamine or urea resins. Coating formulations based on A-3621-E5 have excellent exterior durability. Because of this, A-3621-E5 is suitable for light metallics as well as straight shades.

Physical Characteristics:
Nonvolatile: 70+-1
Solvent:
Ethyl 3-ethoxypropionate: 100%

Advantages:
Excellent Exterior Durability
Excellent Abrasion Resistance
Good Detergent Resistance
Excellent Salt Spray and Humidity Resistance
Good Cure Response
Good Color Retention and Non-Yellowing
Good Stain Resistance
Not Recommended for Finishes Requiring High Degrees of Flexibility

Suggested Uses:
Outdoor Vending Machines
Household Appliances
Metal Outdoor Furniture
Metal Industrial Building
Aluminum Extrusions
Metal Shelving
Office Furniture
Laboratory Furniture
Exterior Lighting

COOK PAINT AND VARNISH CO.: COROC Resins and Additives:
Acrylics(Continued):

A-3866-A2
High Solids Thermosetting Acrylic
(Spray Application)

A-3866-A2 is a thermosetting acrylic copolymer designed for high solids, low V.O.C. coatings. Enamels based on A-3866-A2 exhibit good durability, gloss, and exterior exposure. A-3866-A2 requires a melamine crosslinker to develop good film properties. A-3866-A2 has shown excellent compatibility with polyester resins, especially COROC R-1645-H6. A-3866 is also available at 80% NV in an 80/20//MAK/Xylene blend.

Suggested Uses:
A-3866-A2 can be used in spray enamels for:
- Appliances
- Shelving
- Aluminum Extrusions

Advantages:
Spray systems can be made with this resin that are:
- Low in V.O.C.
- High in Gloss
- Good in Durability
- Good in Exterior Exposure

A-3866-M
High Solids Thermosetting Acrylic
(Spray Application)

A-3866-M is a thermosetting acrylic copolymer designed for high solids, low V.O.C. coatings. Enamels based on A-3866-M exhibit good durability, gloss, and exterior exposure. A-3866-M requires a melamine crosslinker to develop good film properties. A-3866-M has shown excellent compatibility with polyester resins, especially COROC R-1645-H6.

Physical Characteristics:
Nonvolatile: 80+-1
Solvent:
Methyl Amyl Ketone: 80%
Xylene: 20%
Viscosity: Z2-Z4

Suggested Uses:
A-3866-M can be used in spray enamels for:
Appliances--Shelving--Aluminum Extrusions

Advantages:
Low in V.O.C. High in Gloss
Good in Durability Good in Exterior Exposure

COOK PAINT AND VARNISH CO.: COROC Resins and Additives:
Acrylics(Continued):

A-3866-M
High Solids Thermosetting Acrylic
(Coil Application)

A-3866-M is a thermosetting acrylic copolymer designed for high solids, low V.O.C. coatings. Enamels based on A-3866-M exhibit excellent durability, gloss, and exterior exposure. A-3866-M requires a melamine crosslinker to develop good film properties. A-3866-M has shown excellent compatibility with polyester resins, especially COROC R-1645-H6.

Physical Characteristics:
- Nonvolatile: 80+-1
- Solvent:
  - Methyl Amyl Ketone: 80%
  - Xylene: 20%
- Viscosity: Z2-Z4

Suggested Uses:
A-3866-M can be used in coil coatings for:
- Steel and Aluminum Siding
- Over the Road Trailers

Advantages:

Low in V.O.C.	Excellent in Durability
High in Gloss	Excellent in Exterior Exposure

A-509-V3
Water Reducible Acrylic Resin

A-509-V3 is a hydroxy functional thermosetting acrylic which can be utilized as a water-reducible resin by proper neutralization with an organic amine; followed by the addition of deionized water. A-509-V3 will crosslink efficiently with water soluble melamines and urea-formaldehyde resins. This resin is suggested for use as a full gloss baking enamel for coil coating or spray application.

Physical Characteristics:
- Nonvolatile: 65+-1
- Solvent:
  - Ethylene glycol monopropyl ether: 100%
- Viscosity: Z4-Z5

Properties:

Excellent exterior durability	Excellent flexibility
Good fabrication and forming	Excellent impact resistance
Excellent gloss	

Suggested Uses:

Aluminum siding	Metal furniture
Coating Whites	Lighting Fixtures

COOK PAINT AND VARNISH CO.: COROC Resins and Additives:
Acrylics (Continued):

A-1536-V8
Thermosetting Water Reducible Acrylic Resin

A-1536-V8 is a hydroxy functional thermosetting water reducible acrylic resin, which can be made water reducible by proper neutralization with an organic amine followed by the addition of de-ionized water. A-1536-V8 can be cross-linked with water soluble melamines and urea-formaldehyde resins. This resin is suggested for use in interior or exterior industrial enamel applications. A-1536-V8 is very good for appliance coatings where water systems are needed to minimize the problems encountered with solvent base systems.

Physical Characteristics:
- Nonvolatile: 70+-1
- Solvent:
  - Butyl Cellosolve: 100%
- Viscosity: Z3-Z5

Properties:
- Good Exterior Durability
- Excellent Color
- Good Chemical and Stain Resistance
- Excellent Flexibility
- Excellent Impact Resistance

Suggested Uses:
- Interior and Exterior Baking Enamels
- General Coil Coatings
- Appliance Finishes
- Small Appliance Finishes

A-1744-V3
Water Reducible Thermosetting Acrylic Resin

A-1744-V3 is an acid functional thermosetting acrylic resin, supplied in butanol. This resin can be made water soluble by proper neutralization with an organic amine and the addition of de-ionized water. This resin can be cross-linked with water soluble melamines and urea-formaldehyde resins. This resin is suggested for use as full gloss baking enamels for coil coating or spray applications.

Properties:
- Excellent Flexibility
- Good Form and Fabrication
- Excellent Solvent Resistance
- Excellent Impact Resistance
- Excellent Gloss

Suggested Uses:
- Pigment Grinding Vehicle
- Interior Industrial Baking Enamels
- Appliance and Small Appliance Coatings

COOK PAINT AND VARNISH CO.: COROC Resins and Additives:
Acrylic(Continued):

A-1892-V3
Water Reducible Acrylic Resin

A-1892-V3 can be utilized as a water-reducible acrylic resin by proper neutralization with an organic amine followed by the addition of de-ionized water. A-1892-V3, a hydroxy functional thermosetting acrylic, will crosslink efficiently with water soluble melamines and urea-formaldehyde resins. This resin is suggested for use as a full gloss baking enamel for coil coating or spray application. Enamels based on this resin should be considered where exterior durability is required.

Physical Characteristics:
Nonvolatile: 72+-1
Solvent:
propylene glycol monoethel ether: 100%
Viscosity: Z4-Z6

Properties:
Excellent exterior durability
Excellent flexibility
Good fabrication and forming
Excellent impact resistance
Excellent gloss

Suggested Uses:
Aluminum Siding
Coating Whites
Metal Furniture
Lighting Fixtures

A-2588-V8
Water Reducible Thermosetting Acrylic Resin

A-2588-V8 is a thermosetting acrylic resin which can be reduced in de-ionized water after the resin has been properly neutralized with an organic amine. The acrylic resin can be efficiently crosslinked with water soluble melamine and urea resins. A-2588-V8 is suggested for use as a general line, spray or roller application, full gloss, baking enamel. A-2588-V8 can also be used as a near high solids coating when cured with low viscosity melamines.

Properties:
Excellent flexibility
Excellent gloss
Fair color retention
Excellent Impact Resistance
Good Fabrication and Forming

Suggested Uses:
Industrial Baking Enamel
Coating Whites
Near High Solids Coatings
Coil Coatings
Metal Household Goods

COOK PAINT AND VARNISH CO.: COROC Resins and Additives:
Acrylics(Continued):

A-2500-M1
Thermosetting Acrylic Resin

COROC Resin A-2500-M1 is a hydroxyl functional acrylic which can be cured at elevated temperatures with melamine and urea resins.

A-2500-M1 has also been satisfactorily cured with aliphatic isocyanates, in a two package system, to give a light stable polyurethane coating. Curing A-2500-M1 with isocyanates may be done at room temperature or at low oven temperatures, 150-200F.

Enamels which are prepared from this resin exhibit excellent flexibility, adhesion and exterior durability.

Physical Characteristics:

Nonvolatile: 60+-1
Solvent:
  Xylene: 75%
  Propylene Glycol Monomethyl Ether Acetate: 25%
Viscosity: Z1-Z3

Properties:

Excellent exterior durability
Very good flexibility
Very good impact resistance
Excellent color retention
Very good adhesion

Suggested Uses:

Automotive and Transportation Finishes
Heavy Duty Maintenance
Farm Implements

COOK PAINT AND VARNISH CO.: COROC Resins and Additives:
Acrylics(Continued):

XA-3977-M
Experimental Hydroxy Functional
Isocyanate-Reactive Acrylic Resin

XA-3977-M is a hydroxy functional acrylic that is designed for durable coatings when formulated with aliphatic or aromatic isocyanates. Coatings made from this system are designed for room temperature cure and are ideal for high performance application where baking is unsatisfactory. Examples of such are heavy construction equipment, buses and substrates like wood.

The higher hydroxy equivalent weight permits the formulator to use less isocyanate to achieve proper crosslinking, which allows for a more economical coating.

Physical Characteristics:
- Nonvolatile: 60+-1
- Solvent:
  - xylene: 75%
  - EKTASOLVE EP*: 25%
- Viscosity: U-W

* ethyl 3 ethoxypropionate (Eastman)

Properties:
- Excellent exterior durability
- Good chemical resistance
- Excellent color retention
- Good impact resistance

Suggested Uses:
- Farm implements
- Construction equipment
- Heavy duty maintenance
- Clear coat urethane automotive finishes

COOK PAINT AND VARNISH CO.: COROC Resins and Additives:
Acrylic(Continued):

A-222-A1
Thermoplastic Acrylic Resin

COROC resin A-222-A1 is a thermoplastic acrylic that is supplied in toluene at 50 nonvolatile. Enamels made with A-222-A1 exhibit outstanding durability, extremely high gloss and color retention.

A-222-A1 exhibits excellent resistance to vegetable oils, grease and chemical fumes.

A-222-A1 can be applied in either clear or pigmented coatings which can be air or forced dried.

Properties:
- Outstanding durability
- Extremely high gloss
- Excellent resistance to water, alcohols, and chemical fumes.
- Good color retention
- Exceptional resistance to oils and greases
- Water clear transparency

A-2081-A1
Thermoplastic Acrylic Resin

A-2081-A1 is a general purpose thermoplastic acrylic resin for use in clear and pigmented finishes. In air dry systems, A-2081-A1 will increase hardness and shorten tack free times. In baking systems; film hardness and adhesion will be increased, as well as resistance to chemical fumes. This can be very useful where a high gloss nonyellowing white is needed. Modification of existing resin systems with A-2081-A1 will increase detergent and soapy water resistance.

Physical Characteristics:
- Nonvolatile: 50+-1
- Solvent:
  - toluene: 100%
- Viscosity: Y-Z1

COOK PAINT AND VARNISH CO.: COROC Resins and Additives:
Acrylic(Continued):

A-468-A2
Thermoplastic Acrylic Resin

COROC A-468-A2 is a thermoplastic acrylic supplied in xylene at 50% nonvolatile. A-468-A2 displays excellent compatibility with many alkyd and acrylic resins and is very useful in pigment dispersion. Systems modified with A-468-A2 displayed reduced dry times, improved exposure properties and improved chemical resistance.

Typical Physical Characteristics:
Nonvolatile: 50+-1
Solvent:
Xylene: 100%
Viscosity: Y-Z1

Properties:
Excellent Gloss
Excellent Dispersion Characteristics
Good Flexibility
Good Adhesion

Suggested Uses:
Automotive and Transportation Finishes
Farm Implements
Office Furniture and Cabinets

XA-3201-M
Thermoplastic Acrylic Additive

XA-3201-M is a high solids, low viscosity acrylic additive developed to improve the dry times of high solids alkyds. Coatings modified with XA-3201-M display reduced tack-free times, improved gloss and color retention and improved chemical resistance.

Typical Physical Characteristics:
Nonvolatile: 64+-1
Solvent:
MPK: 80%
Toluene: 20%
Viscosity: U-W

Advantages:
Coatings modified with COROC XA-3201-M display:
Reduced tack-free time
Increased gloss
Increased gloss and color retention
Improved chemical resistance
Improved weatherability
Suggested Uses: Office furniture/Farm implements/
Appliances/Construction equipment

COOK PAINT AND VARNISH CO.: COROC Resins and Additives:
Additives:

A-2201-M
Acrylic Flow Control Agent

A-2201-M is a surface tension modifier for nonaqueous coating systems. A-2201-M is compatible with alkyds, oil free polyesters, acrylics, and certain epoxy modified resins. Because A-2201-M contains no silicone, it should not affect recoatability.

When A-2201-M is incorporated at levels of 0.1-2%, the following film properties will be enhanced:
- Reduction of fish-eyes, craters, and orange peel
- Improved flow and leveling
- Improved gloss
- Improved air release in high solids systems
- Improved wetting of the substrate

Physical Characteristics:
- Nonvolatile: 70+-1
- Solvent:
  - Xylene: 56%
  - Butyl Alcohol: 44%

A-2678-M
Acrylic Flow Agent for Water-Borne Systems

COROC A-2678-M is a non-silicone flow and leveling control agent for water reducible systems. Its ability to promote leveling and flow is based upon its ability to modify surface tensions. A-2678-M has no effect upon other properties of the film, especially adhesion and recoatability. It can be incorporated into a formulation by direct addition either prior to or after pigment dispersion. A-2678-M is already neutralized and reduced in water.

Physical Characteristics:
- Nonvolatile: 50+-1
- Solvent:
  - water: 55.0
  - dipropylene glycol monomethyl ether: 23.5
  - propylene glycol monomethyl ether: 2.6
  - dimethylamino ethanol: 18.9
- Viscosity: W-Y

Suggested Uses:

A-2678-M has been found to be compatible with most water reducible resins: acrylics, alkyds, epoxy esters, and polyesters.

COOK PAINT AND VARNISH CO.: COROC Resins and Additives:
Additives (Continued):

A-468-A2
Thermoplastic Acrylic Resin

COROC A-468-A2 is a thermoplastic acrylic supplied in xylene at 50% nonvolatile. A-468-A2 displays excellent compatbility with many alkyd and acrylic resins and is very useful in pigment dispersion. Systems modified with A-468-A2 displayed reduced dry times, improved exposure properties and improved chemical resistance.

Typical Physical Characteristics:
- Nonvolatile: 50+-1
- Solvent:
  - Xylene: 100%
- Viscosity: Y-Z1

Properties:
- Excellent Gloss
- Excellent Dispersion Characteristics
- Good Flexibility
- Good Adhesion

Suggested Uses:
- Automotive and Transportation Finishes
- Farm Implements
- Office Furniture and Cabinets

XA-3201-M
Thermoplastic Acrylic Additive

XA-3201-M is a high solids, low viscosity acrylic additive developed to improve the dry times of high solids alkyds. Coatings modified with XA-3201-M display reduced tack-free times, improved gloss and color retention and improved color resistance.

Typical Physical Characteristics:
- Nonvolatile: 64+-1
- Solvent:
  - MPK: 80%
  - Toluene: 20%
- Viscosity: U-W

Advantages:
Coatings modified with COROC XA-3201-M display:
- Reduced tack-free time
- Increased gloss
- Increased gloss and color retention
- Improved chemical resistance
- Improved weatherability

COOK PAINT AND VARNISH CO.: COROC Resins and Additives:
Additives(Continued):

A-3661-E3
Acrylic Wetting and Dispersing Agent

A-3661-E3 is an acrylic grinding agent designed to aid the formulator in pigment wetting and dispersion. A-3661-3 may be used in acrylic and alkyd formulation, but is especially useful in oil free polyesters where pigment dispersion can be a challenge. Optimum results are obtained by grinding the pigment in the primary resin with A-3661-E3. Poorer quality grinds (concerning viscosity and film gloss) are obtained when the melamine is incorporated in the grind.

Physical Characteristics:
Nonvolatile: 48+-1
Solvent:
propylene glycol mono-methyl ether acetate: 100%
Viscosity: Z-Z2

XA-4031-A2
Experimental Acrylic Air Release Agent

COROC XA-4031-A2 is a silicone free surface tension modifier for nonaqueous coating systems. As well as being an excellent air release agent, XA-4031-A2 significantly improves flow and leveling in a variety of systems. A stepwise addition is recommended to determine the optimum level for a particular system. Experience has shown concentrations varying from 0.1% - 2.0% to be most effective.

Typical Physical Characteristics:
Nonvolatile: 60+-1
Solvent:
Xylene: 100%
Viscosity: Alt.-A

COOK PAINT AND VARNISH CO.: COROC Resins and Additives:
High Solids:

M-1117-B
High Solids Butylated Melamine
Hexa-Butoxymethyl Melamine
(HBMM)

COROC Resin, M-1117-B, is a fully butylated melamine compound supplied at 96% solids which chemically is hexa-butoxymethyl melamine(HBMM). M-1117-B has a very high tolerance for naptha solvents and exhibits compatibility with a wide range of polymers including alkyds, polyesters and the thermoset acrylics. M-1117-B is an ideal candidate replacement for the conventional polymeric butylated melamines in the formulation of low VOC enamels and primers.

Properties:
Good exterior durability
Good humidity resistance
Good gloss and gloss retention
Very good flexibility
Low VOC
Excellent package stability
Good electrodeposition characteristics
Good intercoat adhesion
Excellent compatibility
Increased hydrophobicity

Suggested Uses:
M-1117-B should be considered as a replacement for conventional polymeric butylated melamine cross-linkers in the formulation of low VOC coatings based on alkyds of all types, polyesters, and thermoset acrylics.

COOK PAINT AND VARNISH CO.: COROC Resins and Additives: High Solids(Continued):

R-1369-E3
High Solids Polyester Resin

COROC resin R-1369-E3 is an oil-free polyester formulated for blending with melamine or urea resins for high-solids application and cured at moderately elevated temperatures. R-1369-E3 is supplied in propylene glycol monomethyl ether acetate.

This resin should be considered for use in general industrial applications. Enamels based on R-1369-E3 have shown good exterior durability with adequate adhesion and flexibility.

Coating formulations at 60-65% volume solids have been formulated with R-1369-E3 for application with conventional spray equipment. Higher volume solids are possible when hot spray application is used.

Physical Characteristics:
- Nonvolatile: 77+-1
- Solvent:
  - propylene glycol monomethyl ether acetate: 100%
- Viscosity: X-Z

Properties:
- Coatings with high volume solids
- Excellent gloss
- Good adhesion and flexibility
- Adequate salt spray and humidity resistance

Suggested Uses:
- General Industrial Use
- Coil Coatings

R-1645-H6
High Solids Polyester

R-1645-H6 is a flexible high solids coil coating polyester designed to be cross-linked with melamine resins and cured at elevated temperatures.

Physical Characteristics:
- Nonvolatile: 80+-1
- Solvent:
  - AROMATIC 100: 100%
- Viscosity: Z1-Z2

Suggested Uses:
- General line coil products

Advantages:
- Very Good Flexibility
- Higher Solids
- Good Exposure Properties

COOK PAINT AND VARNISH CO.: COROC Resins and Additives:
High Solids (Continued):

RS-296-H6
High Solids Vinyl Toluene Modified Oil

RS-296-H6 is a high solids, vinyl toluene modified oil which is supplied at 72 volatile in AROMATIC #100 solvent.

In addition to having excellent snap dry at ambient or force dry temperatures, RS-296-H6 has excellent pigment wetting characteristics and good flow. Systems using RS-296-H6 exhibit good hardness and gloss.

RS-296-H6 offers many of the advantages of styrenated alkyds while offering a high degree of tolerance for naphtha and aliphatic solvents.

RS-296-H6 should not be considered for systems requiring good exterior durability.

Properties:
- High Solids-low VOC
- Excellent flow
- Good pigment wetting characteristics
- Good naphtha tolerance
- Excellent gloss
- Excellent hardness
- Limited exterior durability

Suggested Uses:
- Floor sealers
- Fast dry industrial enamels
- Hardware enamels
- Toy enamels

S-1661-A2
High Solids Alkyd Resin

COROC S-1661-A2 is a high solids soya alkyd resin designed to be cross-linked with high solids urea or melamine resins to yield low VOC coatings. S-1661-A2 is supplied at 90 non-volatile in xylene for use in spray application.

Properties:
- Low VOC coating (Below 2.0 lbs./gal.)
- Good hardness
- Excellent color
- Good exterior durability
- Good mar resistance

Suggested Uses:
- Office Furniture
- Aluminum Extrusion
- Shelving
- Interior Lighting
- Lawn Furniture
- Toys
- File Cabinets
- General Baking Finishes

COOK PAINT AND VARNISH CO.: COROC Resins and Additives: High Solids(Continued):

S-2080-E3
High Solids Alkyd Resin

S-2080-E3 is a short oil phthalic alkyd formulated for blending with melamine or urea resins and cured at elevated temperatures. S-2080-E3 can also be cured with aromatic or aliphatic isocyanates at room temperature or at elevated temperatures. S-2080-E3 is supplied at 85% nonvolatile in propylene glycol monomethyl ether acetate.

S-2080-E3 should be considered for use in general purpose baking enamels where maximum durability is not a requirement. Enamels based on this resin have shown excellent gloss, adhesion, and flexibility.

Coating formulations at 80% volume solids have been formulated with this resin. At these solids, the coating was applied by hot spray.

Properties:
- Coatings with high volume solids
- Excellent gloss
- Very good adhesion and flexibility
- Good durability and color retention
- Good salt spray and humidity resistance

Suggested Uses:
- Metal furniture
- Small kitchen appliances
- Electrical equipment
- Lawn care equipment

SL-51-K2
High Solids Air Dry Chain Stopped Alkyd

COROC Resin SL-51-K2 is an air dry high solids chain stopped alkyd supplied at 80% non-volatile in MPK (methyl propyl ketone). SL-51-K2 is based on the most advanced technology specifically designed for development of low VOC coatings having very rapid dry and excellent exterior durability. SL-51-K2 should also be considered for low bake, low VOC industrial coatings, either modified or unmodified.

Suggested Uses:
- Implement Enamels
- Automotive Refinishing Enamels
- Drum Enamels
- Office and Lawn Furniture Enamels
- Recreational Items
- Lighting Fixtures

COOK PAINT AND VARNISH CO.: COROC Resins and Additives:
High Solids(Continued):

T-185-A1 and T-185-H8
High Solids Alkyd Resin

T-185 is a high solids tall oil modified alkyd resin designed to be crosslinked with high solids urea or melamine resins to produce low VOC coatings. T-185-A1 is supplied at 85 nonvolatile in toluene for spray applied coatings while T-185-H8 is supplied at 85 nonvolatile in high boiling aromatic solvent for roller coating enamels.

Properties:
- Low V.O.C. Coatings
- Good humidity and salt spray resistance
- Excellent color
- Very good flexibility and impact resistance
- Good mar resistance
- Good exterior durability

Suggested Uses:
- Aluminum Extrusion
- Shelving
- Lawn Furniture
- Small Implement Enamel
- Office Furniture

XA-3201-M
Thermoplastic Acrylic Additive

XA-3201-M is a high solids, low viscosity acrylic additive developed to improve the dry times of high solids alkyds. Coatings modified with XA-3201-M display reduced tack-free times, improved gloss and color retention and improved chemical resistance.

Typical Physical Characteristics:
- Nonvolatile: 64+-1
- Solvent:
  - MPK: 80%
  - Toluene: 20%
- Viscosity: U-W

Advantages:
- Reduced tack-free time
- Increased gloss
- Increased gloss and color retention
- Improved chemical resistance
- Improved weatherability

Suggested Uses:
- Office furniture
- Appliances
- Farm implements
- Construction equipment

COOK PAINT AND VARNISH CO.: COROC Resins and Additives:
High Solids(Continued):

XA-4031-A2
Experimental Acrylic Air Release Agent

COROC XA-4031-A2 is a silicone free surface tension modifier for nonaqueous coating systems. As well as being an excellent air release agent, XA-4031-A2 significantly improves flow and leveling in a variety of systems. A stepwise addition is recommended to determine the optimum level for a particular system.

Typical Physical Characteristics:
- Nonvolatile: 60+-1
- Solvent:
  - Xylene: 100%
- Viscosity: Alt.-A

XRC-360-E2X6A
High Solids Epoxy Ester Resin

COROC XRC-360-E2X6A is a high solids epoxy ester resin which can be blended with melamines for excellent baking systems or used in combination with driers for air dry coatings.

While XRC-360-E2X6A is handled much the same way as an alkyd, its properties with respect to chemical resistance and wear resistance are superior to alkyds. Its color retention and durability are good, but it may show early chalking on exterior exposure work. The latter is generally true of most epoxy coatings.

Typical Physical Characteristics:
- Nonvolatile: 75+-1
- Solvent:
  - Butyl Acetate: 100%
- Viscosity: Z1-Z3

Properties:
- Good Color Retention
- Good Exterior Durability
- Good Salt Spray and Humidity Resistance
- Very Good Adhesion to Metals
- Excellent Chemical Resistance

Suggested Uses:

XRC-360-E2X6A is recommended for use as a primer vehicle although it also performs well in topcoat applications

May be blended with melamine crosslinkers at 10% to 30% for optimum performance in baking systems.

COOK PAINT AND VARNISH CO.: COROC Resins and Additives:
Urea and Melamines:

M-1117-B
High Solids Butylated Melamine
Hexa-Butoxymethyl Melamine
(HBMM)

COROC Resin, M-1117-B, is a fully butylated melamine compound supplied at 98% solids which chemically is hexa-butoxymethyl melamine(HBMM). M-1117-B has a very high tolerance for naphtha solvents and exhibits compatibility with a wide range of polymers including alkyds, polyesters and the thermoset acrylics. M-1117-B is an ideal candidate replacement for the conventional polymeric butylated melamines in the formulation of low VOC enamels and primers.

Physical Characteristics:
Nonvolatile: 98+-2
Viscosity: Z1-Z3
Equivalent Weight: 215

Properties:
Good exterior durability
Good humidity resistance
Good gloss and gloss retention
Very good flexibility
Low VOC
Excellent package stability

Suggested Uses:
M-1117-B should be considered as a replacement for conventional polymeric butylated melamine cross-linkers in the formulation of low VOC coatings based on alkyds of all types, polyesters, and thermoset acrylics. This would include enamels requiring exterior exposure and humidity resistance.

COOK PAINT AND VARNISH CO.: COROC Resins and Additives:
Urea and Melamines(Continued):

M-281-M, M-281-V3
Melamine-Formaldehyde Resin

M-281-M is a melamine resin supplied either 100% butanol or in a combination of xylene and butanol. It is an excellent resin for modifying hydroxyl functional thermoset acrylics. M-281-M(V3) was designed to minimize overspray cratering and solvent popping with fast cure speed to reach maximum weatherability and resistance to humidity. The broad compatibility of M-281-M makes it a useful modifier for a variety of resins: short to medium oil alkyds, epoxies, and styrenated alkyds.

M-281-M and its Rule 66 exempt counterpart M-281-V3 are compatible with alkyds, thermosetting acrylics, styrenated alkyds, and epoxies.

Physical Characteristics:
M-281-M
- Nonvolatile: 60+-1
- Solvent:
  - butanol: 50%
  - xylene: 50%
- Viscosity: X-Z

M-281-V3:
- Nonvolatile: 60+-1
- Solvent:
  - butanol: 100%
- Viscosity: X-Z

Properties:
- Good exterior durability
- Excellent gloss and gloss retention
- Automotive quality finishes
- Good color and color retention
- Low temperature cure

COOK PAINT AND VARNISH CO.: COROC Resins and Additives:
Urea and Melamines (Continued):

M-8506-M
Melamine-Formaldehyde Resin

COROC Resin M-8506-M is a melamine resin designed to be used in roller-coating formulations where very short bakes and high oven temperatures are encountered. This fast curing speed may be used advantageously in many baking enamels at regular baking temperatures to give hard, mar resistant, high gloss films. M-8506-M has good compatibility with short oil alkyds and thermosetting acrylics.

Physical Characteristics:
- Nonvolatile: 50+-1
- Solvent:
  - butanol: 95%
  - xylene: 5%
- Viscosity: J-O

Properties:
- Very fast curing speed
- Very good mar resistance
- Excellent color and color retention
- Excellent gloss and gloss retention
- Very good exterior durability

Suggested Uses:
The very fast curing speed of M-8506-M makes it a very useful melamine in formulating roller coatings for high-speed coil coating lines.

M-8784-M
Melamine-Formaldehyde Resin

COROC Resin M-8784-M is a butylated melamine resin supplied in butanol-xylene at 50% nonvolatile. It may be used in baking formulations with either alkyd or acrylic resins to produce metallic or colored enamels. Metallic formulations containing M-8784-M are characterized by their richness in depth of gloss and luster.

Properties:
- Good curing speed
- Excellent exterior durability
- Excellent color and color retention
- Very good resistance to water soak
- Very good resistance to humidity

Suggested Uses:
M-8784-M, when blended with short oil alkyds or thermosetting acrylics, produces baking finishes which may be used for painting many metal goods.

COOK PAINT AND VARNISH CO.: COROC Resins and Additives:
Urea and Melamines(Continued):

U-4834-M
Urea-Formaldehyde Resin

COROC Resin U-4834-M is a fast curing urea resin which has been widely used in many different kinds of baking finishes. It offers to the formulator a lower cost conversion resin while maintaining many of the performance properties of the more expensive melamine resins.

It may be used as the sole modifier resin or in combination with melamine resins to produce coatings having good hardness and flexibility along with outstanding adhesion. The compatibility characteristics of U-4834-M indicate it may be used in short oil alkyds, epoxy esters, and vinyls.

Physical Characteristics:
- Nonvolatile: 50+-1
- Solvent:
  - butanol: 50%
  - xylene: 50%
- Viscosity: W-Y

Suggested Uses:
The outstanding performance of COROC Resin U-4834-M makes it an extremely useful conversion resin for oxidizing or non-oxidizing alkyds, epoxies, or epoxy esters.

U-5185-M
Urea-Formaldehyde Resin

COROC Resin U-5185-M is a butylated urea resin designed to give improved compatibility with medium to long oil alkyds and acrylic resins. It is slower curing than COROC U-4834-M and gives slightly softer films at equal baking schedules. This may be an advantage since a higher concentration of low cost urea resin may be incorporated into the formulation. U-5185-M may be used to modify all types of resins to give higher solids at reduced viscosity in all baking applications.

Physical Characteristics:
- Nonvolatile: 60+-1
- Solvent:
  - butanol: 90%
  - xylene: 10%
- Viscosity: S-V

Suggested Uses:
U-5185-M may be used to modify a broad range of resins: short to long oil alkyds, thermosetting acrylics, epoxies, styrenated alkyds, and epoxy esters.

COOK PAINT AND VARNISH CO.: COROC Resins and Additives:
Water Dispersible:

Z-27-M
Water Dispersible Alkyd Resin

COROC Z-27-M is a low cost, chain stopped, water dispersible alkyd supplied at 70% NV in a 50/50 blend of butanol and ethylene glycol monobutyl ether. It is a lower solids version of Z-1727-M. Z-27-M is recommended for industrial metal finishes with an air dry or force dry cure schedule. COROC resins Z-1726-M and Z-1727-M1 are recommended for those applications which require a faster dry than obtained with Z-27-M.

Suggested Uses:
- Automotive under the hood(chassis, shocks, water pumps, etc.)
- Implements and hardware
- General industrial finishes

Advantages:
- Low VOC
- Good exterior durability
- Good corrosion and humidity resistance
- Excellent application properties

Z-1722-M
Water Dispersible Chain Stopped Alkyd Resin

COROC Resin Z-1722-M is a water dispersible alkyd designed to give exceptional salt spray and humidity resistance coupled with outstanding exterior durability. Coatings formulated with Z-1722-M have excellent application properties at low VOC values with moderately fast dry. These finishes exhibit outstanding adhesion and impact-resistance.

Z-1722-M may be used in primers and topcoats for spray, dip or flowcoat application and may be air dried or baked.

Physical Characteristics:
- Nonvolatile: 75+-1
- Volatile:
  - 2-Butoxyethanol: 50%
  - n-butanol: 50%

Properties:
- Excellent humidity resistance
- Excellent salt spray resistance
- Very good exterior durability
- Excellent application properties
- Low VOC
- Good package stability
- Excellent impact resistance
- Excellemt adhesion

Suggested Uses:
- Automotive
- Plastics
- Metal furniture
- Implement and heavy equipment
- Appliances

COOK PAINT AND VARNISH CO.: COROC Resins and Additives:
Water Dispersible(Continued):

Z-1726-M
Water Dispersible Chain Stopped Alkyd

COROC Z-1726-M is a chain stopped water dispersible alkyd resin supplied at 75 nonvolatile in a 1:1 mixture of 2-butoxyethanol and sec-butanol.

COROC Z-1726-M was designed for air-dry and low-bake coatings. It may be used in high or low gloss enamels and primers for spray, dip or flow coat application. COROC Z-1726-M has outstanding corrosion and humidity resistance coupled with excellent exterior durability.

Properties:
- Very good gloss
- Excellent flexibility
- Very good humidity resistance
- Very good salt spray resistance
- Very good durability
- Good stability

Suggested Uses:
- Automotive under the hood parts
- Small appliances
- Recreational equipment
- Implement and hardware

Z-1726-M1
Water Dispersible Chain-Stopped Alkyd

COROC Z-1726-M1 is a chain-stopped water dispersible alkyd resin supplied at 75 nonvolatile in a 1:1 mixture of PROPASOL P and sec-butanol.

COROC Z-1726-M1 was designed for air-dry and low-bake coatings. It may be used in high or low gloss enamels and primers for spray, dip or flow coat application. COROC Z-1726-M1 has outstanding corrosion and humidity resistance coupled with excellent exterior durability.

Properties:
- Very good gloss
- Excellent flexibility
- Very good humidity resistance
- Very good salt spray resistance
- Very good durability
- Good stability

Suggested Uses:
Automotive under the hood parts/Small appliances/Recreational equipment/Implement and hardware

COOK PAINT AND VARNISH CO.: COROC Resins and Additives:
Water Dispersible(Continued):

Z-1727-M
Water Dispersible Chain Stopped Alkyd Resin

COROC Z-1727-M is a low cost, chain stopped, water dispersible alkyd supplied at 75% NV in a 50/50 blend of butanol and BUTYL CELLOSOLVE. Z-1727-M is recommended for industrial metal finishes with an air dry or force dry cure schedule. For those applications requiring faster dry, COROC Z-1726-M and Z-1727-M1 are recommended.

Suggested Uses:

- Automotive under the hood parts
- Lawn furniture
- Appliances
- Hardware

Advantages:

- Low VOC
- Good salt spray and humidity resistance
- Excellent application properties-spray, dip or flow coat
- Good impact resistance
- Good exterior durability

Z-1727-M1
Water Dispersible Chain Stopped Alkyd

COROC Z-1727-M1 is a chain stopped, water dispersible alkyd resin supplied at 75% NV in a 50/50 blend of sec-butanol and PROPASOL P. Z-1727-M1 is recommended for general industrial or automotive under the hood parts requiring good corrosion and humidity resistance as well as fast dry. For those applications requiring a higher degree of corrosion resistance, COROC Z-1726-M is recommended.

Suggested Uses:

- Automotive under the hood
- Implements and hardware
- General maintenance coatings
- Scaffolding
- Meter housings

Advantages:

- Low VOC
- Good salt spray and humidity resistance
- Good application properties

COOK PAINT AND VARNISH CO.: COROC Resins and Additives:
Water Dispersible(Continued):

Z-2055-M
Water Reducible Chain Stopped Alkyd Resin

Z-2055-M is a medium oil alkyd which when neutralized with a suitable amine, is reducible in water. The resin is supplied at 70% nonvolatile in a 1:1 solvent blend of propylene glycol monomethyl ether and butanol.

Z-2055-M may be used in an air dry application where very good drying characteristics are required. This resin may also be blended with water soluble melamines and ureas for use in off-white and light colored baking enamels.

Z-2055-M offers good flexibility, excellent gloss, and durability.

Properties:
- Excellent exterior durability
- Very good gloss
- Excellent color and color retention
- Good impact resistance
- Good humidity resistance

Suggested Uses:
- Implement and hardware
- Primers
- Automotive accessories
- Metal signs and building products
- Household appliances

XZ-128-MXA
Water Dispersible Alkyd

COROC XZ-128-MXA is an experimental water dispersible chain-stopped alkyd. It is supplied in a 50/50 blend of butoxyethanol and sec-butanol at 70% nonvolatile. The material displays excellent humidity resistance and application properties.

Properties:
- Excellent Package Stability
- Good Corrosion and Humidity Resistance
- Excellent Application Properties
- Good Tack-Free Times

Suggested Uses:
- Automotive Under the Hood Parts
- Small Appliances
- Recreational Equipment
- Implement and Hardware

COOK PAINT AND VARNISH CO.: COROC Resins and Additives:
Water Dispersible(Continued):

XZ-128-M1X2
Water Dispersible Alkyd

COROC XZ-128-M1X2 is a low cost version of XZ-128-MXA Experimental Water Dispersible Chain-Stopped Alkyd. It is supplied in a 50/50 blend of butoxyethanol and sec-butanol at 70% nonvolatile. Like XZ-128-MXA, the material displays excellent humidity resistance and application properties.

Properties:
- Excellent Package Stability
- Good Corrosion and Humidity Resistance
- Excellent Application Properties
- Good Tack-Free Times
- Low Cost Relative to Similar Systems

Suggested Uses:
- Automotive under the Hood Parts
- Small Appliances
- Recreational Equipment
- Implement and Hardware

COOK PAINT AND VARNISH CO.: COROC Resins and Additives:
Water Reducible:

A-509-V3
Water Reducible Acrylic Resin

A-509-V3 is a hydroxy functional thermosetting acrylic which can be utilized as a water-reducible resin by proper neutral-iazation with an organic amine; followed by the addition of deionized water. A-509-V3 will crosslink efficiently with water soluble melamines and urea-formaldehyde resins. This resin is suggested for use as a full gloss baking enamel for coil coating or spray application. Enamels based on this resin should be considered where exterior durability is required.

Physical Characteristics:
- Nonvolatile: 65+-1
- Solvent: Ethylene glycol monopropyl ether: 100%
- Viscosity: Z4-Z5

Properties:
- Excellent exterior durability
- Excellent flexibility
- Good fabrication and forming
- Excellent impact resistance
- Excellent gloss

Suggested Uses:
- Aluminum siding
- Coating Whites
- Metal furniture
- Lighting Fixtures

A-1536-V8
Thermosetting Water Reducible Acrylic Resin

A-1536-V8 is a hydroxy functional thermosetting water reducible acrylic resin, which can be made water reducible by proper neutralization with an organic amine followed by the addition of de-ionized water. A-1536-V8 can be cross-linked with water soluble melamines and urea-formaldehyde resins. This resin is suggested for use in interior or exterior industrial enamel applications: A-1536-V8 is very good for appliance coatings where water systems are needed to minimize the problems encountered with solvent base systems.

Physical Characteristics:
- Nonvolatile: 70+-1
- Solvent:
  - BUTYL CELLOSOLVE: 100%
- Viscosity: Z3-Z5

Properties:
- Good Exterior Durability
- Good Chemical and Stain Resistance
- Excellent Impact Resistance
- Excellent Color
- Excellent Flexibility

Suggested Uses:
- Interior and Exterior Baking Enamels
- Small Appliance Finishes
- General Coil Coatings
- Appliance Finishes

COOK PAINT AND VARNISH CO.: COROC Resins and Additives:
Water Reducible(Continued):

A-1744-V3
Water Reducible Thermosetting Acrylic Resin

A-1744-V3 is an acid functional thermosetting acrylic resin, supplied in butanol. This resin can be made water soluble by proper neutralization with an organic amine and the addition of de-ionized water. This resin can be cross-linked with water soluble melamines and urea-formaldehyde resins. This resin is suggested for use as full gloss baking enamels for coil coating or spray applications.

Physical Characteristics:
- Nonvolatile: 75+-1
- Solvent:
  - butanol: 100%
- Viscosity: Z5-Z6

Properties:
- Excellent Flexibility
- Good Form and Fabrication
- Excellent Solvent Resistance
- Excellent Impact Resistance
- Excellent Gloss

Suggested Uses:
Pigment Grinding Vehicle/Interior Industrial Baking Enamels/ Appliance and Small Appliance Coatings

A-1892-V3
Water Reducible Acrylic Resin

A-1892-V3 can be utilized as a water-reducible acrylic resin by proper neutralization with an organic amine followed by the addition of de-ionized water. A-1892-V3, a hydroxy functional thermosetting acrylic, will crosslink efficiently with water soluble melamines and urea-formaldehyde resins. This resin is suggested for use as a full gloss baking enamel for coil coating or spray application. Enamels based on this resin should be considered where exterior durability is required.

Physical Characteristics:
- Nonvolatile: 72+-1
- Solvent:
  - Ethylene Glycol Monoethyl Ether: 100%
- Viscosity: Z4-Z6

Properties:
Excellent exterior durability/Excellent flexibility/Good fabrication and forming/Excellent impact resistance/Excellent gloss

Suggested Uses:Aluminum Siding/Metal Furniture/Coating Whites/Lighting Fixtures

COOK PAINT AND VARNISH CO.: COROC Resins and Additives:
Water Reducible(Continued):

A-2588-V8
Water Reducible Thermosetting Acrylic Resin

A-2588-V8 is a thermosetting acrylic resin which can be reduced in water after the resin has been properly neutralized with an organic amine. The acrylic resin can be efficiently crosslinked with water soluble melamine and urea resins. A-2588-V8 is suggested for use as a general line, spray or roller application, full gloss, baking enamel. A-2588-V8 can also be used as a near high solids coating when cured with low viscosity melamines.

Physical Characteristics:
- Nonvolatile: 80+-1
- Solvent:
  - BUTYL CELLOSOLVE: 100%
- Viscosity: Z3-Z5

Properties:
- Excellent flexibility
- Fair color retention
- Good Fabrication and Forming
- Excellent gloss
- Excellent Impact Resistance

Suggested Uses:
- Industrial Baking Enamel
- Coating Whites
- Near High Solids Coatings
- Coil Coatings
- Metal Household Goods

A-2678-M
Acrylic Flow Agent for Waterborne Systems

COROC A-2678-M is a non-silicone flow and leveling control agent for water reducible systems. Its ability to promote leveling and flow is based upon its ability to modify surface tensions. A-2678-M has no effect upon other properties of the film, especially adhesion and recoatability. It can be incorporated into a formulation by direct addition either prior to or after pigment dispersion. A-2678-M is already neutralized and reduced in water.

Physical Characteristics:
- Nonvolatile: 50+-1
- Solvent:
  - water: 55.5
  - dipropylene glycol monomethyl ether: 23.2
  - propylene glycol monomethyl ether: 2.6
  - dimethylamino ethanol: 18.7
- Viscosity: W-Y

Suggested Uses:
- Compatible with most water reducible resins.

COOK PAINT AND VARNISH CO.: COROC Resins and Additives:
Water Reducible(Continued):

B-453-V8
Water Reducible Styrenated Epoxy Resin

COROC Resin B-453-V8 is a styrenated epoxy ester which may be used for both air dry and bake applications. When neutralized this resin can readily be reduced with water. B-453-V8 can be cured with water dispersible melamine or urea resins at elevated temperatures. B-453-V8 can also be air dried with metallic driers at room temperature. Superior humidity and corrosion resistance can be attained with B-453-V8 when used in either air dry or bake primers. This resin exhibits good cure response at 220 to 350F and air dry to tack free in pigmented films in about one hour. This resin also has good gloss, good adhesion, and good compatibility with other water reducible resins.

Physical Characteristics:

Nonvolatile: 70+-1
Solvent:
Ethylene Glycol Monobutyl Ether: 100%
Viscosity: Z5-Z6+

Properties:

Excellent Adhesion
Excellent Humidity Resistance
Excellent Corrosion and Salt Spray Resistance
Good Hydrolytic Stability
Very Good Gloss
Excellent Early Water Resistance
Good Compatibility with other Resins
Good Cure Response

Suggested Uses:

B-453-V8 is recommended to be used where fast dry and superior corrosion resistance are required. It should be considered as a primer vehicle for farm implements, automotive parts, outdoor furniture, and appliances. B-453-V8 gives excellent hardness, adhesion to the substrate, and intercoat adhesion. B-453-V8 is not recommended for exterior use where chalk resistance and gloss retention are critical.

COOK PAINT AND VARNISH CO.: COROC Resins and Additives:
Water Reducible(Continued):

B-1537-M
Water Reducible Acrylic Modified Alkyd

COROC resin B-1537-M water reducible acrylic modified alkyd was designed for high quality, rapid dry coatings. When neutralized with a suitable amine, B-1537-M will give excellent film properties and may be used in air or force dry systems. This resin may also be blended with water reducible melamines to obtain full gloss enamels which are hard and chemical resistant.

B-1537-M is supplied at 75% nonvolatile in ethylene glycol monobutyl ether and butanol.

Properties:
- Excellent Spray Characteristics
- Good Humidity and Salt Spray Resistance
- Excellent Mar Resistance
- Excellent Color and Gloss Retention
- Very Good Exterior Durability
- Rapid Dry
- Good Chemical Resistance

Suggested Uses:
- Small Appliances
- Garden Tools and Small Implement Enamels
- Engine Enamels
- Primers for Structural Steel
- Toy Enamels
- Computer Hardware

R-1782-V8
Water Reducible Polyester

COROC resin R-1782-V8 is a high solids water soluble oil free polyester that is supplied at 81 nonvolatile in Ethylene Glycol Monobutyl Ether. Upon neutralization with a suitable amine and reduced with water, R-1782-V8 has good hydrolytic stability over a long period of time.

Enamels based on R-1782-V8 should be considered for coil coatings and general industrial bake systems. R-1782-V8 will crosslink with water soluble melamines or ureas to form a hard thermoset film that exhibits excellent flexibility, gloss, and color retention.

Properties:
- Excellent Color Retention
- Excellent Exterior Durability
- Good Humidity Resistance
- Excellent Gloss Retention
- Good Mar Resistance

Suggested Uses:
Coil Coatings/Small Appliances/Office Equipment/Lawn Furniture/Lighting Fixtures(exterior and interior/Metal Shelving

COOK PAINT AND VARNISH CO.: COROC Resins and Additives:
Water Reducible(Continued):

S-432-M
Water Reducible Alkyd

COROC Resin S-432-M is rapid dry, water reducible, soya alkyd. The resin is supplied at 71% nonvolatile in a 45:55 mixture of isobutyl alcohol and BUTYL CELLOSOLVE, respectively. In order to reduce S-432 in water, the resin must be first neutralized with an appropriate amine.

S-432-M may be used in baking enamels by blending the resin with water soluble melamine or urea resins. S-432-M may also be force air dried at elevated temperatures using metallic driers.

Enamels prepared with S-432-M have high gloss retention and very good durability. Some yellowing may occur at high bake temperatures. However, increasing the melamine level may improve this discoloration.

Properties:

- Excellent gloss
- Good flexibility
- Excellent durability
- Very good hardness and mar resistance
- Good humidity resistance

Suggested Uses:

- Automotive under the hood parts
- Small appliances
- Recreational equipment
- Implement and hardware
- Toy enamels

T-1070-M
Water Reducible Alkyd Resin

COROC Resin T-1070-M is a water reducible medium tall oil alkyd resin designed especially for use in industrial baking finishes. T-1070-M may be blended with water soluble melamine or urea-formaldehyde resin to obtain full gloss enamels.

Enamels prepared from T-1070-M produce hard, tough, mar resistant finishes having good resistance to salt spray and humidity.

Physical Properties:

- Nonvolatile: 75+-1
- Solvent:
  - propylene glycol monomethyl ether: 50%
  - butanol: 50%
- Viscosity: Z2-Z4

Properties:

Excellent spray characteristics/Good humidity and salt spray resistance/Excellent mar resistance/Excellent color and gloss/Very good exterior durability

Suggested Uses:

Water heater covers/Metal cabinets/Heat registers/Metal signs

DANIEL PRODUCTS CO.: SYNOTEX 800 and ALSYNOL RS-47:

Cyclized Rubber Resins for Coatings

SYNOTEX 800 and ALSYNOL RS-47 are hard resins derived from synthetic rubber by isomerization to form ring structures. They are classified as cyclized synthetic isoprene or, more simply, cyclized rubber.

Properties:

SYNOTEX 800 and ALSYNOL RS-47 have excellent solubility in aliphatic as well as aromatic solvents and good compatibility with drying oils, long- and medium-oil alkyds and some acrylics and short oil alkyds. They are insoluble in alcohols, esters, glycols and ketones and incompatible with nitrocellulose and chlorinated rubber. SYNOTEX 800 and ALSYNOL RS-47 have good solvent release and dry rapidly by solvent evaporation.

SYNOTEX 800 is a high viscosity grade of cyclized rubber which forms hard films when unmodified. ALSYNOL RS-47 is a low viscosity grade which produces softer, more flexible films. Both cyclized rubber resins contain double bonds which will crosslink on aging, causing a gradual increase in solvent resistance.

Specifications:

SYNOTEX 800:

Acid Value: 5 Max.
Color (Gardner) 50% in Toluene: 9 Max.
Melting Range (Capillary Tube): 115-125C
Typical Melting Point (Ring and Ball): 130-140C
Wt./Gal. (Sp.Gr.): 8.3 Lbs. (1.00)
Viscosity, Brookfield Spindle #4 @ 60 rpm
50% in Toluene: 600+-50 cps
50% in Mineral Spirits: 5,000+-500 cps

ALSYNOL RS-47:

Acid Value: 5 Max.
Color (Gardner) 50% in Toluene: 9 Max.
Melting Range (Capillary Tube): 80-90C
Typical Melting Point (Ring and Ball): 110-125C
Wt./Gal. (Sp.Gr.): 8.3 Lbs. (1.00)
Viscosity, Brookfield Spindle #4 @ 60 rpm
50% in Toluene: 350+-30 cps
50% in Mineral Spirits: 1,300+-200 cps

DOCK RESINS CORP.: Acrylic Resin Solutions:

Thermoplastic Acrylics:

DORESCO Products:

AC2-25:
% Solids: 40
Solvents: Toluol
Viscosity: Z-Z2
Type: Terpolymer
Remarks: Adhesion promoted, soft, impact resistant film

AC2-50:
% Solids: 40
Solvents: Toluol
Viscosity: 700-1700 cps
Type: Terpolymer
Remarks: Mar resistant cement sealer, metalized plastics, automotive refinishing vehicle

AC4-26:
% Solids: 40
Solvents: Toluol
Viscosity: W-Y
Type: Copolymer
Remarks: Adhesion to glass, metals

AC4-64:
% Solids: 45
Solvents: Toluol
Viscosity: 4000-8000 cps
Type: Copolymer
Remarks: Adhesion to glass, metals

AC4-74:
% Solids: 45
Solvents: Toluol
Viscosity: 6000-10000 cps
Type: Copolymer
Remarks: Adhesion to glass, metals

AC7-4:
% Solids: 30
Solvents: Toluol-Butanol-MEK
Viscosity: 210-280 cps
Type: Homopolymer
Remarks: Maximum solids, hardness, automotive refinish vehicle

DOCK RESINS CORP.: Acrylic Resins Solutions(Continued):

Thermoplastic Acrylics(Continued):

DORESCO Products:

AC13-1:
% Solids: 30
Solvents: Toluol, Butanol
Viscosity: 235-365 cps
Type: Homopolymer
Remarks: Maximum hardness homopolymer

AC22-2:
% Solids: 51
Solvents: Toluol
Viscosity: 4000-6000 cps
Type: Terpolymer
Remarks: Compatible with vinyls, cellulosics

AC24-15:
% Solids: 50
Solvents: Toluol
Viscosity: W-Y
Type: Copolymer
Remarks: Medium hardness, N/C compatible

AC24-15MIBK:
% Solids: 50
Solvents: MIBK
Viscosity: U-W
Type: Copolymer
Remarks: Medium hardness, N/C compatible

AC25-14:
% Solids: 45
Solvents: Aliphatics
Viscosity: V-X
Type: Homopolymer
Remarks: Alkyd compatible; maintenance paints, auto refinish

AC25-22:
% Solids: 45
Solvents: VM&P
Viscosity: 800-1600 cps
Type: Homopolymer
Remarks: Alkyd compatible; maintenance paints, auto refinish

DOCK RESINS CORP.: Acrylic Resins Solutions(Continued):

Thermoplastic Acrylics(Continued):

DORESCO Products:

AC25-25:
% Solids: 45
Solvents: Mineral Spirits
Viscosity: 800-1600 cps
Type: Homopolymer
Remarks: Softer than AC25-22 type, soluble in low KB solvents

AC27-2:
% Solids: 50
Solvents: Toluol
Viscosity: 3700-6200 cps
Type: Copolymer
Remarks: Softer film than L91-1 types; aerosol paint vehicle

AC27-14:
% Solids: 50
Solvents: Toluol
Viscosity: 1350-2250
Type: Copolymer
Remarks: Lower viscosity version of AC27-2 for easier spraying

AC32-17:
% Solids: 50
Solvents: S-100
Viscosity: Y-Z2
Type: Terpolymer
Remarks: Silk screen ink vehicle; similar to AC32-18

AC32-18:
% Solids: 50
Solvents: Xylol, Toluol
Viscosity: 2400-3800 cps
Type: Terpolymer
Remarks: Compatible with short oil alkyds; maintenance paints

AC32-22:
% Solids: 50
Solvents: Xylol, Toluol
Viscosity: 2400-2800 cps
Type: Terpolymer
Remarks: Compatible with short oil alkyds, epoxies

DOCK RESINS CORP.: Acrylic Resin Solutions(Continued):

Thermoplastic Acrylics(Continued):

DORESCO Products:

AC86-1:
% Solids: 45
Solvents: Toluol
Viscosity: 5000-9000 cps
Type: Copolymer
Remarks: High molecular weight medium hardness film

AC86-4:
% Solids: 45
Solvents: MEK
Viscosity: 3000-6000 cps
Type: Copolymer
Remarks: High molecular weight, nitrocellulose compatible

AC92-9:
% Solids: 51
Solvents: Toluol
Viscosity: 2000-5000 cps
Type: Copolymer
Remarks: Medium hard film. Low cost, general purpose acrylic

AC92-9X:
% Solids: 50
Solvents: Xylol
Viscosity: 5000-8000 cps
Type: Copolymer
Remarks: AC92-9 in Xylol

AC102-8:
% Solids: 40
Solvents: Mineral Spirits
Viscosity: 1400-2600 cps
Type: Homopolymer
Remarks: Soft aliphatic soluble acrylic

AC110-8:
% Solids: 40
Solvents: Ethyl Acetate-Toluol
Viscosity: Z-Z3
Type: Copolymer
Remarks: Medium hard, nitrocellulose compatible

DOCK RESINS CORP.: Acrylic Resin Solutions(Continued):

Thermoplastic Acrylics(Continued):

DORESCO Products:

AC129-3:
% Solids: 40
Solvents: Ethyl Acetate-Toluol
Viscosity: 3000 cps
Type: Copolymer
Remarks: Transfer printing vehicle

AC152-19:
% Solids: 25
Solvents: IPAc-Toluol
Viscosity: O-T
Type: Terpolymer
Remarks: Flexible, transparent-film primer and ELECTROFAX binder

L91-1:
% Solids: 51
Solvents: Toluol
Viscosity: 2000-5000 cps
Type: Copolymer
Remarks: Medium hard film. General purpose acrylic

L91-1X:
% Solids: 50
Solvents: Xylol
Viscosity: 5000-8000 cps
Type: Copolymer
Remarks: L91-1 in Xylol

L91-17:
% Solids: 40
Solvents: MEK
Viscosity: Z1-Z5
Type: Homopolymer
Remarks: Maximum film hardness

Styrene Acrylates:

SA1-3:
% Solids: 65
Solvents: SOLVESSO 100
Viscosity: 40,000-60,000 cps
Type: Copolymer
Remarks: Clear acrylic caulk base

DOCK RESINS CORP.: Acrylic Resins Solutions(Continued):

Styrene Acrylates(Continued):

DORESCO Products:

SA2-5:
% Solids: 50
Solvents: SOLVESSO 100
Viscosity: U-Z
Type: Copolymer
Remarks: Cement sealer, fusible to brick

SA2-10:
% Solids: 50
Solvents: Mineral Spirits
Viscosity: U-Z
Type: Copolymer
Remarks: Maximum aliphatic compatibility

Thermosetting Acrylics

AE49-2:
% Solids: 53
Solvents; Water
Viscosity: 30-100 cps
Type: Copolymer
Remarks: Soft Alkali catalyzed self cross linking polymer

AE49-7:
% Solids: 55
Solvents: Water, Coalascing Solvent
Viscosity: 300-500 cps
Type: Copolymer
Remarks: Harder Alkali catalyzed self cross linking polymer

TA1-15:
% Solids: 55
Solvents: Xylol, BUTYL CELLOSOLVE
Viscosity: X-Z
Type: Terpolymer
Remarks: OH Modified, high quality acrylic baking enamel vehicle

TA1-45:
% Solids: 50
Solvents: Xylol, BUTYL CELLOSOLVE
Viscosity: N-V
Type: Terpolymer
Remarks: Melamine containing high quality acrylic baking enamel vehicle

DOCK RESINS CORP.: Acrylic Resin Solutions(Continued):

Thermosetting Acrylics(Continued):

DORESCO Products:

TA19-7:
% Solids: 50
Solvents: Xylol, Butanol
Viscosity: 1,000-3,000 cps
Type: Terpolymer
Remarks: OH Modified, high quality acrylic baking enamel vehicle

TA22-1:
% Solids: 40
Solvents: Isopropanol, MEK
Viscosity: Z-Z3
Type: Terpolymer
Remarks: Self cross linking general purpose acrylic vehicle

TA23-2:
% Solids: 40
Solvents: Isopropanol, MEK
Viscosity: Z5-Z7
Type: Terpolymer
Remarks: High molecular weight self cross linking acrylic vehicle

High Solids Enamel Vehicles

TA7-7:
% Solids: 77
Solvents: MEK
Viscosity: 5,000-10,000 cps
Type: Multipolymer
Remarks: High solids acrylic enamel vehicle for baking with melamines

TA13-23:
% Solids: 85
Solvents: MAK
Viscosity: 3,000-6,000 cps
Type: Multipolymer
Remarks: High solids acrylic enamel vehicle for baking with melamines or reacting with isocyanates

DOCK RESINS CORP.: Acrylic Resin Solutions(Continued):

Water Reducible:

DORESCO Products:

ACW8-6:
% Solids: 30
Solvents: Water
Viscosity: 300 cps max.
Type: Terpolymer
Remarks: Gravure-Flexo water-based resin

AQUAHYDE 100:
% Solids: 30
Solvents: Water, Alcohol
Viscosity: 1000-1800 cps
Type: Terpolymer
Remarks: Heat Resistant, ink vehicle

ACW25-1:
% Solids: 33
Solvents: Water
Viscosity: 600-1200 cps
Type: Copolymer
Remarks: Modifier to improve dry and performance of air dry water alkyds. Experimental.

ACW25-2:
% Solids: 33
Solvents: Water
Viscosity: 320 cps
Type: Copolymer
Remarks: Adhesion promoted ACW25-1

DOCK RESINS CORP.: DORESCO Alkyd Resins:

Products:

DORESCO AKD34-3:

DORESCO AKD34-3 is a tall oil fatty acid modified alkyd formulated especially for the emulsion pigment printing portion of the textile processing industry. It forms stable emulsions that provide good color value.

Generally, DORESCO AKD34-3 can be formulated into quick air drying enamels suitable for application on toys, buses, trucks, and farm implements.

Typical Physical Constants:

Solids: 50+-1%
Solvent: Xylol
Acid No.: 6-12
Color: 11 Max.
Lbs./Gal.: 8.02
PA Content: 39%
Modification: Rosin

DORESCO AKD51-3:

DORESCO AKD51-3 is a short oil rosin modified alkyd that is designed to provide a cured film that will accept leaf for decorative printing effects. It is normally blended with amine resins; just before coating over printed stock the finish is catalyzed with DORESCO G-17, a para toluene sulfonic acid solution @ 50% solids. After curing, bronze, aluminum or other leaf is hot stamped to complete the job.

Typical Physical Constants:

Solids: 50+-1%
Solvents: Xylol, Toluol
Viscosity (Gardner): T-X
Acid No.: 40-45
Color: 10 max.
Lbs./Gal.: 8.4
PA Content: 38%
Rosin: Present

DORESCO AKD55-5:

DORESCO AKD55-5 is a pure tall oil fatty acid alkyd that is short in oil length. It is a high quality alkyd used to formulate thermosetting clear overprint varnishes. These finishes are very responsive to DORESCO G-17, a 50% solution of para toluene sulfonic acid. Properly cured finishes retain color, resist blocking, marring, and abrasion.

In general, DORESCO AKD55-5 can be formulated into baking enamels that provide excellent gloss, gloss retention, film properties, high hiding and durability.

Typical Physical Constants:

Solids: 60+-1%	Color: 4 Max.
Solvents: Xylol, Toluol	Lbs./Gal.: 8.4
Acid No.: 5 approx.	PA Content: 40%

DOCK RESINS CORP.: DORESCO Alkyd Resins(Continued):

Product:

DORESCO L20-13:

DORESCO L20-13 is an alkyd resin solution designed for lacquers. It should not be used by itself with nitrocellulose, but with maleic resin and plasticizers. It is recommended as a grinding vehicle. Raw umber and other difficult to disperse pigments grind readily in DORESCO L20-13, providing excellent paste stability and gloss.

DORESCO L20-13 has relatively low viscosity at high solids. It is ideal for hot spray lacquers or other systems where high solids are required.

Typical Physical Constants:
Solids: 75%
Solvents: Xylol and Isopropanol
Viscosity: X-Z
Color: 6 max.
Acid No.: 6.0 max.
Lbs./Gal.: 9.0

DORESCO L20-35:

DORESCO L20-35 is a medium oil length alkyd that has somewhat better color retentive properties than soya alkyds but has poorer drying properties.

It is recommended for quick curing bake systems where color retention is important but the high cost and bake cycle required of a coconut alkyd is not warranted.

It can be used in nitrocellulose lacquers without lifting the undercoat. It is in Rule 66 solvents and can be used in flexo ink systems. In catalyzed systems with amine resins, stability and response are excellent.

Typical physical constants:
Solids: 70+-2%
Solvents: VM&P Naphtha, Isopropanol
Viscosity: Z-Z2
Acid No.: 8 Max.
Color: 4-6
Lbs./Gal.: 8.35

DORESCO L226-1:

DORESCO L226-1 is a medium short alkyd designed to provide relatively high solids baking systems. It is a good grinding vehicle and is excellent for stability in alkyd-amine baking enamels.

In catalyzed alkyd-amine systems, DORESCO L226-1 is excellent for response and stability.

Typical physical constants:
Solids: 60+-1%
Solvents: Xylol-Toluol
Viscosity: Z3-Z5
Acid No: 7-11
Color: 5-7
Lbs./Gal.: 8.48

DOCK RESINS CORP.: DORESCO Alkyd Resins(Continued):

Product:

DORESCO M99-14:

DORESCO M99-14 is a non-drying alkyd resin solution recommended for use with lacquer, and with butylated urea and melamine resins for enamels.

Enamels formulated with DORESCO M99-14 and amine resins have excellent color retention, gloss, durability, humidity and alkali resistance, even on overbake schedules.

In lacquers, DORESCO M99-14 provides excellent exterior durability, gloss, fullness, and elasticity. However, it is not recommended as the sole plasticizer and some additional plasticizer and lacquer gum should be included in lacquer formulations.

Typical physical constants:

Solids: 60%
Solvents: Xylol
Viscosity: Z2-Z4
Acid No.: 3-8
Color: 1-4
Oil: Coconut
Oil Content: Short
Lbs./Gal.: 8.0

DORESCO AKO4-1:

DORESCO AKO4-1 is a saturated polyester resin solution supplied at 60% solids in toluene. It is a highly specialized product made of FDA approved, carefully polymerized raw materials. As a result, upon decomposition, negligible odor or fumes evolve.

DORESCO AKO4-1 is a base for coatings for remote facsimile transmission paper. The molecular weight is high enough to carry black pigment and provide color intensity but low enough to prevent paper curl.

DORESCO AKO4-1 by itself is not a complete vehicle. Plasticizers and conducting additives are used along with conductive carbon black.

Typical physical constants:

Solids: 60+-1%
Viscosity (Gardner): T-X
Acid No.: 12 Max.
Ring & Ball Melting Point (solids): 225-250F
Lbs./Gal.: 7.4

DOW CORNING CORP.: Silicone Paint Resins:

Dow Corning silicone paint resins feature resistance to temperature extremes, moisture, corrosion, electrical discharge and weathering. They are compatible with many organic resins, and many combinations of silicone resins can be formulated for specific applications and film properties.

Selection by Application and Use:

DOW CORNING 804 resin:
Application: A hard resin which air dries tack free. Can be used as an additive to improve paint flowout. Also used in conjunction with DOW CORNING Q1-9770 Release Additive for release coatings.
How to Use: Can be used by itself or blended with organic resins to improve heat stability and water resistance.

DOW CORNING 805 resin:
Application: A soft, flexible resin used primarily with metallic pigments such as aluminum and zinc.
How to Use: Can be used by itself or sometimes blended with DOW CORNING 806A to achieve intermediate hardness, especially in aluminum paint formulations.

DOW CORNING 806A resin:
Application: A moderately hard resin for decorative finishes and as an additive for paints to improve initial gloss, flow-out and wetting.
How to Use: Can be used by itself or sometimes blended with DOW CORNING 805 resin to achieve intermediate hardness, especially in aluminum paint formulations.

DOW CORNING 808 resin:
Application: Flexible film resin used in colored baking enamels for space heaters, clothes driers, stoves, and similar applications.
How to Use: A lower cost resin than DOW CORNING 805 and often used in similar formulations.

DOW CORNING 840 resin:
Application: A moderately hard resin for improving heat resistance and weathering of organic resins. Also used as an additive to improve flow-out and the initial gloss of paint.
How to Use: Often blended up to 50% with organic resins to improve heat resistance as measured by gloss and color retention and exterior durability as measured by gloss, color retention, and chalk resistance.

DOW CORNING CORP.: Silicone Paint Resins(Continued):

Selection by Application and Use(Continued):

Product:

DOW CORNING Q1-2529 resin:
Application: Same as DOW CORNING 805 resin but formulated to comply with Los Angeles County's Rule 66 Standards.
How to Use: Same as DOW CORNING 805 resin.

DOW CORNING Q1-2530 resin:
Application: Same as DOW CORNING 840 but at 80% solids for lower VOC coatings.
How to Use: Same as DOW CORNING 840 resin, but adjust for higher solids.

DOW CORNING Z-6018 resin:
Application: A hard resin that is primarily used as a reactive intermediate but can be blended with the other silicone resins to improve hardness. Also used as a flow additive for epoxy coatings.
How to Use: Compatible with the other silicone resins up to 50 percent solids. Soluble in ketones, esters, chlorinated solvents, alcohols, aromatic hydrocarbons, and solvent blends having a kauri-butanol value greater than 50.

DOW CORNING 6-2230 resin:
Application: A hard resin with excellent hot hardness. Can be used to increase the hardness of other silicone resins and to improve heat stability of powder coatings. Also used in conjunction with DOW CORNING Q1-9770 Release Additive for release coatings.
How to Use: Unmodified it produces coatings that are very hard and brittle. Soluble in ketones, esters, aromatic and chlorinated solvents.

Selection by Typical Properties:
DOW CORNING Z-6018 resin:
- Meets requirements of Los Angeles County's Rule 66: Yes
- Resin Solids, percent: 100
- Specific Gravity at 77F (25C): 1.07*
- Viscosity at 77F (25C),*
  - centipoises: 30*
  - Gardner-Holt: A-3 to A*

DOW CORNING 804 resin:
- Meets requirements of Los Angeles County's Rule 66: No
- Resin Solids, percent: 60
- Solvent: toluene
- Specific Gravity at 77F (25C): 1.07
- Viscosity at 77F (25C),
  - centipoises: 30
  - Gardner-Holt: A-3 to A

* 60% solids in toluene

DOW CORNING CORP.: Silicone Paint Resins(Continued):

Selection by Typical Properties(Continued):

DOW CORNING 805 resin:
Meets requirements of Los Angeles County's Rule 66: No
Resin Solids, percent: 50
Solvent: xylene
Specific Gravity at 77F (25C): 1.01
Viscosity at 77F (25C),
centipoises: 115
Gardner-Holt: C to F

DOW CORNING 806A resin:
Meets requirements of Los Angeles County's Rule 66: No
Resin Solids, percent: 50
Solvent: xylene, toluene
Specific Gravity at 77F (25C): 1.03
Viscosity at 77F (25C),
centipoises: 150
Gardner-Holt: D to H

DOW CORNING 808 resin:
Meets requirements of Los Angeles County's Rule 66: No
Resin Solids, percent: 50
Solvent: xylene
Specific Gravity at 77F (25C): 1.01
Viscosity at 77F (25C),
centipoises: 150
Gardner-Holt: D to H

DOW CORNING 840 resin:
Meets requirements of Los Angeles County's Rule 66: No
Resin Solids, percent: 60
Solvent: toluene
Specific Gravity at 77F (25C): 1.06
Viscosity at 77F (25C),
centipoises: 17
Gardner-Holt: A-1 to A-3

DOW CORNING Q1-2529 resin:
Meets requirements of Los Angeles County's Rule 66: Yes
Resin Solids, percent: 50
Solvent: Solvent Blend:

Solvent Blend	Wt.%
Toluene	17.6
VM&P Naphtha (C8+0.1%)	49.4
Oxygenated Solvents	33.0

Specific Gravity at 77F (25C): 0.96
Viscosity at 77F (25C),
centipoises: 115
Gardner-Holt: C to F

DOW CORNING CORP.: Silicone Paint Resins(Continued):

Selection by Typical Properties(Continued):

DOW CORNING Q1-2530 resin:
- Meets requirements of Los Angeles County's Rule 66: No
- Resin Solids, percent: 80
- Solvent: toluene
- Specific Gravity at 77F (25C): 1.14
- Viscosity at 77F (25C):
  - centipoises: 800
  - Gardner-Holt: U to W

DOW CORNING 6-2230 resin:
- Meets requirements of Los Angeles County's Rule 66: Yes
- Resin Solids, percent: 100
- Specific Gravity at 77F (25C): 1.08++
- Viscosity at 77F (25C):
  - centipoises: 80++
  - Gardner-Holt: B to C++

++ 60% solids in 75/25 toluene/n-butyl acetate

DUPONT CO.: ELVACITE Acrylic Resins:

ELVACITE acrylic resins are polymers or copolymers of methyl methacrylate or other acrylic monomers for use in solvent-applied lacquers, inks, adhesives and specialty coatings. The ELVACITE resins offer the distinctive combination of properties associated with high-quality acrylic coating resins:

* Outstanding resistance to ultraviolet degradation
* Excellent weatherability
* Superior abrasion resistance
* Excellent resistance to water, alcohol, dilute acids and alkalies, chemical fumes and corrosive and oxidizing environments
* High gloss and hardness
* Generally low pigment reactivity
* Water-white color and transparency
* Resistance to discoloration on heating

Grade Numbers:

2008:
Polymer Type: Methyl methacrylate
Typical Inherent Viscosity: 0.19
Molecular Weight: Low
Lowest molecular weight grade. Used in lacquers and gravure inks requiring low viscosity at high solids. Highest tack temperature. Good pigment dispersion.

2009:
Polymer Type: Methyl methacrylate
Typical Inherent Viscosity: 0.39
Molecular Weight: Medium
A more soluble, softer version of ELVACITE 2010. Shows unusual chip resistance for a hard polymer.

2010:
Polymer Type: Methyl methacrylate
Typical Inherent Viscosity: 0.40
Molecular Weight: Medium
General purpose grade for lacquer coatings such as barrier topcoatings for vinyl fabric. Highest tensile and hardness.

2013:
Polymer Type: Methyl/n-butyl methacrylate copolymer
Typical Inherent Viscosity: 0.17
Molecular Weight: Low
Fast-dissolving, low viscosity grade with quick solvent release for use in industrial lacquers, aerosols, inks, and coatings for plastics. Softer than the methyl methacrylate polymers.

DUPONT CO.: ELVACITE Acrylic Resins(Continued):

<u>Grade:</u>

2014:
Polymer Type: Methyl methacrylate copolymer
Typical Inherent Viscosity: 0.40
Molecular Weight: Medium
Softest and most flexible of the ELVACITE copolymer resins. Good pigment wetting. Useful for clear, translucent or pigmented metal lacquers, e.g., for aircraft. Also useful for ELECTROFAX copy paper coatings and seamless flooring.

2016:
Polymer Type: Methyl/n-butyl methacrylate copolymer
Typical Inherent Viscosity: 0.25
Molecular Weight: Low
Fast dissolving, low viscosity general purpose resin for lacquer and toner use.

2021:
Polymer Type: Methyl methacrylate
Typical Inherent Viscosity: 0.50
Molecular Weight: High
A higher molecular weight version of ELVACITE 2010 for better abrasion resistance in vinyl topcoats.

2028:
Polymer Type: Methacrylate copolymer
Typical Inherent Viscosity: 0.22
Molecular Weight: Low
Unique in its broad solubility and compatibility. Designed for flexographic inks, general purpose lacquers and metal coatings.

2041:
Polymer Type: Methyl methacrylate
Typical Inherent Viscosity: 1.25
Molecular Weight: Very High
The highest molecular weight grade. Used in barrier topcoating of vinyl fabric for dielectric heat sealability, maximum abrasion resistance, block resistance, and slip.

2042:
Polymer Type: Ethyl methacrylate
Typical Inherent Viscosity: 0.80
Molecular Weight: Very High
Tough, alcohol-tolerant, broadly compatible grade for use in abrasion-resistant coatings such as high-gloss clear lacquers for decals and outdoor signs. Slightly softer than ELVACITE 2013.

DUPONT CO.: ELVACITE Acrylic Resins(Continued):

Grade:

2043:
Polymer Type: Ethyl methacrylate
Typical Inherent Viscosity: 0.21
Molecular Weight: Low
Fast-dissolving, low-viscosity resin with alcohol solubility for flexographic inks. Most compatible grade. Has excellent pigment wetting ability and broad solubility for use in lacquers for solvent-sensitive substrates. Equal in hardness to ELVACITE 2042.

2044:
Polymer Type: n-Butyl methacrylate
Typical Inherent Viscosity: 0.50
Molecular Weight: High
Softest, most flexible bead grade. Useful in adhesives for smooth plastic films and aluminum, and in silk screen inks. Will plasticize and improve adhesion of harder butyl grades (ELVACITE 2045 and 2046) and nitrocellulose. Improves outdoor durability of vinyl chloride resins in pigmented lacquers.

2045:
Polymer Type: Isobutyl methacrylate
Typical Inherent Viscosity: 0.63
Molecular Weight: High
Hardest of the weak-solvent-soluble butyl grades. Useful in topcoats for vacuum metallized plastics and in lacquers for other solvent-sensitive substrates. Softer than ELVACITE 2042 and 2043.

2046:
Polymer Type: n-Butyl/isobutyl methacrylate, 50/50 copolymer
Typical Inherent Viscosity: 0.54
Molecular Weight: High
Medium-hardness butyl grade. Soluble (like ELVACITE 2044 and 2045) in mineral spirits, VM&P naphtha, and some alcohols. Useful in aluminum finishes, ceramic coatings, and aerosol snows.

DUPONT CO.: HYPALON Chlorinated Polyolefins:

Curable, flexible resins...for use in high-performance specialty and OEM coatings

HYPALON chlorinated polyolefin resins are chlorinated and chloro sulfonated ethylene polymers for use in solvent-borne high performance coatings. HYPALON offers the properties traditionally valued in other chlorinated binders such as solution vinyls and chlorinated rubber; these benefits include:

* broad chemical resistance
* impermeability to moisture and oxygen
* toughness
* flame retardance

Like most chlorinated binders, HYPALON is particularly useful for anticorrosion and waterproofing paints.

The outstanding physical property balance of HYPALON offers many performance advantages in properly formulated coatings. Specific attributes depend on grade and formulations, but generally include:

* superior corrosion protection in harsh envrionments
* excellent weatherability and color stability
* outstanding impact and abrasion resistance
* convenient application characteristics
* excellent substrate and intercoat adhesion

HYPALON 20 and 30 Chlorosulfonated Polyethylene

HYPALON 20 and 30 are the traditional elastomeric HYPALON chlorinated polyolefin grades for solvent-borne coating systems. Relatively high in molecular weight, these chlorosulfonated polyethylene resins are most useful where the ultimate in weatherability, chemical resistance, elasticity and durability are needed.

Chemical and Physical Characteristics:

HYPALON 20:
- Chemical composition
  - Chlorine, wt%: 29
  - Sulfur, wt%: 1.4
  - Sulfonyl chloride, WPE: 2,286
- Molecular weight (N): 50,000
- Specific Gravity: 1.14

HYPALON 30:
- Chemical composition
  - Chlorine, wt%: 43
  - Sulfur, wt%: 1.1
  - Sulfonyl chloride, WPE: 2,910
- Molecular weight (N): 23,000
- Specific Gravity: 1.26

DYNAMIT NOBEL CHEMICALS: DYNACOLL A Radiation Curable Polyesters for the Formulation of Pressure Sensitive Adhesives:

DYNACOLL A is a class of liquid, saturated copolyesters carrying a certain amount of terminal acrylic double bonds. They can be coated onto suitable substrates and subsequently crosslinked by means of energy-rich radiation (i.e. UV or EB) to permanently tacky films. The average molecular weight of the copolyesters ranges from 3000 to 6000. Their glass transition temperatures are between -10 and -50C.

The following features are characteristic:
- low odor
- water clear to amber colored
- free of solvents
- free of monomers or reactive diluents
- low oligomer content
- low toxicity
- little or no skin irritation
- attractive environmental aspects
- good shelf life

DYNACOLL A:

1530 VP:
Tg, C: -50
Reactivity: moderate
Brookfield Viscosity, Pa.s: 50C: 70

1550 VP:
Tg, C: -50
Reactivity: high
Brookfield Viscosity, PA.s: 50C: 120

2330 VP:
Tg, C: -30
Reactivity: moderate
Brookfield Viscosity, Pa.s: 50C: 90

2350 VP:
Tg, C: -30
Reactivity: high
Brookfield Viscosity, Pa.s: 50C: 130

2530 VP:
Tg, C: -50
Reactivity: moderate
Brookfield Viscosity, Pa.s: 50C: 20

DYNAMIT NOBEL CHEMICALS: DYNACOLL A Radiation Curable Polyesters for the Formulation of Pressure Sensitive Adhesives (Continued):

DYNACOLL:

2550 VP:
Tg, C: -50
Reactivity: high
Brookfield Viscosity, Pa.s: 30

3130 VP:
Tg, C: -10
Reactivity: moderate
Brookfield Viscosity, Pa.s: 90

3150 VP:
Tg, C: -10
Reactivity: high
Brookfield Viscosity, Pa.s: 120

3230 VP:
Tg, C: -20
Reactivity: moderate
Brookfield Viscosirty, Pa.s: 90

3250 VP:
Tg, C: -20
Reactivity: high
Brookfield Viscosity, Pa.s: 140

3330 VP:
Tg, C: -30
Reactivity: moderate
Brookfield Viscosity, Pa.s: 80

3350 VP:
Tg, C: -30
Reactivity: high
Brookfield Viscosity, Pa.s: 120

DYNAMIT NOBEL CHEMICALS: DYNACOLL RP Hydroxyl Terminated Polyesters for the Production of Reactive Adhesives and Sealants:

DYNACOLL RP comprises a class of linear, saturated copolyesters with medium molecular weights between 2000 and 6000. Due to their terminal hydroxyl groups they are processable with isocyanates to reactive hot-melts or polyurethanes.

DYNACOLL RP copolyesters form a building block system of three polyester groups, which are compatible with each other to a large extent:

1. solid, amorphous copolyesters with glass transition temperatures above 0C and with flow points above 50C (series 100),
2. liquid copolyesters with glass transition temperatures between -10C and -60C (series 200),
3. solid, partially crystalline copolyesters with glass transition temperatures below 0C and flow points above 60C (series 300).

<u>DYNACOLL</u> <u>type:</u>

RP 110:
Tg (C): +10
OH number (mg KOH/g): 55
Viscosity (Pa.s): 100c: 10
colour: yellowish

RP 130:
Tg (C): +30
OH number (mg KOH/g): 35
Viscosity (Pa.s): 100C: 230
colour: yellowish

RP 140:
Tg (C): +40
OH number (mg KOH/g): 20
Viscosity (Pa.s): 120C: 270
colour: yellowish

RP-210:
Tg (C): -10
OH number (mg KOH/g): 30
Viscosity (Pa.s): 100C: 5
colour: brown

DYNAMIT NOBEL CHEMICALS: DYNACOLL RP(Continued):

DYNACOLL type:

RP 220:
Tg (C): -20
OH number (mg KOH/g): 30
viscosity (Pa.s): 10C: 7
colour: yellowish

RP 230:
Tg (C): -30
OH Number (mg KOH/g): 30
Viscosity (Pa.s): 100C: 4
colour: yellowish

RP 250:
Tg (C): -50
OH Number (mg KOH/g): 20
Viscosity (Pa.s): 100C: 3
colour: colourless

RP 320:
Tg (C): -20
OH number (mg KOH/g): 30
Viscosity (Pa.s): 100C: 30
colour: white

RP 340:
Tg (C): -40
OH number (mg KOH/g): 30
Viscosity (Pa.s): 120C: 4
colour: white

RP 350:
Tg (C): -50
OH number (mg KOH/g): 30
Viscosity (Pa.s): 100C: 2
colour: white

RP 360:
Tg (C): -60
OH number (mg KOH/g): 30
Viscosity (Pa.s): 80C: 3
colour: white

High-molecular weight DYNAPOL copolyesters:

S 1402:
Tg (C): -10
OH number (mg KOH/g): <5
Viscosity (pa.s): 120C: 330C
colour: white

1420:
Tg (C): -10
OH number (mg KOH/g): <5
Viscosity (pa.s): 120C: 660
colour: yellowish

DYNAMIT NOBEL CHEMICALS: DYNAPOL Copolyester Raw Materials for Industrial Bondings and Coatings:

Product:

high crystalline:

DYNAPOL S 242:
Colour: White
Density: 1,30
Viscosity number: dl/g: 0,7
Flow point: C: 200
Melting point: C: 200
Possible applications: Bonding in the shoe industry

DYNAPOL S 243:
Colour: white
Density: 1,30
Viscosity number: dl/g: 0,7
Flow point: C: 192
Melting point: C: 192
Possible applications: Different applications

DYNAPOL S 390:
Colour: yellowish
Density; 1,17
Viscosity number: dl/g: 0,65
Flow point: C: 185
Melting point: C: 178
Possible applications: different applications, e.g. bonding of filters

DYNAPOL S 360:
Colour: white
Density: 1,26
Viscosity number: 0.85
Flow point: C: 180
Melting point: C: 176
Possible applications: bonding in the shoe industry

DYNAPOL S 350:
Colour: white
Density: 1,27
Viscosity number: 0,8
Flow point: C: 180
Melting point: C: 175
Possible applications: Different applications

DYNAMIT NOBEL CHEMICALS: DYNAPOL Copolyester Raw Materials for Industrial Bondings and Coatings(Continued):

medium crystalline:

DYNAPOL S 50:
colour: greenish
Density: 1,35
Viscosity number: dl/g: 0.95
Flow point: C: 202
Melting point: C:
Possible applications: fibre industry

DYNAPOL P 1500:
Colour: white
Density: 1,28
Viscosity numner: dl/g: 0,9
Flow point: C: 175
Melting point: C: 170
Possible applications: Rwa material for producing powders, for fluidized bed coating and elecrostatic powder coating

DYNAPOL S 341:
Colour: white
Density: 1,25
Viscosity number: dl/g: 0,65
Flow point: C: 160
Melting point: C: 148
Possible applications: Different applications

DYNAPOL S 1249:
Colour: white
Density: 1,28
Viscosity number: dl/g: 0,8
Flow point: C: 145
Melting point: C: 128
Possible applications: Bonding of textiles

DYNAPOL S 1272:
Colour: white
Density: 1,28
Viscosity number: dl/g: 0,85
Flow point: C: 140
Melting point: C: 130
Possible applications: Different applications

DYNAPOL S 1218:
Colour: white
Density: 1,26
Viscosity number: dl/g: 0,8
Flow point: C: 132
Melting point: C: 107
Possible applications: Bonding of textiles

DYNAMIT NOBEL CHEMICALS: DYNAPOL Copolyester Raw Materials for Industrial Bondings and Coatings(Continued):

medium crystalline:

DYNAPOL S 1227:

Colour: white
Density: 1,19
Viscosity number: dl/g: 0.8
Flow point: C: 120
Melting Point: C: 99
Possible applications: Different applications

DYANPOL S 1228:

Colour: white
Density: 1,25
Viscosity number: dl/g: 0,65
Flow point: C: 115
Melting Point: C: 95
Possible applications: Different applications

low crystalline:

DYNAPOL S 1440:

Colour: greenish
Density: 1,26
Viscosity number: dl/g: 0,9
Flow point: C: 153
Possible applications: Hot melt adhesive for laminatings, e.g. in the electrical industry

DYNAPOL S 1450:

Colour: greenish
Density: 1,30
Viscosity number: dl/g: 0,9
Flow point: C: 140
Possible applications: Hot melt adhesive for laminatings, e.g. in the electrical industry

DYNAPOL S 1401:

Colour: white
Density: 1,19
Viscosity number: dl/g: 0,8
Flow point: C: 100
Melting point: C: 58
Possible applications: For the laminating of sheets, e.g. in the furniture and automobile industry

DYNAMIT NOBEL CHEMICALS: DYNAPOL Copolyester Raw Materials for Industrial Bondings and Coatings(Continued):

low crystalline:

DYNAPOL S 1402:
Colour: white
Density: 1.24
Viscosity number: dl/g: 0,8
Flow point: C: 100
Melting point: C: 56
Possible applications: For laminating of sheets, e.g.in the furniture and automobile industry

DYNAPOL S 1225:
Colour: white
Density: 1,24
Viscosity number: dl/g: 0,65
Flow point: C: 96
Melting point: C: 58
Possible applications: Hot melt and solvent-based-adhesive

DYNAPOL S 342:
Colour: yelloow/brown
Density: 1,29
Viscosity number: dl/g: 0,3
Flow point: C: 140
Melting point: C: 134
Possible applications: Different applications

DYNAMIT NOBEL CHEMICALS: DYNAPOL Copolyester Raw Materials for Industrial Bondings and Coatings(Continued):

amorphous:

DYANPOL L 912:
Colour: Transparent
Density: 1,20
Viscosity number: dl/g: 0,55
Flow point: C: 173
Possible applications: Raw material for varnishes and adhesives

DYNAPOL L 207:
Colour: Transparent
Density: 1,23
Viscosity number: dl/g: 0,75
Flow point: C: 160
Possible applications: Raw material for varnishes and adhesives

DYNAPOL L 205:
Colour: Transparent
Density: 1,23
Viscosity number: dl/g: 0,53-0,58
Flow point: C: 145
Possible applications: Raw material for varnishes and adhesives

DYNAPOL L 206:
Colour: yellowish/transparent
Density: 1,23
Viscosity number: dl/g: 0,62-0,65
Flow point: C: 155
Possible applications: Raw material for varnishes and adhesives

DYNAPOL L 208:
Colour: yellowish/transparent
Density: 1,22
Viscosity number: dl/g: 0,65-0,80
Flow point: C: 150
Possible applications: Raw material for varnishes and adhesives

DYNAPOL L 411:
Colour: yellowish/transparent
Density: 1,25
Viscosity number: dl/g: 0,60-0,65
Flow point: C: 135
Possible applications: Raw material for varnishes and adhesives

DYNAMIT NOBEL CHEMICALS: DYNAPOL Copolyester Raw Materials for Industrial Bondings and Coatings(Continued):

amorphous:

DYNAPOL L 600:
Colour: yellowish/transparent
Density: 1,18
Viscosity number: dl/g: 0,65
Flow point: C: 0.65
Posssible applications: Raw material for varnishes and adhesives

DYNAPOL L 860:
Colour: yellowish/transparent
Density: 1,19
Viscosity: dl/g: 0,60-0,70
Flow point: C: 120
Possible applications: Raw material for varnishes and adhesives

DYNAPOL S 1313:
Colour: yellowish
Density: 1,16
Viscosity: dl/g: 0,85
Flow point: C: 110
Possible applications: For laminated sheets, e.g. in the packaging industry

DYNAPOL S 1430:
Colour: greenish
Density: 1,23
Viscosity: dl/g: 0,9
Flow point: C: 100
Possible applications: For laminated sheets, e.g. in the packaging industry

DYNAPOL S 1420:
Colour: yellowish
Density: 1,22
Viscosity number: dl/g: 0,9
Flow point: C: 85
Possible applications: For laminated sheets, e.g. in the packaging industry

EASTMAN CHEMICAL PRODUCTS, INC.: EASTMAN Water Dispersible Polymers:

For coatings, adhesives, and aqueous colloids for personal care and household products.

EASTMAN AQ 29S:
Form: Pellets
lb/gal: 11.1
Acid No. : <2
Molecular Weight: 14,000
Hydroxyl No.: 6
Tg, C: 29

EASTMAN AQ 29D:
Form: Aqueous Dispersion 30% NV
lb/gal: 8.8
Viscosity, cP: 40
pH: 6-7

EASTMAN AQ 38S:
Form: Pellets
lb/gal: 11.2
Acid No.: <2
Molecular Weight: 18,000
Hydroxyl No.: 5
Tg, C: 38

EASTMAN AQ 38D:
Form: Aqueous Dispersion 25% NV
lb/gal: 8.8
Viscosity, cP: 18
pH: 7

EASTMAN AQ 55S:
Form: Pellets
lb/gal: 11.0
Acid No.: <2
Molecular Weight: 18,000
Hydroxyl No.: 5.3
Tg, C: 55

EASTMAN AQ 55D:
Form: Aqueous Dispersion 28% NV
lb/gal: 9.0
Viscosity, cP: 40
pH: 5-6

FRP CO.: FILTREZ Resins:

Ester Gums:

HMP Ester Gum:
Glycerol ester of Gum Rosin offering good viscosity and color retention.

Specifications:
- Melting Point (A.S.T.M. Ring & Ball): 90-100C
- Acid Value: 8 Max.
- Viscosity (60% in Mineral Spirits, Gardner-Holdt): D-G
- Color (60% in Mineral Spirits, Gardner Standards): 8-10
- Pounds per gallon: 8.9-9.2
- Flash Point (C.O.C.): 405F.

Solubility:

Aliphatic Hydrocarbons:	Excellent
Aromatic Hydrocarbons:	Excellent
Alcohols:	Insoluble

Typical Uses:

Interior Coatings	Lacquers
Varnishes	Hot-melt adhesives
Phenolic modifier	

Improves brushability, gloss and resistance to sagging in quality interior coat formulations.

HMP Ester Gum 2010:
A premium lacquer grade ester gum, especially designed for good resistance to film hazing at high resin-nitro-cellulose ratios. Exhibits high alcohol tolerance, and lower acid value than standard grade ester gums.

Specifications:
- Melting Point (A.S.T.M. Ring & Ball): 85-95C
- Acid Value: 8 Max.
- Viscosity (60% in Mineral Spirits, Gardner-Holdt): E-H
- Color (60% in Mineral Spirits, Gardner Standards): 10 Max.
- Pounds per gallon: 8.9-9.2
- Flash Point (C.O.C.): 490F.

Solubility:

Aliphatic Hydrocarbons:	Excellent
Aromatic Hydrocarbons:	Excellent
Alcohols:	Limited

Typical Uses:

Interior lacquers	Exterior lacquers
Metal coatings	Polishing-type lacquers

Excellent for metal lacquers exposed to the weather.

FRP CO.: FILTREZ Resins(Continued):

Ester Gums(Continued):

FILTREZ 13:

Pentaerythritol ester of rosin widely soluble in drying oils, especially raw, cooked and blown. It is also compatible with hydrocarbon waxes, ester waxes and chlorinated rubber. It is heat stable, resistant to oxidation and exhibits a high refractive index, making it desirable for use in gloss finishes.

Specifications:
- Melting Point (A.S.T.M. Ring & Ball): 98-104C.
- Acid Value: 10-15
- Viscosity (60% in Mineral Spirits, Gardner-Holdt): E-H
- Color (60% in Mineral Spirits, Gardner Standards): 10 Max.
- Pounds per gallon: 9.0-9.2
- Flash Point (C.O.C.): 505F.

Solubility:
- Aliphatic Hydrocarbons: Excellent
- Aromatic Hydrocarbons: Excellent
- Alcohols: Insoluble

Typical Uses:
- Adhesives
- Emulsion Paints
- Wax Compositions
- Antifouling Paints
- Oleoresinous Varnishes

FILTREZ 15:

A pentaerythritol ester gum offering a high melting point, viscosity and low acid value.

Specifications:
- Melting Point (A.S.T.M. Ring & Ball): 106-113C.
- Acid Value: 10-15
- Viscosity (60% in Mineral Spriits, Gardner-Holdt): H-K
- Color (60% in Mineral Spirits, Gardner Standards): 10 max.
- Pounds per gallon: 9.0-9.15
- Flash Point (C.O.C.): 440F.

Solubility:
- Aliphatic Hydrocarbons: Excellent
- Aromatic Hydrocarbons: Excellent
- Alcohols: Insoluble

Typical Uses:
- Interior enamels
- Varnishes
- Hot-melt adhesives
- Gloss paints
- Phenolic modifier
- Waxes

Improves brushability, non-sagging and resistance to water spotting in interior enamel and gloss paint formulation.

FRP CO.: FILTREZ Resins(Continued):

Isomerized Rosins & Esters:

ISOESTER B:

An ester of isomerized rosin offering good resistance to oxidation, light and crystallization.

Specifications:
- Melting Point (A.S.T.M. Ring & Ball): 76-83C
- Acid Value: 9-12
- Viscosity (70% in Mineral Spirits, Gardner-Holdt): R-V
- Color (70% in Mineral Spirits, Gardner Standards): 10 Max.
- Pound per gallon: 8.9-9.1
- Flash Point (C.O.C.): 435F.

Solubility:

Aliphatic Hydrocarbons:	Excellent
Aromatic Hydrocarbons:	Excellent
Alcohols:	Insoluble

Typical Uses:
- Water-dispersed adhesives
- Heat-sensitive adhesives
- Lacquers
- Pressure-sensitive adhesives
- Embossing inks

FRP Maleic Resins:

FILTREZ 329:

A maleic-modified glycerol ester, possessing a high melting point and exceptionally high alcohol tolerance. Exhibits very good lacquer compatibility and solvent release.

Specifications:
- Melting Point (A.S.T.M. Ring & Ball): 130-140C
- Acid Value: 25-35
- Viscosity (60% in Toluol, Gardner-Holdt): Q-T
- Color (60% in Toluol, Gardner Standards): 10 Max.
- Pounds per gallon: 9.3-9.6
- Flash Point (C.O.C.): 575F

Solubility:

Aliphatic Hydrocarbons:	Limited
Aromatic Hydrocarbons:	Good
Alcohols:	Limited

Typical Uses:
- Lacquers
- Enamels
- Furniture finishes
- Varnishes
- Hot-melt adhesives
- Gravure printing inks

FRP CO.: FILTREZ Resins(Continued):

Maleic Resins(Continued):

FILTREZ 330:

A maleic-modified glycerol ester possessing a high melting point, good alcohol tolerance and excellent lacquer compatibility. Lacquers prepared with FILTREZ 330 have excellent color retention and adhesion.

Specifications:
- Melting Point (A.S.T.M. Ring & Ball): 135-145C.
- Acid Value: 25-35
- Viscosity (60% in Toluol, Gardner-Holdt): R-V
- Color (60% in Toluol, Gardner Standards): 10 Max.
- Pounds per gallon: 9.3-9.6
- Flash Point (C.O.C.): 575F.

Solubility:

Aliphatic Hydrocarbons:	Limited
Aromatic Hydrocarbons:	Good
Alcohols:	Limited

Typical Uses:
- Lacquers
- Lacquer primers and sealers
- Hot-melt coatings
- Gravure printing inks
- Varnishes
- Tin and furniture finishes
- Enamels

FILTREZ 330A:

A maleic-modified glycerol ester possessing a high melting point, good alcohol tolerance and excellent lacquer compatibility. It has fast solvent release and excellent color stability.

Specifications:
- Melting Point (A.S.T.M. Ring & Ball): 130-140C
- Acid Value: 35-42
- Viscosity (60% in Toluol, Gardner-Holdt): J-M
- Color (60% in Toluol, Gardner Standards): 10 Max.
- Pounds per gallon: 9.3-9.6
- Flash Point (C.O.C.): 575F.

Solubility:

Aliphatic Hydrocarbons:	Limited
Aromatic Hydrocarbons:	Good
Alcohols:	Limited

Typical Uses:
- Lacquers
- Varnishes
- Baking enamels
- Printing inks
- Lacquer primers and sealers
- Tin and furniture finishes
- Hot-melt coatings

FRP CO.: FILTREZ Resins(Continued):

Maleic Resins(Continued):

FILTREZ 338:

A maleic-modified glycerol ester with a high melting point and good alcohol tolerance. Lacquers incorporating this resin have high gloss, hardness and good adhesion.

Specifications:
- Melting Point (A.S.T.M. Ring & Ball): 135-145C.
- Acid Value: 25-35
- Viscosity (60% in Toluol, Gardner-Holdt): S-W
- Color (60% in Toluol, Gardner Standards): 10 Max.
- Pounds per gallon: 9.3-9.6
- Flash Point (C.O.C.): 575F.

Solubility:
- Aliphatic Hydrocarbons: Limited
- Aromatic Hydrocarbons: Good
- Alcohols: Limited

Typical Uses:

Lacquers	Varnishes
Lacquer primers & sealers	Tin & furniture finishes
Hot-melt coatings	Enamels
Gravure inks	

FILTREZ 339:

A maleic-modified glycerol ester possessing a high melting point and very good alcohol tolerance. Exhibits excellent lacquer compatibility and solvent release.

Specifications:
- Melting Point (A.S.T.M. Ring & Ball): 130-140C.
- Acid Value: 35-42
- Viscosity (60% in Toluol, Gardner-Holdt): K-O
- Color (60% in Toluol, Gardner Standards): 10 Max.
- Pounds per gallon: 9.3-9.6
- Flash Point (C.O.C.): 575F.

Solubility:
- Aliphatic Hydrocarbons: Limited
- Aromatic Hydrocarbons: Good
- Alcohols: Limited

Typical Uses:

Lacquers	Lacquer primers and sealers
Varnishes	Tin and furniture finishes
Baking enamels	Hot-melt coatings
Gravure inks	

FRP CO.: FILTREZ Resins(Continued):

Maleic Resins(Continued):

FILTREZ 345:

Maleic-modified-penta ester with a wide solubility range.

Specifications:
Melting Point (A.S.T.M. Ring & Ball): 130-140C.
Acid Value: 15-25
Viscosity (50% in Mineral Spirits, Gardner-Holdt): W-Z
Color (50% in Mineral Spirits, Gardner Standards): 11 Max.
Pounds per gallon: 9.3-9.5
Flash Point (C.O.C.): 470F.

Solubility:

Aliphatic Hydrocarbons:	Good
Aromatic Hydrocarbons:	Good
Alcohols:	Insoluble

Typical Uses:
Heat-set printing inks
Enamels
Paints
Varnishes
Machinery finishes
Metal primers

FILTREZ 360:

Maleic-pentaerythritol rosin ester possessing very high melting point and low acid value. Also it exhibits light color and excellent color retention.

Specifications:
Melting Point (A.S.T.M. Ring & Ball): 145-155C.
Acid Value: 12-25
Viscosity (60% in Xylene, Gardner-Holdt): Z-Z3
Color (50% in Mineral Spirits, Gardner Standards): 10 max.
Pounds per gallon: 9.3-9.5

Solubility:

Aliphatic Hydrocarbons:	Excellent
Aromatic Hydrocarbons:	Excellent
Alcohols:	Insoluble
Varnish Oils:	Good
Esters:	Good

Typical Uses:
Heat-set inks
Gloss letter-press inks
Overprint varnishes

FRP CO.: FILTREZ Resins(Continued):

FILTREZ 3790:

An amber colored, hard, thermoplastic rosin based resin with good solubility in ink oils. As a modified rosin pentaester with unique solution properties, this product may find utility in a number of application areas.

Product Specifications:
Melting Point (A.S.T.M. Ring & Ball): 115-125C.
Acid Value: 25 Max.
Viscosity (60% solids in Toluene): C-E
Color: 12 Max.
Pounds per gallon: 9.2-9.4

Solubility:

Aromatic hydrocarbons:	Soluble
Aliphatic hydrocarbons:	Soluble
Chlorinated hydrocarbons:	Soluble
C1-C4 alcohols:	Insoluble

Suggested Uses:
Varnishes, flush colors, enamels, heat set inks, gel vehicles, hot melt coatings, machinery finishes, textile print coatings, hardware finishes.

FILTREZ 3791:

An amber colored, hard, thermoplastic rosin based resin with intermediate solubility in ink oils. As a modified rosin pentaester with unique solution properties, this product may find utility in a number of application areas.

Product Specifications:
Melting Point (A.S.T.M. Ring & Ball): 125-135C.
Acid Value: 25 Max.
Viscosity (60% solids in Toluene): G-K
Color: 12 Max.
Pounds per Gallon: 9.2-9.4

Solubility:

Aromatic hydrocarbons:	Soluble
Aliphatic hydrocarbons:	Soluble
Chlorinated hydrocarbons:	Soluble
C1-C4 alcohols:	Insoluble

Suggested Uses:
Varnishes, flush colors, enamels, heat set inks, gell vehicles, hot melt coatings, machinery finishes, textile print coatings, hardware finishes.

FRP CO.: FILTREZ Resins(Continued):

Maleic Resins(Continued):

FILTREZ 3792:

An amber colored, hard, thermoplastic rosin based resin with limited solubility in ink oils. As a modified rosin pentaester with unique solution properties, this product may find utility in a number of application areas.

Product Specifications:
Melting Point (A.S.T.M. Ring & Ball): 135-145C.
Acid Value: 25 Max.
Viscosity (60% solids in Toluene): O-S
Color: 12 Max.
Pounds per gallon: 9.2-9.4

Solubility:

Aromatic hydrocarbons:	Soluble
Aliphatic hydrocarbons:	Soluble
Chlorinated hydrocarbons:	Soluble
C1-C4 alcohols:	Insoluble

Suggested Uses:
Varnishes, flush colors, enamels, heat set inks, gel vehicles, hot melt coatings, machinery finishes, textile print coatings, hardware finishes.

FILTREZ 3795:

An exceptionally light colored, hard, thermoplastic rosin based resin with excellent solubility in ink oils. As a modified rosin pentaester with unique solution properties, this product may find utility in a number of application areas.

Product Specifications:
Melting Point (A.S.T.M. Ring & Ball): 115-125C
Acid Value: 25 Max.
Viscosity (60% solids in Toluene): B-G
Color: 12 Max.
Pounds per gallon: 9.2-9.4

Solubility:

Aromatic hydrocarbons:	Soluble
Aliphatic hydrocarbons:	Soluble
Chlorinated hydrocarbons:	Soluble
C1-C4 alcohols:	Insoluble

Suggested Uses:
Varnishes, flush colors, enamels, heat set inks, gel vehicles, hot melt coatings, machinery finshes, textile print coatings, hardware finishes.

FRP CO.: FILTREZ Resins(Continued):

Maleic Resins:

FILTREZ 3796:

An exceptionally light colored, hard, thermoplastic rosin based resin with good solubility in ink oils. As a modified rosin ester pentaester with unique solution properties, this product may find utility in a number of application areas.

Product Specifications:
Melting Point (A.S.T.M. Ring & Ball): 125-135C
Acid Value: 25 Max.
Viscosity (60% solids in Toluene): C-E
Color: 12 Max.
Pounds per gallon: 9.2-9.4

Solubility:

Aromatic hydrocarbons:	Soluble
Aliphatic hydrocarbons:	Soluble
Chlorinated hydrocarbons:	Soluble
C1-C4 alcohols:	Insoluble

Suggested Uses:
Varnishes, flush colors, enamels, heat set inks, gel vehicles, hot melt coatings, machinery finishes, textile print coatings, hardware finishes.

FILTREZ 3797:

An exceptionally light colored, hard, thermoplastic rosin based resin with intermediate solubility in ink oils. As a modified rosin penta-ester with unique solution properties, this product may find utility in a number of application areas.

Product Specifications:
Melting Point (A.S.T.M. Ring & Ball): 135-145C
Acid Value: 25 Max.
Viscosity (60% solids in Toluene): G-K
Color: 12 max.
Pounds per gallon: 9.2-9.4

Solubility:

Aromatic hydrocarbons:	Soluble
Aliphatic hydrocarbons:	Soluble
Chlorinated hydrocarbons:	Soluble
C1-C4 alcohols:	Insoluble

Suggested Uses:
Varnishes, flush colors, enamels, heat set inks, gel vehicles, hot melt coatings, machinery finishes, textile print coatings, hardware finishes.

FRP CO.: FILTREZ Resins(Continued):

Maleic Resins(Continued):

FILTREZ 380:

Maleic-pentaerythritol rosin ester possessing very high melting point and low acid value. Also, it exhibits light color and excellent color retention.

Specifications:
- Melting Point (A.S.T.M. Ring & Ball): 140-150C.
- Acid Value: 12-25
- Viscosity (60% in Xylene, Gardner-Holdt): X-Z
- Color (60% in Xylene, Gardner Standards): 10 max.
- Pounds per gallon: 9.3-9.5

Solubility:

Aliphatic Hydrocarbons:	Excellent
Aromatic Hydrocarbons:	Excellent
Alcohols:	Insoluble
Varnish Oils:	Good
Esters:	Good

Typical Uses
- Heat-set inks
- Gloss letter-press inks
- Overprint varnishes

FILTREZ 3000:

Maleic-glycerol rosin ester.

Specifications:
- Melting Point (A.S.T.M. Ring & Ball): 125-135C.
- Acid Value: 35-42
- Viscosity (60% in Toluol, Gardner-Holdt): F-K
- Color (60% in Toluol, Gardner Standards): 10 Max.

Solubility:

Aliphatic Hydrocarbons:	Limited
Aromatic Hydrocarbons:	Good
Alcohols:	Insoluble

Typical Uses:
- Gravure Inks
- Lacquers
- Coatings

FRP CO.: FILTREZ Resins(Continued):

FILTREZ 5014:

FILTREZ 5014 is an alcohol, aqueous ammonia soluble resin that produces more durable film forming properties than similar rosin based resins. FILTREZ 5014 is compatible with most of the resins now being used to produce water type flexographic inks. It is also compatible with SS nitrocellulose.

Specifications:
- Melting Point (A.S.T.M. Ring & Ball): 135-145C.
- Acid Value: 160-180
- Viscosity (50% in Ethanol, Gardner-Holdt): B-E
- Color (50% in Ethanol, Gardner Standards): 10 max.
- Pounds per gallon: 9.8-10.0
- Flash Point (C.O.C.): 425F.

Solubility:

Aliphatic Hydrocarbons:	Insoluble
Aromatic Hydrocarbons:	Insoluble
Alcohols:	Soluble

Typical Uses:
- Water type flexographic inks
- Gravure inks

Fumaric Resins:

FILTREZ 521:

An alcohol soluble, fumaric modified ester of rosin. FILTREZ 521 is used in Flexographic and Gravure printing inks and coating applications.

Specifications:
- Melting Point (A.S.T.M. Ring & Ball): 130-140C.
- Acid Value: 210-225
- Viscosity (50% in Ethanol, Gardner-Holdt): A3-A1
- Color (50% in Ethanol, Gardner Standards): 9 Max.
- Pound per gallon: 9.5-9.7
- Flash Point (open cup): 500F.

Solubility:

Aliphatic Hydrocarbons:	Insoluble
Aromatic Hydrocarbons:	Good
Alcohols:	Good

Typical Uses:
- Alcohol flexographic inks
- Moisture-set inks
- Gravure inks
- Lacquers

FRP CO.: FILTREZ Resins(Continued):

Fumaric Resins(Continued):

FILTREZ 525:

Fumaric-modified rosin, alcohol soluble resin, especially formulated to exhibit fast dry and solvent release in water and alcohol flexographic inks.

Specifications:
- Melting Point (A.S.T.M. Ring & Ball): 140-155C
- Acid Value: 100-115
- Viscosity (50% in Ethanol, Gardner-Holdt): B-G
- Color (50% in Ethanol, Gardner Standards): 10 Max.
- Pounds per gallon: 9.9-10.1

Solubility:

Aliphatic Hydrocarbons:	Insoluble
Aromatic Hydrocarbons:	Insoluble
Alcohols:	Good

Typical Uses:
- Flexograpic Inks
- Spirit Varnishes
- Gloss steam-set inks
- Aniline Inks
- Floor Waxes

FILTREZ 526:

Fumaric-modified alcohol soluble resin. FILTREZ 526 is useful in alcohol and water flexographic ink systems. It exhibits fast dry and low amine requirements in aqueous systems.

Specifications:
- Melting Point (A.S.T.M. Ring & Ball): 125-135C.
- Acid Value: 115-130
- Viscosity (50% in Ethanol, Gardner-Holdt): A-A2
- Color (50% in Ethanol, Gardner Standards): 10 Max.
- Pounds per gallon: 9.8-10.0

Solubility:

Aliphatic Hydrocarbons:	Insoluble
Aromatic Hydrocarbons:	Insoluble
Alcohols:	Good

Typical Uses:
- Flexographic inks
- Aniline inks
- Floor waxes

FRP CO.: FILTREZ Resins(Continued):

FUMARIC Resins(Continued):

FILTREZ 591:

A fumaric-modified resin especially developed for steam-set printing inks. Color, viscosity and water tolerance assure easy formulation and operation in standard steam-set ink use.

Specifications:
- Melting Point (A.S.T.M. Ring & Ball): 135-145C
- Acid Value: 290-310
- Viscosity (50% in Diethylene Glycol, Gardner-Holdt): Z1-Z2
- Color (50% in Diethylene Glycol, Gardner Standards): 11 Max.
- Pounds per gallon: 9.6-9.7
- Flash Point: 490F.

Solubility:

Aliphatic Hydrocarbons:	Insoluble
Aromatic Hydrocarbons:	Insoluble
Alcohols:	Soluble

Typical Uses:
- Steam-set printing inks
- Shellac substitutes
- Floor waxes

FILTREZ 591 Solutions:

Solutions of FILTREZ 591 in Diethylene Glycol

Specifications: 50% Soln.:
- Color (Gardner Standards): 11 Max.
- Acid Value: 124-134
- Viscosity (Gardner-Holdt @ 25C): Z1-Z2
- Pounds per gallon @ 25C: 9.55-9.65
- Flash Point (C.O.C.): 290F.

Specifications: 60% Soln.:
- Color (Gardner Standards): 11 Max.
- Acid Value: 135-155
- Viscosity (Gardner-Holdt @ 25C): Z7+-1
- Pounds per gallon @ 25C: 9.65-9.75
- Flash Point (C.O.C.): 290F.

Specifications: 65.5% Soln.:
- Color (Gardner Standards): 11 Max.
- Acid Value: 145-165
- Viscosity (Gardner-Holdt @ 25C): 26-32 min.
- Pounds per gallon @ 25C.: 9.65-9.75
- Flash Point (C.O.C.): 290F.

Typical Uses:
- Steam-set printing inks

FRP CO.: FILTREZ Resins(Continued):

Fumaric Resins(Continued):

FILTREZ 593A:

Fumaric modified resin designed for low cost alcohol flexographic and moisture-set ink systems. This resin offers good resistance to crystallization in alcohol solutions.

Specifications:
- Melting Point (A.S.T.M. Ring & Ball): 125-135C.
- Acid Value: 230-250
- Viscosity (50% in Ethanol, Gardner-Holdt): A3-A1
- Color (50% in Ethanol, Gardner Standards): 10 Max.
- Pounds per gallon: 9.6-9.8

Solubility:

Aliphatic Hydrocarbons:	Insoluble
Aromatic Hydrocarbons:	Good
Alcohols:	Good

Typical Uses:
- Alcohol flexographic inks
- Moisture-set inks
- Gravure inks
- Lacquers

FILTREZ 597:

FILTREZ 597 is a fumaric modified rosin ester designed for low cost water based inks and alcohol based inks. FILTREZ 597 possesses good pigment wetting properties and is compatible with shellacs, acrylics, proteins and SS nitrocellulose.

Specifications:
- Melting Point (A.S.T.M. Ring & Ball): 125-135C.
- Acid Value: 145-160
- Viscosity (50% in Ethanol, Gardner-Holdt): B-E
- Color (50% in Ethanol, Gardner Standards): 10 Max.
- Pounds per gallon: 9.8-10.0
- Flash Point (C.O.C.): 450F.

Solubility:

Aliphatic Hydrocarbons:	Insoluble
Aromatic Hydrocarbons:	Slightly
Alcohols:	Soluble

FRP CO.: FILTREZ Resins(Continued):

Phenolic Resins:

FILTREZ 600:

A phenolic modified glycerol ester, soluble in aromatic solvents and all vegetable oils. Possesses maximum durability, color retention and high elasticity. Recommended for short and medium length tung oil varnishes and spar varnish films, such as floor varnish and primers. Miscible with chlorinated rubber and ethylcellulose.

Specifications:
- Melting Point (A.S.T.M. Ring & Ball): 130-140C.
- Acid Value: 12-20
- Viscosity (60% in Toluol, Gardner-Holdt): F-J
- Color (60% in Toluol, Gardner Standards): 12 Max.
- Pounds per gallon: 9.1-9.2
- Flash Point (C.O.C.): 470F.

Solubility:
- Aliphatic Hydrocarbons: Good
- Aromatic Hydrocarbons: Good
- Alcohols: Poor

Typical Uses:
- Tung oil varnishes
- Floor varnishes
- Varnishes primers

FILTREZ 608:

A high melting phenolic resin designed especially for the ink manufacturer. This resin is characterized by its excellent solubility in low KB solvents and pigment wetting properties.

Specifications:
- Melting Point (A.S.T.M. Ring & Ball): 145-155C.
- Acid Value: 12-25
- Viscosity (50% in 38 KB Mineral Spirits): V-Y
- Color (50% in 38 KB Mineral Spirits): 11 Max.
- Pounds per gallon: 9.1-9.2

Solubility:
- Aliphatic Hydrocarbons: Good
- Aromatic Hydrocarbons: Good
- Alcohols: Insoluble

Typical Uses:
- Heat-set inks

FRP CO.: FILTREZ Resins(Continued):

Phenolic Resins(Continued):

FILTREZ 623:
A high phenolic-modified resin which exhibits hardness, solubility, color retention and durability in a wide range of varnish products. It is very adaptable to short oil varnishes where rapid through drying is desired or in medium and long oil varnishes for use in printing inks and floor coverings.

Specifications:
Melting Point (A.S.T.M. Ring & Ball): 135-145C.
Acid Value: 12-25
Viscosity (60% in Xylol, Gardner-Holdt): V-X
Color (60% in Xylol, Gardner Standards): 11 Max.
Pounds per gallon: 9.15-9.25
Flash Point (C.O.C.): 545F.

Solubility:

Aliphatic Hydrocarbons:	Good
Aromatic Hydrocarbons:	Good
Alcohols:	Insoluble

Typical Uses:
Printing inks
Floor coverings
Protective coatings

FILTREZ 624:
A high phenolic-modified resin which exhibits hardness, solubility, color retention and durability in a wide range of varnish products. It is very adaptable to short oil varnishes where rapid thorough drying is desired or in medium and long oil varnishes for use in printing inks and floor coverings.

Specifications:
Melting Point (A.S.T.M. Ring & Ball): 140-150C
Acid Value: 12-25
Viscosity (60% in Xylol, Gardner-Holdt): X-Z1
Color (60% in Xylol, Gardner Standards): 11 Max.
Pounds per gallon: 9.15-9.25
Flash Point (C.O.C.): 545F.

Solubility:

Aliphatic Hydrocarbons:	Good
Aromatic Hydrocarbons:	Good
Alcohols:	Insoluble

Typical Uses:
Printing inks
Floor coverings
Protective coatings

FRP CO.: FILTREZ Resins(Continued):

Phenolic Resins(Continued):

FILTREZ 625N:

A phenolic-modified pentaerythritol ester possessing good resistance to after-yellowing, hot water and alkali and excellent pigment stability.

Specifications:
- Melting Point (A.S.T.M. Ring & Ball): 145-155C
- Acid Value: 12-25
- Viscosity (56% in Toluol, Gardner-Holdt): P-T
- Color (56% in Toluol, Gardner Standards): 12 Max.
- Pounds per gallon: 9.15-9.25
- Flash Point (C.O.C.): 545F.

Solubility

Aliphatic Hydrocarbons:	Good
Aromatic Hydrocarbons:	Good
Alcohols:	Insoluble

Typical Uses:
- Paints
- Enamels
- Varnishes
- Metal Primers

FILTREZ 640:

A phenolic-modified resin used in the preparation of industrial varnishes, maintenance paints, rust protective coatings and implement coatings. Used to modify waxes and alkyds.

Specifications:
- Melting Point (A.S.T.M. Ring & Ball): 130-140C.
- Acid Value: 10-20
- Viscosity (60% in Toluol, Gardner-Holdt): N-T
- Color (60% in Toluol, Gardner Standards): 12 Max.
- Pounds per gallon: 9.1-9.2
- Flash Point (C.O.C.): 490F.

Solubility:

Aliphatic Hydrocarbons:	Good
Aromatic Hydrocarbons:	Good
Alcohols:	Limited

Typical Uses:
- Agriculture implement enamels
- Enamels
- Wax modifier
- Paints
- Industrial varnishes
- Floor and deck varnishes
- Protective coatings

FRP CO.: FILTREZ Resins(Continued):

Phenolic Resins(Continued):

FILTREZ 640X:

A phenolic-modified resin used in the preparation of industrial varnishes, maintenance paints, rust protective coatings and implement coatings. Used to modify waxes and alkyds.

Specifications:
- Melting Point (A.S.T.M. Ring & Ball): 140-150C
- Acid Value: 10-25
- Viscosity (60% in Toluol, Gardner-Holdt): V-Y
- Color (60% in Toluol, Gardner Standards): 12 Max.
- Pounds per gallon: 9.1-9.2
- Flash Point (C.O.C.): 490F.

Solubility:
- Aliphatic Hydrocarbons: Good
- Aromatic Hydrocarbons: Good
- Alcohols: Limited

Typical Uses:
- Agriculture implement varnishes
- Enamels
- Wax modifier
- Paints
- Industrial varnishes
- Floor and deck varnishes
- Protective coatings

FILTREZ 651:

Phenol-formaldehyde modified ester of rosin exhibiting high melting point, viscosity and excellent solubility in low KB solvents and vegetable oils.

Specifications:
- Melting Point (A.S.T.M. Ring & Ball): 150-160C.
- Acid Value: 12-25
- Viscosity (60% in Toluol, Gardner-Holdt): Z-Z3
- Color (60% in Toluol, Gardner Standards): 12 max.
- Pounds per gallon: 9.1-9.3

Solubility:
- Aliphatic Hydrocarbons: Good
- Aromatic Hydrocarbons: Good
- Alcohols: Insoluble

Typical Uses:
- Paints
- Lacquers
- Printing Inks

FRP CO.: FILTREZ Resins(Continued):

Phenolic Resins(Continued):

FILTREZ 652:

Phenol-formaldehyde modified ester of rosin exhibiting high melting point, viscosity and excellent solubility in low KB solvents and vegetable oils.

Specifications:
- Melting Point (A.S.T.M. Ring & Ball): 150-165C.
- Acid Value: 12-25
- Viscosity (46% in Toluene, Gardner-Holdt): H-M
- Color (46% in Toluene, Gardner Standards): 12 Max.
- Pounds per gallon: 9.1-9.3

Solubility:
- Aliphatic Hydrocarbons: Good
- Aromatic Hydrocarbons: Good
- Alcohols: Insoluble

Typical Uses:
- Paints
- Lacquers
- Printing Inks

FILTREZ 659:

A high phenolic resin possessing fast body, fast dry and excellent durability.

Specifications:
- Melting Point (A.S.T.M. Ring & Ball): 130-140C
- Acid Value: 10-20
- Viscosity (60% in Toluol, Gardner-Holdt): M-P
- Color (60% in Toluol, Gardner Standards): 12 Max.
- Pounds per gallon: 9.0-9.1
- Flash Point (C.O.C.): 490F.

Solubility:
- Aliphatic Hydrocarbons: Good
- Aromatic Hydrocarbons: Good
- Alcohol: Insoluble

Typical Uses:
- Varnishes
- Enamels
- Paints

FRP CO.: FILTREZ Resins(Continued):

Phenolic Resins(Continued):

FILTREZ 660:

Phenol-formaldehyde modified ester of rosin exhibiting high melting point, viscosity and excellent solubility in low KB solvents and vegetable oils.

Specifications:
Melting Point (A.S.T.M. Ring & Ball): 160-170C.
Acid value: 12-25
Viscosity (60% in Toluol, Gardner-Holdt): Z-Z3
Color (60% in Toluol, Gardner Standards): 12 Max.
Pounds per gallon: 9.1-9.3

Solubility:

Aliphatic Hydrocarbons:	Good
Aromatic Hydrocarbons:	Good
Alcohols:	Insoluble

Typical Uses:
Paints
Lacquers
Printing Inks

FILTREZ 693:

Phenol-formaldehyde modified ester of rosin exhibiting high melting point, viscosity and excellent solubility in low KB solvents and vegetable oils.

Specifications:
Melting Point (A.S.T.M. Ring & Ball): 165-175C.
Acid Value: 12-25
Viscosity (46% in Toluene, Gardner-Holdt): H-M
Color (46% in Toluene, Gardner Standards): 12 Max.
Pounds per gallon: 9.1-9.3

Solubility:

Aliphatic Hydrocarbons:	Good
Aromatic Hydrocarbons:	Good
Alcohols:	Insoluble

Typical Uses:
Paints
Lacquers
Printing Inks

FRP CO.: FILTREZ Resins(Continued):

Phenolic Resins(Continued):

FILTREZ 6190:

A phenolic modified ester having excellent solubility in low KB ink oils.

Specifications:
- Melting Point (A.S.T.M. Ring & Ball): 140-150C.
- Acid Value: 15-27
- Viscosity (50% in M-470): Z2-Z4
- Color (50% in M-470): 14 Max.
- Pounds per gallon: 9.1-9.3

Solubility:

Aliphatic Hydrocarbons:	Good
Aromatic Hydrocarbons:	Good
Alcohols:	Insoluble

Typical Uses:
- Heat-set inks
- Flushes

Special Resins:

FILTREZ 895:

A styrene ester that exhibits good adhesion and compatibility when used as a replacement for up to 30% of the nitrocellulose in lacquers and inks.

Specifications:
- % Total Solids: 100%
- Melting Point: 87-93C. (Ring & Ball)
- Acid Value: 4.0 Max.
- Viscosity (75% in Butyl Acetate, Gardner-Holdt): Z4-Z6
- Color (Gardner Standards): 6 Max.
- Flash Point (Tag Closed Cup): 250F.
- Density at 23C.: 9.1 lbs./gal.
- Molecular Weight (Number Average): 1500
- Glass Temperature (By Differential Scanning Calorimetry): 32C.

Solubility

Excellent solubility in esters, ketones, and aromatic hydrocarbons

Aliphatic Hydrocarbons:	Insoluble
Alcohols:	Insoluble

FREEMAN CHEMICAL CORP.: CHEMPOL Conventional Solids Resins:

CHEMPOL:

11-2426:
Type: Long oil soya alkyd
Remarks: Good flow, compatibility, gloss; meets TT-R-266, Type 1, Class A.

11-3407:
Type: Medium soya alkyd
Remarks: Excellent salt spray and exterior durability; used with urea or melamine resins.

11-3548:
Type: Medium soya alkyd
Remarks: General purpose primers and enamels; excellent color retention.

11-2932:
Type: Short oil chain stopped alkyd
Remarks: Durable, fast air dry, good compatibility, excellent aliphatic solubility.

11-3068:
Type: Short oil chain stopped alkyd
Remarks: Air dry or bake with urea or melamine resins; excellent exterior durability.

11-3830:
Type: Short oil chain stopped alkyd
Remarks: Rapid dry and good compatability with medium oil alkyds; good aliphatic solubility.

11-4119:
Type: Short oil soya alkyd
Remarks: Fast set and through dry; good exterior durability; can be used with melamine or urea resins.

11-4218:
Type: Short tall oil alkyd
Remarks: General purpose baking vehicle; hardness, mar resistance; compatible with most melamine and urea resins.

11-4272:
Type: Short tall oil alkyd
Remarks: Economical baking alkyd.

FREEMAN CHEMICAL CORP.: CHEMPOL Conventional Solids Resins (Continued):

CHEMPOL:

11-4300:
Type: Short DCO Alkyd
Remarks: Exceptional adhesion, flexibility; excellent salt spray, moisture and detergent resistance in bake applications.

11-4321:
Type: Short coconut oil alkyd
Remarks: Good color retention in melamine or urea-cured bake finishes; nitrocellulose resin plasticizer.

11-4423:
Type: Short tall oil alkyd
Remarks: Fast dry, high gloss, good color retention and exterior durability.

11-3819:
Type: Polyester
Remarks: Overbake resistance; outstanding hardness, flexibility, impact and exterior durability.

11-3915:
Type: Polyester
Remarks: Rapid cure response; good flexibility, adhesion and exterior durability.

12-2101:
Type: Silicone alkyd
Remarks: Air dry; meets TT-E-490E.

12-4102:
Type: Rosin modified alkyd
Remarks: Fast drying; excellent gasoline and corrosion resistance.

12-4008:
Type: Phenolic modified tung oil alkyd
Remarks: Designed for textured/wrinkle finishes; good moisture resistance.

13-1214:
Type: Acrylic alkyd copolymer
Remarks: Fast air dry or bake; excellent mar resistance, adhesion, exterior durability, flexibility and compatability.

FREEMAN CHEMICAL CORP.: CHEMPOL Conventional Solids Resins (Continued):

CHEMPOL:

13-1527:
Type: Acrylic alkyd copolymer
Remarks: Excellent humidity resistance and plastic substrate adhesion.

13-1501:
Type: Acrylic alkyd copolymer
Remarks: Fast dry and good compatibility and exterior durability.

13-1402:
Type: Vinyl toluene alkyd copolymer
Remarks: Fast drying, excellent recoatability, good color retention and salt spray resistance.

13-1905:
Type: Vinyl toluene alkyd copolymer
Remarks: Very fast dry, high gloss, good pigment wetting; excellent for aerosol use.

13-2211:
Type: Vinyl toluene alkyd copolymer
Remarks: Fast dry, good compatability; can be used in aerosols.

13-2502:
Type: Vinyl toluene alkyd copolymer
Remarks: Excellent pigment wetting, toughness, flexibility; wide compatibility.

14-0141:
Type: Vinyl toluene alkyd copolymer
Remarks: Very fast dry, compatible with low solvency aliphatics; aerosol and plastic substrate coatings.

13-2228:
Type: Styrenated alkyd copolymer
Remarks: Fast dry, excellent initial gloss; used for hammer finishes.

15-2518:
Type: Phenolic modified tung oil varnish
Remarks: Designed for textured/wrinkle finishes under adverse conditions; air dry or bake.

FREEMAN CHEMICAL CORP.: CHEMPOL Conventional Solids Resins (Continued):

CHEMPOL:

16-0001:
Type: Epoxy ester
Remarks: Air dry or bake with melamine or urea resins; high chemical and corrosion resistance; excellent nonferrous substrate adhesion.

16-0113:
Type: Epoxy ester
Remarks: For use in weld through, zinc rich primers; excellent corrosion resistance.

17-3210:
Type: Thermoset acrylic copolymer
Remarks: Hydroxyl functional; excellent gloss, hardness and overbake resistance.

18-0089:
Type: Oil modified urethane alkyd
Remarks: Air dry, good abrasion resistance on wood substrates.

18-0094:
Type: Oil modified urethane alkyd
Remarks: Quick development of hardness, mar resistance; clear or pigmented wood coatings.

18-0103:
Type: Urethane alkyd
Remarks: Fast dry, economical resin.

21-7018:
Type: Proprietary resin blend
Remarks: Bake and force dry gloss paper coatings with high mar and soiling resistance; excellent blocking resistance; excellent hot foil stamping acceptance.

50-0068:
Type: Overprint varnish
Remarks: Water reducible, high gloss; fast cure response.

51-4055:
Type: Nonoxidizing alkyd
Remarks: High gloss conversion coatings for paper; excellent color and gloss retention.

FREEMAN CHEMICAL CORP.: CHEMPOL High Solids Resins:

Alkyd (Drying or Semi-Drying Oil Type):

CHEMPOL 11-2831:
75% solids "Chain Stopped" alkyd. Combines good handling properties with fast dry. Will meet 3.5 pound "VOC" coating requirement.

CHEMPOL 11-3071:
75% solids, the fastest drying "Chain Stopped" alkyd. For quality industrial finishes having excellent exterior durability. Designed for 3.5 pound "VOC" coating systems.

CHEMPOL 11-3121:
80% solids "Chain Stopped" alkyd thinned in MAK. Provides fast viscosity reduction, outstanding gloss and gloss retention. Also available in PM Acetate as 11-3135.

CHEMPOL 11-3132:
Lower molecular weight version of 11-3121 supplies at 75% solids in xylene.

CHEMPOL 11-3135:
This is 11-3121 cut in PM Acetate.

CHEMPOL 11-3618:
86% solids short tall oil baking alkyd. Rapid viscosity reduction and good low temperature cure response for drum enamel and general industrial applications.

CHEMPOL 11-3630:
80% solids, low rosin tall oil baking alkyd cut in xylene. Requires a minimum of 30% urea or melamine resin to achieve full cure film properties. Recommended for applications where metal preparation may be marginal. Will provide coatings meeting 2.8-3.0 pound "VOC" requirements. Designed to be a general purpose "workhorse" higher solids baking vehicle.

CHEMPOL 11-3631:
90% solids version of 11-3630 supplied in MIBK.

CHEMPOL 11-3903:
85% solids coconut alkyd in MPK. Has excellent color retention on overbake.

FREEMAN CHEMICAL CORP.: CHEMPOL High Solids Resins(Continued):

Modified Alkyds:

CHEMPOL 12-1816:
75% solids silicone modified chain-stopped alkyd. Designed for fast drying enamels with excellent exterior durability. Recommended for transportation finishes. 12-1816 is supplied in xylol for electrostatic spray applications.

CHEMPOL 12-2204:
75% solids phenolic modified alkyd supplied in xylene. Suggested for use in high solids (3.5 lb. "VOC") air dry, force dry applications or 3.0 pound "VOC" baking primers and enamels. Outstanding corrosion resistance where color is not of prime importance.

CHEMPOL 12-3030:
High solids rosin modified alkyd with fast dry and excellent corrosion resistance. Recommended for primers. Can be recoated with enamels containing ketone and ester solvents.

Alkyd Copolymers:

CHEMPOL 13-1428:
75% solids, very fast drying styrene, alkyd copolymer supplied in xylene/MPK solvents.

CHEMPOL 13-1516:
70% solids, fast drying alkyd acrylic copolymer supplied in toluene. Compatible with thermoplastic acrylic and vinyl resins. Performance properties similar to conventional solids alkyd copolymers 13-1501 and 13-1502. 3.5 pound "VOC" coatings can be realized when heat is used to achieve lower viscosities at point of application.

CHEMPOL 13-1604:
70% solids, fast drying acrylic alkyd copolymer supplied in a solvent blend designed to provide flow and leveling without sagging. Excellent adhesion to treated and untreated metal substrates and plastic surfaces. Compatible with thermoplastic acrylic and vinyl resins. Will provide 3.5 pound "VOC" coatings when the coatings are applied at elevated temperatures(120F-160F).

CHEMPOL 13-1845:
75% solids, fast drying acrylic alkyd copolymer. Outstanding initial gloss and gloss retention can be achieved. Excellent pigment wetting properties and adhesion to treated and untreated metal substrates and plastic surfaces. Can be force dried (with driers) or modified with 10-15% melamine or urea resins to provide a true thermoset finish. 3.0-3.5 pound "VOC" finishes are possible when heat is used to lower viscosity at point of application.

FREEMAN CHEMICAL CORP.: CHEMPOL High Solids Resins(Continued):

Alkyd Copolymers(Continued):

CHEMPOL 14-0149:
75% solids vinyl toluene alkyd which provides rapid lacquer-like, tack-free times for a wide variety of general industrial uses.

Polyesters(oil free alkyds):

CHEMPOL 11-2339:
100% solids resin designed for 2.8-3.0 pound "VOC" baking application. Contains no reactive diluents. Can be reduced with all aromatic solvents. Requires 30% minimum melamine or urea resin to achieve maximum cure film performance properties.

CHEMPOL 11-2821:
80% solids polyester designed to produce tough, hard films when crosslinked with urea and/or melamine resins. Cured films exhibit exterior durability, abrasion, stain and detergent resistance. Finishes formulated with 11-2821 will meet 2.8-3.0 pound "VOC" at point of application.

CHEMPOL 11-3361:
80% solids polymer cut in xylene for spray application where faster evaporating solvents are required. 2.8-3.0 pound "VOC" can be obtained with proper melamine urea selections.

CHEMPOL 11-3369:
70% solids polyester designed for coil coating. Provides excellent flexibility.

CHEMPOL 11-3640:
High solids enamels formulated for 3.0 or lower VOC are achievable. Provides greater coverage, with hardness, flexibility and color retention comparable or superior to conventional alkyd baking enamels. 11-3640 also demonstrates excellent electrostatic spray characteristics.

CHEMPOL 11-4517:
High solids, carboxyl functional polymer.

FREEMAN CHEMICAL CORP.: CHEMPOL High Solids Resins(Continued):

Thermosetting Acrylic Resins:

CHEMPOL 17-3708:

70% solids, carboxyl functional acrylic resin specifically designed for roller coat, metal deco application. Designed to cure with epoxy resins. Will also cure with melamine or urea resins, but at a much slower rate. Chemical resistance, ability to take "pasteurization" plus a high degree of adhesion over tin plate are properties acrylic epoxy coatings possess.

CHEMPOL 17-4001:

75% solids, hydroxyl functional acrylic in MAK. Requires melamine or urea resin to achieve cure. Cured films provide high hardness, excellent stain and chemical resistance. When blended with higher solids polyester resin, 17-4001 will impart these properties to the total blend depending upon degree of modification. Recommended for appliance coatings.

Epoxy Esters:

CHEMPOL 16-1200:

75% solids, thermosetting epoxy ester. Requires melamine or urea resin to achieve cure. Versatile functionality also permits curing with polyamines or polyamides. Resin modifier for polyester baking finishes. Improves adhesion, flexibility, water and corrosion resistance at 10-15% additions based on total resin solids. When used as a prime vehicle in bake applications, provides excellent detergent resistance. When incorporated as a modifier for polyester baking enamels, it will lower viscosities.

CHEMPOL 16-1202:

The same higher performance polymer as 16-1200, only supplied in high boiling aromatic solvent SC-150 for coil, metal decorating and dip applications.

FREEMAN CHEMICAL CORP.: CHEMPOL Resins for 2-Component Urethanes:

Freeman Chemical offers a complete line of both polyester and acrylic resins designed for formulating state-of-the-art two component urethane finishes for the most demanding applications. Clear and pigmented coatings for aircraft, maintenance, transportation and plastic substrate applications can be formulated to provide superior toughness and resistance to weathering, abrasion, corrosion and chemical attack.

CHEMPOL 18-1705:
Acrylic copolymer for high performance exterior applications requiring hardness, gloss retention and chemical resistance.

CHEMPOL 18-1708:
Acrylic copolymer providing high hardness, gloss retention and chemical resistance.

CHEMPOL 18-2217:
High solids polyester for applications where hardness and chemical resistaance are of primary importance over exterior durability.

CHEMPOL 18-2219:
Higher solids polyester which improves gloss retention and flexibility when blended with CHEMPOL 18-2230.

CHEMPOL 18-2222:
High solids polycaprolactone which provides the highest order of chemical and corrosion resistance.

CHEMPOL 18-2230:
Polyester resin for uses where chemical resistance and hardness are primary concerns.

CHEMPOL 18-2231:
Polyester for applications where gloss retention and chemical resistance are of concern.

CHEMPOL 11-2339:
100% solids polyester compatible with acrylic and other polyesters. Can be used to lower viscosity and achieve lower VOC coatings.

FREEMAN CHEMICAL CORP.: Water Reducible Resins:

CHEMPOL 10-0091:
Oxidizing Alkyd
Short oil water reducible fast air drying alkyd for coatings requiring a high order of hardness, water resistance, gloss retention and corrosion resistance.

CHEMPOL 10-0503:
Polyester
For high performance coil and general industrial applications. Exhibits rapid cure and provides hardness, flexibility and excellent color and gloss retention.

CHEMPOL 10-1300:
Acrylic Alkyd Copolymer
For very fast air dry or baking enamels. Recommended especially for use on heavy castings and other rigid substrates.

CHEMPOL 10-1313:
Acrylic Alkyd Copolymer
For flow coat and dip tank application. Excellent stability, corrosion resistance and toughness. For use in bake or air dry application. Exhibits excellent hardness and adhesion to various substrates.

CHEMPOL 10-1602:
Epoxy Ester
For flow coat and dip tank primer application. Outstanding salt spray resistance and humidity resistance. Can be used for both bake and air dry application.

CHEMPOL 10-1706:
Thermoset Acrylic
Fast cure at low temperatures (200-250F). Must be crosslinked with melamines or urea films to achieve cured films. Outstanding hardness and adhesion to various substrates.

CHEMPOL 10-1105:
Coconut Alkyd
Melamine or urea curable baking resin offering outstanding color retention on overbake.

GENERAL ELECTRIC CO.: Silicone Resins:

Silicone Resin:

SR 112:

Key Characteristic: Heat Resistance, Weather Durability
Typical Application: Suitable for high heat and weather-resistant coatings.
Silicone Content %: 50+-1
Specific Gravity 25C (77F): 1.01
Solvent System: Xylol
Viscosity-Bkfld. (cps): 120-220
Gardner: D-H
Color/Gardner: Light straw/0-3
Flash Point Pensky-Martens: 27C (80F)
Density lbs/gal: 8.4

SR 125:

Key Characteristic: Toughness, Resistance to Thermal Shock Heat Resistance
Typical Application: Suitable for coatings requiring flexibility, toughness, high heat resistance, good gloss retention, leafing properties, thermal shock resistance, weather resistance
Silicone Content: 50+-1
Specific Gravity 25C (77F): 1.00
Solvent System: Xylol
Viscosity Bkfld (cps): 110-320
Color/Gardner: Light straw/0-1
Flash Point Pensky-Martens: 24C(75F)
Density, lbs/gal: 8.3

SR 141:

Key Characteristic: Excellent Resistance to Extreme Temperatures
Typical Application: Suitable for coatings requiring optimum performance characteristics above 424C(800F).
Silicone Content %: 50+-1
Specific Gravity 25C (77F): 1.02
Solvent System: Toluol
Viscosity Bkfld. cps: 100-200
Gardner: D-H
Color/Gardner: Light straw/0-2
Flash Point Pensky-Martens: 5C (41F)
Density, lbs/gal: 8.5

GENERAL ELECTRIC CO.: Silicone Resins(Continued):

Silicone Resins:

SR 165:

Key Characteristic: Toughness, Resistance to Thermal Shock Heat Resistance

Typical Application: High solids in VM&P naphtha solvent systems. Flexibility, toughness, thermal shock resistance, weather resistance.

Silicone Content %: 86+-1
Specific Gravity 25C (77F): 1.09
Solvent System: VM&P Naphtha
Viscosity-Bkfld (cps): 300-1,200
Gardner: H-V
Color/Gardner: Light straw/0-2
Flash Point Pensky Martens: 11C (54F)
Density, lbs/gal: 9.1

SR 240:

Key Characteristic: Resistance to Extreme Temperatures

Typical Application: Suitable for coatings requiring optimum performance characteristics above 424C (800F):

Silicone Content %: 50+-1
Specific Gravity 25C (77F): 1.02
Solvent System: Toluol
Viscosity Bkfld. (cps): 35-90
Gardner: A-1 to C
Color/Gardner: Light straw/0-3
Flash Point Pensky-Martens: 10C (50F)
Density, lbs/gal: 8.5

SR 882:

Key Characteristic: Compatibility with Organic Materials

Typical Applications: Recommended for use as blending resins to upgrade heat service life of organic coatings and improved weather durability of exterior finishes.

Silicone Content %: 80+-1
Specific Gravity 25C (77F): 1.14
Solvent System: Toluol
Viscosity Bkfld. (cps): 200-1,000
Gardner: F-W
Color/Gardner: Light straw/0-3
Flash Point Pensky-Martens: 11C(52F)
Density lbs/gal: 9.5

Typical General Electric Silicone Resin Applications:

Boiler Stacks
Jet Engine Components
Furnace Fronts
Gun Components (High Temperature)
Incinerators
Exhaust Manifolds
Missile Coating
High Temperature Appliances
Metal Primer (High Temperature)

HENKEL CORP.: G-CURE Acrylic Resins:

G-CURE Acrylic Resins are an extraordinary family of products for the formulation of two-pak acrylic urethane coatings. These hydroxyl functional acrylics will ambient cure with aliphatic polyisocyanates to provide finishes witn excellent performance characteristics.

Gloss Retention:
G-CURE Acrylic Resins show low chalk rates and excellent gloss retention.

Chemical Resistance:
In the area of chemical resistance, acrylic urethane coatings formulated with G-CURE Acrylic Resins offer performançe comparable to or better than epoxy/polyamide coatings and are far superior to air dry alkyd finishes.

Adhesion:
Coatings based on G-CURE Acrylic Resins provide superior adhesion to plastics and other problem substrates.

Energy Savings:
Though G-CURE Acrylic Resin based coatings attain full cure at room temperatures, as little as 10 minutes oven time at 150F. is sufficient oven energy to provide "handling hardness," permitting immediate movement of coated items to storage for full cure.

Formulation Savings:
Significant additional savings result from the fact that coatings based on G-CURE Acrylic Resins use a lower ratio of the isocyanate than is required in conventional two-pak polyester coatings.

Broad Application Range:
The many advantages of tough urethane coatings formulated ted with G-CURE Acrylic Resins make them ideal for a wide variety of applications, and their excellent performance without high temperature cure is especially important in items too large for the oven. Areas where these coatings are finding use include:

* Maintenance--tanks, plant equipment, off-shore rigs.
* Transportation--trucks, buses, railroad, ships, light aircraft, motor homes, fire trucks
* Recreational Equipment--dune buggies, sports equipment
* Factory Finished Products--business machine plastic parts, appliances, automotive parts.

HENKEL CORP.: G-CURE Acrylic Resins(Continued):

G-CURE 867:

* Over-all premium urethane performance properties
* Durability, gloss retention and chemical resistance unsurpassed by anything in the market today
* Requires about half as much isocyanate as typical polyesters.

G-CURE 105:

* Higher volume solids to reduce solvent emissions
* Outstanding gloss retention properties--best of the G-CURE resin family
* Excellent flow and leveling properties allow for brushable urethane coatings and high build topcoats over inorganic zinc

G-CURE 868:

* Use with harder polyisocyanates for top urethane performance and gloss retention
* Highly flexible coatings when formulated with DESMODUR N with excellent adhesion to a variety of plastic and metal substrates
* Modifier for the other G-CURE resins to increase flexibility and maintain gloss retention

G-CURE 869:

* Very fast dry times for industrial coating applications
* Lowest isocyanate demand of the G-CURE resin family
* Gloss retention similar to premium baked acrylic systems

G-CURE 105:
Hydroxyl Equivalent Weight: 530
G-CURE: Isocyanate Demand: 2.8:1

G-CURE 867:
Hydroxyl Equivalent Weight: 600
G-CURE: Isocyanate Demand: 3:1

G-CURE 868:
Hydroxyl Equivalent Weight: 800
G-CURE: Isocyanate Demand: 4:1

G-CURE 869:
Hydroxyl Equivalent Weight: 1000
G-CURE: Isocyanate Demand: 5:1

HERCULES INC.: HERCOFLEX 500 Synthetic Resin:

A Liquid Resinous Plasticizer and Tackifier

HERCOFLEX 500, a resinous plasticizer derived from rosin, is a light-amber-colored liquid resin. It has marked resistance to aging and minimum odor. HERCOFLEX 500 can be used as a resinous plasticizer or tackifier in finished products such as lacquers, inks, adhesives, and floor tiles. Its broad compliance with FDA requirements makes it an excellent safe candidate for replacing phthalate plasticizers.

Product Specifications:
Color, Gardner, max: 5
Acid Number, max: 4-12
Brookfield viscosity at 25C, poises: 25-65

Outstanding Characteristics:
Resinous nature, high refractive index, low odor, low vapor pressure, and high boiling point; excellent pigment-wetting and surface-wetting properties; wide-range solubility and compatibility.

Typical Uses:
When used as a part of the resin and plasticizer constituents of nitrocellulose lacquers, HERCOFLEX 500 provides exceptionally clear films with high gloss and durability. These features are attributable to its resinous nature and its unusually high refractive index. As a result, it is recommended for use in lacquers and overprint varnishes.

HERCOFLEX 500 is an excellent modifier for pressure-sensitive and hot-melt adhesives. In chlorinated rubber and similar rubber-based paints, the alkali resistance of HERCOFLEX 500 makes it a suitable plasticizing material. It can also be used as a replacement for castor oil as a plasticizer, and as a resin component in inks, floor tile, vinyl plastics, rubber compositions, and related applications.

FDA Status:
HERCOFLEX 500 is in compliance with requirements of the U.S. Food and Drug Administration for use in materials contacting foods as specified in the Code of Federal Regulations, Title 21.

HERCULES INC.: Hydrocarbon and Modified Hydrocarbon Synthetic Resins:

Resin Classification: Aliphatic:

Products:

ADTAC B10 BHT:
(Petroleum Monomers)
Softening point (R&B), C: 10
Color Gardner: 5
ADTAC B25 BHT:
(Petroleum Monomers)
Softening point (R&B), C: 25
Color Gardner: 5.5
ADTAC L100 BHT:
(Petroleum Monomers)
Softening point (R&B), C: 100
Color Gardner: 6
Melt viscosity, C: 1 poise: 225
PICCODIENE 2215:
(Dicyclopentadiene)
Softening point (R&B), C: 102
Color Gardner: 11
Melt viscosity, C: 1 poise: 212
PICCOPALE 70:
(Dienes and Other Olefins)
Softening point (R&B), C: 70
Color Gardner: 9
Melt viscosity, C: 1 poise: 185
PICCOPALE 100:
(Dienes and Other Olefins)
Softening point (R&B), C: 100
Color Gardner: 8
Melt viscosity, C: 1 poise: 230
PICCOTAC A:
(Petroleum Monomers)
Softening point (R&B), C: 118
Color Gardner: 6
Melt viscosity, C: 1 poise: 224
PICCOTAC B:
(Petroleum Monomers)
Softening point (R&B), C: 100
Color Gardner: 5
Melt viscosity, C: 1 poise : 213
PICCOTAC C:
(Petroleum Monomers)
Softening point (R&B), C: 100
Color Gardner: 6
Melt viscosity, C: 1 poise: 212
PICCOTAC 95:
(Petroleum Monomers)
Softening point (R&B), C: 95
Color Gardner: 5
Melt viscosity, C: 1 poise: 205

HERCULES INC.: Hydrocarbon and Modified Hydrocarbon Synthetic Resins(Continued):

Resin Classification: Aliphatic(Continued):

PICCOVAR AB165:
(Petroleum Monomers)
Softening Point (R&B), C: 165
Color, Gardner: 3
Melt Viscosity, C: 1 poise: 237

PICCOVAR AB180:
(Petroleum Monomers)
Softening Point (R&B), C: 178
Color, Gardner: 3
Melt Viscosity, C: 1 poise: 250

Resin Classification: Aromatic:

AD4100 BHT:
(Modified Aromatic)
Softening point (R&B), C: 100
Color, Gardner: 8
Melt Viscosity, C: 1 poise: 190

AD1100 BHT:
(Modified Aromatic)
Softening point (R&B), C: 97
Color, Gardner: 8
Melt Viscosity, C: 1 poise: 190

AD1115 BHT:
(Modified Aromatic)
Softening point (R&B), C: 115
Color, Gardner: 8
Melt Viscosity, C: 1 poise: 208

LA95 BHT:
(Petroleum Feedstock)
Softening Point (R&B), C: 94
Color, Gardner: 8

PICCOVAR:
(Petroleum Feedstock)

AP10:
Softening Point (R&B), C: 11
Color, Gardner: 13
Melt Viscosity, C: 1 poise: 85

AP25:
Softening Point (R&B), C: 32
Color, Gardner: 13
Melt Viscosity, C: 1 poise: 108

L30:
Softening Point (R&B), C: 30
Color, Gardner: 11
Melt Viscosity, C: 1 poise: 93

L60:
Softening Point (R&B), C: 58
Color, Gardner: 11
Melt Viscosity, C: 1 poise: 129

HERCULES INC.: Hydrocarbon and Modified Hydrocarbon Synthetic Resins (Continued):

Resin Classification: Aromatic (Continued):

PICCO 5000 Series:
(Petroleum Monomer)

5070:
Softening Point (R&B), C: 70
Color, Gardner: 11
Melt Viscosity, C: 1 poise: 165
5100:
Softening Point (R&B), C: 100
Color, Gardner: 11
Melt Viscosity, C: 1 poise: 191
5110:
Softening Point (R&B), C: 110
Color, Gardner: 11
Melt Viscosity, C: 1 poise: 200
5120:
Softening Point (R&B), C: 120
Color, Gardner: 11
Melt Viscosity, C: 1 poise: 222
5130:
Softening Point (R&B), C: 130
Color, Gardner: 11
Melt Viscosity, C: 1 poise: 229
5140:
Softening Point (R&B), C: 140
Color, Gardner: 11
Melt Viscosity, C: 1 poise: 245

PICCO 6000 Series:
(Petroleum Monomer)

6070:
Softening Point (R&B), C: 70
Color, Gardner: 11
Melt Viscosity, C: 1 poise: 170
6100:
Softening Point (R&B), C: 100
Color, Gardner: 9
Melt Viscosity, C: 1 poise: 195
6110:
Softening Point (R&B), C: 110
Color, Gardner: 9
Melt Viscosity, C: 1 poise: 205
6115:
Softening Point (R&B), C: 115
Color, Gardner: 9
Melt Viscosity, C: 1 poise: 217

HERCULES INC.: Hydrocarbon and Modified Hydrocarbon Synthetic Resins(Continued):

Resin Classification: Hydrogenated:

REGALREZ:
(Pure Monomer)

3102:
Softening Point (R&B), C: 102
Color: Crystal-clear
Melt viscosity, poises: 460 at 125C

6108:
Softening Point (R&B), C: 108
Color: Crystal-clear
Melt viscosity, poises: 170 at 130C

1126:
Softening Point (R&B), C: 126
Color: Crystal-clear
Melt viscosity, poises: 520 at 145C

1094:
Softening Point (R&B), C: 94
Color: Crystal-clear
Melt viscosity, poises: 250 at 120C

1018:
Softening Point (R&B), C: 18
Color: Crystal-clear
Melt viscosity, poises: 600 at 35C

1033:
Softening Point (R&B), C: 33
Color: Crystal-clear
Melt viscosity, poises: 900 at 45C

1065:
Softening Point (R&B), C: 65
Color: Crystal-clear
Melt viscosity, poises: 720 at 80C

1078:
Softening Point (R&B), C: 78
Color: Crystal-clear
Melt viscosity, poises: 880 at 90C

REGALITE 7070:
(Mixed Monomer)
Softening Point (R&B), C: 71
Color: <1
Melt viscosity, poises: 530 at 85C

HERCULES INC.: Hydrocarbon and Modified Hydrocarbon Synthetic Resins(Continued):

Resin Classification: Pure Monomer:

PICCOLASTIC:
(Styrene and Modified Styrenes)

A5:
Softening Point (R&B), C: 5
Color, Gardner: 3
Melt viscosity, C: 1 poise: 62

A25:
Softening Point (R&B), C: 25
Color, Gardner: 4
Melt Viscosity, C: 1 poise: 86

A50:
Softening Point (R&B), C: 50
Color, Gardner: 2
Melt Viscosity, C: 1 poise: 109

A75:
Softening Point (R&B), C: 75
Color, Gardner: 2
Melt Viscosity, C: 1 poise: 150

D100:
Softening Point (R&B), C: 100
Color, Gardner: 2
Melt Viscosity, C: 1 poise: 250

D125:
Softening Point (R&B), C: 125
Color, Gardner: 2

D150:
Softening Point (R&B), C: 146
Color, Gardner: 3

E50:
Softening Point (R&B), C: 50
Color, Gardner: 3
Melt Viscosity, C: 1 poise: 180

E100:
Softening Point (R&B), C: 100
Color, Gardner: 3

E125:
Softening Point (R&B), C: 125
Color, Gardner: 3

HERCULES INC.: Hydrocarbon and Modified Hydrocarbon Synthetic Resins(Continued):

Resin Classification: Pure Monomer(Continued):

KRISTALEX:
(alpha-Methylstyrene)

3025:
Softening Point (R&B), C: 25
Color, Gardner: <1

3070:
Softening Point (R&B), C: 69
Color, Gardner: <1
Melt viscosity, C: 1 poise: 147

3085:
Softening Point (R&B), C: 85
Color, Gardner: <1
Melt viscosity, C: 1 poise: 180

3100:
Softening Point (R&B), C: 100
Color, Gardner: <1
Melt Viscosity, C: 1 poise: 190

1120:
Softening Point (R&B), C: 120
Color, Gardner: <1
Melt Viscosity, C: 1 poise: 236

PICCOTEX:
(Vinyltoluene/alpha-Methylstyrene)

75:
Softening Point (R&B), C: 75
Color, Gardner: <1
Melt Viscosity, C: 1 poise: 153

LC:
Softening Point (R&B), C: 90
Color, Gardner: <1
Melt Viscosity, C: 1 poise: 175

100:
Softening Point (R&B), C: 98
Color, Gardner: <1
Melt Viscosity, C: 1 poise: 195

120:
Softening Point (R&B), C: 118
Color, Gardner: <1
Melt Viscosity, C: 1 poise: 230

HERCULES INC.: Hydrocarbon and Modified Hydrocarbon Synthetic Resins(Continued):

Resin Classification: Terpene:

PICCOLYTE:
(alpha-Pinene)

A115:
Softening point (R&B), C: 115
Color, Gardner: 6
Melt Viscosity, C: 1 poise: 200

A125:
Softening point (R&B), C: 125
Color, Gardner: 5
Melt Viscosity, C: 1 poise: 210

A135:
Softening point (R&B), C: 135
Color, Gardner: 5
Melt Viscosity, C: 1 poise: 220

PICCOLYTE:
(d-Limonene)

C10:
Softening point (R&B), C: 10
Color, Gardner: 4

C100:
Softening point (R&B), C: 100
Color, Gardner: 4
Melt Viscosity, C: 1 poise: 189

C115:
Softening point (R&B), C: 115
Color, Gardner: 4
Melt Viscosity, C: 1 poise: 204

C125:
Softening point (R&B), C: 125
Color, Gardner: 4
Melt Viscosity, C: 1 poise: 214

C135:
Softening point (R&B), C: 131
Color, Gardner: 4
Melt Viscosity, C: 1 poise: 224

HERCULES INC.: Hydrocarbon and Modified Hydrocarbon Synthetic Resins(Continued):

Resin Classification: Terpene:

PICCOLYTE:
(beta-pinene)

S10:
Softening Point (R&B), C: 10
Color, Gardner: 2
Melt Viscosity, C: 1 poise: 110

S25:
Softening Point (R&B), C: 25
Color, Gardner: 2
Melt Viscosity, C: 1 poise: 125

S55:
Softening Point (R&B), C: 55
Color, Gardner: 2
Melt Viscosity, C: 1 poise: 150

S70:
Softening Point (R&B), C: 70
Color, Gardner: 2
Melt Viscosity, C: 1 poise: 165

S85:
Softening Point (R&B), C: 85
Color, Gardner: 2
Melt Viscosity, C: 1 poise: 191

S100:
Softening Point (R&B), C: 100
Color, Gardner: 2
Melt Viscosity, C: 1 poise: 202

S115:
Softening Point (R&B), C: 115
Color, Gardner: 2
Melt Viscosity, C: 1 poise: 220

S125:
Softening Point (R&B), C: 125
Color, Gardner: 2
Melt Viscosity, C: 1 poise: 225

S135:
Softening Point (R&B), C: 135
Color, Gardner: 2
Melt Viscosity, C: 1 poise: 232

HERCULES INC.: Hydrocarbon and Modified Hydrocarbon Synthetic Resins(Continued):

Resin Classification: Functional Resins:

PICCOFYN:
(Terpene Phenolic)

A135:
Softening Point (R&B), C: 135
Color, Gardner: 7
Melt Viscosity, C: 1 poise: 220

T125:
Softening Point (R&B), C: 120
Color, Gardner: 7
Melt Viscosity, C: 1 poise: 168

HERCULES INC.: PICCO 6000 Aromatic Hydrocarbon Resins:

Low- to Medium-Softening-Point Series

PICCO 6000 aromatic hydrocarbon resins are a series of low molecular weight, nonpolar, amber-colored thermoplastic resins produced from petroleum-derived monomers. Ranging in softening point from low to medium, these resins are characterized by their excellent resistance to acids, alkalies, and moisture; by their good initial color and color stability, and, when formulated with elastomers, by their good balance of flex, tear, tack, and adhesion properties they provide.

PICCO 6000 series resins are particularly indicated as tackifiers in sealant and adhesive systems, and as processing and reinforcing agents in rubber compounds.

PICCO 6100 is also offered in solution in mineral spirits.

Product Specifications: PICCO Resin:

6070:
- Softening point, R&B, C: 68-73
- Color, Gardner, max: 12

6100:
- Softening point, R&B, C: 98-102
- Color, Gardner, max: 11

6110:
- Softening point, R&B, C: 108-112
- Color, Gardner, max: 11

6115:
- Softening point, R&B, C: 113-118
- Color, Gardner, max: 11

Outstanding Characteristics:

PICCO 6000 resins have excellent solvent release, tack, and tack retention; wide solubility and compatibility; and good resistance to oxygen and UV light. They are extremely water repellent; provide excellent pigment wetting; and can be used with both reinforcing and nonreinforcing fillers. They are viscous fluids at rubber-milling temperatures, and aid compounding by reducing viscosity and adding tack while on the rolls. These advantages, combined with improvement in mold flow during cure, are obtained without lowering the cured hardness of the finished product.

Typical Uses:

The highly aromatic nature and wide range of thermoplastic properties offered by PICCO 6000 resins suggest their use in a variety of industrial applications. They should be considered as tackifiers in various elastomer-based solvent, water, and hot-melt adhesive systems. They can be used as tackifiers, reinforcing agents, and extenders in rubber compounding and extruding applications, and they show excellent leafing and solvent-release properties in metallic cold-cut paints.

ICI AMERICAS INC.: LUMIFLON Fluoropolymer Resins:

LUMIFLON is a fluoropolymer engineered to combine the weatherability and chemical resistance features of polyvinylidene fluoride (PVDF) with the gloss, adhesion, and ambient cure characteristics of acrylics. It offers a unique set of properties which produces the finest architectural coatings and meets the stringent requirements of many of today's--and tomorrow's--specialty applications.

Versatile Products for a Variety of Applications:

Architectural:

* roof and wall materials
* doors and windows
* decks, handrails, fences

Specialty:

* plastic films
* aerospace
* solar energy hardware
* fluorescent and incandescent light coatings
* traffic signs and fixtures
* outdoor plastic structures

Maintenance:

* industrial plants
* marine facilities
* remote structures

Transportation:

* automotive
* aircraft
* ships

LUMIFLON resins mean superior weatherability

LUMIFLON offers a choice of gloss

LUMIFLON resins give you a choice of curing conditions.

Clear transparent films with LUMIFLON

LUMIFLON resins are solvent soluble

LUMIFLON coatings adhere to most substrates

LUMIFLON coatings show good chemical resistance

ICI AMERICAS INC.: LUMIFLON Fluoropolymer Resins(Continued):

Four products are available to meet a range of applications:

Grade:

LF 100:
Feature: High viscosity (5000 cps*)
Principal Uses: Brush and roll applied coatings

LF 200:
Feature: Low viscosity (800 cps*)
Principal Uses: Spray applied coatings

LF 302:
Features: Absorbs UV light
Principal Uses: Clear coatings for plastics

LF 400:
Features: Improved pigment compatibility
Principal Uses: Dispersing organic and carbon black pigments

* 50% solution in xylene

Each of these grades is crosslinkable with either isocyanate or melamine hardeners and shows cure rates similar to acrylic-urethane/acrylic-melamine systems.

LUMIFLON Grades:

Grade:

LF 100:
% Wt. Solids: 50
Viscosity: cps: 5,000
Solvent: Xylene
Features: High viscosity
Uses: Brush coatings

LF 200:
% Wt. Solids: 60
Viscosity: cps: 4,000
Solvent: Xylene
Features: Low viscosity
Uses: Spray coatings

LF 302:
% Wt. Solids: 50
Viscosity: cps: 800
Solvent: Xylene
Features: UV absorbing
Uses: Clear coatings

ICI AMERICAS, INC.: LUMIFLON Fluoropolymer Resins(Continued):

Grade:

LF 400:
% Wt. Solids: 50
Viscosity: cps: 800
Solvent: Xylene
Features: Improved pigment compatibility
Uses: Spray coatings

LF 510:
% Wt. Solids: 40
Viscosity: cps: 7,500
Solvent: SOLVESSO 150/cyclohexanone

LF 515:
% Wt. Solids: 40
Viscosity: cps: 13,000
Solvent: SOLVESSO 150/cyclohexanone

LF 502:
% Wt. Solids: 40
Viscosity: cps: 650
Solvent: SOLVESSO 150/cyclohexanone

LF 552:
% Wt. Solids: 40
Viscosity: cps: 750
Solvent: SOLVESSO 150/cyclohexanone

LF 504:
% Wt. Solids: 40
Viscosity: cps: 5,700
Solvent: SOLVESSO 150/cyclohexanone

LF 554:
% Wt. Solids: 40
Viscosity: cps: 8,600
Solvent: SOLVESSO 150/cyclohexanone

Features: Improved flexibility; slower evaporating solvents
Uses: Coil coatings

ICI AMERICAS INC.: LUMIFLON Fluoropolymer Resins(Continued):

Grade:

LF 601:

% Wt. Solids: 50
Viscosity: cps: 800
Solvent: X/T5
Features: Improved Elongation
Uses: Coatings for flexible substrates

LF 916:

% Wt. Solids: 65
Viscosity: cps: 4,500
Solvent: Xylene
Features: Improved leveling
Uses: Automotive topcoats

LF 936:

% Wt. Solids: 65
Viscosity: cps: 5,000
Solvent: Xylene
Features: Higher solids
Uses: High build coatings

LF 110:

% Wt. Solids: 40
Viscosity: cps: 3,000
Solvent: Mineral Spirits
Features: Not cross-linkable
Uses: Lacquer coatings

O.G. INNES CORP.: Natural and Processed Resins:

Gradings:

Manila:
- Copal A
- Loba A Bold
- Loba B Nubs
- Loba C Small Nubs
- Loba D Chips
- CBB Nubs
- DBB Chips
- CNE Nubs
- DK Chips
- Loba Seeds & Dust
- WS Extra Pale Soft Sorts
- MA Amber Soft Sorts
- Philippine Pale Bold
- Philippine Pale Nubs
- Philippine Pale Chips
- Philippine Seeds & Dust

Pontianak:
- Genuine Bold Scraped
- Genuine Split Chips
- Genuine Nubs
- Genuine Chips

Property Ranges:

Manilas:
- Refractive Index: 1.538-1.540
- Softening Point: 81-90C. (178-194F.)
- Melting Point: 114-125C. (237-266F.)
- Specific Gravity: 1.06-1.08
- Acid No.: 110-141
- Saponification No.: 140-190
- Iodine No.: 113-125

Pontianaks:
- Refractive Index: 1.540
- Softening Point: 99-135C. (210-275F.)
- Melting Point: 136-169C. (277-336F.)
- Specific Gravity: 1.07-1.08
- Acid No.: 112-120
- Saponification No.: 150-180
- Iodine No.: 106-125

The alcohol soluble resins are used for paper and packaging coatings of all types, label varnishes, wall paper finishes, floor finishes, wall sizes, primers, lacquers, adhesives, inks and cements.

Manila Gum may be rendered water soluble in amine or caustic solutions for inclusion in liquid floor polishes and similar materials.

Pontianaks are similar to the alcohol soluble Manilas but have higher melting points and are therefore used in more exacting specifications particularly in the specialty coating and adhesive fields. While the Manilas are not generally cooked, Pontianak has been used in cooked oil vehicles.

FDA approved.

O.G. INNES CORP.: Natural and Processed Resins(Continued):

DAMAR

Gradings:

Singapore Damar:
- No. 1
- No. 2
- No. 3
- Seeds
- Dust

Siam Damar

Batavia Damar:
- Standard A/E Mixed
- Standard A/D Mixed
- Standard A/C Mixed
- Bold A
- Medium B
- Nubs C
- Chips D
- Grain E
- Seeds F
- Dust

Property Ranges:
- Refractive Index: 1.535-1.538
- Softening Point: 67-75C.
  153-167F.
- Melting Point: 99-115C.
  210-238F.
- Specific Gravity: 1.04-1.06
- Acid No.: 17-34
- Saponification No.: 30-40
- Iodine No.: 103-123

Damars provide unique depth of gloss, superior color and gloss retention, adhesion and flexibility and resistance to greases, cold checking and scuffing. A general all-around varnish resin for cold cuts, blends or cooked oil vehicles. Its fume resistance and binding value recommend Damar for white enamels and pastels. Recommended for adhesives and inks. A toner resin with miscibility and versatility.

FDA approved.

Dewaxed Damar:

An outstanding lacquer resin with interesting compatibilities.

Raw Damar is not compatible with alcohols and must be "dewaxed" to render it soluble in a broad range of lacquer solvents and other materials.

Solutions:

TA-44 Dewaxed Damar Solution - 43% solids in 50/50 toluol and ethyl alcohol.

Solids: #101 Pale Solid Dewaxed Damar:

Softening point: 77-81C or 170-178C

Melting Point: 79-85C or 174-185C

O.G. INNES CORP.: Natural and Processed Resins(Continued):

East Indias

Types:

Pale:
Bold
Nubs
Chips
Dust

Property Ranges:
Refractive Index: 1.541
Softening Point: 109-128C.
228-262F.
Melting Point: 140-156C.
284-313F.
Specific Gravity: 1.04-1.05
Acid No.: 20-27
Saponification No.: 35-40
Iodine No.: 73-87

Characteristics:
High melting point Damar type resins with adhesion, flexibility and color retention. Scuff and fume resistant, tough, adhesive and flexible.

Pale East India may be controlled for attractive sheen by adjustment of solvents and thinners.

Batu:

Gradings:
Bold Scraped
Bold Unscraped
Nubs & Chips
Dust

Property Ranges:
Refractive Index: 1.538
Softening Point: 111-132C.
232-270F.
Melting Point: 168-180C.
334-356F.
Specific Gravity: 1.00-1.05
Acid No.: 18-30
Saponification No.: 33-50
Iodine No.: 71-81

O. G. INNES CORP.: Natural and Processed Resins(Continued):

East Indias(Continued):

Batu East India cut cold yields a flat. As heats are increased, if cooked in oils, the flat may be reduced until a high temperature provides sturdy, high gloss vehicles. This resin is tough, scuff resistant, has fast solvent release, is adhesive and shows resistance to water, greases and temperature changes. Its water resistance and adhesion under tough wearing conditions particularly recommend it for undercoats on wood and metals, traffic marking paints and similar outlets. It has good compatibility with oils, resins, waxes, cellulose derivatives, chlorinated rubber, stearic acid, asphalt and pitches. Batu is used for adhesives and cements.

Black:

Gradings:
- Bold Scraped
- Bold Unscraped
- Nubs & Chips

Property Ranges:
- Refractive Index: 1.541
- Softening Point: 114-125C.
237-257F.
- Melting Point: 160-164C.
320-327F.
- Specific Gravity: 1.04
- Acid No.: 20-25
- Saponification No.: 27-40
- Iodine No.: 78-84

Black East India has adhesion and flexibility and is particularly interesting in metal undercoats. With brine and scuff resistance, it may be used in a range of outlets. An interesting ingredient in undercoats for ship bottom paints and automotive as well as can coatings where flexibility is needed.

The East Indias cover broad usages for wood, metal and cement surfaces, inks, stains, sizings, adhesives, gasket compounds, outdoor as well as inside finishes, undercoats and tough wearing vehicles.

FDA approved.

O. G. INNES CORP.: Natural and Processed Resins (Continued):

Gum Elemi

Property Ranges:
- Softening Point: Plastic
- Specific Gravity: 1.03
- Acid No.: 20-35
- Saponification No.: 21-44
- Iodine No.: 113-123

Gum Elemi is a film forming plasticizing resin with superior water resistance and excellent color and gloss retention, adhesion and flexibility, especially recommended for lacquers. FDA Approved.

Sandarac

Property Ranges:
- Refractive Index: 1.545
- Softening Point: 100-130C.
212-266F.
- Melting Point: 135-150C.
275-302F.
- Specific Gravity: 1.05-1.09
- Acid No.: 117-155
- Saponification No.: 145-157
- Iodine No.: 112-141

This is one of the oldest known resins in the artist and coating fields. Used in spirit varnishes, lacquers, dental, art and specialty furniture products it gives a hard white spirit varnish with excellent color retention and stability and adhesion as well as flexibility.

FDA approved.

Mastic

Property Ranges:
- Softening Point: 55C. approx. (131F.)
- Melting Point: 76C. approx. (168F.)
- Specific Gravity: 1.04-1.06
- Acid No.: 50-70
- Saponification No.: 62-90
- Iodine No.: 64-142

This relatively soft spirit soluble resin is also of ancient usage. Used in spirit varnish lacquers and adhesives, printing inks and specialty purposes particularly in the artists field it is compatible with a wide range of vegetable oils, cellulose derivatives and both nitrocellulose and ethylcellulose and is readily miscible with asphalts and pitches on heating.

FDA approved.

INTEREZ: NOVACURE Radiation Curable Oligomers:

Resin:

3200:
Comments: Low viscosity, good pigment wetting
Viscosity, cps: 2,000
Weight/Epoxide (minimum): 16,000

3201:
Comments: Low viscosity, good flexibility & chemical resistance
Viscosity, cps: 4,000
Weight/Epoxide (minimum): 15,000

3500:
Comments: Very good flexibility
Viscosity, cps: 55,000

3600:
Comments: Very fast UV cure & very good chemical resistance
Viscosity, cps: 1,500

3700:
Comments: Very good chemical resistance & hardness
Viscosity, cps: 2,500
Weight/Epoxide (minimum): 16,000

3700-15S:
Comments: Low viscosity, good chemical resistance & hardness (contains 15% styrene)
Viscosity, cps: 7,000
Weight/Epoxide (minimum): 16,000

3700-20T:
Comments: Very fast UV cure & very good chemical resistance (contains 20% NOVAMER TMPTA Monomer)
Viscosity, cps: 72,000
Weight/Epoxide (minimum): 16,000

3700-25R:
Comments: Low viscosity, high gloss finish & good flexibility (contains 25% NOVAMER TRPGDA Monomer)
Viscosity, cps: 15,000
Weight/Epoxide (minimum): 20,000

3701:
Comments: Very good chemical resistance & hardness
Viscosity, cps: 4,000
Weight/Epoxide (minimum): 10,000

INTEREZ: NOVACURE Radiation Curable Oligomers(Continued):

Resin:

3702:
Comments: Very good pigment wetting
Viscosity, cps: 2,500
Weight/Epoxide (minimum): 16,000

3703:
Comments: Very good UV cure rate, flexibility, chemical resistance & hardness
Viscosity, cps: 2,000

6700:
Comments: Acrylated aromatic urethane, very good flexibility, abrasion resistance & color
Viscosity, cps: 6,000

7100:
Comments: Acrylated oligomeric amine, low viscosity & very fast cure
Viscosity, cps: 1,000

8800-20R:
Comments: Acrylated aliphatic urethane; very good flexibility & abrasion resistance (contains 20% NOVAMER TRPGDA Monomer)
Viscosity, cps: 1,750

INTEREZ: NOVAMER Radiation Curable Multifunctional Monomers:

Monomer:

TMPTA:
Description: Trimethylolpropane triacrylate
Viscosity, cps: 50-150
Ester Rank (min): 2.7

TRPGDA:
Description: Tripropyleneglycol triacrylate
Viscosity, cps: 10-20
Ester Rank (min): 1.8

HDODA:
Description: Hexanediol diacrylate
Viscosity, cps: 5-8
Ester Rank (min): 1.85

TTEGDA:
Description: Tetraethylene glycol diacrylate
Viscosity, cps: 15-30
Ester Rank (min): 1.8

PETA:
Description: Pentaerythritol triacrylate
Viscosity, cps: 600-800
Ester Rank (min): 3.2

GPTA:
Description: Glycerol propoxy triacrylate
Viscosity, cps: 70-110
Ester Rank (min): 2.7

TMPEOTA:
Description: Trimethylol propane ethoxy triacrylate
Viscosity, cps: 30-90
Ester Rank (min): 2.8

TMPTMA:
Description: Trimethylolpropane trimethacrylate
Viscosity, cps: 25-75
Ester Rank (min): 2.6

ODA:
Description: Octyl/Decyl acrylate
Viscosity, cps: 2-3
Ester Rank (min): 0.96

INTEREZ: POLY/TEX Acrylic Resins:

Resin:

950:
Comments: Hydroxypropyl methacrylate monomer
Viscosity: A4
Solvent: N/A
Nonvolatiles, percent weight: 100
Acid Value: 34

975:
Comments: Hydroxyl functional resin for Acrylic/Urethane finishes (Hydroxyl Eq. Wt.=365 on solids)
Viscosity: Z4
Solvent: NBA
Nonvolatiles, percent weight: 70
Acid Value: 14

CMD 9790:
Comments: Dispersion of CMD 979
Viscosity: 12,000 cps
Solvent: Water/X/EP
Nonvolatiles, percent weight: 42

611Q:
Comments: Epoxy ester emulsion EPI-TEX 199 type epoxy ester emulsion
Viscosity: 100 K.U.
Solvent: Water/Aromatic
Nonvolatiles, percent weight: 60
Acid Value: 1

INTEREZ: EPI-TEX Epoxy Ester Resins:

183:
Comments: General purpose air dry and forced dry finishes
Nonvolatiles, percent weight: 50
Solvent: X
Viscosity: X
Acid Value (maximum): 2.5

199:
Comments: General purpose with slow solvent for brushing
Nonvolatiles, percent weight: 60
Solvent: M
Viscosity: Z1
Acid Value (maximum): 10

INTEREZ: EPI-TEX Epoxy Ester Resins(Continued):

Resin:

1662:
Comments: Exempt version of EPI-TEX 183
Nonvolatiles, percent weight: 50
Solvent: Contains xylene, VM&P naphtha, n-butyl alcohol and n-butyl acetate
Viscosity: X
Acid Value (maximum): 2.5

1663:
Comments: Exempt version of EPI-TEX 199
Nonvolatiles, percent weight: 60
Solvent: Contains high solvency hydrocarbons and mineral spirits.
Viscosity: Z2
Acid Value (maximum): 10

Solvent Codes:
A=acetone
EB=2-butoxyethanol
EE=2-ethoxyethanol
EEA=2-ethoxyethyl acetate
EP=2-propoxyethanol
M=mineral spirits
MAK=methyl amyl ketone
MEK=methyl ethyl ketone
MIBK=methyl isobutyl ketone
MPA=1-methoxy-2-propyl acetate
NBA=n-butyl acetate
Tol=toluene
X=xylene

INTEREZ: Miscellaneous Products:

Flat Base:

720:
Comments: A nonpigmented flatting agent
Viscosity: J
Nonvolatiles, percent weight: 35
720E:
Comments: Exempt version of 720 Flat Base
Viscosity: H
Nonvolatiles, percent weight: 35
SYNTEX 3928:
Comments: Exempt modified phthalic alkyd
Viscosity: G
Max. acid value=20
Nonvolatiles, percent weight:60
SYNTEX 3981:
Comments: Accelerator for amino resin-containing thermosetting systems/ Nonvolatiles, percent weight: 23

INTEREZ: Waterborne/Reducible Resins:

Resin:

EPI-REZ WD 510:
Comments: EPI-REZ 510 for use in waterborne applications
Viscosity: 11,000
Nonvolatile percent weight: 100

CMD 35201:
Comments: EPI-REZ 522-C epoxy dispersion
Viscosity: 12,000
Solvent: Water
Nonvolatile percent weight: 60

CMD WJ55-3520:
Comments: EPI-REZ 520-C epoxy dispersion for coatings
Viscosity: 12,000
Solvent: Water/EP
Nonvolatile percent weight: 55

CMD WJ55-3540:
Comments: EPI-REZ 540-C epoxy dispersion for bake coatings
Viscosity: 12,000
Solvent: Water/EP
Nonvolatile percent weight: 55

CMD W55-5003:
Comments: Polyfunctional aromatic epoxy dispersion, average functionality of 3
Viscosity: 10,000
Solvent: Water
Nonvolatile percent weight: 57

CMD WJ55-5522:
Comments: Polyfunctional epoxy dispersion for coatings
Viscosity: 12,000
Solvent: Water/EP
Nonvolatile percent weight: 55

CMD W60-5520:
Comments: Epoxy urethane dispersion
Viscosity: 15,000
Solvent: Water
Nonvolatile percent weight: 60

POLY-TEX 611Q:
Comments: EPI-TEX 199 type epoxy ester emulsion
Viscosity: 100 K.U.
Solvent: Water/Aromatic
Nonvolatile percent weight: 60

INTEREZ: Waterborne/Reducible Resins(Continued):

Resin:

CMD 9790:
Comments: Dispersion of CMD 979
Viscosity: 12,000
Solvent: Water/X/EP
Nonvolatile percent weight: 42

EPI-CURE W50-8535:
Comments: Ambient curing water soluble amine
Viscosity: Z5
Solvent: Water
Nonvolatile percent weight: 50

CMD JT60-8536:
Comments: Polyamidoamine for aqueous maintenance coatings
Viscosity: Y
Solvent: Tol/EB/EP
Nonvolatile percent weight: 60

CMD WJ60-8537:
Comments: Ambient curing water soluble amine
Viscosity: Z3
Solvent: Water/EE
Nonvolatile percent weight: 60

CMD J60-8290:
Comments: Amine adduct for maintenance coatings
Viscosity: Z4
Solvent: EP
Nonvolatile percent weight: 60

Solvent Codes:
A=acetone
EB=2-butoxyethanol
EE=2-ethoxyethanol
EEA=2-ethoxyethyl acetate
EP=2-propoxyethanol
M=mineral spirits
MAK=methyl amyl ketone
MEK=methyl ethyl ketone
MIBK=methyl isobutyl ketone
MPA=1-methoxy-2-propyl acetate
NBA=n-butyl acetate
Tol=toluene
X=xylene

JOHNSON WAX: Polymers/Resins:

CDX-588:
Developmental Acrylic Polymer for Higher Solids Urethane Coatings

CDX-588 is an acrylic polymer designed to be used as a polyol for higher solids, isocyanate-cross-linked coatings. It is supplied as a 68% non-volatile solution in n-butyl acetate. High performance, higher solids acrylic-urethane coatings can be produced from CDX-588 crosslinked with a high solids aliphatic polyisocyanate such as DESMODUR L-2291A (Mobay).

CDX-588 should be considered as a candidate in formulating the following types of acrylic-urethane coatings:

* High performance maintenance topcoats for steel structures, factory equipment, and marine applications.
* High film build maintenance coatings.
* Higher solids acrylic-urethane coatings for finishing plastic business machine cabinets and other plastic items using 2-component, mixing head spray guns.
* Higher solids acrylic-urethane OEM transportation finishes (truck, bus, RV, rail, aircraft).

Key Features and Benefits:
* High Solids. Acrylic-urethane coatings based on CDX 588 and a high solids polyisocyanate can be applied at higher solids than conventional urethane finishes.
* Competitive Performance. Compared to low solids acrylic-urethane coatings, higher solids finishes based on CDX-588 offer a competitive performance profile.
* Acrylic Composition. The optimized composition of CDX-588 offers the excellent exterior durability and tint retention expected of acrylics. This is far superior to polyester polyols and with a much lower isocyanate demand for cost savings.
* Legislation-Compliant, Low Toxicity Solvent. CDX-588 is supplied in a Rule 66-compliant solvent with low toxicity.

Typical Physical Constants of CDX-588:
Viscosity, Brookfield: 6500 cps
Non-Volatile: 68% (in Butyl Acetate)
Hydroxyl Equivalent Weight (Solids): 460
Hydroxyl Equivalent Weight (As Supplied): 676
Density as Supplied: 8.8 lb/gal
Density of Solids: 9.7 lb/gal
Color, Gardner: <1

JOHNSON WAX: Polymers/Resins(Continued):

JONCRYL 95:

Acrylic Emulsion Polymer for Aqueous Wood Stains and Sealers

JONCRYL 95 is an acrylic emulsion polymer designed for use as a vehicle for water-based stains and clear sealers for wood. When formulated with water soluble dyes, JONCRYL 95 provides fast drying, sandable wood stains and sealers, with good penetration and non-grain-raising characteristics.

Key Features & Benefits:

1) Small particle Size Emulsion. This property allows good penetration into wood substrates with minimum grain raising and excellent wood sealing properties.
2) Excellent Film Formation Characteristics without the need for organic co-solvents. This results in rapid dry time and hardness development, and permits sanding and topcoating after only 15 to 30 minutes air dry.
3) Wide Viscosity Range Controlled by pH. Viscosity can be increased by addition of ammonia or volatile amines which allows formulation for a variety of application methods ranging from spray to brushing.

Typical Physical Constants of Joncryl 95:
Appearance: Transparent Emulsion
Viscosity: 100 cps
Non-Volatile by Weight: 30%
Non-Volatile by Volume: 27%
Density: 8.7 lb/gal
Tg: 43C
Minimum Filming Temp.: 20C
Freeze/Thaw Stability: Stable
Acid Value: 65
pH: 8.3

JOHNSON WAX: Polymers/Resins(Continued):

JONCRYL 500:

Acrylic Oligomer for High Solids Industrial Coatings

JONCRYL 500 is an innovative acrylic oligomer for use in high solids thermosetting industrial coatings which comply with the Clean Air Act. High solids coatings containing less than 2.8 lb VOC (volatile organic compounds) per gallon can be formulated at viscosities which allow application by conventional spray at room temperature--without the problems of viscosity management, film thickness control, or severe sag previously associated with high solids coatings.

JONCRYL 500 should be considered as a candidate for high solids coatings in any application where low solids thermosetting acrylics are currently used and where performance requirements such as exterior durability, detergent or chemical resistance exclude polyesters.

Key Features & Benefits:

1) Optimum Molecular Weight with Narrow Molecular Weight Distribution.
2) Outstanding Application Properties. High solids coatings can be formulated at less than 2.8 lb VOC per gallon, even for conventional air spray application...without heaters...without sag problems...without film thickness control problems. JONCRYL 500 formulas can also be electrostatically sprayed by either No. 1 or No. 2 process ...with high transfer efficency...with excellent wrap... with recoatability.
3) Acrylic Composition. JONCRYL 500 provides the outstanding performance expected of acrylics. It contains no reactive diluents or backbone ester linkages.
4) Broad Compatibility. JONCRYL 500 is compatible with a wide variety of solvents, resins, polymers, reactive diluents and additives giving maximum formulating versatility.
5) Excellent Stability. Shelf and accelerated stability of formulated coatings are excellent. Overbake color stability is excellent.

Typical Physical Constants of JONCRYL 500:
Solids by Weight, 1 hr at 300F: 80%
Solvent: Methyl n-Amyl Ketone
Viscosity, Brookfield: 40 poise
Density as supplied: 8.55 lb/gal
Density of Solids: 9.15 lb/gal
Hydroxyl Number of Solids: 140
Acid Number of Solids: 3
Flash Point, Pensky-Martens: 120F
Appearance: Clear Syrup
Color, Gardner: 1

JOHNSON WAX: Polymers/Resins(Continued):

JONCRYL 530:

Low Cost Acrylic Emulsion Polymer

JONCRYL 530 is a hard, rheology-controlled acrylic emulsion for use in formulating economical industrial coatings for wood or metal. Being a rheology-controlled emulsion, JONCRYL 530 offers the formulator superb application properties, low foam, inherent pigment dispersing ability without the use of conventional surfactants, good adhesion, and high gloss potential.

Key Features and Benefits:

* Household Chemical Resistance. JONCRYL 530 exhibits excellent resistance to common household chemicals and detergents.
* Water Resistance. JONCRYL 530 is hydrophobic in nature and exhibits good water and humidity resistance.
* Rheology-Controlled Emulsion Characteristics: Excellent gloss and DOI. Excellent shear stability as well as inherent pigment dispersing ability.
* Hardness and Hot Block Resistance. JONCRYL 530 is an inherently hard thermoplastic emulsion which, when used as a sole vehicle or modifier, can enhance a coating's block resistance.
* Good Compatibility with Alkyds and Other Emulsions. JONCRYL 530 has good compatibility with other JONCRYL emulsions. This allows a formulator to optimize properties by blending.
* Excellent Performance and Cost Savings. The high solids (49%) and improved economics of JONCRYL 530 allow the formulator to develop coatings with improved hardness, water resistance, and application properties while enjoying raw material savings.

Typical Phyiscal Constants of JONCRYL 530:
Appearance: Translucent Emulsion
Viscosity: 700 cps
Non-Volatile by Weight: 49%
Non-Volatile by Volume: 47%
Density as Supplied: 8.73 Lbs./Gallon
Density of Solids: 9.19 Lbs./Gallon
pH: 8.5
Freeze/Thaw Stability: Passes 5 Cycles
Mechanical Stability: No Effect
Heat Stability (14 Days/120F): Passes
Volatile Organic Compounds by Weight: None
Volatile: Water
Tg: 75C

JOHNSON WAX: Polymers/Resins(Continued):

JONCRYL 538:

Acrylic Emulsion Polymer for Solvent-Resistant Industrial Lacquers

JONCRYL 538 is a Rheology Controlled, acrylic emulsion polymer for producing gasoline and solvent-resistant waterborne lacquers. As the only thermoplastic acrylic emulsion polymer which provides solvent-resistance without acrylonitrile, JONCRYL 538 is an innovative development. As a Rheology Contolled emulsion, JONCRYL 538 provides superb application properties, low foam, inherent pigment dispersing ability, adhesion to difficult substrates, and high gloss potential.

JONCRYL 538 should be considered as a candidate in formulating the following types of waterborne coatings:

* General metal finishes.
* Lacquers for automotive parts and accessories.
* Farm and home implement coatings.
* Business machine finishes.
* Wood flat line primers and topcoats.
* Concrete sealers.

Key Features & Benefits:

Compared to conventional solvent-resistant acrylic emulsion polymers, JONCRYL 538 provides the following key features and benefits:

* Improved Solvent Resistance, especially to ketones, butyl acetate, and 1,1,1-trichloroethane.
* Acrylonitrile-Free Formula.
* Superior Applications Properties. As a Rheology Controlled emulsion, JONCRYL 538 provides low foaming, excellent leveling, and freedom from film defects by spray, dip, roll-coater, or curtain-coater application.
* Higher Gloss and Distinctness of Image, equivalent to many solvent-based coatings.
* Inherent Pigment Dispersing Ability, which allows the use of JONCRYL 538 in the grind as a partial or total replacement for pigment dispersants and wetting agents. This results in coatings with improved resistance properties.
* Improved Formulating Versatility. With higher solids content, superior solvent tolerance, and improved stability, JONCRYL 538 offers wider latitude in formulating parameters, choice of additives, and manufacturing techniques.
* Excellent Adhesion to a Variety of Substrates, including treated steel, oily metal, plastics, and engineered wood products.

Typical Physical Constants of Joncryl 538:

Appearance: Translucent Emulsion
Viscosity, Brookfield: 250 cps
Non-Volatile by Weight: 45%
Non-Volatile by Volume: 42%

JOHNSON WAX: Polymers/Resins(Continued):

JONCRYL 540:

Acrylic Emulsion Polymer for Thermosetting Industrial Coatings

JONCRYL 540 is a Rheology Controlled acrylic emulsion polymer for producing thermosetting industrial coatings. As a rheology Controlled emulsion, JONCRYL 540 provides superb application properties, particularly in direct- or reverse-roll coating operations. Rheology Controlled emulsions are also low foaming, have inherent pigment dispersing ability, promote adhesion to difficult substrates, and have high gloss potential.

JONCRYL 540 should be considered in formulating the following types of waterborne, thermosetting coatings:

* Wood flat line finishes.
* Metal decorating.
* Coil decorating.
* General metal finishing.

Key Features & Benefits:

* Superior Application Properties. Rholoogy Controlled emulsions are much less pseudoplastic than conventional polymer emulsions. This controlled rheology means that JONCRYL 540 has superior application properties, particularly in direct- or reverse-roll coat operations. Add low foam and excellent leveling and you have a product with truly superior application properties.
* Highest Gloss and Distinctiveness of Image, fully equivalent to solvent-based systems.
* Competitive Mechanical and Resistance Properties. Corrosion resistance...flexibility...impact resistance...JONCRYL 540 has it.
* Excellent Adhesion to a Variety of Substrates, including treated steel and aluminum, oily metal, and engineered wood products.
* Inherent Pigment Dispersing Ability, which allows the use of JONCRYL 540 in the grind as a partial or total replacement for pigment dispersants and wetting agents. This results in coatings with improved resistance properties, especially to water.

Typical Physical Constants of JONCRYL 540:

Appearance: Translucent Emulsion
Viscosity, Brookfield: 200 cps
Non-Volatile by Weight: 45%
Non-Volatile by Volume: 43%
Density as supplied: 8.70 lb/gal
Density of Solids: 9.15 lb/gal
pH: 9.0

JOHNSON WAX: Polymers/Resins(Continued):

JONCRYL 554

Acrylic Emulsion Polymer for Wood Coatings

JONCRYL 554 is a styrenated acrylic emulsion polymer for use in high gloss, fast dry air coatings for wood products. JONCRYL 554 is recommended for use as a replacement for solvent-borne lacquers in applications where gloss, dry time and block resistance are of primary importance.

Key Features & Benefits:

1) Fast Air Dry and Hardness Development. JONCRYL 554 provides high gloss, block resistant films under ambient drying conditions.
2) Unique Rheology. JONCRYL 554 exhibits rheology more nearly Newtonian than conventional emulsion polymers. This property allows excellent flow and leveling with roll coat, dip or spray applications.
3) High Solvent Tolerance and Compatibility. A wide variety of solvents can be used in formulating with JONCRYL 554 to adjust the drying rate of the polymer.
4) Excellent Adhesion to natural wood and engineered wood products.

Typical Physical Constants of JONCRYL 554:

Appearance: Milky Emulsion
Non-Volatile by Weight: 46%
Non-Volatile by Volume: 44%
Density as supplied: 8.70 lb/gal
Density of Solids: 9.16 lb/gal
Viscosity: 500 cps
pH: 8.5
Minimum Filming Temperature, C: 27
Freeze/Thaw Stability: Stable

JOHNSON WAX: Polymers/Resins(Continued):

JONCRYL 587:

Acrylic Polymer Flake for High Performance Urethane Coatings

JONCRYL 587 is an acrylic polymer designed to be used as a polyol for high performance isocyanate-cross-linked coatings. It is supplied as an easy dissolving solid flake for maximum latitude in the selection of a wide variety of formulation solvents. And JONCRYL 587 offers economic advantages available only from a solid flake resin.

High performance acrylic-urethane coatings can be produced when JONCRYL 587 is crosslnked with aliphatic polyisocyanates. JONCRYL 587 acrylic-urethane coatings offer an outstanding balance of properties which make them candidates for the following uses:

* Transportation finishes (truck, bus, RV, auto refinish, ship, rail, aircraft).
* Maintenance coatings for steel structures and factory equipment.
* Coatings for heat-sensitive substrates such as plastic business machine cabinets, plastic automotive parts, and wood products.
* High performance OEM industrial coatings with low-temperature curing.

Key Features and Benefits:

* Versatility. As a solid flake resin, JONCRYL 587 allows the formulator to choose his own solvents.
* Value. Polymer flake is the most economical way to buy.
* Performance. JONCRYL 587 provides the high performance necessary for maintenance coatings in harsh chemical and marine environments, the excellent appearance over a long service life demand of transportation finishes, and the quick dry/handling propeties needed for finishing plastic business machine cabinets.

Typical Physical Constants of JONCRYL 587:

Non-Volatile: 100%
Equivalent Weight: 600
Density: 9.7 lb/gal
Specific Gravity: 1.16
Color, Gardner: Less than 1

JOHNSON WAX: Polymers/Resins(Continued):

SCX-501:

Acrylic Oligomer for High Solids Coatings

SCX-501 is an acrylic oligomer for use in high solids thermoset coatings which comply with the Clean Air Act. High solids coatings containing less than 2.8 lb. VOC (volatile organic compounds) per gallon can be formulated at viscosities which allow application by conventional spray at room temperature--without the problems of viscosity management, film thickness control, or severe sag.

SCX-501 is an alternative choice to JONCRYL 500 and is supplied in a 75/25 blend of PM Acetate/Xylene. This choice of an alternative solvent will offer the formulator the following incentives:

* Higher electrical resistance for better electrostatic spray properties.
* Low odor, for better handling in plant and on the finishing line.

SCX-501 should be considered as a candidate for high solids coatings in any application where low solids thermoset acrylics are currently used and where performance requirements such as exterior durability, detergent or chemical resistance are needed.

Typical Physical Constants of SCX-501:
Solids by Weight, 1 hr. at 300F: 80%
Solvent: PM Acetate/Xylene
Viscosity, Brookfield: 6300 cps
Viscosity, Gardner Holdt: Z-1 to Z-3
Density as Supplied: 8.84 lb/gal
Density of Solids: 9.15 lb/gal
Hydroxyl Number of Solids: 140
Equivalent Weight: 500
Acid Number of Solids: 3
Flash Point - Pensky Martens: 104F
Appearance: Clear Syrup
Color, Gardner: 1

JOHNSON WAX: Polymers/Resins(Continued):

SCX-504:

Acrylic Oligomer for High Solids Coatings

SCX-504 is an acrylic oligomer for use in high solids thermoset coatings which comply with the Clean Air Act. High solids coatings can be formulated at viscosities which allow application by conventional spray at room temperature--without the problems of film thickness control, or severe sag.

SCX-504 is an alternative choice to JONCRYL 500 and is supplied in Xylene. This choice of an alternative solvent will offer the formulator the incentive of having:

* A lower cost, high solids acrylic oligomer with similar performance profile to that of JONCRYL 500.
* Higher electrical resistance for better electrostatic spray properties.
* Low odor, for better handling in-plant and on the finishing line.

SCX-504 should be considered as a candidate for high solids coatings in any application where low solids thermoset acrylics are currently used and where performance requirements such as exterior durability, detergent or chemical resistance are needed.

Typical Physical Constants of SCX-504:

Solids by Weight, 1 hr. at 300F: 80%
Solvent: Xylene
Viscosity, Brookfield: 6500 cps
Viscosity, Gardner Holdt: Z-1 to Z-3
Density as Supplied: 8.70 lb/gal
Density of Solids: 9.15 lb/gal
Hydroxyl Number of Solids: 140
Acid Number of Solids: 3
Flash Point, Tag Closed Cup: 83F
Appearance: Clear Syrup
Color, Gardner: 1

NL CHEMICALS, INC.: Alkyd and Polyester Resins: High Solids Resins:

Non-Oxidizing Alkyds:

This section includes resins with semi-drying or nondrying characteristics. They range in oil length from short to long. Two primary application types are the market targets for these resins: industrial finishes when modified with amino for automotive topcoats and appliance finishes, and nitrocellulose lacquers when used as modifiers as part of the plasticizing system.

AROPLAZ 1351:

High quality castor-based alkyd plasticizer, 100% solids.

AROPLAZ 2477-X-65:

Semi-drying plasticizing alkyd for lacquers and bake enamels. Makes outstanding furniture lacquers.

AROPLAZ 2575-X-60:

Light colored resin for economical enamels and lacquers.

1351:
- Percent Solids: 97-100
- Viscosity: G-H: W-Y

2477-X-65:
- Percent Solids: 65
- Solvent: X
- Viscosity: G-H: X-Z

2575-X-60:
- Percent Solids: 60
- Solvent: X
- Viscosity: G-H: Z2-Z4

Polyesters/Oil-Free Alkyds:

This category contains non-oxidizing alkyds which are oil-free and which usually are cross-linked with amino resins during baking. Generally they are extremely flexible and have outstanding adhesion to metal surfaces lending themselves to interior/exterior decorative enamels and coil coatings. They also serve in low temperature cure isocyanate crosslinked finishes and as lacquer plasticizers. Higher solids is a feature of these finishes.

AROPLAZ 6025-Z-70:

Highest flexibility (O-1T) for coil coatings, excellent durability.
- Percent Solids: 70
- Solvent: R/X/Y1
- Viscosity: G-H: W-X+

NL CHEMICALS, INC.: Alkyd and Polyester Resins: High Solids Resins(Continued):

Silicone Modified Resins:

Alkyds and oil-free polyesters in this group are modified with silicone monomers to provide exterior durability, gloss and color retention. Resins in this group include oxidizing and non-oxidizing resin systems. The oxidizing systems are excellent vehicles for architectural and industrial maintenance coatings. The non-oxidizing systems find use in coil coatings and other postforming applications.

AROPLAZ 1700-M-60:

Premium air-dry siliconized alkyd having excellent gloss and color retention. For TT-E-490E type uses.

AROPLAZ 1710-Z-60:

High silicone 50% oil-free polyester having unsurpassed color and gloss retention on weathering. Used for coil coatings.

AROPLAZ Silicone Modified Resins:

1700-M-60:
Percent Solids: 60
Solvent: M
Viscosity: G-H: V-X

1710-Z-60:
Percent Solids: 60
Solvent: R/A6/B
Viscosity: G-H: W-Y

Polyester Hard Resin:

AROPLAZ 642:

Pale polyester hard resin for lacquers, inks, paper coatings, industrial coatings. May be used as diluent in UV-cure inks. Softening range 85-95C (Mercury method). Soluble in aromatics, partially soluble in alcohol, insoluble in aliphatics and glycols.

Percent Solids: 100

KELPOL High Solids VT Alkyds:

3710-X-80:
Percent Solids: 80
Solvent: X
Viscosity: G-H: Z6-Z7

3755-X-80:
Percent Solids: 80
Solvent: X
Viscosity: G-H: Z7-Z8

NL CHEMICALS, INC.: Alkyd and Polyester Resins: High Solids Resins(Continued):

High Solids Polyesters and Alkyds:

This series includes polyesters and alkyds supplied at solids levels of 75% by weight and higher, which can be used to formulate high solids air dry and bake finishes for various industrial applications.

AROPLAZ 3667-Z-80:

Rosin phenolic modified short oil alkyd for lift-resistant primers and interior enamels.

AROPLAZ 6025-A6-80:

Higher solids version of AROPLAZ 6025-Z-70 resin, may be formulated for spray application.

AROPLAZ 6231-K2-90:

Premium alkyd resin for high solids amino crosslinked general industrial baking finishes with good adhesion, hardness and solvent resistance.

AROPLAZ 6232-A4-85:

Economical short oil alkyd for use in economical high solids general industrial baking finishes.

AROPLAZ 6233-Y1-90:

Alkyd for very high solids general industrial baking finishes

AROPLAZ 6234-G4-85:

High quality short soya alkyd for superior performance in salt spray and weathering.

AROPLAZ 6410-K3-75:

Short oil alkyd based on TOFA for fast air dry high solids industrial coatings. Can also be force dried or baked with or without amino crosslinker.

AROPLAZ 6420-K3-85:

Short oil chain stopped alkyd for low bake or air dry high solids industrial coatings.

AROPLAZ 6440-A4-85:

Medium oil chain stopped alkyd designed for durable high solids air dry industrial finishes.

AROPLAZ 6503-K2-90:

Economical high solids polyester resin designed for low VOC industrial finishes with an excellent cost/performance ratio.

AROPLAZ 6755-A6-80:

High solids polyester resin for appliance and similar premium quality industrial coatings.

AROPLAZ 6768-X-84:

Permium quality polyester for low VOC baking finishes. Aromatic solvent reduced formulas for low conductivity H.S. systems and solvent resistance.

NL CHEMICALS, INC.: Alkyd and Polyester Resins: High Solids Resins (Continued):

<u>AROPLAZ High Solids Polyesters and Alkyds:</u>

3667-Z-80:
Percent Solids: 80
Solvent: K2/K4/X
Viscosity: G-H: Z6-Z7
6025-A6-80:
Percent Solids: 80
Solvent: A6
Viscosity: G-H: Z5-Z6
6231-K2-90:
Percent Solids: 90
Solvent: K2
Viscosity: G-H: Z4+-
6232-A4-85:
Percent Solids: 85
Solvent: A4
Viscosity: G-H: Z4+-
6233-Y1-90:
Percent Solids: 90
Solvent: Y1/B
Viscosity: G-H: Z5+-
6234-G4-85:
Percent Solids: 85
Solvent: G4
Viscosity: G-H: Z4+-
6410-K3-75:
Percent Solids: 75
Solvent: K3/T
Viscosity: G-H: Z2-Z4
6420-K3-75:
Percent Solids: 75
Solvent: K3
Viscosity: G-H: Z+-
6440-A4-85:
Percent Solids: 85
Solvent: A4
Viscosity: G-H: Z4-Z5
6503-K2-90:
Percent Solids: 90
Solvent: K2
Viscosity: G-H: Z6+-
6755-A6-80:
Percent Solids: 80
Solvent: A6/T
Viscosity: G-H: Z5+-
6768-X-84:
Percent Solids: 84
Solvent: X
Viscosity: G-H: Z1-Z3

NL CHEMICALS, INC.: Alkyd Resins: AROPLAZ-Oxidizing (Drying Oil-Modified) Alkyds:

Short Oil Alkyds:

Resins in this section generally have less than 40 percent fatty acid modification. They are usually composed of a dibasic acid such as phthalic anhydride or isopthalic acid, a polyol or blend of polyols; an oil; fatty acid or a blend with synthetic acids. These resins are primarily soluble in strong aromatic solvents, and have only limited tolerance to aliphatic solvents. Drying oil-type short alkyds are used for baking and air-dry industrial applications. The short alkyds may also be considered as plasticizing resins for other high molecular weight polymer systems.

AROPLAZ 1453-X-50:
TOFA medium solids air-dry and bake.
AROPLAZ 1453-6X2-50:
"Exempt" solvent version of AROPLAZ 1453.
AROPLAZ 3252-X-50:
Economical, very fast air-dry chain-stopped for industrials.
AROPLAZ 6008-X-50:
Very fast air dry with outstanding exterior durability on metal; excellent in aerosol paints.
AROPLAZ 6065-X-50:
Fast air-dry TOFA resin for industrial enamels.
AROPLAZ 7310-X-50:
Flexible, durable soya for low bake industrials.

1453-X-50:
Percent Solids: 50
Solvent: X
Viscosity: G-H: Z1-Z3
1453-6X2-50:
Percent Solids: 50
Solvent: 6X2
Viscosity: G-H: Z1-Z3
3252-X-50:
Percent Solids: 50
Solvent: X
Viscosity: G-H: Z1-Z3
6008-X-50:
Percent Solids: 50
Solvent: X
Viscosity: G-H: U-X
6065-X-50:
Percent Solids: 50
Solvent: X
Viscosity: G-H: Z1-Z3
7310-X-50:
Percent Solids: 50
Solvent: X
Viscosity: G-H: T-V

NL CHEMICALS, INC.: Alkyd Resins: AROPLAZ-Oxidizing (Drying Oil-Modified) Alkyds(Continued):

Medium Oil Alkyds:

Alkyd resins with 40 to 55 percent fatty acid modification are classed as medium oils. They generally consist of a dibasic acid such as phthalic anhydride, a polyol or blend of polyols; and drying oils or blends of drying oils or fatty acids. These resins are usually soluble in weaker aliphatic solvents, but are sometimes offered in aromatics to lower the viscosity or improve compatibility. Medium alkyds air dry more positively than short oils and are recommended for air dry industrial or trade sales applications.

AROPLAZ 310-M-50:
"High drink" soya with fast set and dry, cut in mineral spirits.

AROPLAZ 310-V-50:
"High drink" soya with fast set and dry, cut in mineral spirits.

AROPLAZ 1081-M-50:
"High drink" soya/tung with snappy dry for mineral spirits cut.

AROPLAZ 1082-M-50:
TT-R-266D Type IV soya resin serving in trade sales and industrial finishes.

AROPLAZ 1082-V-60:
Same as above but cut in VM&P naphtha. Meets specifications for fast drying traffic paints.

AROPLAZ 1247-T-70:
Grinding vehicle, vehicle for tinting colors, let down with other medium oil alkyd.

AROPLAZ 1365-X-60:
Fast setting linseed primer resin; meets TT-P-1757 and MIL-P-7962D.

AROPLAZ 1365-6X3-60:
"Exempt" solvent version of AROPLAZ 1365.

AROPLAZ 1430-M-50:
High viscosity AROPLAZ 1082 type.

AROPLAZ 1433-M-50:
Medium viscosity soya resin for trade sales enamels.

AROPLAZ 1445-M-50:
Linseed-soya resin meeting TT-R-266D Type III.

NL CHEMICALS, INC.: Alkyd Resins: AROPLAZ-Oxidizing (Drying Oil-Modified) Alkyds(Continued):

310-M-50:
Percent Solids: 50
Solvent: M
Viscosity: G-H: Z4-Z6
310-V-50:
Percent Solids: 50
Solvent: V
Viscosity: G-H: Z2-Z4
1081-M-50:
Percent Solids: 50
Solvent: M
Viscosity: G-H: Z4-Z6
1082-M-50:
Percent Solids: 50
Solvent: M
Viscosity: G-H: W-Y
1082-V-60:
Percent Solids: 60
Solvent: V
Viscosity: G-H: Z3-Z5
1247-T-70:
Percent Solids: 70
Solvent: T
Viscosity: G-H: Z2-Z4
1365-X-60:
Percent Solids: 60
Solvent: X
Viscosity: G-H: Z-Z2
1365-6X3-60:
Percent Solids: 60
Solvent: 6X3
Viscosity: G-H: Z-Z2
1430-M-50:
Percent Solids: 50
Solvent: M
Viscosity: G-H: Z-Z2
1433-M-50:
Percent Solids: 50
Solvent: M
Viscosity: G-H: X+-Z
1445-M-50:
Percent Solids: 50
Solvent: M
Viscosity: G-H: U1/2-Y

NL CHEMICALS, INC.: Alkyd Resins: AROPLAZ-Oxidizing (Drying Oil-Modified) Alkyds(Continued):

Long Oil Alkyds:

Long-oil Alkyds are those with more than 55 percent fatty acid modification. They contain dibasic acid such as phthalic anhydride or isophthalic acid; a polyol or blend of polyols; and a drying oil or fatty acid. These resins are readily soluble at low viscosity in aliphatic solvents. Long oils are used almost exclusively in architectural brushing type coatings for both interior and exterior use. The longer oils are used as vehicles for trim and exterior housepaints, and as modifiers for latex.

AROPLAZ 111-M-70:
General purpose, very long marine oil resin. Enamels, house and trim, aluminum paints, metal primers.

AROPLAZ 317:
Latex modifier. Meets TT-P-102E.

AROPLAZ 1241-M-60HV, 70HV:
Exterior durable resin with good gloss and gloss retention. Architectural enamels with good nonhazing properties.

AROPLAZ 1266-M-60:
Meets TT-R-266D, Type 1, Class B.

AROPLAZ 1266-M-70:
Exterior durable pentaerthryitol type resin that meets TT-R-266D, Type 1, Class A.

AROPLAZ 1271:
100 percent solids linseed resin with high compatibility with other film formers.

AROPLAZ 1272:
100 percent solids soya version of AROPLAZ 1271.

NL CHEMICALS, INC.: Alkyd Resins: AROPLAZ-Oxidizing (Drying Oil-Modified) Alkyds(Continued):

Long Oil Alkyds(Continued):

111-M-70:
Percent Solids: 70
Solvent: M
Viscosity: G-H: Z-Z2

317:
Percent Solids: 100
Viscosity: G-H: Z-Z3

1241-M-60HV:
Percent Solids: 60
Solvent: M
Viscosity: G-H: X-Z

1241-M-70HV:
Percent Solids: 70
Solvent: M
Viscosity: G-H: Z4-Z6

1266-M-60:
Percent Solids: 60
Solvent: M
Viscosity: G-H: T-W

1266-M-70:
Percent Solids: 70
Solvent: M
Viscosity: G-H: Z-Z2

1271:
Percent Solids: 97 min.
Viscosity: G-H: W-Y

1272:
Percent Solids: 97 min.
Viscosity: G-H: W-Y

NL CHEMICALS, INC.: AROFLINT Polyester-Epoxy Resins:

A special class of two-package polyester-epoxy reactive systems is included in this section. These two-component systems produce high performance coatings with excellent color, stain resistance, gloss, color retention, chemical resistance, gloss retention and abuse resistance. They are recommended for use in tile-like, maintenance, product finishing, and industrial enamel-type coatings.

Polyester Components:

202-A6X-60:
The "original" acidic polyester developed for use with AROFLINT 303 component in AROFLINT 505 system.
252-Z1-60:
Low odor acidic polyester developed primarily for trade sales use. Combined with AROFLINT component 607 produces AROFLINT 859 system for tile-like home and office finishes.
404-XX-60:
The "improved" acidic polyester developed for use with AROFLINT component 607 to produce the AROFLINT 1011 system (fastest air-dry). May be used with AROFLINT component 303, where it replaces AROFLINT 202, producing an AROFLINT 707 system.

Epoxy Components:

303-X-90:
The "original" oxirane or epoxy component may be used with AROFLINT 202 in the AROFLINT component 505 system. May be used with AROFLINT component 404 or 252 also.

607:
The oxirane or epoxy component normally used with AROFLINT component 404 or AROFLINT 252 to produce AROFLINT 1011 or 859 systems, respectively. May also be used with AROFLINT component 202.

202-A6X-60:
Percent Solids Weight: 60
Solvent: A6X
Viscosity: G-H: V 1/2 - Y 1/2
252-Z1-60:
Percent Solids Weight: 60
Solvent: M/S/Isobutyl Isobutyrate/S
Viscosity: G-H: G-J
404-XX-60:
Percent Solids Weight: 60
Solvent: A4/V/B/Ethyl Benzene
Viscosity: G-H: R-U
303-X-90:
Percent Solids Weight: 90
Solvent: X
Viscosity: G-H: E-1
607:
Percent Solids Weight: 99
Viscosity: G-H: V-X 2/3

NL CHEMICALS, INC.: AROPLAZ Modified Alkyds:

AROPLAZ-Alkyds Modified with Rosin, Hard Resins and/or Phenolic Resin

Alkyds of varying oil length and phthalic content modified with rosin, hard resins or phenolics, are designated as Modified Alkyds. Modification improves drying speed, hardness, adhesion, and chemical resistance. Modified alkyds are excellent for primers and specialty coatings.

AROPLAZ 123-V-50:
Low cost rosin modified alkyd based on marine oil.
AROPLAZ 156-M-50:
Top quality industrial primer, farm equipment and machine enamels.
AROPLAZ 161-M-55:
Porch and floor, machinery, and rapid dry enamels. Used in TT-P-620 C and TT-C-530 B.
AROPLAZ 181-M-65:
Excellent leafing aluminum paints. Steel maintenance. Floor and deck enamels.
AROPLAZ 663-X-50:
Lift-resistant, fast dry resin meeting TT-E-515A and TT-P-664C specifications.
AROPLAZ 663-6X3-50:
"Exempt" solvent version of AROPLAZ 663-X-50

AROPLAZ:

123-V-50:
Percent Solids: 50
Solvent: X
Viscosity: G-H: Z1-Z3
156-M-50:
Percent Solids: 50
Solvent: M
Viscosity: G-H: Z1-Z3
161-M-55:
Percent Solids: 55
Solvent: M
Viscosity: G-H: Z2-Z4
181-M-65:
Percent Solids: 65
Solvent: M
Viscosity: G-H: X-Z
663-X-50:
Percent Solids: 50
Solvent: X
Viscosity: G-H: W-Y
663-6X3-50:
Percent Solids: 50
Solvent: 6X3
Viscosity: G-H: W-Y

NL CHEMICALS, INC.: AROPLAZ Modified Alkyds: AROFLAT and AROTHIX Flat Wall Alkyds:

AROFLAT and AROTHIX Flat and Thixotropic Wall Alkyds:

Composition varies for the individual product in this section, but generally they share certain properties--high viscosity in solutions at low solids. The high molecular weight ensures low penetration into porous surfaces such as plaster or dry wall. Primary applications for this group of resins include interior brushing flats, primers, undercoats and semi-gloss enamels when used as the sole vehicle or with long oil alkyds.

AROFLAT 3113-P-30:
Fast dry resin for highly pigmented flats, also for semi-gloss paints.

AROFLAT 3114-MO-30:
Similar to AROFLAT 3113-P-30 but is cut in odorless mineral spirits.

AROFLAT 301-M-60:
Ester gum modified alkyd supplied as a firm gel for use as puffing and pigment suspending vehicle in easy brushing contractor paints, architectural semigloss enamels, primers and undercoaters.

AROTHIX 4000-P-40:
Gelled modified alkyd with very high gel strength for use in flat wall paints, primers and undercoaters, or as a modifier in semigloss enamels.

3113-P-30:
Percent Solids: 30
Solvent: P
Viscosity: G-H: U-Y

3114-MO-30:
Percent Solids: 30
Solvent: MO
Viscosity: G-H: T-X+

301-M-60:
Percent Solids: 60
Solvent: M
Viscosity: G-H: Dry Gel

4000-P-40:
Percent Solids: 40
Solvent: P
Viscosity: G-H: Med Gel

NL CHEMICALS, INC.: KELPOXY Elastomer Modified Epoxy Concentrates:

Concentrates of carboxy terminated butadiene nitrile rubber (CTBN) reacted into standard epoxy resins, containing 40% CTBN. The CTBN epoxies produce their effects by forming particles of elastomer about 0.1 to 10 microns in diameter which inhibit crack propagation, resulting in improved toughness, flexibility. and interlaminar shear strength. The KELPOXY resins are generally used as concentrates by diluting with standard epoxy resins so that the CTBN concentration is about 10 to 20 percent. Curable with all of the conventional epoxy curing agents.

G202:
Liquid epoxy resin internally modified to improve toughness, impact strength and adhesion. Its unique modification gives relatively low viscosity for a resin of this type.

G272:
Epoxy concentrate containing 40% of a carboxy terminated acrylonitrile rubber. Improves toughness, impact resistance, crack propagation resistance and flexibility. Has found its major use in adhesives and composites.

G293:
Epoxy concentrate containing 40% of a carboxy terminated acrylonitrile rubber. Similar to the above but based on a CTBN designed to give lower viscosity and better compatibility with more polar resins such as epoxy novolacs. Used in adhesives, composites and coatings.

519-K2-70:
Elastomer modified epoxy resin solution designed to give flexible, impact resistant films when cured with conventional epoxy hardeners and curing agents.

G202:
Percent Solids: 100
Viscosity: G-H: Z6+

G272:
Percent Solids: 100
Viscosity: G-H: Z10+

G293:
Percent Solids: 100
Viscosity: G-H: Z10+

519-K2-70:
Percent Solids: 70
Solvent: K2
Viscosity: G-H: Z3-Z5

NL CHEMICALS, INC.: KELTROL Oil Copolymers/KELPOL Alkyd Copolymers:

Resins in this group are drying oils or drying oil alkyds modified with unsaturated monomers, vinyl toluene, styrene or acrylic materials which are copolymerized with the oil or alkyd. Economical fast dry enamels and general industrial formulations are their prime markets.

KELTROL 1001 and 1074:
Fast drying, economical, clear and pigmented finishes for aerosol coatings, utility sanding sealers, production-line prefinishing of paneling and furniture, prepriming lumber and hardboard, machinery enamels, hammer-like finishes, tub caulking compounds and coating for plastic and vacuum metalizing.

KELPOL 835-50M:
Fast drying resin ideally suited for brushing applications.

KELPOL 890-60X:
High viscosity styrenated alkyd resin with good hardness for hammer finishes.

KELPOL 3710:
Vinyl toluenated soya alkyd copolymer for high solids general purpose utility enamels.

KELPOL 3755-X-80:
Faster drying version of KELPOL 3710 above. VT copolymer with exceptionally fast dry.

<u>KELTROL VT Oil Copolymers:</u>

1001:
Percent Solids: 60
Solvent: M
Viscosity: G-H: Y+-1B

1074:
Percent Solids: 60
Solvent: V
Viscosity: G-H: Z+-1B

<u>KELPOL Alkyd Copolymers:</u>

835-50M:
Percent Solids: 50
Solvent: M
Viscosity: G-H: W-Z

890-60X:
Percent Solids: 60
Solvent: X
Viscosity: G-H: Z3-Z5

3710-X-80:
Percent Solids: 80
Solvent: X
Viscosity: G-H: Z6-Z7

3755-X-80:
Percent Solids: 80
Solvent: X
Viscosity: G-H: Z7-Z8

NL CHEMICALS, INC.: Polyols for Reaction with Urethane Prepolymers:

Used as the second component with aromatic and aliphatic urethane prepolymers for coatings, potting compounds, encapsulants, moldings, fabric impregnants, and sealants.

The polyesters offer improved chemical resistance, gloss retention and durability, and are especially useful in exterior applications when used with aliphatic urethane prepolymers.

The product finishing polyols are designed to give much faster tackfree and cure times when air dried or force cured as on finishing lines. They are also suitable for other applications where fast dry is important.

The acrylic polyol is used for hard, light color gloss retentive films where high chemical resistance is not a requirement.

Polyester Polyols:

POLYOL 4294:
High hydroxyl value polyester for high hardness, chemical and solvent resistance, and outstanding exterior gloss and color retention.

POLYOL 4295:
Lower hydroxyl value polyester for more flexible coatings with greater impact resistance as well as excellent exterior gloss and color retention.

Product Finishing Polyols:

POLYOL 4357:
Very fast drying polyol for production line finishing with either aromatic or aliphatic prepolymers. Supplied in methyl ethyl ketone solvent.

POLYOL 8100:
Polyol for fast drying, higher solids two component urethanes, both aromatic and aliphatic. Supplied in n-butyl acetate.

Acrylic Polyol:
POLYOL 4573:
Polyol for very light color two-component aliphatic urethanes with excellent exterior gloss retention and durability.

Product:

4294:
Percent Solids: 100
Viscosity: G-H: Z8+-

4295:
Percent Solids: 100
Viscosity: G-H: Z10+-

4357:
Percent Solids: 50
Viscosity: G-H: U-W

8100:
Percent Solids: 80
Viscosity: G-H: Z2+

4573:
Percent Solids: 60/Solvent: A6X/Viscosity: G-H: T-X

NL CHEMICALS, INC.: Resins for Water-Reducible Coatings: KELSOL Industrial Resins:

A series of unique water-dispersible resins designed for air-dry, force-dry or low-bake applications. These resins are for use in all colors including whites and pastels and in primers. Most are supplied as base resin and require neutralization before adding water. A water/cosolvent blend is usually required for reduction to application solids. These resins produce enamels with outstanding corrosion resistance for air-dry, force-dry and low-bake applications. Each resin has its own distinct advantage as follows:

KELSOL 3900:
A general purpose resin with a good combination of drying speed, corrosion and humidity resistance.

KELSOL 3902:
A general purpose resin, slightly slower in dry than KELSOL 3900 but superior to it in most other performance properties, especially corrosion resistance and early water resistance.

KELSOL 3904:
An economical general purpose resin, slightly slower in dry rate than KELSOL 3900 but similar in most other properties.

KELSOL 3905:
A general purpose resin with very good stability, corrosion resistance, early water resistance and exterior durability.

KELSOL 3906:
Excellent corrosion resistance and faster dry rate than KELSOL 3902, similar in most other properties.

KELSOL 3907:
Very fast air dry rate and excellent dip tank stability and humidity resistance, with very good corrosion resistance and early water resistance.

KELSOL 3909:
Best corrosion and humidity resistance of the series, and especially good in very thin films. Very good early water resistance.

KELSOL 3910:
Economical short oil alkyd with good corrosion resistance, good humidity resistance and excellent stability for low cost shopcoats and primers.

NL CHEMICALS, INC.: Resins for Water-Reducible Coatings (Continued):

KELSOL Industrial Resins (Continued):

KELSOL 3911:
Economical short oil alkyd with good corrosion resistance and low cost shopcoats and primers.

KELSOL 3950:
Phenolic modified alkyd designed to be used in lift-resistant primers, under topcoats such as two-component urethanes, high solids topcoats, etc.

KELSOL 3960:
Chain stopped alkyd designed to give excellent corrosion resistance, humidity and early water resistance and exterior durability with moderate dry rate.

KELSOL 3961:
Chain stopped alkyd with very fast air dry rate and corrosion resistance but less durability, water and humidity resistance than KELSOL 3960.

KELSOL 3962:
Similar performance compared to KELSOL 3960 and 3961 but better early wet adhesion. Supplied at lower NV and viscosity.

KELSOL 3963:
A fast dry chain stopped alkyd with excellent corrosion and humidity resistances.

KELSOL 3970:
Silicone modified alkyd for the best durability and heat resistance of the series, similar to solvent thinned silicone alkyds.

KELSOL 3980:
Coconut oil alkyd for light color melamine crosslinked baking finishes.

KELSOL 3900-BG4-75:
Percent Solids Weight: 75
Solvent: B/G4
Viscosity: G-H: Z6+-

KELSOL 3902-BG4-75:
Percent Solids Weight: 75
Solvent: B/G4
Viscosity: G-H: Z6+-

NL CHEMICALS, INC.: Resins for Water-Reducible Coatings (Continued):

KELSOL:

KELSOL 3904-BG4-75:
Percent Solids Weight: 75
Solvent: B/G4
Viscosity: G-H: Z6+-

KELSOL 3905-B2G-75:
Percent Solids Weight: 75
Solvent: B2/G4
Viscosity: G-H: Z5+-

KELSOL 3906-B2G-75:
Percent Solids Weight: 75
Solvent: B2/G4
Viscosity: G-H: Z6+-

KELSOL 3907-G2B-75:
Percent Solids Weight: 75
Solvent: B2/G2
Viscosity: G-H: Z6+-

KELSOL 3909-BG4-75:
Percent Solids Weight: 75
Solvent: B/G4
Viscosity: G-H: Z5+-

KELSOL 3910-B2G-75:
Percent Solids Weight: 75
Solvent: B2/G4
Viscosity: G-H: Z6+-

KELSOL 3911-B2G-70:
Percent Solids Weight: 70
Solvent: B2/G4
Viscosity: G-H: Z6+-

KELSOL 3922-G-80:
Percent Solids Weight: 80
Solvent: G2/G5
Viscosity: G-H: Z5+-

KELSOL 3931-WG4-45:
Percent Solids Weight: 45
Solvent: W/G4
Viscosity: G-H: T-W

NL CHEMICALS, INC.: Resins for Water Reducible Coatings (Continued):

KELSOL Industrial Resins(Continued):

KELSOL 3950-B2G-70:
Percent Solids Weight: 70
Solvent: B2/G4
Viscosity: G-H: Z3+-

KELSOL 3960-B2G-75:
Percent Solids Weight: 75
Solvent: B2/G4
Viscosity: G-H: Z6+-

KELSOL 3961-B2G-75:
Percent Solids Weight: 75
Solvent: B2/G4
Viscosity: G-H: Z6+-

KELSOL 3962-B2G-70:
Percent Solids Weight: 70
Solvent: B2/G4
Viscosity: G-H: Z5+-

KELSOL 3963-B2G-70:
Percent Solids Weight: 70
Solvent: B2/G4
Viscosity: G-H: Z5+-

KELSOL 3970-G4-75:
Percent Solids Weight: 75
Solvent: G4
Viscosity: G-H: Z5+-

KELSOL 3980-G4-75:
Percent Solids Weight: 75
Solvent: G4
Viscosity: G-H: Z6+-

AROLON Industrial Resins:

AROLON 465:
Unneutralized, oil-free durable, color retentive polyester for dip, spray, flow or rollercoat application. Must be cross-linked with amino resin. Excellent hardness, flexibility and durability.

AROLON 557:
Unneutralized, thermosetting water dispersible acrylic, which must be crosslinked with aminoplast for best results--may be used in hybrid systems with acrylic emulsions.

NL CHEMICALS, INC.: Resins for Water-Reducible Coatings (Continued):

AROLON Industrial Resins(Continued):

AROLON 559:
Lower viscosity version of AROLON 557 for higher solids coatings.

AROLON 580:
Versatile light colored alkyd dispersion for use in varnishes, gloss and semi-gloss trade sales enamels, exterior house paints and industrial air-dry or baked finishes. Excellent modifier for acrylic emulsions to improve flow, leveling and adhesion.

AROLON 585:
Very fast dry modified alkyd dispersion. Excellent modifier for AROLON 580 or soluble air-dry types to enhance speed of dry and hardness. Like AROLON 580, it contains no organic coupling agent.

AROLON 820:
Unique air-dry acrylic emulsion for industrial maintenance finishes. Offers excellent long-term protection. Corrosion resistance is outstanding.

AROLON 840:
Unique fast air-dry acrylic emulsion for moisture and solvent resistant gloss enamels.

AROLON 850:
Unique air-dry acrylic emulsion with excellent adhesion to plastics and oily metals.

AROLON 921:
Fast air-dry and force dry for industrial primer/shop coat.

AROLON 465-G4-80:
Percent Solids Weight: 80
Solvent: G4
Viscosity: G-H: Z5+-

AROLON 557-B-70:
Percent Solids Weight: 70
Solvent: B2
Viscosity: G-H: Z6+

AROLON 559-G4-70:
Percent Solids Weight: 70
Solvent: G4
Viscosity: Z6+-

NL CHEMICALS, INC.: Resins for Water-Reducible Coatings (Continued):

AROLON Industrial Resins(Continued):

AROLON 580-W-42:
Percent Solids Weight: 42
Solvent: W*
Viscosity: 700-2500 cps

AROLON 585-W-43:
Percent Solids Weight: 43
Solvent: W*
Viscosity: 300-1500 cps

AROLON 820-W-49:
Percent Solids Weight: 49
Solvent: W**
Viscosity: 100-350 cps

AROLON 840-W-39:
Percent Solids Weight: 39
Solvent: W**
Viscosity: 50-200 cps

AROLON 850-W-45:
Percent Solids Weight: 45
Solvent: W**
Viscosity: 700-1500 cps

AROLON 921-A8-70:
Percent Solids Weight: 70
Solvent: G4
Viscosity: G-H: Z3-Z5

AROLON 970-G4-70:
Percent Solids Weight: 70
Solvent: G4
Viscosity: G-H: Z6+

* Supplied neutralized with TEA
**Supplied neutralized with ammonia

NL CHEMICALS, INC.: Urethanes: SPENKEL and SPENLITE:

Urethanes

This group includes oil modified, moisture cured, two-component and lacquer resins. Applications include wood finishing, industrial maintenance, product finishing, plastics and fabric coatings. These are high performance coatings providing abrasion resistance, fast dry and outstanding chemical resistance. All the products are energy efficient in that they air dry or provide low temperature curing finishes, and can be used over substrates that are heat sensitive.

ASTM Type 1
One-Package Stable, Oil Modified, Cured by Oxidation at Ambient or Slightly Elevated Temperatures.

SPENKEL Oil Modified Urethanes:

F19-50MS:
Safflower rapid drying resin for architectural clear and pigmented finishes.

F19-55P:
Easy brushing, rapid drying resin for architectural and pigmented finishes.

F58-60MS:
Higher build than F48. Can be used for exterior finishes.

F34:
Polyurethane resin at 100% nonvolatile developed principally for use as a printing ink vehicle.

F35:
Solvent free oil modified urethane for high solids air dry brushing finishes with low VOC.

F48-50MS and F48-50X:
Quick dry interior, color retentive clears and enamels for furniture and paneling finishes, floors, cabinets, woodwork, antiquing, liquid embroidery. Mineral spirits cut for trade sales finishes and xylol cut for industrial finishes. Compatible with alkyds. Xylol cut can be used with CAB or nitrocellulose for extremely fast dry.

F77-60MS and F77-60X:
Premium high build clear and pigmented floor and exterior finishes, zinc-rich paints, maintenance primers and enamels, implement finishes, utility varnishes.

NL CHEMICALS, INC.: Urethanes: SPENKEL and SPENLITE(Continued):

SPENKEL Oil Modified Urethanes(Continued):

F78-50MS, F78-50X & F78-50T:
Faster dry and harder than F77--same end uses plus production line finishes for paneling, flooring, cabinets, furniture, vacuum metalizing, plastics, aerosols and traffic marking paints. Use toluol cut for fastest dry when modifying with nitrocellulose or CAB.

SPENSOL Water Dispersible Oil Modified Urethane:

F74:
Water dispersible oil modified urethane for clears and enamels.

Product:

F19-50MS:
Percent Solids: 50
Solvent: M
Viscosity: G-H: Z-Z2

F19-55P:
Percent Solids: 55
Solvent: P
Viscosity: G-H: Z-Z2

F58-60MS:
Percent Solids: 60
Solvent: M
Viscosity: G-H: Z2+-1

F34:
Percent Solids: 100
Viscosity: G-H: Z10+

F35:
Percent Solids: 100
Viscosity: G-H: Z3-Z5

F48-50MS:
Percent Solids: 50
Solvent: M
Viscosity: G-H: Z2+-1

F48-50X:
Percent Solids: 50
Solvent: X
Viscosity: G-H: Z+-1

NL CHEMICALS, INC.: Urethanes: SPENKEL and SPENLITE(Continued):

SPENKEL Oil Modified Urethanes(Continued):

F77-60MS:
Percent Solids: 60
Solvent: M
Viscosity: G-H: Z+-1

F77-60X:
Percent Solids: 60
Solvent: X
Viscosity: G-H: S+-1

F78-50MS:
Percent Solids: 50
Solvent: M
Viscosity: G-H: Z1-Z3

F78-50X:
Percent Solids: 50
Solvent: X
Viscosity: G-H: Z1-Z3

F78-50T:
Percent Solids: 50
Solvent: T
Viscosity: G-H: Z1-Z3

SPENSOL Water Dispersible Oil Modified Urethane:

F74-70:
Percent Solids: 70
Solvent: A5/G2
Viscosity: G-H: Z7+-

ASTM Type 2
One-Package Moisture-Cure or Two-Package with Catalyst

SPENKEL Aromatic Moisture Curing Urethanes:

M21-40X:
Highly abrasion resistant coating for maintenance of wood, concrete floors, gymnasiums, masonry sealers, anti-skid floors, linings for box cars and hopper cars, plywood concrete forms, and many other high-service requirements.

NL CHEMICALS, INC.: Urethanes: SPENKEL and SPENLITE(Continued):

SPENKEL Aromatic Moisture Curing Urethanes(Continued):

M37-A6X-42:
Slightly faster dry and harder than M21-40X, with slightly less abrasion resistance for seamless floors and similar applications where higher hardness and faster dry-time are important.

M86-A6X-50:
Very fast drying for high-service production-line finishes on paperboard, bowling alleys, concrete forms, electrical coatings, furniture, prefinished plywood panels.

SPENLITE Aliphatic Moisture Curing Urethane:

M22-X-50:
High-performance clear coating for exterior finishes. Label and overprint varnish. Vinyl and polycarbonate coating. Glaze coat for interior-exterior seamless floors, tile-like gloss coat. Wear layer over deck coatings, slip resistant coatings, clear protective coat over aluminum, brass, copper brightwork. Can be catalyzed and heat cured.

M21-40X:
Percent Solids: 40
Solvent: X
Viscosity: G-H: B-E
M37-A6X-42:
Percent Solids: 42
Solvent: A6X
Viscosity: G-H: C-H
M-65:
Percent Solids: 100
Viscosity: G-H: Z3
M-26:
Percent Solids: 64
Solvent: X
Viscosity: G-H: A-H
M86-A6X-50:
Percent Solids: 50
Solvent: A6X
Viscosity: G-H: B-G

M22-X-40:
Percent Solids: 40
Solvent: X
Viscosity: G-H: A-D
M-27:
Percent Solids: 64
Solvent: X
Viscosity: G-H: H

NL CHEMICALS, INC.: Urethanes: SPENKEL and SPENLITE(Continued):

ASTM Type 5
Polyisocyanate Adducts, Cured with Castor, Polyester, and Polyether Polyols, With and Without Heat.

SPENKEL Aromatic Urethane Prepolymers:

Film properties for the Spenkel materials are based on reaction with a 275 OH modified castor oil.

P49-A6-60:
Chemical, abrasion and impact-resistant finishes for chemical plants, hopper and tank cars, aircraft, concrete, leather, rubber, and marine equipment. Also laboratory benches, bar tops, desk tops, pre-finished flooring. Coatings for cold, damp surfaces: barges, breweries, dairies, food plants, water and sewage units, tank farms, marine, basement, and subways. Also used in adhesives, especially for lamination of film and foil for packaging.

P81-A6-75:
For use with POLYOL 8100 in high solids aromatic enamels and production line high performance finishes.

P82-K4-75:
Optimum film properties with POLYOL 8100 but slightly lower volume solids in finished coating.

SPENLITE Aliphatic Urethane Prepolymers:

P25-A6X-60:
Highly durable color and gloss-retentive clears and enamels for metal, wood, masonry and flexible substrates; as clear glaze coats over other finishes to improve overall performance. Outstanding exterior durability. Compatible with hydroxyl terminated polyesters and acrylics. P25-Z-60 is exempt solvent version of P25-A6X-60. Used as additive for alkyd automotive refinishing enamels to improve dry time, hardness and early water resistance. Compatible with nitrocellulose or CAB to decrease dry time.

P52-A6X-65:
Outstanding exterior durability when reacted with POLYOL 4573.

P74-A6X-70:
Used as an additive in alkyd and acrylic automotive enamels to improve hardness, gloss, water and gasoline resistance, and resistance to tape marking.

NL CHEMICALS, INC.: Urethanes: SPENKEL and SPENLITE(Continued):

Spenkel Aromatic Urethane Prepolymers:

P49-A6-60:
Percent Solids: 60
Solvent: A6
Viscosity: G-H: H-P

P49-A6X-60:
Percent Solids: 60
Solvent: A6X
Viscosity: G-H: F-N

P82-K4-75:
Percent Solids: 75
Solvent: K4
Viscosity: G-H: Z1-Z3

SPENLITE Aliphatic Urethane Prepolymers:

P25-A4-60:
Percent Solids: 60
Solvent: A4
Viscosity: G-H: H-V

P25-A6X-60:
Percent Solids: 60
Solvent: A6X
Viscosity: G-H: U-W

P52-A6X-65:
Percent Solids: 65
Solvent: A6X
Viscosity: G-H: V-Y

P70:
Percent Solids: 70
Solvent: A4
Viscosity: G-H: Z2-Z3

P74-A6X-70:
Percent Solids: 70
Solvent: A6X
Viscosity: G-H: Y-Z

NL CHEMICALS, INC.: Urethane Solution Elastomers/WD Colloidal Urethanes:

ASTM Type 6
Solvent Thinned Urethane Elastomers

SPENKEL Aromatic Urethanes:
Coatings, impregnants, adhesives for fabrics, foam, rubber, plastics, paper and leather.

L66-40MP:
Used primarily in primers for RIM, TPR and EPDM.

L91-40K:
General adhesive; primer for plastics.

SPENKEL Aliphatic Urethanes:
High performance, color-retentive coatings for flexible substrates such as fabrics, foams, leather, rubber, plastics, printing inks, clear overprint coatings, strippable coating.

L06-30S:
Very soft with good hydrolytic stability, low temperature flexibility and abrasion resistance. For use in laminating adhesives and film/fabric adhesives.

L89-30S:
Fast dry surface coatings, medium hardness and good abrasion resistance. Excellent adhesion to ABS, vinyl and leather.

L90-20A:
Supplied in isopropanol and dilutable with alcohols only. Can be used on solvent sensitive substrates, polycarbonate, styrofoam and paper. Compatible with 1/4 second nitrocellulose.

L66-40MP:
Percent Solids: 40
Solvent: MP
Viscosity: G-H: Y-Z
L91-40K:
Percent Solids: 40
Solvent: K1/D
Viscosity: G-H: Z3-Z5
L06-30S:
Percent Solids: 30
Solvent: D/T
Viscosity: G-H: Z5
L89-30S:
Percent Solids: 30
Solvent: D/T
Viscosity: G-H: Z4-Z5
L90-20A:
Percent Solids: 20
Solvent: D
Viscosity: G-H: W-Y

NL CHEMICALS, INC.: Urethane Solution Elastomers/WD Colloidal Urethanes(Continued):

Water Thinned Urethane Elastomers

SPENSOL Water Dispersible Urethanes:

A series of anionic colloidal urethane elastomer dispersions in a water/solvent mixture, and dilutable with water only. All are fully reacted air dry polymer systems that are mechanically and freeze-thaw stable. The films are all high in tensile strength and have excellent water resistance.

SPENSOL Aromatic Urethanes:
L41:
Moderately hard MDI polyester polymer for interior use in primers and darker colored topcoats for plastic, rubber and metal.

L44:
TDI polyether elastomer, softer than L41 in hardness and lower in cost. Inferior to L41 in chemical resistance but excellent hydrolytic stability.

SPENSOL Aliphatic Urethanes:
L51:
High surface hardness and abrasion resistance. Clear and pigmented topcoat finishes on flexible and semi-rigid plastics.

L52:
Good alcohol tolerance, wet rub and alkali resistance. Abrasion resistant surface coatings on plastics requiring low temperature impact properties.

L53:
Soft. General Purpose. Primers and surface coatings on plastics and rubbers.

L53-40:
Supplied at 40% solids and slightly higher viscosity.

L54:
Moderately hard, best compatibility with extender resins and solvent modifiers. Higher initial viscosity, so more adaptable to non-spray applications such as flow, dip and roller coating.

L55:
Clear and pigmented coatings on plastics, fabrics and metal. Excellent cold flexibility and superior wet adhesion.

L56:
Softest, most flexible resin of the series. For excellent hand in textile coatings and binders, leather finishes and adhesives.

NL CHEMICALS, INC.: Urethane Solution Elastomers/WD Colloidal Urethanes(Continued):

Water Thinned Urethane Elastomers:

SPENSOL Water Reducible Aromatic Urethanes:

L41:
Percent Solids: 32
Solvent: W/MP
Viscosity: G-H: S-Z3

L44:
Percent Solids: 32
Solvent: W/MP
Viscosity: G-H: A-X

SPENSOL Water Reducible Aliphatic Urethanes:

L51:
Percent Solids: 30
Solvent: W/MP
Viscosity: G-H: A1-F

L52:
Percent Solids: 30
Solvent: W/MP
Viscosity: G-H: A1-D

L53:
Percent Solids: 30
Solvent: W/MP
Viscosity: G-H: A1-D

L53-40:
Percent Solids: 40
Solvent: W/MP
Viscosity: G-H: H-W

L54:
Percent Solids: 30
Solvent: W/MP
Viscosity: G-H: W-Z3

L55:
Percent Solids: 35
Solvent: W/MP
Viscosity: G-H: D-W

L56:
Percent Solids: 36
Solvent: W/MP
Viscosity: G-H: C-W

NL CHEMICALS, INC.: Solvent Legend:

A1	Ethoxy Ethyl Acetate
A2	Ethyl Acetate
A4	n-Butyl Acetate
A5	Butoxy Ethyl Acetate
A6	Methoxy Propyl Acetate
B	n-Butyl Alcohol
B2	sec-Butyl Alcohol
D	Isopropyl Alcohol
DMF	Dimethyl Formamide
G	Mixed Glycol/Glycol Ether
G2	Propoxy Propanol
G4	2-Butoxy Ethanol
G5	Diethylene Glycol n-Butyl Ether
K1	Methyl Ethyl Ketone
K2	Methyl Isobutyl Ketone
K3	Methyl n-Propyl Ketone
K4	Methyl Amyl Ketone
K5	Methyl Heptyl Ketone
MP	n-Methyl-2-Pyrrolidone
M	Mineral Spirits (Exempt)
MO	Odorless Mineral Spririts (Exempt)
P	ALIPHATIC 140 Type (Exempt)
R	AROMATIC 150 Type
S	AROMATIC 100 Type
T	Toluene (Toluol)
V	VM&P Naphtha (Exempt)
W	Water
X	Xylene (Xylol)
XX	Exempt Blends
Y1	Ethyl Ethoxy Propionate
6X2	Exempt Solvent Blend with Ester
6X3	Exempt Solvent Blend-No Ester Present

C. J. OSBORN: Copolymers:

These products are modified alkyds. Their modifications may be VT, styrene, acrylic or silicone resins. The VT and styrene copolymers are known for their outstanding speed of dry under ambient conditions and may be used in aerosol coatings, sanding sealers, furniture, machinery enamels and hammertone finishes. Silicone alkyds are known for high heat resistant coatings and exterior durability on metallic surfaces.

Silicone Modified Polyesters:

CJO Code:

13-A-6119:
Type: Silicone modified polyester
% Non-Volatile: 48-50
Solvent: PM acetate
Viscosity: D-G
Color (Gardner): 7 max.
Acid Number (Solids): 10 max.
Modifying Oil: None
WPG Average: 9.05
Uses: Silicone modified saturated polyester designed for high performance and thermal stability.

13-A-6739:
Type: Silicone modified polyester
% Non-Volatile: 64-66
Solvent: PM acetate
Viscosity: Y-Z2
Color (Gardner): 3 max.
Acid Number (Solids): 10 max.
Modifying Oil: None
WPG Average: 9.35
Uses: Premium industrial coatings, high temperature resistance.

13-A-5523:
Type: Silicone modified polyester
% Non-Volatile: 64-66
Solvent: PM acetate
Viscosity: Z-Z3
Color (Gardner): 3 max.
Acid Number (Solids): 10 max.
Modifying Oil: None
WPG Average: 9.40
Uses: This polyester has a 50% silicone modification. Therefore, it is designed for high performance coatings. High solids should be of help to the formulator.

C. J. OSBORN: Copolymers(Continued):

Silicone Modified Polyesters(Continued):

CJO Code:

13-A-5566:
Type: Silicone polyester
% Non-Volatile: 49-51
Solvent: Methyl propyl ketone
Viscosity: A1-B
Color (Gardner): 3 max.
Acid Number (Solids): 10 max.
Modifying Oil: None
WPG Average: 8.18
Uses: Any application requiring toughness, abrasion resistance color retention at high temp. and good durability.

13-A-6109:
Type: Silicone polyester copolymer
% Non-Volatile: 49-51
Solvent: N-hexyl CARBITOL
Viscosity: G-J
Color (Gardner): 3 max.
Acid Number (Solids): 10 max.
Modifying Oil: None
WPG Average: 8.80
Uses: Any application requiring toughness, abrasion resistance, color retention at high temp. and good durability.

13-A-6114:
Type: Silicone modified polyester
% Non-Volatile: 48-50
Solvent: PM acetate
Viscosity: D-G
Color (Gardner): 4 max.
Acid Number (Solids): 10 max.
Modifying Oil: None
WPG Average: 9.05
Uses: This product is a silicone modified saturated polyester designed for high performance and thermal stability.

13-A-5013:
Type: Silicone polyester resin
% Non-Volatile: 49-51
Solvent: PM acetate
Viscosity: E-H
Color (Gardner): 4 max.
Acid Number (Solids): 10 max.
Modifying Oil: None
WPG Average: 9.07
Uses: High silicone copolymer designed for maximum heat stability.

C. J. OSBORN: Copolymers(Continued):

Silicone Modified Polyesters and Alkyds:

CJO Code:

13-3164:

Type: Silicone modified long oil soya alkyd
% Non-Volatile: 59-61
Solvent: Exempt MS
Viscosity: U-W
Color (Gardner): 7 max.
Acid Number (Solids): 10 max.
Modifying Oil: Soya
WPG Average: 7.93

Uses: This long oil soya alkyd may be used for superior exterior coatings on metal marine finishes, trim and trellis enamels and used to upgrade conventional long oil alkyd based enamels. This may also be used to meet Federal spec. TT-E-490E.

13-A-3491:

Type: Silicone polyester copolymer
% Non-Volatile: 48-50
Solvent: CELLO. ACETATE 98%
Butanol 2%
Viscosity: D-F
Color (Gardner): 4 max.
Acid Number (Solids): 10 max.
Modifying Oil: None
WPG Average: 8.95

Uses: Any application requiring toughness, abrasion resistance and color retention at high temperature and good durability. This product could be used in appliances, outdoor signs & coil coatings. There are some electrical properties.

13-0098:

Type: Silicone modified resin
% Non-Volatile: 49-51
Solvent: Exempt MS
Viscosity: J-L
Color (Gardner): 7 max.
Acid Number (Solids): 10 max.
Modifying Oil: Linseed/Soya
WPG Average: 7.70

Uses: This resin is recommended for high temperature aluminum pigmented paints. It exhibits excellent durability and air-drying characteristics.

C. J. OSBORN: Copolymers(Continued):

Silicone Modified Polyesters and Alkyds(Continued):

CJO Code:

13-0640:
Type: Silicone modified alkyd
% Non-Volatile: 59-61
Solvent: Exempt VM&P Naphtha
Viscosity: U-W
Color (Gardner): 7 max.
Acid Number (Solids): 10 max.
Modifying Oil: Soya
WPG Average: 7.80
Uses: Recommended for high temp. aluminum pigmented paints. Exhibits excellent durability and air-dry characteristics. May also be used with marine finishes and exterior coatings on metal or wood.

13-A-1736:
Type: Silicone modified polyester
% Non-Volatile: 49-51
Solvent: MIBK
Viscosity: D-F
Color (Gardner): 3 max.
Acid Number (Solids): 10 max.
Modifying Oil: None
WPG Average: 8.10
Uses: Designed for heat resistance, hardness and color retention.

13-2700:
Type: Linseed modified silicone copolymer
% Non-Volatile: 49-51
Solvent: Xylol
Viscosity: S-V
Color (Gardner): 8 max.
Acid Number (Solids): 15 max.
Modifying Oil: Linseed
WPG Average: 8.30
Uses: This product meets TT-P-28 compositioning for high heat aluminum paints. It has excellent weatherability.

C. J. OSBORN: Copolymers (Continued):

Vinyl Toluene:

CJO Code:

13-A-6169:
Type: Vinyl toluene copolymer
% Non-Volatile: 59-61
Solvent: Exempt MS
Viscosity: Z-Z2
Color (Gardner): 7 max.
Acid Number (Solids): 10 max.
Modifying Oil: Linseed
WPG Average: 7.54
Uses: This vinyl toluene is a fast drying copolymer dispersed in exempt mineral spirits and is suitable for sanding sealers, aerosol paints & wood finishes.

13-A-6170:
Type: Vinyl toluene copolymer
% Non-Volatile: 59-61
Solvent: Exempt VM&P
Viscosity: X-Y
Color (Gardner): 7 max.
Modifying Oil: Linseed
WPG Average: 7.50
Uses: This vinyl toluene is particularly designed for fast drying sanding sealers.

13-A-5422:
Type: Vinyl toluene alkyd copolymer
% Non-Volatile: 49-51
Solvent: Exempt VM&P
Viscosity: Y-Z1
Color (Gardner): 5 max.
Acid Number (Solids): 10 max.
Modifying Oil: Soya
WPG Average: 7.50
Uses: Very fast drying copolymer suitable for aerosol overprint and sanding sealers.

13-A-5444:
Type: Vinyl toluene alkyd copolymer
% Non-Volatile: 49-51
Solvent: Exempt MS
Viscosity: V-X
Color (Gardner): 6 max.
Modifying Oil: TOFA
WPG Average: 7.48
Uses: Quick dry machinery enamels.

C. J. OSBORN: Copolymers(Continued):

Vinyl Toluene(Continued):

CJO Code:

13-A-4731:
Type: Vinyl toluene modified alkyd
% Non-Volatile: 49-51
Solvent: 66/3 VM&P Naphtha
Viscosity: Z2-Z4
Color (Gardner): 3 max.
Acid Number (Solids): 10 max.
Modifying Oil: TOFA
WPG Average: 7.39
Uses: Sanding sealers, aerosol coatings, paper coatings & quick dry enamels.

13-2916:
Type: Vinyl toluene modified alkyd
% Non-Volatile: 59-61
Solvent: Exempt VM&P Naphtha
Viscosity: Z-Z2
Color (Gardner): 5 max.
Acid Number (Solids): 15 max.
Modifying Oil: TOFA
WPG Average: 7.70
Uses: This vinyl toluene copolymer may be used in quick air dry pigmented coatings. For aerosol coatings, for air dry primers, machinery enamels and drum enamels. It is an excellent product for sanding sealers and top coats for wood finishes, paper coating, gloss, and satin finishes.

Silicone Alkyds:

13-A-7019:
Type: Silicone alkyd
% Non-Volatile: 49-51
Solvent: Exempt MS,
Exempt VM&P Xylol
Viscosity: V-X
Color (Gardner): 6 max.
Acid Number (Solids): 10 max.
Modifying Oil: Linseed
WPG Average: 7.57
Uses: Silicone alkyd finds important usage in high heat aluminum paints. It is prepared for Rule 66 compliance and is designed for the requirements of TT-P-28E.

C. J. OSBORN: Copolymers(Continued):

Silicone Alkyds(Continued):

CJO Code:

13-A-4283:
Type: Silicone epoxy modified alkyd
% Non-Volatile: 49-51
Solvent: Exempt MS
Xylol
Viscosity: Z-Z2
Color (Gardner): 9 max.
Acid Number (Solids): 12 max.
Modifying Oil: Linseed
WPG Average: 7.70
Uses: Exterior durability, increased temperature resistance, and excellent adhesion on metal.

13-A-3979:
Type: Silicone alkyd copolymer
% Non-Volatile: 69-71
Solvent: Diethylene glycol monobutyl ether & acetate
Viscosity: Z4-Z6
Color (Gardner): 9 max.
Acid Number (Solids): 15 max.
Modifying Oil: Linseed
WPG Average: 9.08
Uses: Used in high heat aluminum paints, this product has excellent weatherability.

13-0678:
Type: Silicone alkyd copolymer
% Non-Volatile: 59-61
Solvent: Exempt MS
Viscosity: Y-Z1
Color (Gardner): 7 max.
Acid Number (Solids): 10 max.
Modifying Oil: Soya
WPG Average: 8.00
Uses: Superior exterior coating on metal or wood marine finishes.

C. J. OSBORN: Copolymers (Continued):

Acrylic Copolymers:

CJO Code:

13-A-5390:
Type: Acrylic copolymer resin
% Non-Volatile: 40-51
Solvent: Xylol
Viscosity: U-W
Color (Gardner): 6 max.
Acid Number (Solids): 16 max.
Modifying Oil: Soya, dehydrated castor
WPG Average: 8.07
Uses: An excellent resin for industrial metal coating.

13-A-8625:
Type: Acrylic copolymer
% Non-Volatile: 49-51
Solvent: Xylol
Viscosity: L-S
Color (Gardner): 5 max.
Acid Number (Solids): 16 max.
Modifying Oil: Dehydrated castor
WPG Average: 8.10
Uses: This is an excellent resin for industrial metal coatings with fast dry, outstanding color & color retention. Provides good rapid cure for fast production schedules with good resistance to overbake. It is recommended also for metal furniture, cabinets and similar applications including machinery and implement enamel.

13-A-8633-A:
Type: Acrylic copolymer
% Non-Volatile: 49-51
Solvent: SOLVESSO 150
Viscosity: U-X
Color (Gardner): 5 max.
Acid Number (Solids): 15 max.
Modifying Oil: Dehydrated castor
WPG Average: 8.20
Uses: This acrylic copolymer has been designed for a broad spectre of applications in the metal decorating field. It is characterized by very rapid cure, excellent levelling, fabrication and processing. Gloss, color retention and flexibility are also excellent pigmented systems as well as clear finishing varnishes for regular overprint. Coatings may be formulated using this resin.

C. J. OSBORN: Copolymers (Continued):

Acrylic Copolymers (Continued):

CJO Code:

13-A-5509:
Type: Acrylic copolymer
% Non-Volatile: 74-76
Solvent: Xylol
Viscosity: Z5-Z7
Color (Gardner): 7 max.
Modifying Oil: Dehydrated castor
WPG Average: 8.52
Uses: Industrial metal coatings, machinery and implement enamel.

13-A-5080:
Type: Acrylic copolymer resin
% Non-Volatile: 49-51
Solvent: Xylol
Viscosity: U-W
Color (Gardner): 6 max.
Acid Number (Solids): 16 max.
Modifying Oil: Dehydrated castor
WPG Average: 8.07
Uses: An excellent resin for industrial metal coatings; fast dry, outstanding color and color retention. It provides rapid cure for fast production schedules with good resistance to overbake. Recommended for metal furniture, cabinets and similar applications. Also for machinery and implement enamels.

13-A-5324:
Type: Acrylic copolymer resin
% Non-Volatile: 49-51
Solvent: Xylol
Viscosity: T-V
Color (Gardner): 5 max.
Acid Number (Solids): 16 max.
Modifying Oil: Deydrated castor
WPG Average: 8.10
Uses: Excellent resin for industrial metal coatings, fast dry, outstanding color and color retention. Provides rapid cure for fast production schedules with good resistance to overbake. Recommended for metal furniture, cabinets and similar applications. Also for machine and implement enamels.

C. J. OSBORN: Copolymers(Continued):

Acrylic Copolymer(Continued):

CJO Code:

13-A-0257:
Type: Acrylic copolymer resin
% Non-Volatile: 49-50
Solvent: SOLVESSO 150
Viscosity: U-X
Color (Gardner): 5 max.
Acid Number (Solids): 9-16
Modifying Oil: DCO
WPG Average: 8.25
Uses: More flexible version of 13-8633.

13-1491:
Type: Acrylic silicone copolymer
% Non-Volatile: 49-51
Solvent: PM acetate
High flash naphtha
Viscosity: Z2-Z5
Color (Gardner): 3 max.
Acid Number (Solids): 8 max.
Modifying Oil: None
WPG Average: 8.75
Uses: Acrylic silicone copolymers designed for high heat resistance and color retention.

Styrenated Alkyds:

13-A-4014:
Type: Styrenated alkyd
% Non-Volatile: 54-56
Solvent: Xylol
Viscosity: X-Z
Color (Gardner): 6 max.
Acid Number (Solids): 12 max.
Modifying Oil: Soya
WPG Average: 8.13
Uses: Color, hardness, gloss and speed of drying suggest applications in an industrial and implement vehicle, toy enamels and hammertone finishes.

13-A-4565:
Type: Styrenated alkyd
% Non-Volatile: 49-51
Solvent: Xylol
Viscosity: X-Z
Color (Gardner): 4 max.
Acid Number (Solids): 9 max.
Modifying Oil: Soya
WPG Average: 8.10
Uses: Styrenated alkyd with a low viscosity reduction.

C.J. OSBORN: Epoxy Esters:

Epoxy esters serve the trade and industrial market for superior chemical resistance, adhesion, abrasion resistance, and epoxy esters can be formulated in air-drying finishes or baked with or without urea and melamine resins for maximum cross-linking.

CJO Code:

06-601:

Type: Epoxy ester
% Non-Volatile: 49-51
Solvent: High flash naphtha
Viscosity: W-Y
Color (Gardner): 7 max.
Acid Number (Solids): 8 max.
Modifying Oil: DCO
WPG Average: 8.00
Uses: White baking enamels for appliances demanding a superior chemical resistance, general industrial finishes and primers, clear chemical or preparation coatings and oil resistant machine tool enamels.

06-601-B:

Type: Epoxy ester
% Non-Volatile: 49-51
Solvent: SOLVESSO 150
Viscosity: Z-Z2
Color (Gardner): 7 max.
Acid Number (Solids): 8 max.
Modifying Oil: DCO
WPG Average: 8.10
Uses: White baking enamels for appliances demanding superior chemical resistance, general industrial finishes and primers, clear chemical or preparation coatings and oil resistant machine tool enamels.

06-601-X:

Type: Epoxy ester
% Non-Volatile: 49-51
Solvent: Xylol
Viscosity: S-V
Color (Gardner): 7 max.
Acid Number (Solids): 8 max.
Modifying Oil: DCO
WPG Average: 7.95
Uses; White baking enamels for appliances demanding superior chemical resistance, general industrial finishes and primers, clear chemical or preparation coatings and oil resistance machine tool enamels.

C.J. OSBORN: Epoxy Esters(Continued):

CJO Code:

06-602-6:
Type: Epoxy ester
% Non-Volatile: 59-61
Solvent: Exempt MS
Viscosity: Z1-Z3
Color (Gardner): 10 max.
Acid Number (Solids): 25 max.
Modifying Oil: Linseed fatty acid/rosin ester
WPG Average: 7.75
Uses: General interior varnish including floor and bar type varnishes. Vehicle for general utility enamels. Also used for high quality drum enamels, automotive primers & chassis finishes as well as concrete floor finish.

06-604:
Type: Epoxy ester
% Non-Volatile: 49-51
Solvent: Exempt MS
Viscosity: F-J
Color (Gardner): 8 max.
Acid Number (Solids): 8 max.
Modifying Oil: Soya
WPG Average: 7.40
Uses: Brushing paints for application as finished coats over prime surfaces. This product has good durability and general exposure performance; may also be used for structural steel paints.

06-609-6:
Type: Epoxy ester
% Non-Volatile: 59-61
Solvent: Exempt MS
Viscosity: Z-Z2
Color (Gardner): 6 max.
Acid Number (Solids): 14 max.
Modifying Oil: TOFA
WPG Average: 7.80
Uses: Fast drying enamels and clears characterized by high gloss and chemical resistance floor enamels and varnishes for both wood, concrete and masonry coatings including primers.

06-A-1495:
Type: Epoxy ester
% Non-Volatile: 49-51
Solvent: High flash naphtha
Viscosity: Z-Z2
Color (Gardner): 3 max.
Acid Number (Solids): 8 max.
Modifying Oil: Soya
WPG Average: 8.10
Uses: Designed for abrasion resistion resistance, good color.

C.J. OSBORN: Epoxy Esters(Continued):

CJO Code:

06-2540:
Type: Linseed fatty acid modified epoxy ester
% Non-Volatile: 49-51
Solvent: Exempt MS
Viscosity: V-X
Color (Gardner): 9 max.
Acid Number (Solids): 8 max.
Modifying Oil: Linseed
WPG Average: 7.40
Uses: Strongly recommended as a vehicle in zinc rich primers, red lead primers, and primers in general.

06-A-2680:
Type: Epoxy ester
% Non-Volatile: 51-53
Solvent: Exempt MS
Xylol, Dipentene
Viscosity: T-V
Color (Gardner): 10 max.
Acid Number (Solids): 6 max.
Modifying Oil: Linseed
WPG Average: 7.70
Uses: Excellent chemical resistance for zinc rich primers.

06-3166:
Type: Modified epoxy ester
% Non-Volatile: 49-51
Solvent: Xylol
Viscosity: W-Y
Color (Gardner): 8 max.
Acid Number (Solids): 8 max.
Modifying Oil: Drying oil fatty acid
WPG Average: 7.96
Uses: Industrial finishes both air dry and baking. Suitable for white baking enamels.

06-A-3597:
Type: Epoxy ester
% Non-Volatile: 49-51
Solvent: Xylol
Viscosity: W-Y
Color (Gardner): 8 max.
Acid Number (Solids): 8 max.
Modifying Oil: TOFA
WPG Average: 8.04
Uses: This product was designed for fast drying toughness, hardness, and chemical resistance. Air dry and baking finishes for primers, industrial and machinery enamels.

C. J. OSBORN: Epoxy Esters(Continued):

CJO Code:

06-A-4297:
Type: Rosin fatty acid
Epoxy ester
% Non-Volatile: 48-50
Solvent: Exempt MS
Viscosity: W-Y
Color (Gardner): 10 max.
Acid Number (Solids): 10 max.
Modifying Oil: Vegetable fatty acid
WPG Average: 7.70
Uses: Suitable for floor finishes paneling and furniture refinishing, It produces films with good gloss and chemical Some high flash aromatic solvent is suggested when the non-volatile is reduced to lower solids.

06-A-5598:
Type: Epoxy ester
% Non-Volatile: 59-61
Solvent: MS
VM&P Exempt
Viscosity: Z-Z2
Color (Gardner): 6 max.
Acid Number (Solids): 14 max.
Modifying Oil: TOFA
WPG Average: 7.74
Uses: Fast drying enamels and clears characterized by high gloss and chemical resistance.

06-A-5702:
Type: Epoxy ester
% Non-Volatile: 49-51
Solvent: High flash naphtha
Viscosity: Z3-Z5
Color (Gardner): 5 max.
Acid Number (Solids): 8 max.
Modifying Oil: Semi-oxidizing fatty acid
WPG Average: 8.24
Uses: This product should be considered for its high epoxy content and chemical resistance, color, gloss, adhesion and flexibility. It is used in baking systems wiht a minimum amino-resin modifications.

C. J. OSBORN: Epoxy Esters(Continued):

CJO Code:

06-8953:
Type: Modified Epoxy ester
% Non-Volatile: 49-51
Solvent: Ethylene glycol butyl ether acetate and exempt MS
Viscosity: O-R
Color (Gardner): 6 max.
Acid Number (Solids): 6 max.
Modifying Oil: Linseed fatty acid
WPG Average: 7.50
Uses: Zinc-rich primers, red lead primers and primers in general.

06-9879:
Type: Epoxy ester
% Non-Volatile: 49-51
Solvent: Exempt
Viscosity: S-V
Color (Gardner): 4 max.
Acid Number (Solids): 8 max.
Modifying Oil: DCO
WPG Average: 7.55
Uses: May be formulated to meet EPA Solvent regulations. May be used wherever MIRASOL 06-601 epoxy ester is used.

C. J. OSBORN: High Solids Resins:

These products comprise all of the previous resins outlined from 01 to the 13 series and illustrate the types of high solid resins available. Care should be exercised when reviewing this series, as it contains non-drying, drying and baking systems.

CJO Code:

01-124:
Type: Pure non-drying alkyd
% Non-Volatile: 79-81
Solvent: Xylol
Viscosity: M-P
Color (Gardner): 7 max.
Acid Number (Solids): 20 max.
Modifying Oil: Coconut
WPG Average: 8.47
Uses: This is a resin plasticizer for nitrocellulose. Being longer in oil has a more pronounced plasticizing effect. For use in high solids lacquers.

01-131:
Type: Non-drying plasticizing type alkyd
% Non-Volatile: 79-81
Solvent: Toluol
Viscosity: U-X
Color (Gardner): 4 max.
Acid Number (Solids): 10 max.
Modifying Oil: None
WPG Average: 8.85
Uses: Valuable in formulating high solids lacquers due to low viscosity. Nitrocellulose lacquers for both wood and metal may be formulated with this alkyd.

02-A-4773:
Type: Modified medium oil alkyd
% Non-Volatile: 79-81
Solvent: MEK/PM Glycol ether
Viscosity: Z3-Z5
Color (Gardner): 12 max.
Acid Number (Solids): 40 max.
Modifying Oil: TOFA/Soya
WPG Average: 8.67
Uses: Tough, durable non-lifting primers with excellent adhesion and fast dry.

C.J. OSBORN: High Solids Resins(Continued):

CJO Code:

03-A-3263:
Type: Chain stopped alkyd
% Non-Volatile: 74-76
Solvent: Xylol
Viscosity: Z4-Z6
Color (Gardner): 8 max.
Acid Number (Solids): 12 max.
Modifying Oil: Soya
WPG Average: 8.87
Uses: This chain stopped alkyd is designed for compatibility with a variety of film formers like CAB, NC, vinyls, chlorinated rubbers, urea and melamine resins. It is therefore very useful in a variety of applications, air dry and bake.

03-A-4374:
Type: High solid, short oil alkyd
% Non-Volatile: 89-91
Solvent: Xylol,
DOW BC 300
Viscosity: Y-Z2
Color (Gardner): 8 max.
Acid Number (Solids): 15 max.
Modifying Oil: TOFA
WPG Average: 9.05
Uses: High solid, low viscosity, short oil alkyd suitable for crosslinking with melamine and urea resins. It has good application in industrial coatings.

03-A-4477:
Type: Baking or air-dry industrial finishing alkyds.
% Non-Volatile: 79-81
Solvent: MEK/PM Glycol ether
Viscosity: Z2-Z4
Color (Gardner): 9 max.
Acid Number (Solids): 12 max.
Modifying Oil: TOFA
WPG Average: 8.99
Uses: General purpose industrial baking finish with good urea and melamine compatibility. Has good aliphatic solvent tolerance and is an excellent high solid vehicle.

03-A-5344:
Type: High solid, short oil alkyd
% Non-Volatile: 89-91
Solvent: PM Acetate
Viscosity: Z1-Z3
Color (Gardner): 8 max.
Acid Number (Solids): 10 max.
Modifying Oil: TOFA
WPG Average: 8.90
Uses: Excellent high solid industrial finishes.

C.J. OSBORN: High Solids Resins(Continued):

CJO Code:

03-A-5698:
Type: High solid, short oil alkyd
% Non-Volatile: 89-91
Solvent: PM Acetate
Viscosity: Z4-Z6
Color (Gardner): 6 max.
Acid Number (Solids): 10 max.
Modifying Oil: TOFA
WPG Average: 9.00
Uses: This is a flexible, short oil alkyd designed to impart characteristics that cannot be obtained by the use of conventional short oil alkyds. Coatings may be formulated at high solids that retain their flexibility even after extended baking. This product has excellent pigment wetting, gloss and durability.

03-A-7015:
Type: High solid, short oil alkyd
% Non-Volatile: 89-91
Solvent: BUTYL CELLOSOLVE
Xylol
Viscosity: Z6-Z8
Color (Gardner): 6 max.
Acid Number (Solids): 10 max.
Modifying Oil: TOFA
WPG Average: 8.95
Uses: High solid vehicle noticeable for pigment wetting, hardness and excellent baking properties.

03-A-5885:
Type: High solids baking alkyd
% Non-Volatile: 89-91
Solvent: Butyl acetate
Viscosity: Z2-Z4
Color (Gardner): 6 max.
Acid Number (Solids): 12 max.
Modifying Oil: TOFA
WPG Average: 8.95
Uses: This product is a general purpose, high solid baking alkyd.

03-A-6633:
Type: High solid short oil baking alkyd
% Non-Volatile: 89-91
Solvent: Xylol, butanol
Viscosity: Z4-Z6
Color (Gardner): 8 max.
Acid Number (Solids): 10 max.
Modifying Oil: TOFA
WPG Average: 8.89
Uses: This product is a premium high solid industrial finish.

C. J. OSBORN: High Solids Resins(Continued):

CJO Code:

03-A-7015:
Type: High solid, short oil alkyd
% Non-Volatile: 89-91
Solvent: Xylol
BUTYL CELLOSOLVE
Viscosity: Z6-Z8
Color (Gardner): 6 max.
Acid Number (Solids): 10 max.
Modifying Oil: TOFA
WPG Average: 8.89
Uses: This is a high solid vehicle noticeable for pigment wetting hardness which has excellent baking properties.

03-9899:
Type: High solid, short oil alkyd
% Non-Volatile: 89-91
Solvent: PM Acetate
Viscosity: Z4-Z6
Color (Gardner): 6 max.
Acid Number (Solids): 10 max.
Modifying Oil: TOFA
WPG Average: 9.00
Uses: This is a flexible, short oil alkyd designed to impart a combination of characteristics which may not be obtained with the use of conventional short oil alkyds. Coatings may be formulated at high solids that retain their flexibility even after extended bake schedules. Excellent wetting of pigments & excellent durability and gloss are noticeable characteristics of this product.

05-A-0565:
Type: Long oil alkyd
% Non-Volatile: 82-84
Solvent: Exempt MS
Viscosity: X-Y
Color (Gardner): 6 max.
Acid Number (Solids): 10 max.
Modifying Oil: Specially selected acids
WPG Average: 8.40
Uses: Long oil color retentive architectural type vehicle. premium resin for best interior and exterior enamel formulations.

C.J. OSBORN: High Solids Resins(Continued):

CJO Code:

05-2164:
Type: Long oil latex modifier
% Non-Volatile: 100
Solvent: None
Viscosity: Z3-Z5
Color (Gardner): 9 max.
Acid Number (Solids): 10 max.
Modifying Oil: Linseed
WPG Average: 8.50
Uses: Additive for exterior latex paints to impart improved adhesion and durability over chalky surfaces.

05-2165:
Type: Long oil latex modifier
% Non-Volatile: 100
Solvent: None
Viscosity: Z2-Z4
Color (Gardner): 7 max.
Acid Number (Solids): 10 max.
Modifying Oil: Soya
WPG Average: 8.50
Uses: Additive for exterior latex paints to impart improved adhesion and durability over chalky surfaces.

05-2657:
Type: Long oil alkyd house paint vehicle
% Non-Volatile: 83-85
Solvent: Exempt MS
Viscosity: T-U
Color (Gardner): 7 max.
Acid Number (Solids): 7 max.
Modifying Oil: Soya/Linseed
WPG Average: 7.90
Uses: Alkyd exterior house paint, uniform gloss, excellent flow easy brushing and high viscosities. It is also useful in house paint formulations.

05-2677:
Type: Long oil isophthalic acid alkyd
% Non-Volatile: 89-91
Solvent: Exempt MS
Viscosity: V-X
Color (Gardner): 8 max.
Acid Number (Solids): 9 max.
Modifying Oil: Linseed
WPG Average: 8.00
Uses: This resin meets Federal spec TT-P-102 and may be used in meeting TT-P-81; exhbits excellent gloss retention and flexibility.

C.J. OSBORN: High Solids Resins(Continued):

CJO Code:

05-A-5416:
Type: High solid long oil alkyd
% Non-Volatile: 80
Solvent: Exempt MS
Viscosity: Z3-Z5
Color (Gardner): 10 max.
Acid Number (Solids): 12 max.
Modifying Oil: Linseed
WPG Average: 8.22
Uses: Interior and exterior architectural enamel.

05-A-5632:
Type: Long oil latex modifier
% Non-Volatile: 100
Solvent: None
Viscosity: Z1-Z3
Color (Gardner): 7 max.
Acid Number (Solids): 10 max.
Modifying Oil: Soya
WPG Average: 8.45
Uses: Designed for latex modifications to increase adhesion and chalk resistance.

07-RF-0245:
Type: Emulsifiable resin ester
% Non-Volatile: 97-99
Solvent: None
Viscosity: 120-240 poises
Color (Gardner): 6 max.
Acid Number (Solids): 12 max.
Modifying Oil: TOFA
WPG Average: 8.20
Uses: This product is used to upgrade exterior latex paints-improves resistance to chipping and peeling, imparts better adhesion to chalky and glossy surfaces. Also, better mildew resistance is obtained.

07-A-3423:
Type: High solid wrinkle vehicle
% Non-Volatile: 79-81
Solvent: MIBK
Viscosity: Y-Z1
Color (Gardner): 12 max.
Acid Number (Solids): 36
WPG Average: 8.67
Uses: High solid low VOC applications.

C.J. OSBORN: High Solids Resins(Continued):

CJO Code:

08-A-3702:
Type: Oil modified polyurethane
% Non-Volatile: 100
Solvent: None
Viscosity: 65-100 poises
Color (Gardner): 8 max.
Acid Number (Solids): 5 max.
Modifying Oil: Linseed
WPG Average: 8.47
Uses: High solid, low viscosity--good scuff resistance, suggests applications in stains, trim and trellis.

08-A-3815:
Type: Oil modified polyurethane
% Non-Volatile: 100
Solvent: None
Viscosity: 2,000-2,750 poises
Color (Gardner): 8 max.
Acid Number (Solids): 5 max.
Modifying Oil: Linseed
WPG Average: 8.60
Uses: High solid, excellent drying good scuff resistance suggest application in stain, sealers, house paints, trim and trellis.

08-A-4639:
Type: Oil modified polyurethane
% Non-Volatile: 100
Solvent: None
Viscosity: 500-800 poises
Color (Gardner): 8 max.
Acid Number (Solids): 5 max.
Modifying Oil: Linseed
WPG Average: 8.41
Uses: High solid, excellent drying good scuff resistance suggest application in stain, sealers, house paints, trim and trellis.

09-908:
Type: Oil modified polyester resin plasticizer
% Non-Volatile: 100
Solvent: None
Viscosity: Z1-Z3
Color (Gardner): 8 max.
Acid Number (Solids): 5 max.
Modifying Oil: Castor
WPG Average: 8.23
Uses: This resin is used as a plasticizer for nitrocellulose, ethyl cellulose and polyvinyl butyral. In addition to flexibility, elasticity and adhesion, 09-908 is an excellent pigment grinding vehicle. It is recommended as a plasticizer.

C. J. OSBORN: High Solids Resins(Continued):

CJO Code:

09-3231:
Type: Polyester
% Non-Volatile: 79-81
Solvent: Toluene
Viscosity: U-W
Color (Gardner): 4 max.
Acid Number (Solids): 13 max.
Modifying Oil: None
WPG Average: 8.85
Uses; Polymeric plasticizer contributing to flexibility and adhesion.

09-A-5236:
Type: Saturated polyester
% Non-Volatile: 89-91
Solvent: PM Acetate
Viscosity: Z5-Z7
Color (Gardner): 5 max.
Acid Number (Solids): 15 max.
WPG Average: 9.90
Uses: High solid polyester designed for color, hardness, and good flexibility.

09-A-5877:
Type: Saturated polyester
% Non-Volatile: 84-86
Solvent: PM Acetate
Methyl ethyl ketone
Viscosity: Z-Z3
Color (Gardner): 3 max.
Acid Number (Solids): 10 max.
Modifying Oil: None
WPG Average: 9.20
Uses: High solids, flexible with good color retention.

09-A-6914:
Type: High solid, short oil alkyd
% Non-Volatile: 84-86
Solvent: Xylol
Viscosity: Z6-Z8
Color (Gardner): 10 max.
Acid Number (Solids): 10 max.
Modifying Oil: TOFA
WPG Average: 8.72
Uses: This short oil alkyd is an excellent high solid baking vehicle.

C. J. OSBORN: High Solids Resins(Continued):

CJO Code:

09-9152:
Type: Oil free polyester
% Non-Volatile: 84-86
Solvent: 50% MEK
50% CELL. ACET.
Viscosity: Y-W
Color (Gardner): 2 max.
Acid Number (Solids): 10 max.
WPG Average: 9.10
Uses: This low temperature, heat convertible polyester is a distinctly new type of resin for the highest quality finishes. It forms a tough, non-yellowing film with gloss and color retention. Chemical and physical properties are superior to coventional alkyds. It exhibits exceptional alkali and salt spray resistance.

9-10017:
Type: Solvent free polyester
% Non-Volatile: 100
Solvent: None
Viscosity: Y-Z1
Color (Gardner): 3 max.
Acid Number (Solids): 10 max.
Modifying Oil: None
WPG Average: 9.55
Uses: High solids, industrial baking coatings.

9-10086:
Type: High solids polyester
% Non-Volatile: 100
Solvent: None
Viscosity: Y-Z1
Color (Gardner): 3 max.
Acid Number (Solids): 15 max.
Modifying Oil: None
WPG Average: 9.60
Uses: Solvent-free plasticizing resin for industrial and lacquer type finishes.

9-10104:
Type: 100% solvent-free polyester
% Non-Volatile: 100
Solvent: None
Viscosity: U-W
Color (Gardner): 3 max.
Acid Number (Solids): 10-20
Modifying Oil: None
WPG Average: 9.85
Uses: This solvent-free polyester has excellent gloss and very good flexibility.

C.J. OSBORN: High Solids Resins(Continued):

CJO Code:

10-1516:
Type: Solvent-free, water reducible polyester
% Non-Volatile: 100
Solvent: None
Viscosity: X-Z
Color (Gardner): 3 max.
Acid Number (Solids): 55-65
Modifying Oil: None
WPG Average: 10.00
Uses: High solids, industrial baking finishes.

10-A-4067:
Type: Polyester reducible in water without the aid of amine.
% Non-Volatile: 79-81
Solvent: BUTYL CELLOSOLVE
Viscosity: Z5-Z7
Color (Gardner): 3 max.
Acid Number (Solids): 50-60
Modifying Oil: None
WPG Average: 9.53
Uses: This resin is designed in such a way that it may be introduced in a water system without the help of amines. It may be used with substantial amounts of ureas or melamine resins with which it is capable of crosslinking.

10-A-5426:
Type: Water soluble polyester
% Non-Volatile: 79-81
Solvent: Sec. butanol
BUTYL CELLO.
Viscosity: Z5-Z7
Color (Gardner): 4 max.
Acid Number (Solids): 50-65
Modifying Oil: None
WPG Average: 9.10
Uses: This water soluble polyester is designed to be water-borne with a minimal amount of co-solvent. This resin needs only moderate quantities of melamine or urea crosslinker to produce a tough, glossy film.

10-A-5816:
Type: Oil-free polyester
% Non-Volatile: 79-81
Solvent: BUTYL CELLO.
Viscosity: Z2-Z4
Color (Gardner): 3 max.
Acid Number (Solids): 41-49
Modifying Oil: None
WPG Average: 9.23
Uses: Designed for being water dispersible. Crosslinks with melamine and urea resins. 10-A-5816 produces flexible enamels.

C. J. OSBORN: High Solids Resins(Continued):

CJO Code:

10-A-5995:
Type: Water dispersible epoxy ester
% Non-Volatile: 79-81
Solvent: BUTYL CELLO., CARBITOL, butanol
Viscosity: Z6-Z8
Color (Gardner): 10 max.
Acid Number (Solids): 40-50
Modifying Oil: DCO FA
WPG Average: 8.80
Uses: Low temperature metal baking finish.

10-A-6441:
Type: Water soluble oleoresinous
% Non-Volatile: 100
Solvent: None
Viscosity: Z5-Z7
Color (Gardner): 10 max.
Acid Number (Solids): 100-120
Modifying Oil: Tung
WPG Average: 8.30
Uses: This vehicle produces baked films which are very hard, mar-proof and fairly glossy. They do not need a cross-linker. Air dry films from this resin are flat because of micro-wrinkling. Drying time is about 6 hours.

10-9889:
Type: Oil free, water reducible polyester (not pre-neutralized)
% Non-Volatile: 78-82
Solvent: N-butyl alcohol
Viscosity: Z4-Z6
Color (Gardner): 2 max.
Acid Number (Solids): 50-60
Modifying Oil: None
WPG Average: 9.00
Uses: For exterior as well as interior applications where gloss, flexibility, hardness and color retention is required over metal substrates. Baked films exhibit superior salt-spray resistance and durability.

10-10070:
Type: Oil-free water reducible polyester
% Non-Volatile: 79-81
Solvent: BUTYL CELLOSOLVE
Viscosity: Z4-Z6
Color (Gardner): 3 max.
Acid Number (Solids): 55-65
Modifying Oil: None
WPG Average: 9.35
Uses: May be used for coatings as well as metal decorating.

C.J. OSBORN: Long Oil Alkyds:

Alkyds that have a low phthalic content. These alkyds are generally soluble in aliphatic solvents. They will generally serve the industrial and interior and exterior architectural market.

CJO Code:

05-A-0565:
Type: Long oil alkyd
% Non-Volatile: 82-84
Solvent: Exempt MS
Viscosity: X-Y
Color (Gardner): 6 max.
Acid Number (Solids): 10 max.
Modifying Oil: Specially selected acids
WPG Average: 8.40
Uses: A long oil, color retentive architectural type vehicle. Premium resin for best interior and exterior enamel formulations.

05-2164:
Type: Long oil latex modifier
% Non-Volatile: 100
Solvent: None
Viscosity: Z3-Z5
Color (Gardner): 9 max.
Acid Number (Solids): 10 max.
Modifying Oil: Linseed oil
WPG Average: 8.50
Uses: Additive for exterior latex paints to impart improved adhesion and durability over chalky surfaces.

05-2165:
Type: Long oil latex modifier
% Non-Volatile: 100
Solvent: None
Viscosity: Z2-Z4
Color (Gardner): 7 max.
Acid Number (Solids): 10 max.
Modifying Oil: Soya
WPG Average: 8.50
Uses: Additive for exterior latex paints to impart improved adhesion and durability over chalky surfaces.

C.J. OSBORN: Long Oil Alkyds(Continued):

CJO Code:

05-2263:
Type: Long oil oxidizing
% Non-Volatile: 69-71
Solvent: Exempt MS
Viscosity: Y-Z1
Color (Gardner): 12 max.
Acid Number (Solids): 10 max.
Modifying Oil: Linseed
WPG Average: 8.00
Uses: Meets Federal Spec. TT-R-266D Type II Class A.

05-2271:
Type: Long oil isophthalic alkyd
% Non-Volatile: 69-71
Solvent: Exempt MS
Viscosity: Y-Z
Color (Gardner): 12 max.
Acid Number (Solids): 12 max.
Modifying Oil: Fish oil
WPG Average: 8.00
Uses: Low cost maintenance enamel, trim and trellis, barn paint structural steel and rust-inhibitive paints and primers.

05-2358:
Type: Flat Alkyd
% Non-Volatile: 39-41
Solvent: Odorless MS
Viscosity: Z2-Z4
Color (Gardner): 9 max.
Acid Number (Solids): 14 max.
Modifying Oil: TOFA
WPG Average: 7.20
Uses: For interior flat and semi-gloss paint.

05-2388 EX:
Type: Long oil
% Non-Volatile: 69-71
Solvent: Exempt MS
Viscosity: Z-Z2
Color (Gardner): 7 max.
Acid Number (Solids): 10 max.
Modifying Oil: Soya
WPG Average: 8.00
Uses: Meets Fed. spec. TT-R-266D Type I Class A. Interior architectural enamels, brushing, levelling and non-sagging are outstanding features of this product.

C.J. OSBORN: Long Oil Alkyds(Continued):

CJO Code:

05-2389-60:
Type: Long oil alkyd
% Non-Volatile: 59-61
Solvent: Exempt MS
Viscosity: U-W
Color (Gardner): 7 max.
Acid Number (Solids): 10 max.
Modifying Oil: Soya
WPG Average: 7.65
Uses: Architectural gloss or semi-gloss enamels with good color and color retention.

05-2627:
Type: Long oil alkyd
% Non-Volatile: 49-51
Solvent: Exempt MS
Viscosity: Y-Z1
Color (Gardner) 9 max.
Acid Number (Solids): 10 max.
Modifying Oil: Fish oil
Uses: Low cost vehicle with superior wetting designed for maintenance enamels & primers.

05-2657:
Type: Long oil alkyd
% Non-Volatile: 83-85
Solvent: Exempt MS
Viscosity: T-U
Color (Gardner): 7 max.
Acid Number (Solids): 7 max.
Modifying Oil: Soya/Linseed
WPG Average: 7.90
Uses: Alkyd exterior house paint, uniform gloss, excellent flow, easy brushing at high viscosities.

05-2677:
Type: Long oil isophthalic acid alkyd
% Non-Volatile: 89-91
Solvent: Exempt MS
Viscosity: V-X
Color (Gardner): 8 max.
Acid Number (Solids): 9 max.
Modifying Oil: Linseed
WPG Average: 8.00
Uses: This resin meets Federal specification TT-P-102 and may be used in meeting TT-P-81. Exhibits excellent gloss retention and flexibility.

C. J. OSBORN: Long Oil Alkyds(Continued):

CJO Code:

05-2773:
Type: Pure long oil drying alkyd
% Non-Volatile: 52-54
Solvent: Exempt MS
Viscosity: P-Q
Color (Gardner): 4-6
Acid Number (Solids): 10 max.
Modifying Oil: Soya
WPG Average: 7.60
Uses: Good general purpose enamel vehicle.

05-A-3976:
Type: Long oil alkyd
% Non-Volatile: 69-71
Solvent: Odorless MS
Viscosity: Z6-Z8
Color (Gardner): 8 max.
Acid Number (Solids): 10 max.
Modifying Oil: Soya FA
WPG Average: 7.85
Uses: Interior, exterior architectural enamels brushing leveling, non sagging properties.

05-4336:
Type: Long oil alkyd
% Non-Volatile: 59-61
Solvent: Exempt MS
Viscosity: U-W
Color (Gardner): 7 max.
Acid Number (Solids): 10 max.
Modifying Oil: Soya
WPG Average: 7.65
Uses: Meets TT-R-226 Type I Class B.

05-A-5416:
Type: High solid, long oil alkyd
% Non-Volatile: 79-81
Solvent: Exempt MS
Viscosity: Z3-Z5
Color (Gardner): 10 max.
Acid Number (Solids): 12 max.
Modifying Oil: Linseed
WPG Average: 8.22
Uses: Interior & exterior architectural enamel.

C.J. OSBORN: Long Oil Alkyds(Continued):

CJO Code:

05-A-5632:
Type: Long oil latex modifier
% Non-Volatile: 100
Solvent: None
Viscosity: Z1-Z3
Color (Gardner): 7 max.
Acid Number (Solids): 10 max.
Modifying Oil: Soya
WPG Average: 8.45
Uses: Designed for latex modification to increase adhesion and chalk resistance.

05-A-6419:
Type: Long oil alkyd
% Non-Volatile: 69-71
Solvent: Exempt MS
Viscosity: X-Z
Color (Gardner): 8 max.
Acid Number (Solids): 12 max.
Modifying Oil: Soya FA
WPG Average: 7.98
Uses: Architectural enamel, premium quality, good color retention.

05-A-6610:
Type: Long oil alkyd
% Non-Volatile: 59-61
Solvent: Exempt MS
Viscosity: Z-Z2
Color (Gardner): 7 max.
Acid Number (Solids): 8 max.
Modifying Oil: Soya
WPG Average: 7.85
Uses: High quality architectural enamels.

C.J. OSBORN: Medium Oil Alkyds:

Alkyds that have a medium phthalic content. These alkyds are generally soluble in aromatic and/or aliphatic solvents. These alkyds will generally serve the industrial and architectural interior and exterior coatings market both for bakes and air-dry.

CJO Code:

04-108:
Type: Pure soya medium alkyd
% Non-Volatile: 54-56
Solvent: Exempt MS
Viscosity: Z4-Z6
Color (Gardner): 6 max.
Acid Number (Solids): 12 max.
Modifying Oil: Soya
WPG Average: 7.70
Uses: This product has excellent levelling and flow characteristics along with fast air drying and a fast baking rate. It has excellent gloss and gloss retention as well as excellent color and color retention. This product also has good compatibility with other alkyds, oils, and varnishes. It may be used in the following applications: metal decorating, architectural white enamels, air dry and baked coatings for general use.

04-A-0427:
Type: Flat alkyd
% Non-Volatile: 34-36
Solvent: Odorless MS
Viscosity: T-W
Color (Gardner): 5 max.
Acid Number (Solids): 10 max.
Modifying Oil: Soya
WPG Average: 7.10
Uses: Interior flat and semi-gloss paints.

04-A-0911:
Type: Medium oil alkyd
% Non-Volatile: 59-61
Solvent: 88% SOLVESSO 100
12% Xylol
Viscosity: Y-Z1
Color (Gardner): 5 max.
Acid Number (Solids): 15 max.
Modifying Oil: TOFA
WPG Average: 8.33
Uses: Fast drying industrial enamel and good drying properties with good color and gloss retention.

C. J. OSBORN: Medium Oil Alkyds(Continued):

CJO Code:

04-A-1199:
Type: Medium oil alkyd
% Non-Volatile: 49-51
Solvent: Xylol
Viscosity: Z1-Z3
Color (Gardner): 14 max.
Acid Number (Solids): 25 max.
Modifying Oil: Fish oil
WPG Average: 8.22
Uses: Designed for good pigment wetting and drying products.

04-1871:
Type: Medium oil alkyd
% Non-Volatile: 39-41
Solvent: 75% Exempt VMP Naphtha
25% Exempt MS
Viscosity: V-W
Color (Gardner): 6 max.
Acid Number (Solids): 10 max.
Modifying Oil: TOFA/Linseed
WPG Average: 7.20
Uses: Fast dry implement maintenance enamels with excellent adhesion and durability.

04-2264:
Type: Medium oil alkyd
% Non-Volatile: 49-51
Solvent: Exempt MS
Viscosity: Y-Z1
Color (Gardner): 10 max.
Acid Number (Solids): 10 max.
Modifying Oil: Linseed/Soya
WPG Average: 7.65
Uses: Medium oil alkyd meeting TT-R-266-D Type III.

04-2472 G-60:
Type: Medium oil alkyd
% Non-Volatile: 59-61
Solvent: Aromatic petroleum
Viscosity: V-W
Color (Gardner): 5 max.
Acid Number (Solids): 10 max.
Modifying Oil: TOFA/Castor
WPG Average: 8.35
Uses: Recommmended for roller coat applications, metal deco. areas.

C.J. OSBORN: Medium Oil Alkyds(Continued):

CJO Code:

04-2486:
Type: Alkyd flat vehicle
% Non-Volatile: 29-31
Solvent: Exempt MS
Viscosity: Z2-Z4
Color (Gardner): 4 max.
Acid Number (Solids): 10 max.
Modifying Oil: TOFA
WPG Average: 7.00
Uses: Interior flat and semi-gloss paints.

04-2627:
Type: Pure medium marine oil alkyd
% Non-Volatile: 49-51
Solvent: Exempt MS
Viscosity: Y-Z
Color (Gardner): 9 max.
Acid Number (Solids): 10 max.
Modifying Oil: Fish Oil
WPG Average: 7.45
Uses: Excellent as a low cost trim & trellis enamel. Has excellent outdoor durability as a maintenance enamel vehicle.

04-3052:
Type: Medium length pure soya alkyd
% Non-volatile: 59-61
Solvent: Exempt MS
Viscosity: Z6-Z8
Color (Gardner): 6 max.
Acid Number (Solids): 6 max.
Modifying Oil: Soya
WPG Average: 7.90
Uses: White exterior enamels, implement enamels and metal decorating enamels.

04-A-4295:
Type: Linseed modified alkyd
% Non-Volatile: 59-61
Solvent: Exempt (Xylol, VM&P, Isobutanol)
Color (Gardner): 8 max.
Acid Number (Solids): 30 max.
Modifying Oil: Linseed
WPG Average: 8.10
Uses: Excellent resistance to abrasion, hydrocarbons and water making it a very good vehicle for fast set and quick dry primers and topcoats.

C.J. OSBORN: Medium Oil Alkyds(Continued):

CJO Code:

04-4337:
Type: Medium oil, air-dry and baking alkyd
% Non-Volatile: 49-51
Solvent: Exempt MS
Viscosity: Z1-Z3
Color (Gardner): 9 max.
Acid Number (Solids): 10 max.
Modifying Oil: Linseed
WPG Average: 7.70
Uses: Fast dry implement and maintenance enamels, has excellent adhesion and durability.

04-A-4472:
Type: Medium oil alkyd
% Non-Volatile: 69-71
Solvent: Xylol
Viscosity: Z4-Z6
Color (Gardner): 8 max.
Acid Number (Solids): 15 max.
Modifying Oil: TOFA
WPG Average: 8.64
Uses: This resin finds use in air-dry and baking applications. It has a good range of compatibility with crosslinkers & film formers.

04-A-4509:
Type: Oxidizing alkyd
% Non-Volatile: 69-71
Solvent: Xylol
Viscosity: Z4-Z6
Color (Gardner): 8 max.
Acid Number (Solids): 10 max.
Modifying Oil: Soya
WPG Average: 8.50
Uses: Fast dry, medium oil alkyd compatible with vinyl resins like UC VROH.

04-A-5642:
Type: Medium oil alkyd
% Non-Volatile: 49-51
Solvent: Exempt VMP Naphtha
Viscosity: Z4-Z6
Color (Gardner): 8 max.
Acid Number (Solids): 10 max.
Modifying Oil: Soya
WPG Average: 7.60
Uses: Fast dry implement enamel.

C.J. OSBORN: Medium Oil Alkyds(Continued):

CJO Code:

04-A-5913:
Type: Medium oil alkyd
% Non-Volatile: 49-51
Solvent: Exempt MS
Viscosity: Z4-Z6
Color (Gardner): 6 max.
Acid Number (Solids): 10 max.
Modifying Oil: TOFA/Linseed
WPG Average: 7.64
Uses: Fast dry implement and maintenance enamels.

04-A-6143:
Type: Medium oil alkyd
% Non-Volatile: 49-51
Solvent: Exempt MS
Viscosity: X-Z
Color (Gardner): 7 max.
Acid Number (Solids): 10 max.
Modifying Oil: Soya
WPG Average: 7.63
Uses: Fast dry implement enamel.

04-A-6230:
Type: Medium oil alkyd
% Non-Volatile: 49-51
Solvent: Exempt MS, Xylol
Viscosity: Z1-Z3
Color (Gardner): 6 max.
Acid Number (Solids): 12 max.
Modifying Oil: Soya/Tung Oil
WPG Average: 7.77
Uses: Industrial and farm implement enamels.

04-A-6231:
Type: Medium oil alkyd
% Non-Volatile: 49-51
Solvent: Exempt MS
Viscosity: Z2-Z4
Color (Gardner): 7 max.
Acid Number (Solids): 12 max.
Modifying Oil: Soya
WPG Average: 7.61
Uses: Fast dry implement enamel.

C.J. OSBORN: Medium Oil Alkyds(Continued):

CJO Code:

04-A-6451A:
Type: Pure medium oil alkyd
% Non-Volatile: 69-71
Solvent: Xylol
Viscosity: Z4-Z6
Color (Gardner): 6 max.
Acid Number (Solids): 10 max.
Modifying Oil: Soya
WPG Average: 8.50
Uses: Truck enamels, automotive finishes, rapid dry industrial finishes including applications as pumps, machinery and other industrial equipment. When used alone with amine resins, this product converts more rapidly when baked than other alkyds of similar oil length. This resin merits evaluation for its adhesion, flexibility & compatibility with a wide range of other film formers.

04-7228:
Type: Medium oil alkyd
% Non-Volatile: 49-51
Solvent: Exempt MS
Viscosity: V-W
Color (Gardner): 7 max.
Acid Number (Solids): 10 max.
Modifying Oil: Soya
WPG Average: 7.54
Uses: Fast air-dry enamels excellent color and gloss retention.

04-7378:
Type: Pure medium oil alkyd
% Non-Volatile: 69-71
Solvent: Xylol
Viscosity: Z5-Z6
Color (Gardner): 6 max.
Acid Number (Solids): 15 max.
Modifying Oil: Safflower
WPG Average: 8.50
Uses: Truck enamels, automotive finishes, rapid drying industrial finishes including such applications as pumps, machinery and other industrial equipment. When used alone with amine resins, MIRASOL 04-7378 converts more rapidly when baked than other alkyds of similar oil length. This resin merits evaluation for its adhesion, flexibility and unusual compatibilities with a wide range of other film-formers.

C.J. OSBORN: Medium Oil Alkyds (Continued):

CJO Code:

04-8242:
Type: Pure medium marine oil alkyd
% Non-Volatile: 49-51
Solvent: Exempt MS
Viscosity: Z-Z2
Color (Gardner): 9 max.
Acid Number (Solids): 10 max.
Modifying Oil: Fish Oil
WPG Average: 7.40
Uses: Maintenance enamel vehicle which exhibits excellent outdoor durability; also excellent for use as a low cost trim and trellis enamel vehicle primer.

04-9074:
Type: Pure soya medium oil alkyd
% Non-Volatile: 49-51
Solvent: Exempt MS
Viscosity: U-W
Color (Gardner): 6 max.
Acid Number (Solids): 10 max.
Modifying Oil: Soya
WPG Average: 7.60
Uses: Fast air dry enamels combine good brushing, excellent color and gloss retention. Metal decorating, good application properties, flow and color, and gloss retention.

C.J. OSBORN: Non-Drying Alkyds:

Non-drying alkyds are used as non-migrating plasticizers in lacquers and as a building block in the crosslinking of urea-formaldehyde, melamines and urethanes. They find application in automotive, appliance, furniture and other industrial finishes.

CJO Code:

01-121:
Type: Pure non-drying oil modified alkyd
% Non-Volatile: 49-51
Solvent: Xylol
Viscosity: Z-Z2
Modifying Oil: Cottonseed
WPG Average: 8.40
Uses: A resinous plasticizer for nitrocellulose lacquers both clear and pigmented.

01-124:
Type: Pure non-drying alkyd
% Non-volatile: 79-81
Solvent: Xylol
Viscosity: M-P
Modifying Oil: Coconut
WPG Average: 8.47
Uses: Resin plasticizer for nitrocellulose. Being longer in oil has a more pronounced plasticizing effect. For use in high solids lacquers.

01-126:
Type: Pure non-drying alkyd
% Non-Volatile: 59-61
Solvent: Xylol
Viscosity: Z-Z2
Modifying Oil: Castor
WPG Average: 8.30
Uses: High quality furniture and other wood lacquers. Non-lifting with good adhesion. High quality baking enamels with good toughness, adhesion and gloss.

01-127-65X:
Type: Pure non-drying alkyd
% Non-Volatile: 64-66
Solvent: Xylol
Viscosity: X-Z
Modifying Oil: Castor (partially dehydrated)
WPG Average: 8.55
Uses: Resinous plasticizer for nitrocellulose lacquers, amine baking finishes and vinyl systems. It can be used as the only plasticizer or in combination with chemical or other resinous plasticizers with which it may be compatible.

C.J. OSBORN: Non-Drying Alkyds(Continued):

CJO Code:

01-129A:
Type: Pure non-drying (lauric acid modified alkyd)
% Non-Volatile: 49-51
Solvent: Xylol: 76%
MEK: 6%
Isopropanol:18%
Viscosity: D-G
Modifying Oil: Lauric
WPG Average: 8.33
Uses: Nitrocellulose lacquer where color and adhesion are prime requisites.

01-131:
Type: Non-drying plasticizing type alkyd
% Non-Volatile: 79-81
Solvent: Toluol
Viscosity: U-W
Modifying Oil: None
WPG Average: 8.85
Uses: Valuable in formulating high solids lacquers due to low viscosity. Nitrocellulose lacquers for both wood & metal may be formulated with this alkyd.

01-A-1980:
Type: Non-drying alkyd
% Non-Volatile: 49-51
Solvent: Aromatic blend
Viscosity: E-H
Modifying Oil: Non-oxidizing
WPG Average: 7.75
Uses: Property formulated with MM and/or UF resins, this product makes an excellent wet ink varnish for metal-deco. and roller coat applications. Has excellent color and gloss retention even after processing, adhesion, flexibility and toughness.

01-A-2221:
Type: Non-drying alkyd
% Non-Volatile: 59-61
Solvent: Aromatic hydrocarbon
Viscosity: Z4-Z6
Modifying Oil: Coconut
WPG Average: 8.60
Uses: Baking enamels and quality wood lacquers.

C.J. OSBORN: Non-Drying Alkyds(Continued):

CJO Code:

01-A-2373:
Type: Short oil, non-oxidizing color retentive alkyd
% Non-Volatile: 59-61
Solvent: Xylol
Viscosity: Z2-Z4
Modifying Oil: Pelargonic
WPG Average: 8.71
Uses: For quality baking finishes where maximum color and color retention on overbake are required. Also widely used in lacquers.

01-A-2414:
Type: High solid, non-oxidizing alkyd
% Non-Volatile: 59-61
Solvent: Xylol
Viscosity: Z4-Z6
Modifying Oil: Non-oxidizing
WPG Average: 8.95
Uses: This alkyd finds application in lacquer and baking enamel where high solids, color, gloss and hardness are important goals.

01-2525:
Type: Pure non-drying alkyd
% Non-Volatile: 59-61
Solvent: Xylol
Viscosity: Z1-Z3
Modifying Oil: Coconut
WPG Average: 8.58
Uses: Highest quality furniture and wood lacquers. High quality enamels with excellent toughness and gloss when used in alkyd/amino systems.

01-2525-6T:
Type: Pure non-drying alkyd
% Non-Volatile: 59-61
Solvent: Toluol
Viscosity: V-X
Modifying Oil: Coconut
WPG Average: 8.60
Uses: Highest quality furniture and wood lacquers. High quality baking enamels with excellent toughness and gloss when used in alkyd/amino systems.

C.J. OSBORN: Non-Drying Alkyds (Continued):

CJO Code:

01-2525 EX:
Type: Pure non-drying alkyd
% Non-Volatile: 59-61
Solvent: Exempt VMP Naphtha Toluol/Butanol
Viscosity: Z3-Z5
Modifying Oil: Coconut
WPG Average: 8.50
Uses: Highest quality furniture wood lacquers, high quality baking enamels w/excellent toughness and gloss when used in alkyd/amino systems.

01-A-5306:
Type: Non-oxidizing alkyd
% Non-Volatile: 64-66
Solvent: Xylol/Toluol
N-Propanol
Viscosity: X-Z1
Modifying Oil: None
WPG Average: 8.60
Uses: Low cost plasticizing resin.

01-A-5704:
Type: Non-drying
% Non-Volatile: 69-71
Solvent: BUTYL CELLO.
Viscosity: Z3-Z5
Modifying Oil: Coconut
WPG Average: 9.00
Uses: High quality furniture and wood lacquers. High quality baking enamel with excellent toughness and gloss when used in alkyd/amino systems.

01-7809:
Type: Non-drying alkyd
% Non-Volatile: 49-51
Solvent: High boiling aromatic solvent
Viscosity: 45-55" #4 Ford cup
Modifying Oil: Coconut
WPG Average: 7.70
Uses: This resin is recommended for roller coat application in metal decorating areas. It is a very good wet ink (vehicle). It will take pasteurization and has excellent adhesion and resistance to dry heat.

C.J. OSBORN: Non-Drying Alkyds(Continued):

CJO Code:

01-9589:
Type: Non-drying alkyd
% Non-Volatile: 59-61
Solvent: Xylol
Viscosity: Z1-Z3
Modifying Oil: Coconut
WPG Average: 8.60
Uses: High quality furniture wood lacquers. High quality baking enamels with excellent toughness and gloss when used in alkyd amino systems.

C. J. OSBORN: Polyesters:

Polyesters have a primary use in the industrial baking market. These products are known for their outstanding adhesion and flexibility to metal surfaces. Their use may be interior or exterior. These products are generally oil-free and are available in high solids versions. Polyesters may also be used to plasticize nitrocellulose lacquers, both air-dry and forced. They are available in both solvent and water reducible varieties.

<u>CJO Code:</u>

09-A-0575:

Type: Saturated polyester
% Non-Volatile: 59-61
Solvent: High flash naphtha
Viscosity: Z-Z2
Color (Gardner): 6 max.
Acid Number (Solids): 10 max.
WPG Average: 8.75
Uses: MIRASOL 09-A-0575 is a versatile metal deco. polyester.

09-902:

Type: Oil modified polyester resin plasticizer
% Non-Volatile: 59-61
Solvent: Toluol
Viscosity: T-V
Color (Gardner): 8 max.
Acid Number (Solids): 35 max.
Modifying Oil: Castor
WPG Average: 8.02
Uses: This resin is used solely as a plasticizer for other film formers such as nitrocellulose, ethyl cellulose and polyvinyl butyral. Imparts such properties as permanent flexibility, toughness, adhesion--particularly to rubber.

09-908:

Type: Oil modified polyester resin plasticizer
% Non-Volatile: 100
Solvent: None
Viscosity: Z1-Z3
Color (Gardner): 8 max.
Acid Number (Solids): 5 max.
Modifying Oil: Castor
WPG Average: 8.23
Uses: This resin is used as a plasticizer for nitrocellulose, ethyl cellulose and polyvinyl butyral. In addition to flexibility and elasticity and adhesion, 09-908 is an excellent pigment grinding vehicle. It is recommended as a plasticizer for very flexible lacquers such as fabric finishes.

C. J. OSBORN: Polyesters (Continued):

CJO Code:

09-3231:
Type: Saturated polyester
% Non-Volatile: 79-81
Solvent: Toluene
Viscosity: U-W
Color (Gardner): 4 max.
Acid Number (Solids): 13 max.
WPG Average: 8.85
Uses: A polymeric plasticizer contributing to flexibility and adhesion.

09-A-3768:
Type: Oil-free polyester
% Non-Volatile: 69-71
Solvent: High flash naphtha
Viscosity: Z-Z2
Color (Gardner): 4 max.
Acid Number (Solids): 8 max.
WPG Average: 8.75
Uses: Polyester for use in metal finishing and decorating.

09-A-4029:
Type: Special polyester type non-drying alkyd
% Non-Volatile: 49-51
Solvent: Xylol
Viscosity: U-W
Color: Gardner
Acid Number (Solids): 14 max.
WPG Average: 8.35
Uses: Very high grade appliance finish with a wide range of baking schedules (30' @ 250 to very short times @ 500-550F).

09-A-4060-75:
Type: Oil-free polyester
% Non-Volatile: 74-76
Solvent: Toluol
Viscosity: X-Z
Color (Gardner): 3 max.
Acid Number (Solids): 15 max.
WPG Average: 9.01
Uses: Properly formulated, this polyester has films that are quite flexible and tough. It has a wide range of compatibility making it a suitable resin for lacquer formulations. It also lends itself well for baking applications.

C.J. OSBORN: Polyesters(Continued):

CJO Code:

09-A-4359:
Type: Solvent-free polyester
% Non-volatile: 100
Solvent: None
Viscosity: Z4-Z6
Color (Gardner): 3 max.
Acid Number (Solids): 22-30
WPG Average: 10.48
Uses: This is an unusual polyester because of its tolerance for water and its rapid viscosity reduction. These properties alone make this product an important resin when VOC compliance is one of the objectives. Films crosslinked with melamine or urea resins pass as much as 200 rubs with MEK are very glossy and display good mechanical properties. Industrial applications include metal-deco, industrial enamels and clear finishes.

09-A-4660:
Type: Saturated polyester
% Non-Volatile: 69-71
Solvent: High flash aromatic naphtha
Diacetone alcohol
Viscosity: Z1-Z4
Color (Gardner): 3 max.
Acid Number (Solids): 10 max.
Modifying Oil: None
WPG Average: 8.95
Uses: General maintenance finishes.

09-A-4675:
Type: Oil-free polyester
% Non-Volatile: 74-76
Solvent: Toluol
Viscosity: X-Z
Color (Gardner): 3 max.
Acid Number (Solids): 15 max.
Modifying Oil: None
WPG Average: 9.01
Uses: This oil-free polyester has films that are tough and flexible. It has a wide range of compatibility making it a suitable resin for lacquer formulations. It also lends itself well for baking applications.

C. J. OSBORN: Polyesters(Continued):

CJO Code:

09-A-5204:
Type: High solid polyester
% Non-Volatile: 74-76
Solvent: Toluol
Viscosity: T-W
Color (Gardner): 4 max.
Acid Number (Solids): 20 max.
WPG Average: 9.10
Uses: Flexibility, toughness, wide range of compatibility and color make this polyester an unusual resin for lacquer formulations. A low viscosity and relatively high solids suggests its use in baking finishes.

09-A-5236:
Type: Saturated polyester
% Non-Volatile: 89-91
Solvent: PM Acetate
Viscosity: Z5-Z7
Color (Gardner): 5 max.
Acid Number (Solids): 15 max.
WPG Average: 9.00
Uses: High solid polyester designed for color, hardness and good flexibility.

09-A-5430:
Type: Saturated polyester
% Non-Volatile: 64-66
Solvent: High flash naphtha
Viscosity: Z-Z2
Color (Gardner): 3 max.
Acid Number (Solids): 30 max.
Modifying Oil: None
WPG Average: 8.00
Uses: This polyester is designed for baking requiring superior flexibility and color. It can achieve a good degree of curing with less than 10 percent melamine.

09-A-5848:
Type: Reactive polymer
% Non-Volatile: 69-71
Solvent: SOLVESSO 150
Butanol
Viscosity: Z-Z2
Color (Gardner): 3 max.
Acid Number (Solids): 16 max.
Modifying Oil: None
WPG Average: 8.00
Uses: This reactive polymer has been designed for an application where flexibility and reasonable hardness are important requirements.

C.J. OSBORN: Polyesters(Continued):

<u>CJO</u> <u>Code:</u>

09-A-5877:
Type: Saturated polyester
% Non-Volatile: 84-86
Solvent: PM acetate
Methyl ethyl ketone
Viscosity: Z-Z3
Color (Gardner): 3 max.
Acid Number (Solids): 10 max.
Modifying Oil: None
WPG Average: 9.20
Uses: High solids, flexible good color retention.

09-A-6146:
Type: Saturated polyester
% Non-Volatile: 59-61
Solvent: Xylol
Viscosity: X-Z
Color (Gardner): 4 max.
Acid Number (Solids): 10 max.
Modifying Oil: None
WPG Average: 8.67
Uses: A tough oil free alkyd for baking applications.

09-A-6435:
Type: High solid polyester
% Non-Volatile: 74-76
Solvent: Toluol
Viscosity: T-W
Color (Gardner): 4 max.
Acid Number (Solids): 14 max.
Modifying Oil: None
WPG Average: 9.05
Uses: Flexibility, toughness and a wide range of compatibility in color make this polyester an unusual resin for lacquer formulations. Low viscosity and relatively high solids suggests its use in baking finishes.

09-A-6512:
Type: High solids polyester
% Non-Volatile: 74-76
Solvent: Toluol
Viscosity: T-W
Color (Gardner): 4 max.
Acid Number (Solids): 14 max.
WPG Average: 9.09
Uses: Same applications as 09-A-6435.

C. J. OSBORN: Polyesters(Continued):

CJO Code:

09-A-6914:
Type: High solid, short oil alkyd
% Non-Volatile: 84-86
Solvent: Xylol
Viscosity: Z6-Z8
Color (Gardner): 10 max.
Acid Number (Solids): 10 max.
Modifying Oil: TOFA
WPG Average: 8.72
Uses: This short oil alkyd is an excellent high solid baking vehicle.

09-A-7011:
Type: Saturated polyester
% Non-Volatile: 74-76
Solvent: Toluene
Viscosity: Z5-Z7
Color (Gardner): 2 max.
Acid Number (Solids): 2 max.
Modifying Oil: None
WPG Average: 9.10
Uses: This saturated polyester has been designed for polyester/polyurethane air-dry and baking applications. It features excellent gloss, high impact resistance, good mechanical and chemical resistance.

09-9152:
Type: Oil free polyester
% Non-Volatile: 80-82
Solvent: 50% MEK
50% CELLO. ACETATE
Viscosity: Y-W
Color (Gardner): 2 max.
Acid Number (Solids): 10 max.
Modifying Oil: None
WPG Average: 9.10
Uses: This low-temp. heat convertible polyester is a distinctly new type of resin for the highest quality finishes. It forms a tough non-yellowing film with gloss and color retention. Chemical and physical properties are superior to coventional alkyds. Exhibits exceptional alkali and salt spray resistance.

C.J. OSBORN: Polyesters(Continued):

CJO Code:

09-10017:
Type: Solvent free polyester
% Non-Volatile: 100
Solvent: None
Viscosity: Y-Z1
Color (Gardner): 3 max.
Acid Number (Solids): 10 max.
Modifying Oil: None
WPG Average: 9.55
Uses: High solids, industrial baking coatings.

09-10086:
Type: High solids polyester
% Non-Volatile: 100
Solvent: None
Viscosity: Y-Z1
Color (Gardner): 3 max.
Acid Number (Solids): 15 max.
Modifying Oil: None
WPG Average: 9.60
Uses: High solids baking and lacquers.

09-10104:
Type: 100% solvent free polyester
% Non-Volatile: 100
Solvent: None
Viscosity: U-W
Color (Gardner): 3 max.
Acid Number (Solids): 10-20
Modifying Oil: None
WPG Average: 9.85
Uses: This solvent-free polyester has excellent gloss and very good flexibility.

C. J. OSBORN: Polyurethanes:

ASTM Type I polyurethanes are oil modified, one-package ambient cure polyurethane resins with no free NCO. Their use may be industrial enamels, furniture, wood floors, and paneling as well as requirements for fast dry and highly durable finishes.

CJO Code:

08-A-0013:
Type: Uralkyd
% Non-Volatile: 49-51
Solvent: Xylol
Viscosity: Z2-Z4
Color (Gardner): 8 max.
Acid Number (Solids): 20 max.
Modifying Oil: Linseed
WPG Average: 8.25
Uses: This product's quick drying and early solvent water resistance suggest its use in wood finishes, implement enamels, primers and sealers.

08-A-139:
Type: Oil modified polyurethane
% Non-Volatile: 59-61
Solvent: Exempt MS
Viscosity: Y-Z2
Color (Gardner): 8 max.
Acid Number (Solids): 1 max.
Modifying Oil: Linseed
WPG Average: 7.60
Uses: This is an excellent vehicle for all wood finishes.

08-A-0158:
Type: Oil modified polyurethane
% Non-Volatile: 49-51
Solvent: Exempt MS
Viscosity: X-Y
Color (Gardner): 5 max.
Acid Number (Solids): 1 max.
Modifying Oil: Linseed
WPG Average: 7.49
Uses: This oil modified polyurethane is designed for fast dry and excellent abrasion resistance.

C.J. OSBORN: Polyurethanes(Continued):

<u>CJO</u> <u>Code:</u>

08-A-0617:
Type: ASTM Type I polyurethane
% Non-Volatile: 49-51
Solvent: Exempt MS
Viscosity: Z1-Z3
Color (Gardner): 6 max.
Acid Number (Solids): 2 max.
Modifying Oil: Linseed
WPG Average: 7.45
Uses: Finishes for flooring, paneling and furniture. May be used on certain plastics. A good general purpose varnish with positive drying.

08-A-0895:
Type: ASTM Type I polyurethane
% Non-Volatile: 59-61
Solvent: Exempt MS
Viscosity: Z1-Z3
Color (Gardner): 5 max.
Acid Number (Solids): 2 max.
Modifying Oil: Soya
WPG Average: 7.60
Uses: Vehicle for interior finishes, characterized by optimum color retention with excellent abrasion resistance and good chemical resistance.

08-A-1082:
Type: ASTM Type I polyurethane
% Non-Volatile: 49-51
Solvent: Xylol
Viscosity: Z1-Z3
Color (Gardner): 6 max.
Acid Number (Solids): 2 max.
Modifying Oil: Linseed
WPG Average: 7.80
Uses: Production line paneling, furniture finishes, factory finish flooring and vacuum metallizing.

08-A-1092:
Type: ASTM Type I polyurethane
% Non-Volatile: 49-51
Solvent: Exempt VM&P Naphtha
Viscosity: Z1-Z3
Color (Gardner): 6 max.
Acid Number (Solids): 2 max.
Modifying Oil: Linseed
WPG Average: 7.35
Uses: Aerosol and other applications where fast solvents are required.

C. J. OSBORN: Polyurethanes(Continued):

CJO Code:

08-A-1287:
Type: Oil modified polyurethane
% Non-Volatile: 39-41
Solvent: Exempt MS
Viscosity: E-F
Color (Gardner): 7 max.
Acid Number (Solids): 2 max.
Modifying Oil: Soya/Linseed
WPG Average: 7.30
Uses: Floor finishes and general purpose varnishes. This product may also be reduced for use as a wood sealer. Contains driers.

08-A-2631:
Type: ASTM Type I polyurethane
% Non-Volatile: 49-51
Solvent: Exempt MS
Viscosity: Y-Z2
Color (Gardner): 6 max.
Acid Number (Solids): 2 max.
Modifying Oil: Soya
WPG Average: 7.49
Uses: This polyurethane has fast drying, excellent abrasion and mar resistance. It is a natural for gym floors and other heavy duty applications. Its unusual chemical resistance suggests application in concrete and terrazzo sealers.

08-A-2174:
Type: ASTM Type I polyurethane
% Non-Volatile: 49-51
Solvent: Exempt MS
Viscosity: Z3-Z6
Color (Gardner): 6 max.
Acid Number (Solids): 5 max.
Modifying Oil: Soya
WPG Average: 7.44
Uses: This versatile polyurethane is suitable for floor finishes, panelling, furniture, either by itself or extended with alkyds.

08-A-3702:
Type: Oil modified polyurethane
% Non-Volatile: 100
Viscosity: 65-100 poises
Color (Gardner): 8 max.
Acid Number (Solids): 5 max.
Modifying Oil: Linseed
WPG Average: 8.47
Uses: High solids, low viscosity & good scuff resistance suggest applications in stains, trim & trellis.

C. J. OSBORN: Polyurethanes(Continued):

CJO Code:

08-A-3815:
Type: Oil modified polyurethane
% Non-Volatile: 100
Viscosity: 2000-2750 poises
Color (Gardner): 8 max.
Acid Number (Solids): 5 max.
Modifying Oil: Linseed
WPG Average: 8.60
Uses: High solids, excellent drying & good scuff resistance suggest application in stains, sealers, house paint trim & trellis.

08-A-4309:
Type: Oil modified polyurethane
% Non-Volatile: 59-61
Solvent: Exempt MS
Isobutanol
Viscosity: Z1-Z3
Color (Gardner): 8 max.
Acid Number (Solids): 5 max.
Modifying Oil: Soya
WPG Average: 7.50
Uses: This product finds application in floor finishes, paneling, etc. where speed of dry, color abrasion, mar resistance are important.

08-A-4487:
Type: Safflower modified polyurethane
% Non-Volatile: 49-51
Solvent: Exempt MS
Viscosity: Z-Z2
Color (Gardner): 5 max.
Acid Number (Solids): 2 max.
Modifying Oil: Safflower
WPG Average: 7.54
Uses: This product finds application in floor finishes, panelling, etc. where speed of dry, color abrasion, mar resistance are important.

08-A-4514:
Type: ASTM Type I polyurethane
% Non-Volatile: 49-51
Solvent: Exempt MS
Viscosity: U-W
Color (Gardner): 8 max.
Acid Number (Solids): 5 max.
WPG Average: 7.50
Uses: Versatile polyurethane suitable for floor finishes, paneling, furniture, etc. Clear and pigmented finishes benefit from this polyurethane. It has relatively low viscosity and positive drying.

C. J. OSBORN: Polyurethanes(Continued):

CJO Code:

08-A-4639:
Type: Oil modified polyurethane
% Non-Volatile: 100
Viscosity: 500-800 poises
Color (Gardner): 8 max.
Acid Number (Solids): 5 max.
Modifying Oil: Linseed
WPG Average: 8.41
Uses: High solid, scuff resistance & good drying suggest applications in stain, floor finishes, house paint, trim & trellis.

08-A-5331:
Type: Polyurethane
% Non-Volatile: 54-56
Solvent: Exempt MS
Viscosity: Z4-Z6
Color (Gardner): 6 max.
Acid Number (Solids): 5 max.
Modifying Oil: Soya
WPG Average: 7.65
Uses: Either by itself or extended with alkyds, finds applications for floor finishes, paneling & furniture.

08-A-6248:
Type: Oil modified polyurethane
% Non-Volatile: 49-51
Solvent: Exempt MS
Viscosity: P-R
Color (Gardner): 6 max.
Acid Number (Solids): 2 max.
Modifying Oil: Soya/Tung
WPG Average: 7.45
Uses: Fast drying, mar resistant, finds application in furniture and floor finishes.

08-A-8200M:
Type: Oil modified polyurethane
% Non-Volatile: 49-51
Solvent: Exempt MS
Viscosity: U-X
Color (Gardner): 4 max.
Acid Number (Solids): 2 max.
Modifying Oil: Soya
WPG Average: 7.35
Uses: Clear and satin wood finish aerosol coatings with excellent resistance to weathering and abrasion.

C. J. OSBORN: Polyurethanes(Continued):

CJO Code:

08-9248:
Type: ASTM Type I polyurethane
% Non-Volatile: 59-61
Solvent: Exempt MS
Viscosity: Y-Z2
Color (Gardner): 6 max.
Acid Number (Solids): 2 max.
Modifying Oil: Linseed
WPG Average: 7.50
Uses: Suitable for preparing quick dry varnishes with superior hardness and mar resistance.

08-9963:
Type: ASTM Type I polyurethane
% Non-Volatile: 59-61
Solvent: Exempt MS
Viscosity: X-Z
Color (Gardner): 6 max.
Acid Number (Solids): 2 max.
Modifying Oil: Soya
WPG Average: 7.60
Uses: Interior finishes where color, gloss retention and hardness are important.

C. J. OSBORN: Resin Modified Alkyds:

Alkyds that vary in oil length and phthalic content and are modified with rosin, hard resins, or phenolics. These modifications improve drying speed, hardness, adhesion, and chemical resistance.

CJO Code:

02-101:
Type: Rosin Modified Phthalic-Free alkyd
% Non-Volatile: 49-51
Solvent: VM & P 66/3
MS 66/3
Viscosity: Y-Z1
Color (Gardner): 11 max.
Acid Number (Solids): 45 max.
Modifying Oil: TOFA
WPG Average: 7.55
Uses: Short oil fast drying solution.

02-220:
Type: Short oil phenolic modified alkyd
% Non-Volatile: 49-51
Solvent: Xylol 86%
AMSCO C 14%
Viscosity: V-X
Color (Gardner): 10 max.
Acid Number (Solids): 50 max.
Modifying Oil: Linseed
WPG Average: 8.20
Uses: Highest quality enamel finish for machinery, agricultural equipment, automotive, truck bodies.

02-A-1122:
Type: Resin modified alkyd
% Non-Volatile: 40-45
Solvent: MS VM & P,
Dipentene
Viscosity: A-3
Color (Gardner): 12 max.
Acid Number (Solids): N/A
Modifying Oil: Marine
WPG Average: 7.25
Uses: A versatile exterior sealer marine use. Contains driers ready for use.

C.J. OSBORN: Resin Modified Alkyds(Continued):

CJO Code:

02-1589:
Type: Air-dry, modified short oil alkyd
% Non-Volatile: 49-51
Solvent: Exempt VMP Naphtha
Viscosity: Y-Z
Color (Gardner): 11 max.
Acid Number (Solids): 24 max.
Modifying Oil: Tall oil
WPG Average: 7.50
Uses: Excellent for fast dry low cost enamels. Good compatibility with medium and long oil resins. Can be blended to improve drying times of long oil resins.

02-1826:
Type: Phenolic resin modified alkyd
% Non-Volatile: 49-51
Solvent: Xylol, Toluol
Viscosity: X-Y
Color (Gardner): 15 max.
Acid Number (Solids): 15-20
Modifying Oil: Linseed
WPG Average: 8.40
Uses: Very short linseed alkyd with high phenolic modification.

02-1902:
Type: Modified, medium oil alkyd
% Non-Volatile: 59-61
Solvent: Exempt VM & P Naphtha
Viscosity: Z2-Z4
Color (Gardner): 12 max.
Acid Number (Solids): 30 max.
Modifying Oil: Linseed
WPG Average: 8.00
Uses: Excellent fast dry non-lifting primer vehicle, outstanding salt and spray resistance with excellent pigment compatibility and stability.

02-2318-35:
Type: Alkyd resin solution
% Non-Volatile: 34-36
Solvent: Xylol, VM & P Naphtha
Viscosity: G-I
Color (Gardner): 4-7
Acid Number (Solids): N/A
Modifying Oil: Linseed
WPG Average: 7.45
Uses: General use spar varnish.
Contains driers ready for use.

C. J. OSBORN: Resin Modified Alkyds (Continued):

CJO Code:

02-2369:
Type: Long oil, modified alkyd
% Non-Volatile: 69-71
Solvent: Exempt MS
Viscosity: Z-Z1
Color (Gardner): 14 max.
Acid Number (Solids): 10 max.
Modifying Oil: Selected blend
WPG Average: 8.00
Uses: Excellent porch and deck enamel vehicle, trim & trellis. General purpose maintenance enamels, etc. Good penetration over rusty surfaces.

02-2426:
Type: Modified medium oil alkyd
% Non-Volatile: 49-51
Solvent: Exempt VMP Naphtha
Viscosity: T-V
Color (Gardner): 12 max.
Acid Number (Solids): 40 max.
Modifying Oil: Soya-TOFA
WPG Average: 7.65
Uses: Tough, durable non-lifting primers for adhesion and fast dry.

02-2873:
Type: Modified phenolic resin tung oil varnish
% Non-Volatile: 39-41
Solvent: 93% exempt mineral spirits
7% dipentene
Viscosity: B-D
Color (Gardner): 12 max.
Acid Number (Solids): N/A
Modifying Oil: Tung
WPG Average: 7.20
Uses: Excellent as floor and trim varnish for wood and concrete porches and furniture. Good exterior durability, chemical and salt spray resistance.

02-A-3869:
Type: Phenolic modified alkyd
% Non-Volatile: 49-51
Solvent: Xylol
Viscosity: U-W
Color (Gardner): 10
Acid Number (Solids): 28-32
Modifying Oil: Soya/Linseed/Tung
WPG Average: 8.32
Uses: High quality non-lifting primer vehicle.

C. J. OSBORN: Resin Modified Alkyds(Continued):

CJO Code:

02-A-4196:
Type: Resin modified alkyd
% Non-volatile: 49-51
Solvent: Exempt MS
Viscosity: Z-Z2
Color (Gardner): 11 max.
Acid Number (Solids): 12 max.
Modifying Oil: TOFA
WPG Average: 7.68
Uses: Short oil resin modified alkyd offering a variety of uses.

02-4603:
Type: Universal tinting vehicle
% Non-Volatile: 69-71
Solvent: Xylol
Viscosity: D-G
Color (Gardner): 12 max.
Acid Number (Solids): 32 max.
Modifying Oil: TOFA
WPG Average: 8.26
Uses: This vehicle has been designed for a wide latitude of compatibility in tinting proportions. Low viscosity & super wetting make possible the formulation of high PVC tinting pastes. Useful in trade sale & industrial finishes.

02-A-4773:
Type: Modified medium oil alkyd
% Non-Volatile: 79-81
Solvent: MEK/PM Glycol ether
Viscosity: Z3-Z5
Color (Gardner): 12 max.
Acid Number (Solids): 40 max.
Modifying Oil: Soya/TOFA
WPG Average: 8.67
Uses: Tough, durable non-lifting primers with excellent adhesion and fast dry.

02-A-4963:
Type: Resin modified alkyd
% Non-Volatile: 49-51
Solvent: VM & P 66/3
Isobutyl alcohol
Viscosity: N-P
Color (Gardner): 10 max.
Acid Number (Solids): 28-32
Modifying Oil: Tung oil/linseed
WPG Average: 7.76
Uses: Meets the requirements for MIL 52995 specification.

C. J. OSBORN: Resin Modified Alkyds(Continued):

CJO Code:

02-A-5999:

Type: Rosin phenolic modified alkyd
% Non-Volatile: 59-61
Solvent: Xylol
Viscosity: Z-Z2
Color (Gardner): 11 max.
Acid Number (Solids): 18 max.
Modifying Oil: Linseed
WPG Average: 8.48
Uses: Designed for primers and topcoats where adhesion, gloss and speed of drying are of prime importance.

02-A-6411:

Type: Rosin modified, isophthalic acid
% Non-Volatile: 44-46
Solvent: Exempt MS
Viscosity: J-O
Color (Gardner): 10 max.
Acid Number (Solids): 14 max.
Modifying Oil: Linseed/Fish
WPG Average: 7.42
Uses: Dries exceptionally well making possible the formulation of clear pigmented systems. Baking finishes prepared with this resin modified alkyd are significant for their gloss, hardness and wetting properties.

02-A-8187:

Type: Short oil phenolic modified alkyd resin solution in exempt solvent
% Non-Volatile: 49-51
Solvent: 61% Exempt VMP Naphtha
17% Toluol
22% Isopropanol
Viscosity: V-X
Color (Gardner): 9 max.
Acid Number (Solids): 50 max.
Modifying Oil: Linseed
WPG Average: 7.70
Uses: This alkyd is a special form of MIRASOL 02-220 designed to meet Rule 66.

C. J. OSBORN: Resin Modified Alkyds(Continued):

CJO Code:

02-8394:
Type: Modified alkyd for industrial finishes.
% Non-Volatile: 59-61
Solvent: Xylol
Viscosity: Z1-Z3
Color (Gardner): 10 max.
Acid Number (Solids): 40 max.
Modifying Oil: Castor
WPG Average: 8.55
Uses: This is used for metal primers and one-coat metal decorating applications over various metal substrates including aluminum. It may also be used as a primer on surfaces of wood.

02-8621:
Type: Modified alkyd resin
% Non-Volatile: 49-51
Solvent: 45% Xylol
35% Exempt VMP Naphtha
20% Exempt MS
Viscosity: Z4-Z6
Color (Gardner): 10 max.
Acid Number (Solids): 14 max.
Modifying Oil: TOFA
WPG Average: 7.90
Uses: Low cost enamels and primers where fast air dry and good adhesion are required.

02-9627:
Type: Modified short oil alkyd
% Non-Volatile: 59-61
Solvent: 50% Xylol
50% Exempt Naphtha
Viscosity: Z2-Z4
Color (Gardner): 16 max.
Acid Number (Solids): 30 max.
Modifying Oil: Marine
WPG Average: 8.20
Uses: Lower cost 02-1202, excellent fast dry primer vehicle. Outstanding salt spray resistance with good pigment capability and stability.

C. J. OSBORN: Short Oil Alkyds:

Alkyds that have a high phthalic content. These alkyds are generally soluble in strong, aromatic solvents. These alkyds will generally serve the industrial market for both bake and air-dry.

CJO Code:

03-90:
Type: Pure drying oil modified alkyd
% Non-Volatile: 49-51
Solvent: Xylol
Viscosity: X-Z1
Color (Gardner): 7 max.
Acid Number (Solids): 14 max.
Modifying Oil: Linseed
WPG Average: 8.30
Uses: Industrial formulations for spray coating, both air dry and baking as well as colored enamels based on urea and melamine resins for baking. This is an oxidizing type resinous plasticizer for nitrocellulose lacquers. It may also be used in metal furniture, baking enamels, automotive refinishing enamels and air dry primers and finishes.

03-107:
Type: Pure semi-drying oil alkyd
% Non-Volatile: 49-51
Solvent: Xylol
Viscosity: Y-Z1
Color (Gardner): 6 max.
Acid Number (Solids): 14 max.
Modifying Oil: Soya
WPG Average: 8.20
Uses: For white and colored baking and air drying enamels for interior and exterior use. It is used with nitrocellulose lacquers; this product improves durability, adhesion and toughness.

03-107 EX:
Type: Pure semi-drying oil alkyd
% Non-Volatile: 49-51
Solvent: Exempt
Viscosity: Z2-Z5
Color (Gardner): 6 max.
Acid Number (Solids): 14 max.
Modifying Oil: Soya
WPG Average: 7.80
Uses: For white and colored baking and air dry enamels for interior and exterior use. Used with nitrocelulose lacquers this product improves durability, adhesion and toughness.

C. J. OSBORN: Short Oil Alkyds(Continued):

<u>CJO</u> <u>Code:</u>

03-107 XB:
Type: Pure semi-drying oil alkyd
% Non-Volatile: 59-61
Solvent: 95% Xylol
5% Butanol
Viscosity: Z3-Z5
Color (Gardner): 7 max.
Acid Number (Solids): 14 max.
Modifying Oil: Soya FA
WPG Average: 8.60
Uses: White and colored baking and air dry enamels for interior and exterior use. Used with nitrocellulose lacquers, it improves durability, adhesion and toughness.

03-305:
Type: Pure drying oil modified alkyd
% Non-Volatile: 49-51
Solvent: Xylol
Viscosity: Y-Z1
Color (Gardner): 7 max.
Acid Number (Solids): 14 max.
Modifying Oil: Dehydrated castor
WPG Average: 8.50
Uses: This product has toughness and adhesion and good outdoor durability and may be used as a vehicle for metal finishes, both interior and exterior. Its compatibility with nitrocellulose permits better adhesion, toughness and exterior durability in lacquers.

03-0733:
Type: Short oil pure alkyd
% Non-Volatile: 49-51
Solvent: Xylol
Viscosity: Z2-Z4
Color (Gardner): 3 max.
Acid Number (Solids): 16 max.
Modifying Oil: Special selected acids
WPG Average: 8.30
Uses: For fast dry millwork primers, metal primers and topcoats. Fairly good exterior durability, it's also used in lacquer and conversion coatings.

C. J. OSBORN: Short Oil Alkyds(Continued):

CJO Code:

03-1568:
Type: Baking or air dry industrial finishing alkyd
% Non-Volatile: 49-51
Solvent: Xylol
Viscosity: Z3-Z5
Color (Gardner): 6 max.
Acid Number (Solids): 12 max.
Modifying Oil: TOFA
WPG Average: 8.00
Uses: A general purpose industrial baking finish, good urea and melamine compatibility, good aliphatic solvent tolerance. Also available in exempt solvents.

03-1743:
Type: Short isophthalic industrial roller coating alkyd
% Non-Volatile: 59-61
Solvent: AMSCO G
Viscosity: Z4-Z6
Color (Gardner): 6 max.
Acid Number (Solids): 16 max.
Modifying Oil: TOFA
WPG Average: 8.50
Uses: Outstanding adhesion plus gloss and color retention for coil coatings and roller coat enamels.

03-2314:
Type: Short oil alkyd
% Non-Volatile: 49-51
Solvent: 50% SOLVESSO 100
50% SOLVESSO 150
Viscosity: Z1-Z3
Color (Gardner): 5 max.
Acid Number (Solids): 10 max.
Modifying Oil: TOFA
WPG Average: 8.30
Uses: Excellent flow and leveling roller coating applications fast solvent release, outstanding adhesion and flexibility plus excellent color retention, strongly qualify this product for general roll coat and metal decorating applications.

03-2314-4X:
Type: Medium short oil industrial finish isophthalic alkyd
% Non-Volatile: 49-51
Solvent: Xylol
Viscosity: Z1-Z3
Color (Gardner): 5 max.
Acid Number (Solids): 10 max.
Modifying Oil: TOFA
WPG Average: 8.25
Uses: Excellent industrial baking finish, outstanding adhesion and flexibility, excellent color retention.

C. J. OSBORN: Short Oil Alkyds(Continued):

CJO Code:

03-2318:
Type: Pure short oil linseed alkyd
% Non-Volatile: 59-61
Solvent: Xylol
Viscosity: Z-Z2
Color (Gardner): 9 max.
Acid Number (Solids): 25 max.
Modifying Oil: Linseed
WPG Average: 8.50
Uses: Air dry or baking industrial finishes, good exterior durability.

03-2591:
Type: Short oil pure alkyd
% Non-Volatile: 49-51
Solvent: Xylol
Viscosity: Z2-Z4
Color (Gardner): 4 max.
Acid Number (Solids): 10 max.
Modifying Oil: Special fatty acids
WPG Average: 8.40
Uses; Fast millwork primers, metal primers and topcoats. Fairly good exterior durability, this product may also be used in lacquer.

03-2779:
Type: Short oil industrial finish alkyd
% Non-Volatile: 59-61
Solvent: Xylol
Viscosity: X-Z
Color (Gardner): 4 max.
Acid Number (Solids): 25 max.
Modifying Oil: TOFA
WPG Average: 8.40
Uses: Flexible substrates such as paper coatings, etc. Exhibits good adhesion and flexibility and outstanding wear and mar resistance.

03-2813:
Type: Baking industrial finishing alkyd
% Non-Volatile: 49-51
Solvent: Xylol
Viscosity: Z5-Z6
Color (Gardner): 6 max.
Acid Number (Solids): 10 max.
Modifying Oil: TOFA
WPG Average: 8.30
Uses: This product may be used as a general purpose spray for industrial baking finishes. It has good urea and melamine compatibility as well as good aliphatic solvent tolerance. May be supplied in AMSCO G (MIRASOL 03-2813-G50) for roller coat.

C. J. OSBORN: Short Oil Alkyds(Continued):

CJO Code:

03-3254:
Type: Pure short oil alkyd
% Non-Volatile: 49-51
Solvent: SOLVESSO 100
Viscosity: Z2-Z4
Color (Gardner): 7 max.
Acid Number (Solids): 14 max.
Modifying Oil: Dehydrated castor
WPG Average: 8.40
Uses: This product has toughness & outdoor durability. It may be used as the vehicle for metal finishes, both interior and exterior. Its compatibility with nitrocellulose permits better adhesion, toughness, exterior durability in lacquers. It does not have as fast a set as MIRASOL 03-305.

03-A-3263:
Type: Chain stopped alkyd
% Non-Volatile: 74-76
Solvent: Xylol
Viscosity: Z4-Z6
Color (Gardner): 8 max.
Acid Number (Solids): 12 max.
Modifying Oil: Soya
WPG Average: 8.82-8.92
Uses: This chain stopped alkyd is designed for compatibility with a variety of film formers like CAB, NC, vinyls, chlorinated rubbers, urea & melamine resins. It is therefore very useful in a variety of applications, air dry and bake.

03-A-3882:
Type: Short oil alkyd
% Non-Volatile: 69-71
Solvent: Xylol
Viscosity: Z3-Z5
Color (Gardner): 6 max.
Acid Number (Solids): 3 max.
Modifying Oil: Special selected acids
WPG Average: 8.80
Uses: An excellent product for both air dry and baking applications; low acid numbers, good tolerance for low KB solvents and slow viscosity reduction makes this product an interesting alkyd for the formulator of industrial finishes.

C. J. OSBORN: Short Oil Alkyds(Continued):

CJO Code:

03-A-3297:
Type: Chain stopped alkyd
% Non-Volatile: 59-61
Solvent: Exempt VMP
Isobutanol
Viscosity: Z-Z2
Color (Gardner): 5 max.
Acid Number (Solids): 10 max.
Modifying Oil: TOFA
WPG Average: 8.01
Uses: Industrial bake finishes.

03-A-4374:
Type: High solid, short oil alkyd
% Non-Volatile: 89-91
Solvent: Xylol, DOW BC-300
Viscosity: Y-Z2
Color (Gardner): 8 max.
Acid Number (Solids): 15 max.
Modifying Oil: TOFA
WPG Average: 9.05
Uses: High solid, low viscosity short oil alkyd suitable for crosslinking with melamine and urea resins. It has good application in industrial coatings.

03-A-4477:
Type: Baking or air dry industrial finishing alkyd
% Non-Volatile: 79-81
Solvent: MEK/PM Glycol ether
Viscosity: Z2-Z4
Color (Gardner): 9 max.
Acid Number (Solids): 12 max.
Modifying Oil: TOFA
WPG Average: 8.99
Uses: General purpose industrial baking finish with good urea and melamine compatibility. Has good aliphatic solvent tolerance and is an excellent high solid vehicle.

03-A-4855:
Type: Chain stopped alkyd
% Non-Volatile: 69-71
Solvent: Xylol
Viscosity: Z4-Z6
Color (Gardner): 8 max.
Acid Number (Solids): 15 max.
Modifying Oil: TOFA
WPG Average: 8.60
Uses: This chain stopped alkyd has been designed for compatibility with a variety of film formers like CAB, NC, vinyls, chlorinated rubbers, urea & melamine resins. It is therefore very useful in a variety of applications.

C.J. OSBORN: Short Oil Alkyds(Continued):

CJO Code:

03-A-5766:
Type: Short oil alkyd
% Non-Volatile: 39-41
Solvent: SOLVESSO 150
Viscosity: G-J
Color (Gardner): 6 max.
Acid Number (Solids): 10 max.
Modifying Oil: TOFA
WPG Average: 8.30
Uses: General purpose baking alkyd.

03-A-5344:
Type: High solid, short oil alkyd
% Non-Volatile: 89-91
Solvent: PM Acetate
Viscosity: Z1-Z3
Color (Gardner): 8 max.
Acid Number (Solids): 10 max.
Modifying Oil: TOFA
WPG Average: 8.90
Uses: This is a high solid industrial finish.

03-A-5698:
Type: High solid, short oil alkyd
% Non-Volatile: 89-91
Solvent: PM Acetate
Viscosity: Z4-Z6
Color (Gardner): 6 max.
Acid Number (Solids): 10 max.
Modifying Oil: TOFA
WPG Average: 9.00
Uses: This is a flexible, short oil alkyd designed to impart characteristics that cannot be obtained by the use of conventional short oil alkyds. Coatings may be formulated at high solids that retain their flexibility even after extended baking. This product has excellent pigment wetting, gloss and durability.

03-A-5885:
Type: High solid baking alkyd
% Non-Volatile: 89-91
Solvent: Butyl acetate
Viscosity: Z2-Z4
Color (Gardner): 6 max.
Acid Number (Solids): 12 max.
Modifying Oil: TOFA
WPG Average: 8.95
Uses: This product is a general purpose short oil high solid baking alkyd.

C.J. OSBORN: Short Oil Alkyds(Continued):

CJO Code:

03-A-6633:
Type: High solid short oil alkyd
% Non-Volatile: 89-91
Solvent: Xylol, butanol
Viscosity: Z4-Z6
Color (Gardner): 8 max.
Acid Number (Solids): 10 max.
Modifying Oil: TOFA
WPG Average: 8.89
Uses: This product is a premium high solid industrial finish.

03-A-7015:
Type: High solid short oil alkyd
% Non-Volatile: 89-91
Solvent: BUTYL CELLOSOLVE
Xylol
Viscosity: Z6-Z8
Color (Gardner): 6 max.
Acid Number (Solids): 10 max.
Modifying Oil: TOFA
WPG Average: 8.95
Uses: This is a high solid vehicle noticeable for pigment wetting, hardness which has excellent baking properties.

03-9898:
Type: Short oil alkyd
% Non-Volatile: 49-51
Solvent: Xylol
Viscosity: W-Z
Color (Gardner): 6 max.
Acid Number (Solids): 10 max.
Modifying Oil: Soya
WPG Average: 8.30
Uses: This product is a high quality short oil alkyd designed for maximum gloss & color retention; used to produce air dry and baking finishes with unusual film integrity and adhesion.

C.J. OSBORN: Short Oil Alkyds(Continued):

CJO Code:

03-9899:

Type: High solid, short oil alkyd
% Non-Volatile: 89-91
Solvent: PM Acetate
Viscosity: Z4-Z6
Color (Gardner): 6 max.
Acid Number (Solids): 10 max.
Modifying Oil: TOFA
WPG Average: 9.00
Uses; This is a flexible short oil alkyd designed to impart a combination of characteristics which may not be obtained by the use of conventional short alkyds. Coatings may be formulated at high solids that retain their flexibility even after extended bake schedules. Excellent wetting of pigments and excellent durability & gloss are notable characteristics of this product.

03-A-9938:

Type: Short oil alkyd
% Non-Volatile: 54-56
Solvent: SOLVESSO 150 30%
SOLVESSO 100 10%
Butanol 5%
Viscosity: Z-Z2
Color (Gardner): 5 max.
Acid Number (Solids): 12 max.
Modifying Oil: Soya
WPG Average: 8.45
Uses; Industrial baking finishes.

C.J. OSBORN: Specialties:

The following are special purpose vehicles designed to meet specific requirements. These include wrinkle finishes, both regular and high solid; bronzing liquid; chlorinated alkyds; resin solutions; reactive polymers and emulsifiable resin esters that may be used as latex modifiers.

CJO Code:

07-202 A:
Type: Modified alkyd wrinkle resin
% Non-Volatile: 49-51
Solvent: VMP Naphtha
Xylol
Aromatic solvent
Viscosity: U-W
Color (Gardner): 11 max.
Acid Number (Solids): 36 max.
Modifying Oil: Tung
WPG Average: 7.70
Uses: This product is a standard of synthetic wrinkle vehicles. Formulations may be supplied.

07-RF-0245:
Type: Emulsifiable resin ester
% Non-Volatile: 97-99
Solvent: None
Viscosity: 120-240 poises
Color (Gardner): 6 max.
Acid Number (Solids): 12 max.
Modifying Oil: TOFA
WPG Average: 8.20
Uses: This product is used to upgrade exterior latex paints. It improves resistance to chipping and peeling and imparts better adhesion to chalky and glossy surfaces. Also, better mildew resistance is obtained.

07-501:
Type: Hammer finish alkyd (non-styrenated)
% Non-Volatile: 49-51
Solvent: Xylol
Viscosity: R-T
Color (Gardner): 7-9
Acid Number (Solids): 6-9
Modifying Oil: Blend
WPG Average: 7.88
Uses: This resin has been designed specifically for use in one-coat baking hammer finishes. In combination with amine resins, it has improved solvent resistance and marproofness.

C.J. OSBORN: Specialties(Continued):

CJO Code:

07-1187:
Type: Resin solution
% Non-Volatile: 15-17
Solvent: Toluene ethanol acetone
Viscosity: B-D
Color (Gardner): 1
Modifying Oil: None
WPG Average: 7.45
Uses: CAB solution.

07-2534:
Type: Short oil bronzing liquid
% Non-Volatile: 49-51
Solvent: Exempt MS
Viscosity: A-
Color (Gardner): 18+
Acid Number (Solids): 4 max.
Modifying Oil: Linseed
WPG Average: 7.50
Uses: Bronzing liquid for radiators, burial vaults, etc. It may be used as a general purpose bronzing liquid.

07-A-3423:
Type: High solid, wrinkle vehicle
% Non-Volatile: 79-81
Solvent: MIBK
Viscosity: Y-Z1
Color (Gardner): 12 max.
Acid Number (Solids): 36
WPG Average: 8.67
Uses: High solid, low VOC applications.

07-A-3542:
Type: Chlorinated alkyd
% Non-Volatile: 59-61
Solvent: Exempt MS
Viscosity: U-W
Color (Gardner): 10 max.
Acid Number (Solids): 6 max.
Modifying Oil: Safflower
WPG Average: 8.35
Uses: Safflower version of 07-A-3550.

C.J. OSBORN: Specialties(Continued):

<u>CJO</u> <u>Code:</u>

07-A-3550:
Type: Chlorinated alkyd
% Non-Volatile: 59-61
Solvent: Exempt MS
Viscosity: U-W
Color (Gardner): 10 max.
Acid Number (Solids): 6 max.
Modifying Oil: Soya
WPG Average: 8.35
Uses: This meets MTL-R-21417 Federal specification.

07-A-5836:
Type: Reactive polymer
% Non-Volatile: 69-71
Solvent: SOLVESSO 150
Viscosity: Z2-Z4
Color (Gardner): 3 max.
Acid Number (Solids): 16 max.
Modifying Oil: None
WPG Average: 8.95
Uses: Designed for applications where flexibility and reasonable hardness are important requirements.

07-8438:
Type: Alkyd wrinkle vehicle
% Non-Volatile: 49-51
Solvent: 81.2 Exempt VM&P Naphtha
10.3 Toluol
8.5 Xylol
Viscosity: U-X
Color (Gardner): 10 max.
Acid Number (Solids): 36 max.
Modifying Oil: Tung
WPG Average: 7.60
Uses: 07-A-8438 is a special form of MIRASOL 07-202A designed to permit wrinkle enamel formulations which will meet Rule 66.

C.J. OSBORN: Varnishes:

This series conforms to a variety of end uses such as spar varnish for interior/exterior use, mixing varnish for aluminum paint, requirements for Federal and State Highway Association and uses that specify fast dry, excellent chemical resistance, hardness and high gloss. Its primary use is for wood.

CJO Code:

12-2411:
Type: Gold metal lacquer
% Non-Volatile: 49-51
Solvent: MS 66/3
Viscosity: 50"+-5" in #4 Ford Cup
Color (Gardner): 18
Modifying Oil: Tung/Linseed
WPG Average: 7.40
Uses: Contains driers.

12-A-2659:
Type: Oleoresinous phenolic varnish
% Non-Volatile: 59-61
Solvent: Exempt MS Dipentene
Viscosity: R-T
Color (Gardner): 8 max.
Modifying Oil: Tung
WPG Average: 7.52
Uses: This is a pure phenolic tung oil varnish with excellent toughness, chemical resistance, gloss and weathering characteristics. It may be used in spar varnishes, porch and deck paints and floor finishes.

12-A-3583:
Type: Vacuum metallizing varnish
% Non-Volatile: 29-31
Solvent: Exempt MS
Viscosity: A1-A4
Color (Gardner): 15 max.
Modifying Oil: Soya
WPG Average: 7.08
Uses: May be used as a base coat over plastic in vacuum metallizing finishes.
Contains driers.

12-A-3703:
Type: Tung oil varnish
% Non-Volatile: 53-55
Solvent: Exempt MS
Viscosity: F-H
Color (Gardner): 9 max.
Modifying Oil: Tung oil
WPG Average: 7.44
Uses: General purpose interior/exterior varnish.

C.J. OSBORN: Varnishes (Continued):

<u>CJO</u> <u>Code:</u>

12-A-3922:
Type: Spar varnish
% Non-Volatile: 50-52
Solvent: Exempt
Viscosity: F-H
Color (Gardner): 10 max.
Modifying Oil: Tung/Soya
WPG Average: 7.50
Uses: Exterior and interior wood finishes.

12-A-4294:
Type: Phenolic varnish
% Non-Volatile: 57-59
Solvent: Non photo-chemically reactive
Viscosity: C-F
Color (Gardner): 12 max.
Modifying Oil: Tung/Linseed & castor
WPG Average: 7.40
Uses: This varnish meets the requirements for TTV119D phenolic spar varnish of 90.105 and a SSU at 210-980-1020 SEL. Contains driers.

12-A-4507:
Type: Phenolic varnish
% Non-Volatile: 59-61
Solvent: Exempt MS
Viscosity: L-N
Color (Gardner): 12 max.
Modifying Oil: Tung/Linseed
WPG Average: 7.65
Uses: This phenolic varnish meets the requirements for DOD V15218C Type II, a mixing phenolic varnish.

12-A-4518:
Type: Spar varnish
% Non-Volatile: 50-52
Solvent: Exempt MS
Viscosity: B-E
Color (Gardner): 14 max.
Modifying Oil: Tung oil
WPG Average: 7.25
Uses: This is designed to meet the requirements for State Highway Association, M69-70 Type I aluminum paint.

C.J. OSBORN: Varnishes(Continued):

CJO Code:

12-A-5299:
Type: Black varnish
% Non-Volatile: 45-49
Solvent: MS, aromatic hydrocarbons
Viscosity: 65"-85" #4 Ford cup
Color (Gardner): Black
Modifying Oil: Linseed
WPG Average: 7.39
Uses: This black varnish finds use in the preparation of stain and low cost high gloss black enamels. It is based on Gilsonite asphaltum.

12-A-5396:
Type: Oleoresinous tung oil varnish
% Non-Volatile: 49-51
Solvent: Aliphatic and aromatic hydrocarbons
Viscosity: V-X
Color (Gardner): 8 max.
Modifying Oil: Tung
WPG Average: 7.46
Uses: Oleoresinous tung oil varnish may be used for furniture applications.

12-A-6438:
Type: Phenolic tung oil varnish
% Non-Volatile: 53-55
Solvent: Exempt MS
Viscosity: H-J
Color (Gardner): 8 max.
Modifying Oil: Tung
WPG Average: 7.38
Uses: This is a fairly short oil varnish designed for fast drying, high chemical and abrasion resistance with excellent gloss.

12-A-6500:
Type: Spar varnish
% Non-Volatile: 49-51
Solvent: Exempt MS
Viscosity: A1-A3
Color (Gardner): 6 max.
Acid Number (Solids): 12 max.
Modifying Oil: Soya/Linseed
Uses: Exterior wood finishes.

C.J. OSBORN: Varnishes(Continued):

CJO Code:

12-7933:
Type: Oleoresinous varnish
% Non-Volatile: 57-59
Solvent: Exempt MS
Viscosity: A2-A
Color (Gardner): 13 max.
Modifying Oil: Tung/Linseed
WPG Average: 7.45
Use: Designed for stains and fillers where wood penetration is of prime importance.
Contains driers.

C.J. OSBORN: Water Reducibles:

Water Reducibles have a primary use in the industrial baking market. These products are known for their outstanding adhesion and flexibility to metal surfaces. Their use may be interior or exterior. These products are generally oil-free and are available in high solids versions. Water Reducibles may also be used to plasticize nitrocellulose lacquers, both air-dry and forced.

Polyesters:

CJO Code:

10-A-1559:
Type: Saturated polyester solution
% Non-Volatile: 74-76
Solvent: BUTYL CELLOSOLVE
Viscosity: Z2-Z5
Color (Gardner): 3 max.
Acid Number (Solids): 70-85
Modifying Oil: None
WPG Average: 8.80
Uses: Water reducible for industrial baking finishes.

10-A-2197:
Type: High solids water reducible, oil-free polyester
% Non-Volatile: 74-76
Solvent: BUTYL CARBITOL
Viscosity: Z5-Z7
Color (Gardner): 4 max.
Acid Number (Solids): 40-45
Modifying Oil: None
WPG Average: 9.30
Uses: High gloss, good flow, flexibility and hardness and excellent exterior durability make this a useful resin for aluminum siding, can coatings, automotive and appliance finishes.

10-A-4067:
Type: Polyester, reducible in water without the aid of amine.
% Non-Volatile: 79-81
Solvent: BUTYL CELLOSOLVE
Viscosity: Z5-Z7
Color (Gardner): 3 max.
Acid Number (Solids): 50-60
Modifying Oil: None
WPG Average: 9.53
Uses: This resin is designed in such a way that it may be introduced in a water system without the help of amines. It may be used with substantial amounts of ureas or melamine resins with which it is capable of crosslinking.

C. J. OSBORN: Water Reducibles(Continued):

CJO Code:

10-A-5423:
Type: Water soluble polyester
% Non-Volatile: 77-79
Solvent: Sec-butanol,
BUTYL CELLOSOLVE
Viscosity: Z5-Z7
Color (Gardner): 9 max.
Acid Number (Solids): 50-60
Modifying Oil: None
WPG Average: 8.82
Uses: This polyester is designed to be water-borne with a minimum amount of co-solvent--this resin needs only moderate quantities of melamine or urea crosslinker to produce a tough glossy film. Low temps and short cycles are recommended for better color with this product.

10-A-5426:
Type: Water soluble polyester
% Non-Volatile: 79-81
Solvent: Sec. butanol
BUTYL CELLOSOLVE
Viscosity: Z5-Z7
Color (Gardner): 4 max.
Acid Number (Solids): 50-65
Modifying Oil: None
WPG Average: 9.01
Uses: This water soluble polyester is designed to be water-borne with a minimum amount of co-solvents. The resin needs only moderate quantities of melamine or urea crosslinker to produce a tough, glossy film.

10-A-5816:
Type: Oil-free polyester
% Non-Volatile: 79-81
Solvent: BUTYL CELLOSOLVE, xylol
Viscosity: Z2-Z4
Color (Gardner): 3 max.
Acid Number (Solids): 41-49
Modifying Oil: None
WPG Average: 9.23
Uses: Designed for being water-dispensible. Crosslinked with melamine and urea resin 10-A-5816 produces very flexible enamels.

C.J. OSBORN: Water Reducibles(Continued):

CJO Code:

10-9105:
Type: Oil modified water reducible polyester
% Non-Volatile: 74-76
Solvent: Butanol, BUTYL CELLOSOLVE
Viscosity: Z4-Z6
Color (Gardner): 6 max.
Acid Number (Solids): 50-60
Modifying Oil: Specialty fatty acids
WPG Average: 8.70
Uses: Used for interiors as well as exterior applications where good gloss, flexibility, hardness and color retention are required over metal substrates. The baked films exhibit excellent durability.

10-A-9889:
Type: Oil-free water reducible polyester (not pre-neutralized)
% Non-Volatile: 78-82
Solvent: N-butyl alcohol
Viscosity: Z4-Z6
Color (Gardner): 2 max.
Acid Number (Solids): 50-60
Modifying Oil: None
WPG Average: 9.00
Uses: For exterior as well as interior applications where gloss, flexibility, hardness and color retention is required over metal substrates. Baked films exhibit superior salt resistance and durability.

10-10070:
Type: Oil-free water reducible polyester
% Non-Volatile: 79-81
Solvent: BUTYL CELLOSOLVE
Viscosity: Z4-Z6
Color (Gardner): 3 max.
Acid Number (Solids): 55-65
Modifying Oil: None
WPG Average: 9.35
Uses: This water reducible can be used for industrial baking coatings as well as metal decorating finishes.

C.J. OSBORN: Water Reducibles(Continued):

CJO Code:

Epoxy Esters:

10-A-5763:
Type: Water reducible epoxy ester
% Non-Volatile: 69-71
Solvent: BUTYL CELLOSOLVE
Butanol
Viscosity: Z2-Z5
Color (Gardner): 12 max.
Acid Number (Solids): 50-60
Modifying Oil: None
WPG Average: 8.50
Uses: Water soluble vehicle capable of delivering many of the desired properties of an epoxy ester.

10-A-5795:
Type: Water reducible epoxy ester
% Non-Volatile: 69-71
Solvent: Diethylene glycol ethyl ether, glycol butyl ether, butanol
Viscosity: Z2-Z5
Color (Gardner): 9 max.
Acid Number (Solids): 50-60
Modifying Oil: Nono
WPG Average: 8.61
Uses: Epoxy ester baking vehicle.

10-A-5995:
Type: Water dispersible epoxy ester
% Non-Volatile: 79-81
Solvent: BUTYL CELLOSOLVE
CARBITOL, butanol
Viscosity: Z6-Z8
Color (Gardner): 10 max.
Acid Number (Solids): 40-50
Modifying Oil: DCO FA
WPG Average: 8.80
Uses: Low temperature metal baking finish

C.J. OSBORN: Water Reducibles(Continued):

CJO Code:

Alkyds:

10-A-1084:
Type: Pre-neutralized short oil water reducible alkyd
% Non-Volatile: 50-52
Solvent: Water + co-solvents
Viscosity: S-W
Color (Gardner): 7 max.
Modifying Oil: Soya
WPG Average: 8.65
Uses: This product is unique in that it performs similarly to a conventionally based solvent product, yielding high gloss, industrial bake finishes with good hardness, mar resistance and color retention.

10-A-1086:
Type: Short oil, water reducible alkyd
% Non-Volatile: 74-76
Solvent: Propanol
Viscosity: Z1-Z3
Color (Gardner): 4 max.
Acid Number (Solids): 45-55
Modifying Oil: Soya
WPG Average: 8.60
Uses: This resin is unique in that it performs similar to conventional solvent based products yielding high gloss industrial baking finishes with good hardness, mar resistance and color retention.

10-A-6383:
Type: Water reducible alkyd
% Non-Volatile: 74-76
Solvent: Ethylene glycol butyl ether, butanol
Viscosity: Z5-Z7
Color (Gardner): 8 max.
Acid Number (Solids): 55-70
Modifying Oil: Soya
WPG Average: 8.85
Uses: Architectural or industrial air-dry enamels.

10-A-6441:
Type: Water soluble oleoresinous vehicle
% Non-Volatile: 100
Solvent: None
Viscosity: Z5-Z7
Color (Gardner): 10 max.
Acid Number (Solids): 100-120
Modifying Oil: Tung
WPG Average: 8.30
Uses: This vehicle produces baked films which are very hard, mar-proof and fairly glossy. They do not need a crosslinker. Air-dry films from this resin are flat because of micro-wrinkling.

C.J. OSBORN: Water Reducibles(Continued):

CJO Code:

10-A-7006:
Type: Water dispersible alkyd
% Non-Volatile: 74-76
Solvent: Sec. butanol,
BUTYL CELLO.
Viscosity: Z4-Z6
Color (Gardner): 9 max.
Acid Number (Solids): 35-42
Modifying Oil: None
WPG Average: 8.80
Uses: 10-A-7006 has been designed for good drying, early hardness and good stability.

10-9900:
Type: Short oil, water reducible alkyd
% Non-Volatile: 74-76
Solvent: N-butyl alcohol
Viscosity: Z3-Z5
Color (Gardner): 6 max.
Acid Number (Solids): 45-55
Modifying Oil: Soya
WPG Average: 8.60
Uses: 10-9900 water reducible resin is unique in that it performs to conventional solvent based products yielding high gloss industrial bake finishes with good hardness, mar resistance and color retention.

W.R. Acrylics:

10-A-1268:
Type: High solid, water reducible acrylic
% Non-Volatile: 69-71
Solvent: BUTYL CELLOSOLVE
Viscosity: Z4-Z6
Color (Gardner): 4 max.
Acid Number (Solids): 120-130
Modifying Oil: None
WPG Average: 8.50
Uses: Water reducible industrial baking finishes, metal decorating applications.

C.J. OSBORN: Water Reducibles(Continued):

CJO Code:

W. R. Acrylics(Continued):

10-A-1522:
Type: Water reducible acrylic terpolymer
% Non-Volatile: 69-71
Solvent: BUTYL CELLOSOLVE
BUTYL CARBITOL
Viscosity: Z5-Z7
Color (Gardner): 3 max.
Acid Number (Solids): 120-130
Modifying Oil: None
WPG Average: 8.60
Uses: Water reducible industrial baking finishes, metal decorating applications with good flow and leveling.

10-10026:
Type: Water reducible acrylic
% Non-Volatile: 64-66
Solvent: BUTYL CELLOSOLVE
Viscosity: Z2-Z5
Color (Gardner): 3 max.
Acid Number (Solids): 55-65
Modifying Oil: None
WPG Average: 8.40
Uses: Acrylic for metal deco water reducible industrial baking coating.

10-10096:
Type: High solids, water reducible acrylic
% Non-Volatile: 69-71
Solvent: BUTYL CELLOSOLVE
Viscosity: Z4-Z6
Color (Gardner): 4 max.
Acid Number (Solids): 125-135
Modifying Oil: None
WPG Average: 8.60
Uses: Water reducible acrylic may be used for industrial baking coatings as well as metal decorating application.

POLYVINYL CHEMICALS INC.: NEOCRYL Acrylic Concrete Sealers:

B-725:

Type: MMA/BMA Copolymer Bead
Solids: 100%
Tg: 50C
Viscosity (25C): 350 cps*
Features/Applications: A general purpose polymer with excellent penetration for applications requiring color stability, durability and fast drying as well as chemical resistance.
* 40% Wt solids in Toluene

B-708:

Type: Acrylic Copolymer Bead
Solids: 100%
Tg: 52C
Viscosity (25C): 400 cps*
Features/Applications: An easy-to-use polymer with low moisture vapor transmission. Ideally suited for use as a concrete curing membrane.
* 40% Wt solids in Toluene

A-633:

Type: Acrylic Copolymer
Solids: 42%
Viscosity (25C): 40 cps
Features/Applications: An easy-to-use polymer with wide adhesion latitude for applications requiring rapid hardness development, high ultimate hardness and resistance to gasoline, naphtha, oil, water and corrsoion.

A-645:

Type: Acrylic Polymer
Solids: 40%
Viscosity: 50 cps
Features: An easy-to-use polymer for applications requiring rapid hardness development, high ultimate hardness and resistance to gasoline, naphtha, oil, water and corrosion.

POLYVINYL CHEMICALS INC.: NEOCRYL Solid Acrylic Resins for Coatings:

B-700:
Type: BMA Polymer
Molecular Weight: Medium
Glass Transition Temperature (Tg): 48C
Ring and Ball Softening Point: 172C
Viscosity (40% wt. Solids Toluene, 25C): 500 cps
Features/Applications: Broad solubility and compatibility. Hardtough films provide high gloss, heat resistance, and durability. For use in coatings for wood, metals, and plastics. Excellent adhesion to untreated polypropylene with Eastman's Polyolefin 515-2 modification.

B-723:
Type: BMA Copolymer
Molecular Weight: High
Glass Transition Temperature (Tg): 48C
Ring and Ball Softening Point: 194C
Viscosity (40% wt. Solids Toluene, 25C): 3000 cps
Features/Applications: Excellent adhesion to bare and treated metals. Excellent compatibility, heat resistance, and color retention. For use in clear & pigmented coatings on aluminum, steel, brass, and chrome.

B-723LV:
Type: BMA Copolymer
Molecular Weight: High
Glass Transition Temperature (Tg): 48C
Ring and Ball Softening Point: 180C
Viscosity (40% wt. Solids Toluene, 25C): 2000 cps
Features/Applications: Similar to B-723 in performance properties. Solutions of B-723LV have lower viscosities for better sprayability at higher application solids and pigment loading. Still rather viscous for pigment, spray.

B-725:
Type: MMA/BMA Copolymer
Molecular Weight: Low/Medium
Glass Transition Temperature (Tg): 50C
Ring and Ball Softening Point: 160C
Viscosity (40% wt. Solids Toluene, 25C): 350 cps.
Features/Applications: Fast dissolving with rapid solvent release. Broad compatibility and high solids at low application viscosity. For use in coatings for metal, wood, and plastics. Also for use in traffic and marine paints.

POLYVINYL CHEMICALS INC.: NEOCRYL Solid Acrylic Resins for Coatings(Continued):

B-728:

Type: MMA Polymer
Molecular Weight: Medium
Glass Transition Temperature (Tg): 105C
Ring and Ball Softening Point: >200C
Viscosity (40% wt. Solids Toluene, 25C): 7000 cps*
Features/Applications: Hard and durable with excellent gasoline, oil, and plasticizer resistance. Also, excellent acohol, heat and chemical resistance. For use in coatings for auto refinishing, vinyl, and plastics.
* 40% WT solids in MEK, 25C

B-734:

Type: BMA copolymer
Molecular Weight: Medium
Glass Transition Temperature (Tg): 45C
Ring and Ball Softening Point: 163C.
Viscosity (40% wt. Solids Toluene, 25C): 500 cps
Features/Applications: Compatible with many short and medium oil alkyds. Excellent pigment grinding vehicle. For use in clear & pigmented coatings for auto refinishing, plastics, and for upgrading of alkyd paints.

B-735:

Type: MMA/BMA Copolymer
Molecular Weight: Low/Medium
Glass Transition Temperature (Tg): 72C
Ring and Ball Softening Point: 190C
Viscosity (40% wt. Solids Toluene, 25C): 300 cps
Features: Fast dissolving with rapid solvent release. Good chemical and gasoline resistance. Suggested for automotive plastics, and auto refinshing.

B-736:

Type: MMA Copolymer
Molecular Weight: Medium/High
Glass Transition Temperature (Tg): 95C
Viscosity (40% wt. Solids Toluene, 25C): 800 cps*
Features/Applications: Hard, tough films provide high gloss, mar resistance and gasoline resistance. Excellent exterior durability with resistance to cracking. For use in coatings for clear and pigmented auto refinish systems.
* 36% Wt solids in 60/40 Toluene/MEK, 25C.

POLYVINYL CHEMICALS INC.: NEOCRYL Solid Acrylic Resins for Coatings(Continued):

B-745:
Type: MMA/BMA Copolymer
Molecular Weight: Low/Medium
Glass Transition Temperature (Tg): 53C
Ring and Ball Softening Point: 130C
Viscosity (40% wt. Solids Toluene, 25C): 300 cps
Features/Applications: Excellent adhesion to bare and treated metals. High solids at low application viscosity. For use in Clear and pigmented coatings on aluminum, steel, brass, and chrome. Also, for transportation and tank coatings.

B-1000:
Type: Acrylic Copolymer
Molecular Weight: Low/Medium
Glass Transition Temperature (Tg): 43C
Ring and Ball Softening Point: 110C
Viscosity (40% wt. Solids Toluene, 25C): 350 cps
Features/Applications: Fast dissolving, compatible with nitrocellulose and other acrylic resins. For use in clear and pigmentedcoatings for wood, metal and plastics. Also, as an extender for nitrocellulose lacquers.

B-1019:
Type: MMA/BMA Copolymer
Molecular Weight: Low/Medium
Glass Transition Temperature (Tg): 50C
Ring and Ball Softening Point: 160C
Viscosity (40% wt. Solids Toluene, 25C): 250 cps
Features/Applications: Suggested for clear and metallic aerosols. Excellent durability and fast speed of dry.

POLYVINYL CHEMICAL INC: NEOREZ Water-Borne Polyurethane Dispersions for Coatings:

R-940:
Type: Aromatic
Solids: 31%
Viscosity (25C): 70 cps
pH: 8.3
Features/Applications: Hard films with high tensile and modulus values similar to R-960. For use in clear and pigmented coatings on metal, wood and plastics.

R-960:
Type: Aliphatic
Solids: 33%
Viscosity (25C): 300 cps
pH: 8.0
Features/Applications: Hard films with high tensile and modulus values. For use in tough, UV resistant clear and pigmented coatings on metal, wood, and plastics.

R-962:
Type: Aliphatic
Solids: 33%
Viscosity (25C): 250 cps
pH: 9.0
Features/Applications: Flexible films with high elongation for light stable, abrasion resistant coatings on flexible plastics, elastomers, and foams.

R-963:
Type: Aliphatic
Solids: 33%
Viscosity (25C): 250 cps
pH: 8.0
Features/Applications: High modulus with intermediate hardness and elongation. Suggested for use in coatings on metal, plastic and elastomers, also for use on glass and ceramics.

R-966:
Type: Aliphatic
Solids: 33%
Viscosity (25C): 30 cps
pH: 7.5
Features/Applications: Cosolvent free, UV resistant with ultimate hardness rapidly developed after water evaporation. For use in clear and pigmented coatings on textiles, rubber and plastics.

R-967:
Type: Aliphatic
Solids: 40%
Viscosity (25C): 250 cps
pH: 8.0
Features/Applications: Similar to R-966, but softer and more extensible.

POLYVINYL CHEMICALS INC.: NEOREZ Water-Borne Polyurethanes for Concrete Sealers:

R-940:
Type: Aromatic Dispersion
Solids: 31%
pH: 8.3
Viscosity (25C): 70 cps
Features/Applications: A general purpose polymer designed for chemical resistance and durability. It forms hard and glossy films at ambient temperatures that are tough and flexible with a high degree of abrasion, chemical and standing water resistance. It is crosslinkable.

R-941:
Type: Aromatic Dispersion
Solids: 31%
pH: 7.8
Viscosity (25C): 70 cps
Features/Applications: A durable polymer designed for chemical and water resistance, forming hard and glossy, yet tough and flexible films at ambient temperatures which exhibit a high degree of abrasion, chemical and standing water resistance as well as improved gloss. It is crosslinkable.

R-960:
Type: Aliphatic Dispersion
Solids: 33%
pH: 8.0
Viscosity (25C): 300 cps
Features/Applications: A general purpose polymer with exterior durability and chemical resistance designed for high performance applications where hardness, flexibility, chemical and abrasion resistance are required. The aliphatic backbone results in excellent U.V. resistance. It is cross-linkable, and can also be used to modify floor polishes.

RANBAR TECHNOLOGY INC.: Long Oil Alkyds:

G-4561-X-70 L/LS Alkyd Resin

G-4561-X-70 is a long oil soya/linseed alkyd resin which is supplied at 70% non-volatile in xylene. The combination of soya and linseed oils provides a faster and harder dry than G-4070-X-70. With the exception of the linseed oil content, G-4561-X-70 is similar to a TTR-266D, Type 1, Class C resin.

Typical Properties:
- Non-volatile by weight: 70+-1%
- Weight/gallon, 25C: 8.3 lbs.
- Viscosity, 25C: V-X
- Solvent: Xylene

Solid Resin Properties:
- Phthalic Anhydride: 23% minimum
- Acid Number: 10 maximum
- Oil Type: Soya/Linseed
- Cure: Air Dry, Force Dry

G-4587
Linseed-Penta Grinding Vehicle

G-4587 is a linseed oil-pentaerthryitol vehicle for grinding pigments. A low acid value makes this vehicle suitable for reactive pigments.

Typical Properties:
- Non-volatile by weight: 100%
- Viscosity, Gardner, 25C: C-E
- Weight/gallon, 25C: 7.90
- Color, Gardner: 5-10
- Hydroxyl value: 120+-10

G-1265-M-60 Alkyd Resin

G-1265-M-60 is a long tall oil alkyd supplied at 60% non-volatile in exempt mineral spirits. It is used in enamels and primers requiring good brushability, color retention, and durability.

Solution Properties:
- Non-volatile by weight: 60%
- Viscosity, 25C: T-W
- Color, Gardner: 8 maximum
- Appearance: Clear, light amber liquid
- Solvent: Exempt mineral spirits
- Cure: Air dry

Solids Resin Properties:
- Phthalic Anhydride: 23% minimum
- Oil length: long
- Acid Number: 10 maximum
- Oil type: tall oil

RANBAR TECHNOLOGY INC.: Long Oil Alkyds(Continued):

G-3629-M-60 Alkyd Resin

G-3629-M-60 Alkyd Resin conforms to Federal Specification TT-R-266D, Type II, Class B. It is recommended for use in structural steel primers, such as SSPC Paint Specification No. 2, 11, and 25. Because of its good brushing properties and its ability to dry to a high gloss, it also finds use in clear varnishes.

Solution Properties:
- Non-volatile by weight: 60%
- Weight/gallon: 7.8 pounds
- Viscosity, 25C: T-V
- Color, Gardner: 8 maximum
- Appearance: Clear, light amber liquid
- Solvent: Exempt mineral spirits
- Cure: Air Dry
- Tack Free Time at 20C: 8 hours

Solid Resin Properties:
- Phthalic Anhydride: 23% (minimum)
- Oil length: long
- Acid Number: 10 maximum
- Oil type: linseed

G-4070-M-70 Alkyd Resin

G-4070-M-70 Alkyd Resin conforms to Federal Specification TT-R-266D, Type I, Class A. It is a general purpose resin for interior or exterior use in architectural and trim enamels, structural and machinery paints and primers.

Solution Properties:
- Non-Volatile by weight: 70%
- Weight/gallon, 25C: 8.0 pounds
- Viscosity, 25C: Z2-Z4
- Color, Gardner: 8 maximum
- Appearance: Clear, Amber Liquid
- Solvent: Exempt mineral spirits
- Cure: Air Dry

Solid Resin Properties:
- Phthalic Anhydride: 23% (minimum)
- Oil length: long
- Acid number: 10 maximum
- Oil Type: Soya

RANBAR TECHNOLOGY INC.: Long Oil Alkyds(Continued):

G-4070-M-60 L/S Alkyd Resin

G-4070-M-60 L/S Resin conforms to Federal Specification TT-R-266D, Type I, Class B. It is a general purpose resin for interior or exterior use in architectural and trim enamels, structural and machinery paints, and primers.

The resin is suitable for air dry or force dry applications.

Solution Properties:
- Non-volatile by weight: 60%
- Viscosity, 25C: T-W
- Acid Value: 5-10
- Color, Gardner: 7 maximum
- Appearance: Clear, amber liquid
- Solvent: Exempt mineral spirits
- Cure: Air dry or bake

Solid Resin Properties:
- Phthalic Anhydride: 23% minimum
- Oil length: long
- Acid number: 10 maximum
- Oil type: Soya

G-4070-X-70 L/S Alkyd Resin

G-4070-X-70 is a long oil soya alkyd resin which is supplied at 70% non-volatile in xylene. It conforms to the requirements of Federal Specification TT-R-266D, Type I, Class C and exhibits excellent vinyl compatibility when tested according to TT-R-266D, paragraph 4.2.13.6.

Solution Properties:
- Non-volatile by Weight: 70%
- Weight/gallon, 25C: 8.2 pounds
- Viscosity, 25C: U-W
- Appearance: Clear, light yellow liquid
- Solvent: Xylene
- Cure: Air Dry

Solid Resin Properties:
- Phthalic Anhydride: 23% (minimum)
- Oil Length: long
- Acid number: 10 maximum
- Oil Type: Soya

RANBAR TECHNOLOGY INC.: Long Oil Alkyds(Continued):

G-4070-BC-75 L/S Alkyd Resin

Physical Properties:
- Non-volatile by weight: 75%
- Non-volatile by volume: 72%
- Weight/gallon, 25C: 8.40 pounds
- Viscosity, 25C: X-Z
- Flash Point, Seta: 140F
- Appearance: Amber liquid
- Solvent: Butyl CELLOSOLVE

G-4495 100% L/T Iso Resin

G-4495 is an isophthalic modified resin made from tall oil fatty acids. G-4495 may be used in place of linseed oil/bodied linseed oil in house paints or as a modifier for latex paints. Improvement in dry, yellowing, and durability will be seen when G-4495 is used to replace linseed oil.

Typical Properties:
- Non-volatile by weight: 100%
- Viscosity, Gardner, 25C: Z2-Z4
- Color, Gardner: 10 maximum
- Acid Number: 4-8
- Weight/gallon, 25C: 8.1 lbs.

G-4503-M-75 L/S Iso Resin

G-4503-M-75 is a long oil alkyd based on soya oil and isophthalic acid. The use of isophthalic acid yields resins of better durability and higher thermal endurance than comparable resins using phthalic anhydride. G-4503-M-75 may be used in air dry or force dry coatings or, when combined with amino resins, in baking coatings.

Typical Properties:
- Non-volatile by weight: 75%
- Non-volatile by volume: 69%
- Weight/Gallon, 25C: 8.1 pounds
- Viscosity, Gardner: Z-Z2
- Color, Gardner: 8 max.
- Acid Value: 10 max.
- Flash Point, Seta: 104F
- Solvent: Exempt mineral spirits

RANBAR TECHNOLOGY INC.: Long Oil Alkyds(Continued):

G-4520-V-50 L/T Resin

G-4520-V-50 is a long oil alkyd made from distillated tall oil. Thinned with VM&P Naphtha it can be used where a fast set time is desired. Finished products made with G-4520-V-50 must able to tolerate the presence of rosin.

Typical Properties:
- Non-volatile by weight, %: 50+-1
- Non-volatile by volume, %: 40+-1
- Weight/gallon, 25C: 7.35
- Viscosity, Gardner: Y-Z1
- Color, Gardner: 10 maximum
- Acid Number: 5-10
- Oil Type: Distilled tall oil
- Phthalic Anhydride Content, %: 17

G-4549 100% L/L Alkyd Resin

G-4549 is a long oil linseed alkyd intended to be used as a latex paint modifier. G-4549 will help achieve maximum adhesion and chalk binding when used in acrylic or PVA latex. The recommended usage range is 5 to 15% on the latex solids. A wetting agent and driers are included in G-4549 so that no further additions are necessary. It's composition is as follows:

G-4549 Alkyd	94.46% by weight
Wetting Agent	4.00
6% Manganese Drier	0.54
18% Zirconium Drier	0.60
Anti-skin Agent	0.40
	100.00

Typical Properties:
- Non-volatile by weight: 99+ per cent
- Viscosity, Gardner, 25C: Y-Z1
- Color, Gardner: 13 maximum
- Acid number: 8 maximum
- Weight/gallon, 25C: 8.4 lbs.
- Oil Content: 70%
- Phthalic Content: 15%

G-4549 may be used as a replacement for linseed oil/bodies linseed oil in oil type house paints. Improvement in dry, yellowing, and durability will be seen when replacing linseed oil.

RANBAR TECHNOLOGY INC.: Medium Oil Alkyds:

G-4067-M-50 M/LS Resin

G-4067-M-50 M/LS Resin conforms to Federal Specification TTR-266D, Type III. It produces durable interior or exterior coatings, providing easy brushing, high gloss, and excellent weather resistance. It is suitable for air dry, force dry, or baked coatings.

Physical Properties:
- Non-volatile by weight: 50.0%
- Non-volatile by volume: 41.0%
- Weight/gallon, 25C: 7.60 lbs.
- Viscosity, 25C: X-Z
- Flash Point, Seta: 104F
- Acid Value: 5-12
- Color, Gardner: 10 maximum
- Solvent: Exempt Mineral Spirits
- Oil Type: Soya/Linseed
- Cure: Air Dry or Bake

G-4412-M-50 M/S Alkyd Resin

G-4412-M-50 is a medium soya alkyd resin supplied at 50% non-volatile in mineral spirits. It is designed as an economical vehicle for air dry interior/exterior trade sales enamels, industrial primers and enamels, and maintenance paints. The products made with G-4412-M-50 exhibit good toughness, good exterior durability, and good resistance to water, oil, and gasoline.

G-4412-M-50 exhibits compatibility with vinyl resins when tested according to Federal Specification TT-T-266D, paragraph 4.2.13.6. Also, with either U-F or melamine resins, it forms excellent low temperature cure (250F) coatings.

Solution Properties:
- Non-volatile by Weight: 50+-1%
- Non-volatile by Volume: 41%
- Weight/gallon, 25C: 7.6 pounds
- Viscosity, 25C: Z2-Z4
- Flash Point, Seta: 104F
- Color (Gardner 1933): 8 maximum
- Appearance: Clear, light yellow liquid

Solid Resin Properties:
- Phthalic Anhydride: 35% (minimum)
- Oil length: medium
- Acid number: 10 maximum
- Oil Type: Soya

RANBAR TECHNOLOGY INC.: Medium Oil Alkyds(Continued):

G-4412 M-50 HV Alkyd Resin

G-4412 M-50 HV is a medium oil alkyd resin for interior or exterior use. It is suitable for air drying or baking coatings.

Physical Properties:
- Non-volatile by weight: 50.0%
- Non-volatile by volume: 40.5%
- Weight/gallon, 25C: 7.60 lbs.
- Viscosity, 25C: Z4-Z6
- Flash Point, Seta: 104F
- Acid Value: 5-10
- Color, Gardner: 7 maximum
- Appearance: Clear, amber liquid
- Solvent: Exempt Mineral Spirits
- Oil Type: Soya
- Cure: Air Dry or Bake

G-4412 V-50 Alkyd Resin

G-4412 V-50 is a medium oil alkyd resin for interior or exterior use. It is suitable for air drying or baking coatings.

Physical Properties:
- Non-volatile by weight: 50.0%
- Non-volatile by volume: 40.5%
- Weight/gallon, 25C: 7.3 lbs.
- Viscosity, 25C: Z-Z2
- Flash Point, Seta: 55F
- Acid Value: 5-10
- Color, Gardner: 7 maximum
- Appearance: Clear, amber liquid
- Solvent: Exempt VM&P Naphtha
- Oil Type: Soya
- Cure: Air Dry or Bake

G-4525-M-50 M/L Alkyd Resin

Typical Properties:
- Non-volatile by weight: 50+-1%
- Non-volatile by volume: 41%
- Viscosity, 25C: Z1-Z2
- Flash Point (Seta): 104F
- Acid Value: 5-10
- Color, Gardner: 6
- Solvent: Exempt mineral spirits
- Oil Type: Linseed
- Phthalic content, minimum: 30%
- Cure: Air dry/bake

RANBAR TECHNOLOGY INC.: Oil Free Alkyds:

G-4398-BA-85 High Solids Polyester Resin

G-4398-BA-85 High Solids Polyester Resin is a low molecular weight, oil free, polyester resin. It is designed to be cross linked with amino type resins in high solids baking finishes. As a starting point, 70 pbw of G-4398-BA-85 (solids basis) to 30 pbw CYMEL 303 is recommended.

Typical Properties:
- Non-volatile by weight: 85.0%
- Weight/gallon, 25C: 9.16 pounds
- Viscosity, 25C: Z-Z2
- Acid Value: 13+-2
- Appearance: Clear colorless liquid, no haze
- Solvent: n-Butyl Acetate
- Primary Hydroxyl number: 140 mg KOH/gram
- Average Molecular Weight (ASTM-E-222): 627+-6

Silicone Modified Resins:

G-4489-M-60 L/S Silicone Modified Resin

G-4489-M-60 is a silicone modified, long oil soya alkyd intended for use in durable exterior coatings. Air-drying enamels made with G-4489-M-60 are weather resistant, retain their color and gloss, and have good flexibility. Examples of items to be coated include machinery, freight cars, metal drums, signs, railings, metal doors and siding, and metal structures. G-4489-M-60 may be used to make TT-E-490E enamels.

Typical Properties:
- Non-volatile by weight, %: 60+-1%
- Non-volatile by volume: 52+-1%
- Weight/gallon, 25C: 7.94 lbs.
- Viscosity, Gardner, 25C: X-Z
- Color, Gardner: 10 maximum
- Acid Value: 10 maximum
- Flash Point, Seta: 105F
- V.O.C.: 3.18 pounds/gallon
- Reducing Solvents: VM&P Naphtha or Mineral Spirits or Xylene

RANBAR TECHNOLOGY INC.: Short Oil Alkyds:

G-4401-B-70 High Solids, Short Oil Alkyd

G-4401-B-70 High Solids, Short Oil Alkyd is recommended for air dry coatings or baking coatings when used in combination with amino resins. G-4401-B-70 is supplied at 70% non-volatile to give the formulator greater latitude in the choice of solvents to meet air quality regulations.

Physical Properties:
- Non-volatile by weight: 70%
- Weight per gallon, 25C: 8.60+-.10 pounds
- Viscosity, 25C: Z-Z2
- Flash Point, Seta: 85F
- Color, Gardner: 8 maximum
- Appearance: Free of haze
- Acid Value: 12 maximum
- Solvent: 70:30 n-Butanol: Ethyl Benzene
- Oil Type: Soya

G-4451-X-50 S/T Resin

G-4451-X-50 is a short oil alkyd based on low rosin tall oil fatty acids. It is intended to be used with melamine or urea-formaldehyde resins to formulate baking enamels for interior use. 20 to 30% amino resin is the range suggested as a starting point.

Properties:
- Non-volatile by weight: 50+-1%
- Non-volatile by volume: 43+-1%
- Weight/gallon, 25C: 8.25 lbs.
- Viscosity (Gardner): Z-Z2
- Flash Point, Seta: 80F
- Acid Value: 8 maximum
- Color (Gardner): 6 maximum
- Solvent: Xylene

G-4451-VBT-50 Resin

G-4451-VBT-50 is a short oil alkyd based on low rosin tall oil fatty acids. It is intended to be used with melamine or urea-formaldehyde resins to formulate baking enamels for interior use. The blended solvent replaces xylene and allows the coating formulator to create "exempt" solvent systems.

Properties:
- Non-volatile by weight: 50+-1%
- Non-volatile by volume: 39+-1%
- Viscosity, Gardner: Z1-Z3
- Flash Point, Seta: 50F
- Acid Value: 10 maximum
- Color, Gardner: 8 maximum

RANBAR TECHNOLOGY INC.: Styrene/VT Modified Alkyds & Copolymers:

G-1488E VT/Oil Copolymer Resin

G-1488E is a vinyltoluene/oil copolymer resin supplied at 60% non-volatile in VM&P Naphtha. The resin is used in the manufacture of fast air drying machinery enamels and primers. The products made with this resin dry primarily by evaporation of solvent from the applied coating's surface. To improve the mar resistance of the coating, approximately 10% melamine resin is added and cured for 10 minutes at 250F (121C). G-1488E is lighter in color and faster drying than G-1488D.

Physical Properties:
- Non-volatile by weight: 60+-1%
- Non-volatile by volume: 50+-1%
- Weight/gallon, 25C: 7.65 pounds
- Viscosity, Gardner, 25C: Z1-Z3
- Color, Gardner: 5 max.
- Flash Point (Seta): 52F
- Recommended Drier: 0.03% (metal on resin)
- Tack Free Time: 15-20 minutes

G-1515E VT/Oil Copolymer Resin

G-1515E is a vinyltoluene/oil copolymer resin supplied at 60% non-volatile in mineral spirits. The resin is used in air dry enamels, machinery enamels, floor sealers, primers and surfacers. To improve mar resistance of coatings formulated with G-1515E, add 10% of a melamine resin and cure for 10-20 minutes at 250F (121C).

Typical Properties
- Non-volatile by weight: 60+-1%
- Non-volatile by volume: 53.5%
- Weight/gallon, 25C: 7.45 pounds
- Viscosity, 25C: X-Y
- Flash Point, Seta: 104F
- Appearance: Clear, light yellow liquid
- Solvent: Exempt mineral spirits
- Cure: Solvent evaporation
- Tack Free Time at 19C: 35-45 minutes
- Recommended drier system: 0.03% Cobalt as metal based on resin

RANBAR TECHNOLOGY INC.: Styrene/VT Modified Alkyds & Copolymers (Continued):

G-1781-X-50 Alkyd Resin

G-1781-X-50 is a styrenated alkyd resin with excellent dry-to-handle (15-20 minutes), humidity resistance, film toughness, and gasoline resistance. It is recommended for use as an "over-coat" varnish and as the vehicle for fast drying, pigmented industrial coatings.

Physical Properties:
- Non-volatile by weight: 50+-1%
- Weight/gallon, 25C: 8.0 pounds
- Viscosity, 25C: U-W
- Flash Point, Seta: 76F
- Solvent: Xylene
- Type of Cure: Air drying
- Dry Time at 25C: 15 minutes
- V.O.C.: 3.95 lbs./gallon

RANBAR TECHNOLOGY INC.: Water Reduced Alkyds:

G-4510-BC-75 Water Soluble Alkyd

G-4510-BC-75 is a water soluble alkyd which may be used in air dry, force dry, or baking coatings. As supplied, the resin will require neutralization with an amine and some additional co-solvent before the addition of water. Sufficient co-solvent should be used to maintain an 80:20 ratio of water to co-solvent.

For baking coating, 10-20% of a melamine resin can be used for cross-linking.

Co-solvent, driers and/or melamines, and neutralizer should be added to the resin and mixed completely before the addition of water.

Typical Properties:
- Non-volatile by weight: 75+-1%
- Viscosity, 25C: Z5-Z6
- Acid Value (on non-volatile): 42-47
- Weight/Gallon, 25C: 8.7
- Solvent: 2-butoxy ethanol
- Flash Point, Seta: 140F

Flat Wall Alkyd

G-4594-OMS-35 Flat Wall Alkyd

G-4594-OMS-35 is a high viscosity, low solids alkyd intended for odorless flat paints, primer sealers, and undercoat finishes. G-4594-OMS-35 meets the requirements of Rule 66.

Typical Properties:
- Non-volatile by weight: 35+-1%
- Viscosity, Gardner, 25C: Z3-Z6
- Weight/Gallon, 25C: 7.0+-.10
- Acid Value: Less than 10
- Color, Gardner: 7 maximum
- Solvent: Odorless Mineral Spirits
- Oil: Low Rosin T.O.F.A.

REICHHOLD CHEMICALS, INC.: AMBERLAC Acrylic Monomer Modified Alkyds:

RCI Code:

13-801:

Product: 292-A-150
Type: Acrylic Alkyd Baking
% N.V.: Wt.: 50
Volatile: A 150
Viscosity (Gardner-Holdt): U-W

13-802:

Product: 292-X
Type: Acrylic Alkyd-Air Dry
% N.V.: Wt.: 50
Volatile: X
Viscosity (Gardner-Holdt): N-R

REICHHOLD CHEMICALS, INC.: Amino Resins:

Amino resins are used as the crosslinking agent with various BECKOSOL (alkyd) and EPOTUF (epoxy) resins in baking enamels. In this capacity the amino resins serve to impart hardness, chemical resistance and color retention to the finished product. The three generic types of amino resins supplied by Reichhold are described briefly here.

BECKAMINE and UFORMITE urea-formaldehyde resins provide excellent adhesion and through hardness to baking enamels. They also show good gloss retention and chemical resistance. Urea-formaldehyde resin in combination with alkyds can be used in room temperature cure conversion varnishes when a suitable catalyst is used.

SUPER BECKAMINE and UFORMITE melamine-formaldehyde resins impart excellent surface hardness to baking enamels. They exhibit excellent mar resistance and chemical resistance. Their outstanding durability makes the melamine resins useful in automotive and fluorescent lighting fixture finishes.

UFORMITE benzoguanamine-formaldehyde resins are valued for their chemical and water resistance. They have exceptional resistance to color and gloss change upon high temperature bake or severe overbake. They have poor gloss retention when exposed to the weather.

BECKAMINE Urea-formaldehyde Resins:

<u>RCI Code:</u>

21-500:
- Product: 3250-50
- % N.V.: Wt.: 60
- Volatile: X/BuOH
- Viscosity (Gardner-Holdt): X-Z1

21-505:
- Product: 3530-50
- % N.V.: Wt.: 55
- Volatile: X/I BuOH
- Viscosity (Gardner-Holdt): V-Z

21-506:
- Product: 3531-50
- % N.V.: Wt.: 60
- Volatile: BuOH/I PrOH
- Air Reg.: E
- Viscosity (Gardner-Holdt): T-W

21-508:
- Product: 93-208
- % N.V.: Wt.: 55
- Volatile: X/I BuOH
- Viscosity (Gardner-Holdt): U-Y

E Denotes Exempt Solvent

REICHHOLD CHEMICALS, INC.: Amino Resins(Continued):

RCI Code:

21-510:
Product: P-138-60
% N.V.: Wt.: 60
Volatile: X/BuOH
Viscosity (Gardner-Holdt): R-U

21-511:
Product: P-196-60
% N.V.: Wt.: 60
Volatile: BuOH/EtOH
Air Reg.: E

21-605:
% N.V.: Wt.: 60
Volatile: I BuOH
Air Reg: E
Viscosity (Gardner-Holdt): V-Z

21-625:
% N.V.: Wt.: 70
Volatile: I BuOH
Air Reg.: E
Viscosity (Gardner-Holdt): Z2-Z5

UFORMITE Urea-Formaldehyde Resins:

21-805:
Product: F-240
% N.V.: Wt.: 60
Volatile: X/BuOH
Viscosity (Gardner-Holdt): K-P

21-806:
Product: F-240N
% N.V.: Wt.: 65
Volatile: HFAN/BuOH
Viscosity (Gardner-Holdt): Z1-Z5

E Denotes Exempt Solvent

REICHHOLD CHEMICALS, INC.: Amino Resins(Continued):

SUPER-BECKAMINE Melamine-Formaldehyde Resins:

RCI Code:

27-556:
Product: 3555-60
% N.V.: Wt.: 60
Volatile: BuOH
Air Reg.: E
Viscosity (Gardner-Holdt): V-Y

27-566:
% N.V.: Wt.: 55
Volatile: X/I BuOH
Viscosity (Gardner-Holdt): L-Q

27-568:
Product: 93-241
% N.V.: Wt.: 90*
Volatile: BuOH
Air Reg: E
Viscosity (Gardner-Holdt): Z1-Z4

27-580:
% N.V.: Wt.: 80
Volatile: BuOH
Air Reg: E
Viscosity (Gardner-Holdt): Z2-Z5

UFORMITE Melamine-Formaldehyde Resins:

27-801:
Product: MM-46
% N.V.: Wt.: 60
Volatile: X/BuOH
Viscosity (Gardner-Holdt): I-L

27-802:
Product: MM-47
% N.V.: Wt.: 60
Volatile: X/BuOH
Viscosity (Gardner-Holdt): P-T

27-803:
Product: MM-55
% N.V.: Wt.: 55
Volatile: BuOH
Air Reg: E
Viscosity (Gardner-Holdt): H-L

E Denotes Exempt Solvent
* 90' @ 60C

REICHHOLD CHEMICALS, INC.: Amino Resins(Continued):

RCI Code:

27-804:
Product: MM-57
% N.V.: Wt.: 50
Volatile: BuOH
Air Reg: E
Viscosity (Gardner-Holdt): H-K

27-806:
Product: MM-83
% N.V.: Wt.: 80*
Volatile: I BuOH/I PrOH
Air Reg: E
Viscosity (Gardner-Holdt): T-Z

UFORMITE Benzoguanamine Resins:

27-809:
Product: MX-61
% N. V.: Wt.: 66
Volatile: X/BuOH
Viscosity (Gardner-Holdt): G-M

27-810:
Product: QR-336
% N. V.: Wt.: 96 min.
Volatile: None
Viscosity (Gardner-Holdt): Q-Y**

E Denotes Exempt Solvent
* 90 @ 60C
**Values for 65% solution in 2 parts Xylene/1.3 parts Butanol by weight

REICHHOLD CHEMICALS, INC.: BECKOSOL Unmodified Alkyds with Under 27% Phthalic Content-(Long Oil Alkyds):

<u>RCI Code:</u>

10-024:
Product: P-75
% N.V.: Wt.: 100
Volatile: None
Air Reg.: E
Viscosity (Gardner-Holdt): W-Y
Phthalic Anhydride Min. on N.V.: 17
% and Type Oil: 72.5 Linseed

10-027:
% N.V.: Wt.: 60
Volatile: MS
Air Reg.: E
Viscosity (Gardner-Holdt): V-X
Phthalic Anhydride Min. on N.V.: 23
% and Type Oil: 64 TOFA

10-045:
Product: P-296-70
% N.V.: Wt.: 70
Volatile: MS
Air Reg.: E
Viscosity (Gardner-Holdt): Y-Z2
Phthalic Anhydride Min. on N.V.: 24
% and Type Oil: 65 Soybean

10-046:
Product: P-296-60
% N.V.: Wt.: 60
Volatile: MS
Air Reg.: E
Viscosity (Gardner-Holdt): S-V
Phthalic Anhydride Min. on N.V.: 24
% and Type Oil: 65 Soybean

10-060:
Product: P-470-70
% N.V.: Wt.: 70
Volatile: MS
Air Reg.: E
Viscosity (Gardner-Holdt): Z1-Z4
Phthalic Anhydride Min. on N.V.: 25
% and Type Oil: 65 Soybean

E Denotes Exempt Solvent

REICHHOLD CHEMICALS, INC.: BECKOSOL Unmodified Alkyds With Under 27% Phthalic Content-(Long Oil Alkyds)(Continued):

RCI Code:

10-061:
Product: P-470-60
% N.V.: Wt.: 60
Volatile: MS
Air Reg.: E
Viscosity (Gardner-Holdt): U-W
Phthalic Anhydride Min. on N.V.: 25
% and Type Oil: 65 Soybean

10-088:
% N.V.: Wt.: 70
Volatile: MS
Air Reg.: E
Viscosity (Gardner-Holdt): Z4-Z6
Phthalic Anhydride Min. on N.V.: 23
% and Type Oil: 64 TOFA

10-900:
Product: 97-089
% N.V.: Wt: 100
Volatile: None
Air Reg.: E
Viscosity (Gardner-Holdt): Z-Z2
% and Type Oil: TOFA

10-908:
Product: L-143
% N.V.: Wt.: 100
Volatile: None
Air Reg.: E
Viscosity (Gardner-Holdt): X-Z1
% and Type Oil: Safflower

BECKOSOL Unmodified Alkyds With 27-36% Phthalic Content (Medium Oil Alkyds):

11-010:
Product: 1331-50
% N.V.: Wt.: 50
Volatile: MS
Air Reg.: E
Viscosity (Gardner-Holdt): T-V
Phthalic Anhydride Min. on N.V.: 31
% and Type Oil: 57 Soybean

E Denotes Exempt Solvent

REICHHOLD CHEMICALS, INC.: BECKOSOL Unmodified Alkyds With 27-36% Phthalic Content-(Medium Oil Alkyds):

RCI Code:

11-035:
Product: P-323-50
% N.V.: Wt.: 50
Volatile: MS
Air Reg.: E
Viscosity (Gardner-Holdt): Y-Z2
Phthalic Anhydride Min on N.V.: 35
% and Type Oil: 52 Soybean

11-045:
Product: P-471-50
% N.V.: Wt.: 50
Volatile: MS
Air Reg.: E
Viscosity (Gardner-Holdt): V-Y
Phthalic Anhydride Min. on N.V.: 30
% and Type Oil: 58 Soybean

11-050:
Product: P-531-50
% N.V.: Wt.: 50
Volatile: MS
Air Reg.: E
Viscosity (Gardner-Holdt): Z1-Z3
Phthalic Anhydride Min. on N.V.: 35
% and Type Oil: 52 Mixed oils

11-070:
Product: P-671-50
% N.V.: Wt.: 50
Volatile: MS
Air Reg.: E
Viscosity (Gardner-Holdt): V-Y
Phthalic Anhydride Min. on N.V.: 31
% and Type Oil: 56 Soybean Linseed

11-081:
Product: P-750-50
% N.V.: Wt.: 50
Volatile: VMP
Air Reg.: E
Viscosity (Gardner-Holdt): Z-Z2
Phthalic Anhydride Min. on N.V.: 35
% and Type Oil: 52 Soybean

E Denotes Exempt Solvent

REICHHOLD CHEMICALS, INC.: BECKOSOL Unmodified Alkyds With 27-36% Phthalic Content-(Medium Oil Alkyds)(Continued):

RCI Code:

11-112:
Product: P-893-50
% N.V.: Wt.: 50
Volatile: MS
Air Reg.: E
Viscosity (Gardner-Holdt): Z4-Z6
Phthalic Anhydride Min. on N.V.: 34
% and Type Oil: 53 Soybean

11-630:
% N.V.: Wt.: 50
Volatile: MS
Air Reg.: E
Viscosity (Gardner-Holdt): Z4-Z6
Phthalic Anhydride Min. on N.V.: 35
% and Type Oil: 50 TOFA

REICHHOLD CHEMICALS, INC.: BECKOSOL Unmodified Alkyds with Over 36% Phthalic Content--(Short Oil Alkyds):

12-005:
Product: 1307-50
% N.V.: Wt.: 50
Volatile: X
Viscosity (Gardner-Holdt): N-P
Phthalic Anhydride Min. on N.V. %: 42
% and Type Oil: 41 Soybean

12-006:
Product: 1307-50HV
% N.V.: Wt.: 50
Volatile: X/VMP
Viscosity (Gardner-Holdt): Z1-Z3
Phthalic Anhydride Min. on N.V. %: 42
% and Type Oil: 41 Soybean

12-021:
Product: 1323-60
% N.V.: Wt.: 60
Volatile: T
Viscosity (Gardner-Holdt): X-Z
Phthalic Anhydride Min. on N.V. %: 47
% and Type Oil: 31 Coconut

E Denotes Exempt Solvent

REICHHOLD CHEMICALS, INC.: BECKOSOL Unmodified Alkyds With Over 36% Phthalic Content(Short Oil Alkyds)(Continued):

RCI Code:

12-025:
Product: 1343-50
% N.V.: Wt.: 50
Volatile: X/BuOH
Viscosity (Gardner-Holdt): W-Y
Phthalic Anhydride Min. on N.V. %: 40
% and Type Oil: 32 Oxidizing

12-035:
Product: P-222-60
% N.V.: Wt.: 60
Volatile: X
Viscosity (Gardner-Holdt): Z1-Z3
Phthalic Anhydride Min. on N.V. %: 47
% and Type Oil: 31 Coconut

12-038:
Product: P-372-60
% N.V.: Wt.: 60
Volatile: X
Viscosity (Gardner-Holdt): Z2-Z4
Phthalic Anhydride Min. on N.V. %: 40
% and Type Oil: 45 Linseed

12-054:
Product: P-800-50
% N.V.: Wt.: 50
Volatile: X
Viscosity (Gardner-Holdt): Z2-Z4
Phthalic Anhydride Min. on N.V. %: 41
% and Type Oil: 32 TOFA

12-073:
Product: P-899-60
% N.V.: Wt.: 60
Volatile: X
Viscosity (Gardner-Holdt): T-V
Phthalic Anhydride Min. on N.V. %: 37
% and Type Oil: 42 Mixed Oils

12-093:
Product: P-981-50
% N.V.: Wt.: 50
Volatile: X
Viscosity (Gardner-Holdt): V-Y
Phthalic Anhydride Min. on N.V. %: 40
% and Type Oil: 34 TOFA

E Denotes Exempt Solvent

REICHHOLD CHEMICALS, INC.: BECKOSOL Unmodified Alkyds with Over 36% Phthalic Content-(Short Oil Alkyds)(Continued):

<u>RCI</u> <u>Code:</u>

12-510:
Product: 98-210
% N.V.: Wt.: 90
Volatile: EGMEEA
Viscosity (Gardner-Holdt): Y-Z2
% and Type Oil: TOFA

12-511:
% N.V.: Wt.: 90
Volatile: PGMMEA
Viscosity (Gardner-Holdt): Z3-Z5
% and Type Oil: TOFA

12-512:
% N.V.: Wt.: 95 min.
Volatile: None
Viscosity (Gardner-Holdt): Z3-Z5
% and Type Oil: TOFA

12-513:
% N.V.: Wt.: 75
Volatile: X/IBuOH
Viscosity (Gardner-Holdt): Z-Z2
Phthalic Anhydride Min. on N.V. %: 37
% and Type Oil: Mixed

12-606:
Product: R-66 12-006
% N.V.: Wt.: 50
Volatile: R-66
Viscosity (Gardner-Holdt): Z1-Z3
Phthalic Anhydride Min. on N.V. %: 41
% and Type Oil: 42 Soybean

12-607:
Product: 93-955
% N.V.: Wt.: 80
Volatile: EGMEEA
Viscosity (Gardner-Holdt): Z1-Z3
% and Type Oil: TOFA

12-623:
Product: R-66 12-035
% N.V.: Wt.: 60
Volatile: R-66
Viscosity (Gardner-Holdt): Z3-Z5
Phthalic Anhydride Min. on N.V. %: 47
% and Type Oil: 31 Coconut

E denotes Exempt Solvent

REICHHOLD CHEMICALS, INC.: BECKOSOL Unmodified Alkyds With Over 36% Phthalic Content-(Short Oil Alkyds)(Continued):

RCI Code:

12-645:
Product: R-66 12-054
% N.V.: Wt.: 50
Volatile: R-66
Viscosity(Gardner-Holdt): Z2-Z4
Phthalic Anhydride Min. on N.V. %: 41
% and Type Oil: 32 TOFA

12-681:
Product: R-66 12-093
% N.V.: Wt.: 60
Volatile: R-66
Viscosity (Gardner-Holdt): Z1-Z3
Phthalic Anhydride Min. on N.V. %: 40
% and Type Oil: 34 TOFA

REICHHOLD CHEMICALS, INC.: BECKOSOL Modified Alkyds:

10-001:
Product: 1-50
Type Modifier: Phenol/Rosin
% N.V.: Wt.: 50
Volatile: MS
Air Reg.: E
Viscosity (Gardner-Holdt): Y-Z1
Phthalic Anhydride Min. on N.V. %: 18
% and Type Oil: 52 Linseed/Oiticica

10-010:
Product: 1316-50
Type Modifier: Phenol/Rosin
% N.V.: Wt.: 50
Volatile: T
Viscosity (Gardner-Holdt): D-G
Phthalic Anhydride Min. on N.V. %: 17
% and Type Oil: 52 Linseed/Oiticica

10-015:
Product: 1341-60
Type Modifier: Phenol/Resin
% N.V.: Wt.: 60
Volatile: X/VMP
Viscosity (Gardner-Holdt): Z2-Z4
Phthalic Anhydride Min. on N.V. %: 27
% and Type Oil: 28 Linseed

E Denotes Exempt Solvent

REICHHOLD CHEMICALS, INC.: BECKOSOL Modified Alkyds(Continued):

RCI Code:

10-071:
Product: P-539-50
Type Modifier: Phenol/Resin
% N.V.: Wt.: 50
Volatile: X
Viscosity (Gardner-Holdt): U-W
Phthalic Anhydride Min. on N.V.: 26
% and Type Oil: 27 Tall Oil

10-613:
Product: R-66
P-885-60
Type Modifier: Phenol/Rosin
% N.V.: Wt.: 60
Volatile: T/VMP
Air Reg.: E
Phthalic Anhydride Min. on N.V.: 15
% and Type Oil: 33 TOFA

10-615:
Product: R-66
10-015
Type Modifier: Phenol/Rosin
% N.V.: Wt.: 60
Volatile: R-66
Air Reg.: E
Viscosity (Gardner-Holdt): Z2-Z4
% Phthalic Anhydride Min. on N.V. %: 27
% and Type Oil: 28 Linseed

10-617:
Product: R-66
P-885-50
Type Modifier: Phenol/Resin
% N.V.: Wt.: 50
Volatile: T/VMP
Air Reg.: E
Viscosity (Gardner-Holdt): Y-Z1
Phthalic Anhydride Min. on N.V. %: 15
% and Type Oil: 33 TOFA

11-019:
Product: 1340-50
Type Modifier: Rosin
% N.V.: Wt.: 50
Volatile: VMP
Air Reg.: E
Viscosity (Gardner-Holdt): Z1-Z3
Phthalic Anhydride Min. on N.V.: 33
% and Type Oil: 41 Soybean/Linseed

E Denotes Exempt Solvent

REICHHOLD CHEMICALS, INC.: BECKOSOL Modified Alkyds (Continued):

RCI Code:

11-107:
Product: P-860-50
Type Modifier: Phenol/Rosin
% N.V.: Wt: 50
Volatile: X
Viscosity (Gardner-Holdt): W-Y
Phthalic Anhydride Min. on N.V. %: 33
% and Type Oil: 23 Tall Oil

12-001:
Product: 1320-50
Type Modifier: Rosin
% N.V.: Wt: 50
Volatile: X
Viscosity (Gardner-Holdt): W-Z
Phthalic Anhydride Min. On N.V. %: 43
% and Type Oil: 32 Soybean

12-032:
Product: P-27-50
Type Modifier: Phenol
% N.V.: Wt: 50
Volatile: X
Viscosity (Gardner-Holdt): W-Y
Phthalic Anhydride Min. on N.V. %: 40
% and Type Oil: 41 Soybean

12-079:
Type Modifier: Phenol/Rosin
% N.V.: Wt: 50
Volatile: X
Viscosity (Gardner-Holdt): W-Y
Phthalic Anhydride Min. on N.V. %: 40
% and Type Oil: 32 Linseed/CWO

12-080:
Product: 92-752
Type Modifier: Phenol/Rosin
% N.V.: Wt: 50
Volatile: X
Viscosity (Gardner-Holdt): Y-Z1
Phthalic Anhydride Min. on N.V. %: 39
% and Type Oil: 25 Linseed/CWO

12-090:
Type Modifier: Phenol/Rosin
% N.V.: Wt.: 50
Volatile: X
Viscosity (Gardner-Holdt): V-Y
Phthalic Anhydride Min. on N.V. %: 38
% and Type Oil: 39 Soybean

E Denotes Exempt Solvent

REICHHOLD CHEMICALS, INC.: BECKOSOL Modified Alkyds(Continued):

RCI Code:

13-403*
Product: 92-747
Type Modifier: Phenol/Rosin
% N.V.: Wt.: 75
Volatile: EGMBE
Viscosity (Gardner-Holdt): Z3-Z4

* Design for water reducible coatings

Flat Wall Alkyds:
A special group of alkyd resins with a combination of properties which make them particularly suitable for use in interior flat wall finishes are produced by Reichhold under the trade name WALLKYD.

WALLKYD Special Alkyds for Flat Wall Finishes:

11-024:
Product: 0-1956-35
% N.V.: Wt.: 35
Volatile: OMS
Air Reg.: E
Viscosity (Gardner-Holdt): V-X

11-150:
Product: 1952-40
% N.V.: Wt.: 40
Volatile: MS
Air Reg.: E
Viscosity (Gardner-Holdt): Z-Z3

11-151:
Product: 0-1952-35
% N.V.: Wt.: 35
Volatile: OMS
Air Reg.: E
Viscosity (Gardner-Holdt): Z-Z3

11-153:
% N.V.: Wt.: 30
Volatile: A140
Air Reg.: E
Viscosity (Gardner-Holdt): U-Y

E Denotes Exempt Solvent

REICHHOLD CHEMICALS, INC.: Chain Stopped Alkyds:

Chain-stopped alkyds are vehicles in which a monobasic acid is used to change the degree of polymerization and molecular structure of a conventional alkyd. Prime advantages of chain-stopped alkyds are fast set and tack free dry time along with better durability, color and gloss retention on exterior exposure as compared to fast-dry modified alkyds.

BECKOSOL Chain-Stopped Alkyds:

12-101:
% N.V.: Wt.: 50
Volatile: X
Viscosity (Gardner-Holdt): Z1-Z3

12-102:
% N.V.: Wt.: 50
Volatile: X
Viscosity (Gardner-Holdt): X-Z

12-105:
Product: 92-701
% N.V.: Wt.: 40
Volatile: MS/PGMMEA
Air Reg.: E
Viscosity (Gardner-Holdt): W-Y

12-106:
Product: 92-767
% N.V.: Wt.: 50
Volatile: X
Viscosity (Gardner-Holdt): T-W

12-501:
Product: 93-986
% N.V.: Wt.: 75
Volatile: MAK
Air Reg.: E
Viscosity (Gardner-Holdt): Z4-Z6

12-604:
Product: R-66 12-102
% N.V.: Wt.: 50
Volatile: R-66
Air Reg.: E
Viscosity (Gardner-Holdt): X-Z

12-704:
Product: S-1557-A
% N.V.: Wt.: 50
Volatile: X
Viscosity (Gardner-Holdt): P-R

13-404*:
Product: 92-762
% N.V.: Wt.: 75
Volatile: EGMBE sec-BuOH
Air Reg.: E
Viscosity (Gardner-Holdt): Z5-Z6

E Denotes Exempt Solvent * Design for water reducible coatings

REICHHOLD CHEMICALS, INC.: High Solids Resins:

RCI Code:

12-073:
Product: P899-60
Type: Short Oil Alkyd
% N.V.: Wt.: 60
Volatile: X
Viscosity (Gardner-Holdt): T-V
Type Oil: Mixed Oils

12-501:
Product: 93-986
Type: Chain Stopped Alkyd
% N.V.: Wt.: 75
Volatile: MAK
Air Reg: E
Viscosity (Gardner-Holdt): Z4-Z6
Type Oil: TOFA

12-510:
Product: 98-210
Type: Short Oil Alkyd
% N.V.: Wt.: 90
Volatile: EGMEEA
Air Reg.: E
Viscosity (Gardner-Holdt): Y-Z2
Type Oil: TOFA

12-511:
Type: Short Oil Alkyd
% N.V.: Wt.: 90
Volatile: PGMMEA
Air Reg.: E
Viscosity (Gardner-Holdt): Z3-Z5
Type Oil: TOFA

12-512:
Type: Short Oil Alkyd
% N.V.: wt.: 95 min.
Volatile: None
Air Reg.: E
Viscosity (Gardner-Holdt): Z3-Z5
Type Oil: TOFA

12-513:
Product: 92-732
Type: Short Oil Alkyd
% N.V.: Wt.: 75
Volatile: X/IBuOH
Viscosity (Gardner-Holdt): Z-Z2
Type Oil: Mixed

REICHHOLD CHEMICALS, INC.: High Solids Resins(Continued):

RCI Code:

12-607:
Product: 93-955
Type: Short Oil Alkyd
% N.V.: Wt.: 80
Volatile: EGMEEA
Air Reg.: E
Viscosity (Gardner-Holdt): Z1-Z3
Type Oil: TOFA

13-510:
Type: V.T.
% N.V.: Wt.: 80
Volatile: X
Viscosity (Gardner-Holdt): Z2-Z4

13-511:
Type: Acrylic Alkyd
% N.V.: Wt.: 75
Volatile: X/EGMBE
Viscosity (Gardner-Holdt): Z2-Z4

13-520:
Type: Vinyl Copolymer
% N.V.: Wt.: 75
Volatile: X
Viscosity (Gardner-Holdt): Z1-Z3

13-550:
Type: Polyester
% N.V.: Wt.: 90
Volatile: BuAc
Air Reg.: E
Viscosity (Gardner-Holdt): Z4-Z6
Type Oil: None

13-601:
Product: 93-954
Type: Polyester
% N.V.:Wt.: 80
Volatile (Gardner-Holdt): Z3-Z5
Type Oil: None

27-568:
Product: 93-241
Type: MF
% N.V.: 90*
Volatile: Wt.: BuOH
Viscosity (Gardner-Holdt): Z1-Z4
Type Oil: None

E Denotes Exempt Solvent

REICHHOLD CHEMICALS, INC.: High Solids Resins(Continued):

RCI Code:

27-580:
- Type: MF
- % N.V.: Wt.: 80
- Volatile: BuOH
- Air Reg.: E
- Viscosity (Gardner-Holdt): Z2-Z5
- Type Oil: None

27-806:
- Product: MM-83
- Type: MF
- % N.V.: Wt.: 80
- Volatile: I PrOH/I BuOH
- Air Reg.: E
- Viscosity (Gardner-Holdt): T-Z
- Type Oil: None

E Denotes Exempt Solvent

Government regulations on solvent emissions has mandated that many coatings be applied at higher solids and lower VOC. Reichhold has several alkyd, polyester and amino resins for the formulation of high solids air dry and baking enamels. While the solids of these resins has been increased, coatings can be form ulated with hardness, adhesion, flexibility, and gloss fully comparable to those made with resins of more conventional solids.

REICHHOLD CHEMICALS, INC.: LUSTRASOL Acrylic Monomer Modified Alkyds and Acrylic Copolymers:

RCI Code:

13-046:
Product: 4281-50
Type: Acrylic Alkyd-Air Dry
% N.V.: Wt.: 50
Volatile: X
Viscosity (Gardner-Holdt): T-V

13-076:
Type: Acrylic Thermosetting
% N.V.: Wt.: 50
Volatile: X/BuOH
Viscosity (Gardner-Holdt): L-P

13-077:
Type: Acrylic Alkyd-Air Dry
% N.V.: Wt.: 60
Volatile: X
Viscosity (Gardner-Holdt): W-Y

13-081:
Type: Acrylic Alkyd-Air Dry
% N.V.: Wt.: 50
Volatile: X
Viscosity (Gardner-Holdt): T-V

13-511:
Type: Acrylic Alkyd-Air Dry
% N.V.: Wt.: 75
Volatile: X/EGMBE
Viscosity (Gardner-Holdt): Z2-Z4

13-405*:
Product: 93-996
Type: Acrylic Alkyd-WR
% N.V.: Wt.: 70
Volatile (Gardner-Holdt): Z5-Z7

* Design for water reducible coatings

REICHHOLD CHEMICALS, INC.: Miscellaneous Special Purpose Resins, Additives and Accelerators:

RCI Code:

10-714:
Product: WV-302-T
Type Modifier: Rosin
Type: Wrinkle Finish Vehicle
% N.V.: Wt.: 60
Volatile: T
Viscosity (Gardner-Holdt): M-P
Type Oil: Linseed/CWO

13-003:
Product: 1338-50 MS Type
Type Modifier: Phenol
Type: Hardener for Long & Medium Alkyds
% N.V.: Wt.: 50
Volatile: MS
Air Reg.: E
Viscosity (Gardner-Holdt): W-Y
Type Oil: Oxidizing

13-004:
Product: 1338-50X Type
Type Modifier: Phenol
Type: Hardener for Long & Medium Alkyds
% N.V.: Wt.: 50
Volatile: X
Viscosity (Gardner-Holdt): F-H
Type Oil: Oxidizing

13-030:
Product: 98-600
Type: Chlorinated Alkyd
% N.V.: Wt.: 60
Volatile: MS
Air Reg.: E
Viscosity (Gardner-Holdt): U-W
Type Oil: Safflower

44-444:
Product: LUSTRALITE
Type: Arylsulfonamide Formaldehyde
% N.V.: Wt.: 80
Volatile: BuAc
Air Reg.: E
Viscosity (Gardner-Holdt): Z2 min.
Type Oil: None

10-638:

Type Modifier: Silicone
Type: Alkyd Copolymer
% N.V.: Wt.: 60
Volatile: MS
Air Reg.: E
Viscosity (Gardner-Holdt): V-Y
Type Oil: Soybean

E Denotes Exempt Solvent

REICHHOLD CHEMICALS, INC.: Monomer Modified Resins:

Exceptional dry and film properties can be developed from alkyd resins by incorporating styrene, vinyl toluene, or acrylic monomers into the polymerization reaction of the alkyd. The particular properties of a resin are governed by the type and amount of monomer used.

Styrene modified alkyds are readily soluble in aromatic solvents. These modified alkyds are used to produce fast drying finishes and hammer tones that display high gloss and superior water and alkali resistance.

Vinyl toluene modified alkyds are usually thinned in aliphatic solvents. They are as fast to dry as styrene modified alkyds and have considerably less residual odor. The water and alkali resistance of VT alkyds is better than that of unmodified alkyds. A variation of this type of resin is the vinyl toluenated oil.

Acrylic monomer modifications are supplied both as alkyd modified and as copolymerized acrylates. They may be air drying, thermosetting, or thermoplastic. The acrylic modified resins produce finishes which have excellent gloss, hardness and durability, as well as superior water, chemical, solvent, detergent resistance, and stain resistance.

STYRESOL Styrene Monomer Modified Alkyds:

RCI Code:

13-031:

Product: 4250-50
% N.V.: Wt.: 50
Volatile: X
Viscosity (Gardner-Holdt): T-V

13-041:

Product: 4250-50HV
% N.V.: Wt.: 50
Volatile: X
Viscosity (Gardner-Holdt): U-W

13-055:

Product: 4400-50
% N.V.: Wt.: 50
Volatile: X
Viscosity (Gardner-Holdt): R-U

REICHHOLD CHEMICALS, INC.: Polyurethanes:

Oil-modified, one-component polyurethane resins give the advantages of fast dry, good abrasion resistance, gloss retention (clear films) and hardness. These polyurethanes are commonly used in gymnasiums, on factory floors, and in marine finishes.

UROTUF Oil-Modified One Component Polyurethane Coating Resins (ASTM Type I):

RCI Code

13-307:

% N.V.: Wt.: 60
Volatile: MS
Air Req.: E
Viscosity (Gardner-Holdt): Y-Z1
Type Oil: Linseed

13-308:

% N.V.: Wt.: 60
Volatile: MS
Air Req.: E
Viscosity (Gardner-Holdt): V-X
Type Oil: Soybean/Linseed

13-309:

% N.V.: Wt.: 50
Volatile: MS
Air Req.: E
Viscosity (Gardner-Holdt): Z-Z2
Type Oil: Safflower

E Denotes Exempt Solvent

REICHHOLD CHEMICALS, INC.: RCI Chemically Modified Polyolefins:

64-101 POLYBOND 1001:

A thermoplastic polypropylene-acrylic acid copolymer offering exceptional adhesion to unprimed aluminum, steel, and glass; excellent heat resistance (can be autoclaved), chemical resistance, stiffness and tensile strength; low melt viscosity resulting in good flow properties. Can be fabricated via injection molding, fluidized bed coatings. Commercial applications include: glass fiber and/or mineral reinforced compounds; metal/plastic laminate systems.

64-103 POLYBOND 1016:

Thermoplastic polypropylene-EPDM elastomer-acrylic acid terpolymer offering outstanding adhesion to unprimed aluminum, steel, and glass. Excellent low temperature toughness and impact are obtained as well as good physical properties, heat resistance (can be autoclaved), and processability. Can be fabricated via injection molding, sheet extrusion, blown film, or coextrusion processes. Potential end uses: metal or glass coatings, adhesives for metal and/or glass laminates for flexible and rigid packaging, metal/plastic laminates.

64-106 POLYBOND 106:

Thermoplastic ethylene-vinyl acetate-acrylic acid terpolymer. Can be processed via standard commercial techniques. Commercial applications include hot-melt adhesives.

64-107 POLYBOND 107:

Thermoplastic polyethylene-acrylic acid copolymer offering good adhesion to wide variety of substrates, including unprimed steel and aluminum foil; excellent compatibility with fillers and reinforcing agents. Can be blended via standard commercial compounding techniques. Potential end use: metal and pipe coating systems.

The benefits of polyolefins are already well appreciated: low cost, ease of processing, and excellent properties--but their inherent inertness have often limited their application. POLYBOND technology has generated a versatile family of functional thermoplastic polyolefin copolymers and terpolymers. The ability to adhere a part, to bond it to metal, to electroplate it, to decorate with hot foils, to plasticize it, to reinforce effectively, to fill with a low cost filler, and to bond it to glass, aluminum, steel, etc. will increase the applications for POLYBOND.

REICHHOLD CHEMICALS, INC.: STYRESOL Vinyl-toluene Monomer Modified Alkyds and Copolymers(Continued):

RCI Code:

13-038:
Product: 4256-50LV
% N.V.: Wt.: 50
Volatile: VMP
Air Reg.: E
Viscosity (Gardner-Holdt): X-Z1

13-040:
Product: 4256-50
% N.V.: Wt.: 50
Volatile: VMP
Air Req.: E
Viscosity (Gardner-Holdt): Z3-Z5

13-042:
Product: 4257-50
% N.V.: 50
Volatile: VMP
Air Reg.: E
Viscosity (Gardner-Holdt): Z1-Z3

13-043:
Product: 4257-50LV
% N.V.: Wt.: 50
Volatile: VMP
Air Reg.: E
Viscosity (Gardner-Holdt): W-Z

13-059:
Product: 4445-50
% N.V.: Wt.: 50
Volatile: MS
Air Reg.: E
Viscosity (Gardner-Holdt): Z2-Z4

13-065:
% N.V.: Wt.: 60
Volatile: VMP
Air Reg.: E
Viscosity (Gardner-Holdt): Z-Z2

13-510:
% N.V.: Wt.: 80
Volatile: X
Viscosity (Gardner-Holdt): Z2-Z4

E Denotes Exempt Solvent

REICHHOLD CHEMICALS, INC.: STYRESOL Vinyl-toluene Monomer Modified Alkyds and Copolymers(Continued):

RCI Code:

13-520:
% N.V.: Wt.: 75
Volatile: X
Viscosity (Gardner-Holdt): Z1-Z3

13-716:
% N.V.: Wt.: 60
Volatile: HFVMP/MS
Air Reg.: E
Viscosity (Gardner-Holdt): Y-Z1

REICHHOLD CHEMICALS, INC.: Thermoset Polyesters:

These oil-free polyester resins react under the influence of heat (and acid catalysts) with melamine - or benzoguanamine-formaldehyde resins to form hard, flexible films. They show outstanding chemical resistance and display exceptional color and gloss retention on overbake. When formulated with a melamine resin a finish with superior gloss retention and durability to Florida exposure is produced.

BECKOSOL Polyesters:

RCI Code:

13-542:

% N.V.: Wt.: 60
Volatile: X/EGMBE
Viscosity (Gardner-Holdt): V-X

13-550:

% N.V.: Wt.: 90
Volatile: BuAc
Air Reg.: E
Viscosity (Gardner-Holdt): Z4-Z6

13-601:

Product: 93-954
% N.V.: Wt.: 80
Volatile: EGMEEA
Air Reg.: E
Viscosity (Gardner-Holdt): Z3-Z5

13-642:

Product: 98-202
% N.V.: Wt.: 60
Volatile: BUAc/T/EGMBE
Air Reg.: E
Viscosity (Gardner-Holdt): V-X

E Denotes Exempt Solvent

REICHHOLD CHEMICALS, INC.: Ultra Violet Light Curable Resins:

Ultra violet curable resins contain double bonds that cross-link via free radical polymerization. These double bonds can be either internal ethylene linkages found in unsaturated polyester resins or terminal acrylic groups found in UV curable polyurethanes. With either resin type a photoinitiator must be added to the formulation. This molecule cleaves under the influence of UV light to yield the free radicals which initiate polymerization.

POLYLITE polyester resins are highly reactive unsaturated polyesters supplied in styrene. They cure to produce films with outstanding gloss and very good chemical resistance.

POLYLITE polyester resins are used in filler/sealer coatings for particle board on flat stock lines. They are also used in sanding sealers and topcoats for wood furniture in both flat stock and 3 dimensional applications.

POLYLITE polyurethane resins are acrylic terminated polyurethanes. They find a wide variety of applications including overprint varnishes and "no-wax" floor coatings.

POLYLITE Polyester Resins:

RCI Code:

13-830:
Former Code: 31-039
Type: Rigid, Air Inhibited
Monomer: 29% Styrene
Viscosity, cps: 2600-3600

13-831:
Former Code: 32-703
Type: Rigid, Non Air Inhibited
Monomer: 26% Styrene
Viscosity, cps: 1600-1900

13-832:
Former Code: 31-867
Type: Flexible, Air Inhibited
Monomer: 20% Styrene
Viscosity, cps: 2500-3000

13-834:
Former Code: 31-038
Type: Rigid, Air Inhibited
Monomer: 30% Styrene
Viscosity, cps: 800-1000

13-835:
Former Code: 31-830
Type: Flexible, Air Inhibited
Monomer: 30% Styrene
Viscosity, cps: 700-800

REICHHOLD CHEMICALS, INC.: Ultra Violet Light Curable Resins (Continued):

RCI Code:

13-836:
Type: Rigid, Non Air Inhibited
Monomer: 25% Styrene
Viscosity, cps: 800-1800

13-837:
Former Code: 31-001
Type: Rigid, Air Inhibited
Monomer: 34% Styrene
Viscosity, cps: 850-1100

13-838:
Former Code: 31-820
Type: Flexible, Air Inhibited
Monomer: 20% Styrene
Viscosity, cps: 1100-1400

POLYLITE Polyurethane Resins:

13-850:
Type: Acrylic Terminated
% N.V.: Wt.: 85
Volatile: Toluene
Viscosity, cps: 4500-5000

13-851:
Type: Acrylic Terminated
% N.V.: Wt.: 100
Viscosity, cps: 300-400,000

REICHHOLD CHEMICALS, INC.: Water-Reducible Resins:

<u>RCI</u> <u>Code:</u>

13-400:
Product: BECKOSOL 92-100
Polymer Type: Alkyd
Non-Volatile %: Wt: 75
Volatile: EGMBE
Viscosity: Z5-Z6
Recommended Uses: Air Dry, Force Baking Enamels

13-402:
Product: BECKOSOL
Polymer Type: Alkyd
Non-Volatile %: Wt: 75
Volatile: EGMBE/sec-BuOH
Viscosity: Z5-Z7
Recommended Uses: Air Dry, Force Baking Enamels

13-403:
Product: BECKOSOL 92-747
Polymer Type: Phenol/Rosin Mod. Alkyd
Non-Volatile %: Wt: 75
Volatile: EGMBE
Viscosity: Z3-Z4
Recommended Uses: Air Dry & Bake Primers

13-404:
Product: BECKOSOL 92-762
Polymer Type: Chainstopped
Non-Volatile %: Wt: 75
Volatile: EGMBE/sec-BuOH
Viscosity: Z5-Z7
Recommended Uses: Fast Air Dry Enamels

13-405:
Product: LUSTRASOL 93-996
Polymer Type: Acrylic Mod. Alkyd
Non-Volatile %: Wt: 70
Volatile: PGMPE
Viscosity: Z5-Z7
Recommended Uses: Very Fast Dry Enamels

13-420:
Product: BECKOSOL
Polymer Type: Proprietary
Non-Volatile %: Wt.: 50
Volatile: EGMMEA, EGMBE
Viscosity: W
Recommended Uses: Flow Agent

REICHHOLD CHEMICALS, INC.: Water-Reducible Resins(Continued):

RCI Code:

13-432:
Product: BECKOSOL
Polymer Type: Maleinized Oil
Non-Volatile %: Wt.: 37
Volatile: Water
Recommended Uses: Baking Primers

38-690:
Product: EPOTUF
Polymer Type: Epoxy Ester
Non-Volatile %: Wt: 70
Volatile: EGMBE
Viscosity: Z6-Z7
Recommended Uses: Air Dry & Bake Primers, Enamels

38-691:
Product: EPOTUF
Polymer Type: Epoxy Ester
Non-Volatile %: Wt: 70
Volatile: EGMPE
Viscosity: Z6-Z7
Recommended Uses: Air Dry & Bake Primers, Enamels; FDA approvable

27-568:
Product: SUPER-BECKAMINE
Polymer Type: Methylated MF
Non-Volatile %: Wt: 90*
Volatile: BuOH
Viscosity: Z1-Z4
Recommended Uses: High Solids MF

27-806:
Product: UFORMITE
Polymer Type: Methylated MF
Non-Volatile %: Wt: 80*
Volatile: I PrOH/I BuOH
Viscosity: T-Z
Recommended Uses: Fast Cure MF

* 90 @ 60C

REICHHOLD CHEMICALS, INC.: Solvent Key:

A 140	ALIPHATIC 140
A 150	AROMATIC 150
BuAc	n-Butyl Acetate
BuOH	n-Butanol
DEGMBE	Diethylene Glycol Monobutyl Ether
EEP	Ethyl 3-Ethoxypropionate
EGMBE	Ethylene Glycol Monobutyl Ether
EGMMEA	Ethylene Glycol Monomethyl Ether-Acetate
EGMPE	Ethylene Glycol Monopropyl Ether
EtOH	Ethanol
HFAN	High Flash Aromatic Naphtha
HFVMP	High Flash VM&P Naphtha
I BuOH	iso-Butanol
I PrOH	iso-Propanol
LOMS	Low Odor Mineral Spirits
MAK	Methyl Amyl Ketone
MEK	Methyl Ethyl Ketone
MIBK	Methyl Isobutyl Ketone
MS	Mineral Spirits
PGMME	Propylene Glycol Monomethyl Ether
PGMMEA	Propylene Glycol Monomethyl Ether Acetate
Sec-BuOH	sec-Butanol
T	Toluene
VMP	VM&P Naphtha
X	Xylene

RESINALL CORP.: Hydrocarbon Resins: Aliphatic:

Product Name:

760:
Uses Code: C,P,R
Melting Point: R&B Degrees C: 100
Acid/Iodine Number: <1/130
Density Lbs/Gal: 8.9

762:
Uses Code: C,I
Melting Point: R&B Degrees C: 120
Acid/Iodine Number: <1/145
Density Lbs/Gal: 9.0

764:
Uses Code: C,I
Melting Point: R&B Degrees C: 140
Acid/Iodine Number: <1/150
Density Lbs/Gal: 9.0

766:
Uses Code: C,I,P,R
Melting Point: R&B Degrees C: 105
Acid/Iodine Number: <1/110
Density Lbs/Gal: 8.9

767:
Uses Code: C,P,R
Melting Point: R&B Degrees C: 105
Acid/Iodine Number: <1/170
Density Lbs/Gal: 9.0

768:
Uses Code: C,P,R
Melting Point: R&B Degrees C: 105
Acid/Iodine Number: <1/175
Density Lbs/Gal: 8.9

Code
C Coating
I Ink
P Process Aid
R Rubber

RESINALL CORP.: Hydrocarbon Resins: Aromatic:

Product Name:

700:
Uses Code: A,C,P
Melting Point: R&B degrees C: 100
Acid/Iodine Number: <1/90
Density: Lbs/Gal: 8.8

702:
Uses Code: C,P
Melting Point: R&B degrees C: 120
Acid/Iodine Number: <1/95
Density: Lbs/Gal: 8.8

Code:

A	Adhesives
C	Coating
I	Ink
P	Process Aid
R	Rubber

704:
Uses Code: C,I
Melting Point: R&B degrees C: 140
Acid/Iodine Number: <1/95
Density: Lbs/Gal: 8.9

709:
Uses Code: A,C,R
Melting Point: R&B degrees C: 140
Acid/Iodine Number: <1/110
Density: Lbs/Gal: 9.0

711:
Uses Code; A,C,P,R
Melting Point: R&B degrees C: 100
Acid/Iodine Number: <1/45
Density: Lbs/Gal: 8.9

714:
Uses Code: A,C,R
Melting Point: R&B degrees C: 100
Acid/Iodine Number: <1/40
Density: Lbs/Gal: 9.1

725:
Uses Code: C,I
Melting Point: R&B degrees C: 100
Acid/Iodine Number: <1/40
Density: Lbs/Gal: 8.9

784:
Uses Code: A,I
Melting Point: R&B degrees C: 140
Acid/Iodine Number: <1/90
Density: Lbs/Gal: 9.1

RESINALL CORP.: Hydrocarbon Resins: Aromatic(Continued):

Product Name:

7110:
Uses Code: A,C,P
Melting Point: R&B degrees C: 10
Acid/Iodine Number: <1/25
Density: Lbs/Gal: 8.2

111:
Uses Code: A,C,P
Melting Point: R&B degrees C: 10
Acid/Iodine Number: <1/15
Density: Lbs/Gal: 8.2

Code:
A Adhesives
C Coating
I Ink
P Process Aid

125:
Uses Code: A,C,P
Melting Point: R&B degrees C: 25
Acid/Iodine Number: <1/25
Density: Lbs/Gal: 8.3

RESINALL CORP.: Hydrocarbon-Modified Rosin Esters:

514:
Uses Code: I,P
Melting Point: R&B degrees C: 125
Acid/Iodine Number: 18/125
Density: Lbs/Gal: 9.0

519:
Uses Code: I,P
Melting Point: R&B degrees C: 140
Acid/Iodine Number: 19/120
Density: Lbs/Gal: 9.0

521:
Uses Code: I
Melting Point: R&B degrees C: 155
Acid/Iodine Number: 22/135
Density: Lbs/Gal: 9.0

522:
Uses Code: I
Melting Point: R&B degrees C: 150
Acid/Iodine Number: 25/115
Density: Lbs/Gal: 9.1

531:
Uses Code: I
Melting Point: R&B degrees C: 140
Acid/Iodine Number: 18/120
Density: Lbs/Gal: 9.1

RESINALL CORP.: Modified Rosins:

Product Name:

201:
Uses Code: A,P
Melting Point: R&B degrees C: 100
Acid Number: 150
Density Lbs/Gal: 9.2

202:
Uses Code: R,P
Melting Point: R&B degrees C: 65
Acid Number: 135
Density Lbs/Gal: 9.2

203:
Uses Code: A,R
Melting Point: R&B degrees C: 95
Acid Number: 185
Density Lbs/Gal: 9.2

204:
Uses Code: A,R
Melting Point: R&B degrees C: 75
Acid Number: 155
Density Lbs/Gal: 9.2

212:
Uses Code: A,P
Melting Point: R&B degrees C: 90
Acid Number: 120
Density Lbs/Gal: 9.2

219:
Uses Code: R,P
Melting Point: R&B degrees C: 100
Acid Number: 65
Density Lbs/Gal: 9.3

Code:
A Adhesive
P Process Aid
R Rubber

RESINALL CORP.: Rosin Esters:

Product Name:

601:
   Uses Code: A,I
   Melting Point: R&B degrees C: 160
   Acid Number: 15
   Density: Lbs/Gal: 9.2

605:
   Uses Code: A,P
   Melting Point: R&B degrees C: 80
   Acid Number: 9
   Density: Lbs/Gal: 9.1

608:
   Uses Code: A,P
   Melting Point: R&B degrees C: 98
   Acid Number: 10
   Density: Lbs/Gal: 9.2

610:
   Uses Code: A,P
   Melting Point: R&B degrees C: 105
   Acid Number: 12
   Density: Lbs/Gal: 9.2

611:
   Uses Code: A,C,L
   Melting Point: R&B degrees C: 100
   Acid Number: 15
   Density Lbs/Gal: 9.2

622:
   Uses Code: A,C,L
   Melting Point: R&B degrees C: 90
   Acid Number: 8
   Density Lbs/Gal: 9.0

6200:
   Uses Code: G
   Melting Point: R&B degrees C: 90
   Acid Number: 8
   Density Lbs/Gal: 9.0

Code: A Adhesive
      C Coating
      G Chewing Gum
      I Ink
      L Lacquer

RESINALL CORP.: Rosin Maleics: Alcohol-Insoluble:

Product Name:

417:
Uses Code: C,I
Melting Point: R&B Degrees C: 150
Acid Number: 15
Density: Lbs/Gal: 9.1

434:
Uses Code: C,I
Melting Point: R&B Degrees C: 125
Acid Number: 20
Density: Lbs/Gal: 9.2

435:
Uses Code: C,I
Melting Point: R&B Degrees C: 120
Acid Number: 18
Density: Lbs/Gal: 9.2

436:
Uses Code: C,I
Melting Point: R&B Degrees C: 150
Acid Number: 20
Density: Lbs/Gal: 9.2

440:
Uses Code: C,I
Melting Point: R&B Degrees C: 135
Acid Number: 18
Density: Lbs/Gal: 9.1

470:
Uses Code: C,I
Melting Point: R&B Degrees C: 140
Acid Nunber: 35
Density: Lbs/Gal: 9.2

472:
Uses Code: C,I
Melting Point: R&B Degrees C: 130
Acid Number: 40
Density: Lbs/Gal: 9.2

473:
Uses Code: C,I
Melting Point: R&B Degrees C: 95
Acid Number: 35
Density: Lbs/Gal: 9.2

Code
C Coating
I Ink

RESINALL CORP.: Rosin Maleics: Alcohol-Soluble:

Product Name:

Code:
A Adhesive
C Coating
F Floor Polish
I Ink
V Varnish

802:
Uses: A,F,I
Melting Point: R&B degrees C: 160
Acid Number: 200
Density: Lbs/Gal: 9.8

805:
Uses: F,I
Melting Point: R&B degrees C: 160
Acid Number: 190
Density: Lbs/Gal: 9.8

807:
Uses: F,I
Melting Point: R&B degrees C: 120
Acid Number: 125
Density: Lbs/Gal: 9.8

807A:
Uses: F,I
Melting Point: R&B degrees C: 130
Acid Number: 150
Density: Lbs/Gal: 9.4

807AC:
Uses: C,V
Melting Point: R&B degrees C: 130
Acid Nunber: 155
Density: Lbs/Gal: 9.4

809:
Uses: C,F,I
Melting Point: R&B degrees C: 160
Acid Number: 130
Density: Lbs/Gal: 9.9

833:
Uses: I
Melting Point: R&B degrees C: 135
Acid Number: 235
Density: Lbs/Gal: 9.2

835:
Uses: C,F,I
Melting Point: R&B degrees C: 155
Acid Number: 190
Density: Lbs/Gal: 9.3

836:
Uses: C,F,I
Melting Point: R&B degrees C: 140
Acid Number: 120
Density Lbs/Gal: 9.6

RESINALL CORP.: Rosin-Modified Phenolics:

Product Name:

901:
Uses: C,I
Melting Point: R&B degrees C: 135
Acid Number: 18
Density Lbs/Gal: 9.0

903:
Uses: C,I
Melting Point: R&B degrees C: 170
Acid Number: 20
Density Lbs/Gal: 9.2

905:
Uses: C,I
Melting Point: R&B degrees C: 150
Acid Number: 20
Density Lbs/Gal: 9.1

908:
Uses: C,I
Melting Point: R&B degrees C: 165
Acid Number: 20
Density: Lbs/Gal: 9.1

909:
Uses: C,I
Melting Point: R&B degrees C: 165
Acid Number: 20
Density: Lbs/Gal: 9.1

Code
C Coating
I Ink

ROHM AND HAAS: ACRYLOID Thermoplastic Acrylic Ester Resins:

Grade:

ACRYLOID A-10S:
Hardest of the ACRYLOID thermoplastic resins, resistant to alcohol and water.
Solids %: 30
Chemical Composition: methyl methacrylate polymer
Principal Uses: Heat resistant white enamels with good color retention. Plastisol primers. Plastic barrier coatings.

ACRYLOID A-11:
Similar to ACRYLOID A-10. Slightly softer.
Solids %: 100
Chemical Composition: methyl methacrylate polymer
Principal Uses: Same as ACRYLOID A-10 and ACRYLOID A-101.

ACRYLOID A-101:
ACRYLOID A-10 polymer in methyl ethyl ketone.
Solids %: 40
Chemical Composition: methyl methacrylate polymer
Principal Uses: Vinyl topcoating and printing inks.

ACRYLOID A-21:
Excellent adhesion, hardness and exterior durability
Solids %: 30
100
Chemical Composition: methyl methacrylate polymer
Principal Uses: Automotive finishing and refinishing.

ACRYLOID A-21LV:
Lower viscosity ACRYLOID A-21. Better compatability with cellulosics.
Solids %: 30
Chemical Composition: methyl methacrylate polymer
Principal Uses: Used in bake-sand-bake finishes.

ACRYLOID A-30:
Abrasion resistance, toughness and flexibility. Higher molecular weight.
Solids %: 100
Chemical Composition: methyl methacrylate polymer
Principal Uses: Used in vinyl topcoating with vinyl copolymers and homopolymers.

ACRYLOID B-44S:
ACRYLOID B-44:
Hardness, flexibility and adhesion.
Solids %: 40
100
Chemical Composition: methyl methacrylate copolymer
Principal Uses: Finishes for vacuum-metallized plastics, aluminum sash and chrome plate. Luminescent finishes.

ROHM AND HAAS: ACRYLOID Thermoplastic Acrylic Ester Resins (Continued):

Grade:

ACRYLOID B-48S:
ACRYLOID B-48N:
Toughness, exterior durability, adhesion to bare metal.
Solids %: 45
100
Chemical Composition: methyl methacrylate copolymer
Principal Uses: Clear coating over copper, brass, bronze, zinc; aerosol coatings, plastic coatings.

ACRYLOID B-50S:
Similar to ACRYLOID B-48S but lower cost and poor adhesion to untreated metal.
Solids %: 45
Chemical Composition: methyl methacrylate copolymer
Principal Uses: Coatings for plastics, seamless floors, treated metals, aerosol coatings.

ACRYLOID B-60:
Hardness, fast dry, adhesion to plastics.
Solids %: 100
Chemical Composition: methyl/butyl methacrylate copolymer
Principal Uses: Specialty coatings for automotive and furniture plastics and for plastic films.

ACRYLOID B-66:
Fast air-dry and excellent compatibility with other film formers.
Solids %: 50
51
100
Chemical Composition: methyl/butyl methacrylate copolymer
Principal Uses: Specialty clear and pigmented coatings on metal, wood, plastic and fabric.

ACRYLOID B-67:
Excellent for improving properties of medium and long-oil alkyd and oleoresinous varnishes.
Solids %: 45
100
Chemical Composition: isobutyl methacrylate polymer
Principal Uses: Used in enamels with and without alkyd modification.

ROHM AND HAAS: ACRYLOID Thermoplastic Acrylic Ester Resins (Continued):

Grade:

ACRYLOID B-67MT:
Similar to ACRYLOID B-67 but dissolved in mineral spirits.
Solids %: 45
100
Chemical Composition: isobutyl methacrylate polymer
Principal Uses: Used in plastic coating, printing ink, over print varnishes and refinish lacquers.

ACRYLOID B-72:
Compatible with vinyl and silicone resins. Stable, durable, transparent.
Solids %: 50
100
Chemical Composition: ethyl methacrylate copolymer
Principal Uses: General purpose clear and pigmented coatings, aerosol and inks.

ACRYLOID B-82:
Low cost with typical acrylic properties.
Solids %: 51
100
Chemical Composition: methyl methacrylate copolymer
Principal Uses: General purpose coatings.

ACRYLOID B-84:
Outstanding adhesion to thermoset enamel and plastic sub-substrates.
Solids %: 45
Chemical Composition: methyl methacrylate copolymer
Principal Uses: Refinishing glass-fiber-reinforced plastics and other plastic surfaces. Tie coats. Touch-up enamels for appliances.

ACRYLOID B-99:
Used in pigment grind to improve gloss. Improves short and medium-oil alkyd properties.
Solids %: 50
Chemical Composition: methyl methacrylate copolymer
Principal Uses: Grinding medium, refinish lacquer, and maintenance paints.

ACRYLOID C-10LV:
Softest and most elastic ACRYLOID resin.
Solids %: 40
Chemical Composition: methyl acrylate polymer
Principal Uses: Brushing and dipping coating, thermoplastic adhesive.

ROHM AND HAAS: ACRYLOID Thermoplastic Acrylic Ester Resins (Continued):

Grade:

ACRYLOID F-10:
Good adhesion and elongation. Soluble in petroleum hydrocarbons.
Solids %: 40
Chemical Composition: butyl methacrylate polymer
Principal Uses: Clear and pigmented coatings on metal; luminescent pigment vehicles.

ACRYLOID NAD-10:
Non-aqueous dispersion, fast solvent release, glossy films, good adhesion.
Solids %: 40
Principal Uses: Rotogravure inks, fluorescent inks, general coatings.

Typical Physical Properties of Solvent Grade ACRYLOID Thermoplastic Resins:

Grade:

ACRYLOID A-10S:
Percent Solids: 30
Solvent: PM Acetate
Pounds Per Gallon: 8.6
Viscosity Brookfield cps. at 25C: 800-1,200
Flash Point PMCC, F.: 114

ACRYLOID A-101:
Percent Solids: 40
Solvent: Methyl Ethyl Ketone
Pounds Per Gallon: 7.8
Viscosity Brookfield cps. at 25C: 700 to 1,400
Flash Point PMCC, F: 25

ACRYLOID A-21:
Percent Solids: 30
Solvent: Toluene/Butanol (90/10)
Pounds Per Gallon: 7.8
Viscosity Brookfield cps. at 25C: 235 to 365
Flash Point PMCC, F: 42

ACRYLOID A-21LV:
Percent Solids: 30
Solvent: Toluene/MEK/Butanol (50/40/10)
Pounds Per Gallon: 7.8
Viscosity Brookfield cps. at 25C: 210 to 280
Flash Point PMCC, F: 24

ROHM AND HAAS: ACRYLOID Thermoplastic Acrylic Ester Resins (Continued):

Grade:

ACRYLOID B-44S:
Percent Solids: 40
Solvent: Toluene
Pounds Per Gallon: 8.1
Viscosity Brookfield cps. at 25C: 855 to 1,700
Flash Point PMCC, F: 45

ACRYLOID B-48S:
Percent Solids: 45
Solvent: Toluene
Pounds Per Gallon: 8.1
Viscosity Brookfield cps. at 25C: 6,000 to 10,000
Flash Point PMCC, F: 42

ACRYLOID B-50S:
Percent Solids: 45
Solvent: Toluene
Pounds Per Gallon: 8.1
Viscosity Brookfield cps. at 25C: 5,000 to 9,000
Flash Point PMCC, F: 46

ACRYLOID B-66:
Percent Solids: 51
Solvent: Toluene
Pounds Per Gallon: 8.2
Viscosity Brookfield cps. at 25C: 2,000 to 5,000
Flash Point PMCC, F: 44

ACRYLOID B-67:
Percent Solids: 45
Solvent: VM&P Naphtha
Pounds Per Gallon: 7.3
Viscosity Brookfield cps. at 25C: 900 to 1,600
Flash Point PMCC, F: 54

ACRYLOID B-67MT:
Percent Solids: 45
Solvent: Mineral Spirits
Pounds Per Gallon: 7.4
Viscosity Brookfield cps. at 25C: 800 to 1,600
Flash Point PMCC, F: 106

ACRYLOID B-72:
Percent Solids: 50
Solvent: Toluene
Pounds Per Gallon: 8.3
Viscosity Brookfield cps. at 25C: 3,700 to 6,200
Flash Point PMCC, F: 44

ROHM AND HAAS: ACRYLOID Thermoplastic Acrylic Ester Resins (Continued):

Grade:

ACRYLOID B-82:
Percent Solids: 51
Solvent: Toluene
Pounds Per Gallon: 8.3
Viscosity Brookfield cps. at 25C: 4,000 to 7,000
Flash Point PMCC, F: 40

ACRYLOID B-84:
Percent Solids: 45
Solvent: Toluene/sec. Butanol (85/15)
Pounds per Gallon: 7.9
Viscosity Brookfield cps. at 25C: 4,000 to 8,000
Flash Point PMCC, F: 42

ACRYLOID B-99:
Percent Solids: 50
Solvent: Xylene/Toluene (70/30)
Pounds Per Gallon: 2,100 to 4,200
Viscosity Brookfield cps. at 25C: 2,100 to 4,200
Flash Point PMCC, F: 65

ACRYLOID C-10LV:
Percent Solids: 40
Solvent: Toluene
Pounds Per Gallon: 8.2
Viscosity Brookfield cps. at 25C: 38 to 82
Flash Point PMCC, F: 43

ACRYLOID F-10:
Percent Solids: 40
Solvent: Mineral Spirits/AROMATIC 150 (90/10)
Pounds Per Gallon: 7.4
Viscosity Brookfield cps. at 25C: 1,400 to 2,600
Flash Point PMCC, F: 108

ACRYLOID NAD-10:
Percent Solids: 40
Solvent: VM&P Naphtha
Pounds Per Gallon: 7.1
Viscosity Brookfield cps. at 25C: 290 to 600
Flash Point PMCC, F: 54

ROHM AND HAAS: ACRYLOID Thermoplastic Acrylic Ester Resins (Continued):

Typical Physical Properties of solid grade ACRYLOID thermoplastic resins:

Grade:

ACRYLOID A-11:
- Bulk Density lbs./gallon: 9.8
- Solution Viscosity:
  - Brookfield, cps at 25C: 1200-1900
  - Solvent: 35% in Methyl ethyl ketone

ACRYLOID A-21:
- Bulk Density lbs./gallon: 9.8
- Solution Viscosity:
  - Brookfield, cps at 25C: 235-365
  - Solvent: 30% in Toluene/n-butanol

ACRYLOID A-30:
- Bulk Density lbs./gallon: 9.8
- Solution Viscosity:
  - Brookfield, cps. at 25C: 1500-2500
  - Solvent: 20% in Methyl ethyl ketone

ACRYLOID B-44:
- Bulk Density lbs./gallon: 9.6
- Solution Viscosity:
  - Brookfield, cps. at 25C: 700-1700
  - Solvent: 40% in Toluene

ACRYLOID B-48N:
- Bulk Density lbs./gallon: 9.6
- Solution Viscosity:
  - Brookfield, cps. at 25C: 6000-10000
  - Solvent: 45% in Toluene

ACRYLOID B-60:
- Bulk Density lbs./gallon: 9.4
- Solution Viscosity:
  - Brookfield, cps. at 25C: 1800-2000
  - Solvent: 51% in Methyl ethyl ketone/Toluene

ACRYLOID B-66:
- Bulk Density lbs./gallon: 9.1
- Solution Viscosity:
  - Brookfield, cps. at 25C: 2000-5000
  - Solvent: 50% in toluene

ROHM AND HAAS: ACRYLOID Thermoplastic Acrylic Ester Resins (Continued):

Typical Physical Properties of solid grade ACRYLOID thermoplastic resins(continued):

Grade:

ACRYLOID B-67:
- Bulk Density lbs./gallon: 8.8
- Solution Viscosity:
  - Brookfield, cps at 25C: 250-375
  - Solvent: 45% in Toluene

ACRYLOID B-72:
- Bulk Density lbs./gallon: 9.6
- Solution Viscosity:
  - Brookfield, cps at 25C: 3700-6200
  - Solvent: 50% in toluene

ACRYLOID B-82:
- Bulk Density lbs./gallon: 9.7
- Solution Viscosity:
  - Brookfield, cps at 25C: 4000-7000
  - Solvent: 50% in Toluene

ROHM AND HAAS: Resins for Industrial and Maintenance Coatings:

ACRYLOID AT-400:
Economical, High-Solids, Thermosetting Acrylic Resin for Durable Industrial Baking Finishes

Performance Highlights:
* Outstanding durability; suitable for exterior and interior applications
* High gloss
* Attractive balance of hardness, flexibility, and resistance properties
* Excellent appearance
* Exceptional gloss and color retention; resists overbake color development

Formulation Advantages
* Low VOC
* Cost competitive with conventional systems
* Wide formulating latitude
* Broad compatability with other resins; can be used to modify polyesters and alkyds
* Excellent pigment dispersant

Application Benefits
* Uncatalyzed cure at a low 250F
* Can be crosslinked with monomeric melamine (and external catalyst) for lower VOC
* Excellent spray properties
* Outstanding resistance to sagging, cratering, and rinse-mark telegraphing
* Overspray is less tacky and easier to clean than that of many high-solids polyesters

Typical Properties:
Solids Content, %: 75
Solvent: Methyl n-amyl ketone
Viscosity, cps (Brookfield, 25C): 5000-9000
Density, lbs./U.S. gallon at 25C: 8.54
Specific Gravity at 25C: 1.025
Flash Point, SETA Closed Cup, F(C): 121(49)
Appearance: Clear liquid
Color, APHA: 500 max.
Hydroxyl number (solids): 86
Acid Number (solids): 20-30

ROHM AND HAAS: Resins for Industrial and Maintenance Coatings (Continued):

ACRYLOID AT-410:

High-Solids, Thermosetting Acrylic Resin for Baking Enamels with Excellent Application Properties

Performance Highlights:

* Outstanding hardness
* Exceptional stain and solvent resistance
* High gloss and image clarity
* Resists overbake color development
* Excellent color retention
* Elegant appearance

Formulation Advantages:

* Low VOC
* Broad compatability with other high-solids resins; can be used to modify polyesters and alkyds
* Wide formulating latitude
* Good pigment dispersant
* Baking enamel formulations have excellent viscosity stability

Application Benefits:

* Excellent spray properties
* Outstanding resistance to sagging, cratering, and rinse-mark telegraphing
* Overspray is less tacky and easier to clean than that of many high-solids polyesters
* Uncatalyzed cure at a low 250F
* Can be crosslinked with monomeric melamine (and external catalyst) for lower VOC

Typical Properties:

Solids Content, %: 73
Solvent: Methyl n-amyl ketone
Viscosity, cps. (Brookfield, 25C): 3000-6000
Density, lbs./U.S. gallon at 25C: 8.4
Specific Gravity at 25C: 1.01
Flash Point, SETA Closed Cup, F(C): 113(45)
Appearance: Clear liquid
Color, APHA: 500 max.
Hydroxyl number (solids): 64
Acid number (solids): 20-30

ROHM AND HAAS: Resins for Industrial and Maintenance Coatings: (Continued):

ACRYLOID AT-954:

Thermosetting, Modified-Acrylic Resin for High-Solids Baking Enamels with VOC under 3.0 Lbs./Gal

Performance Highlights:

* High gloss
* Exceptional balance of hardness, flexibility, and resistance properties
* Excellent appearance
* Better outdoor durability than high-solids polyesters
* Higher impact resistance than other high-solids acrylics

Formulation Advantages:

* Low VOC (supplied at 85% solids)
* Excellent as a sole vehicle, or as a performance-improving modifier for other high-solids resins
* Good compatability with high-solids polyesters and alkyds
* Good pigment dispersant

Application Benefits:

* Excellent spray properties
* Outstanding resistance to sagging, cratering, and rinse-mark telegraphing
* Easy-to-remove overspray

Typical Properties:

Solids Content, %: 85
Solvent: Methyl n-amyl ketone
Viscosity, cps. (Brookfield, 25C): 2500-5000
Density, lbs./U.S. gallon at 25C: 8.55
Specific Gravity at 25C: 1.025
Flash Point, SETA Closed Cup, F(C): 128(53)
Appearance: Clear liquid
Color, APHA: 300 max.
Hydroxyl number (solids): 160
Acid number (solids): 9-15

ROHM AND HAAS: Resins for Industrial and Maintenance Coatings (Continued):

ACRYLOID AU-608S and ACRYLOID AU-608X:
Hydroxyl Functional Isocyanate-Reactive Acrylic Resins

Performance Highlights
* Gloss and tint retention, reducing the need to repaint
* Film integrity maintained when exposed to corrosive chemicals
* Long-lasting protection because of abrasion and crack resistance
* Low temperature cure

Typical Properties

ACRYLOID AU-608S:
Solids Content, %: 60
Nonvolatiles: 62.5
Solvent: PM Acetate/Toluene (65/35)
Viscosity, cps 25C: 4500-8500
Wet Density, lbs./U.S. gal.: 8.8
Dry Density, lbs./U.S. gal.: 9.6
Specific Gravity at 25C: 1.06
Hydroxyl Equivalent Weight (solids): 600
Hydroxyl Equivalent Weight (as supplied): 1000
Hydroxyl Number (Solids): 86

ACRYLOID AU-608X:
Solids Content, %: 58
Nonvolatiles: 60
Solvent: Xylene
Viscosity, cps. 25C: 4000-9000
Wet Density, lbs./U.S. gal.: 8.5
Dry Density, lbs./U.S. gal: 9.6
Specific Gravity at 25C: 1.02
Hydroxyl Equivalent Weight (solids): 600
Hydroxyl Equivalent Weight (as supplied): 1032
Hydroxyl Number (Solids): 86

ACRYLOID B-48S:
Thermoplastic Acrylic Resin with Adhesion to Metal

Performance Highlights:
* Excellent adhesion to untreated metal
* Toughness
* Flexibility
* Exterior durability
* Excellent block resistance
* Fast solvent release

Typical Properties:
Solids Content, %: 45
Solvent: Toluene
Viscosity, Brookfield LVF at 45% solids in Toluene: 6000-10000

ROHM AND HAAS: Resins for Industrial and Maintenance Coatings (Continued):

Resin QR-999:

Hydroxyl-Functional, Isocyanate-Reactive Acrylic Resin

Performance Highlights:

* Fast dry
* Excellent durability
* Low temperature cure
* Gloss and tint retention

Typical Properties:
Solids Content, %: 60
Nonvolatiles Content, %: 62
Solvent: n-butyl acetate
Viscosity, cps. at 25C: 3000-6000
Wet Density, lbs./U.S. gal.: 8.6
Dry Density, lbs./U.S. gal.: 9.8
Specific Gravity at 25C: 1.03
Hydroxyl Equivalent Weight (solids): 600
Hydroxyl Equivalent Weight (as supplied): 1000
Color, APHA: 60 max.

Resin QR-946:
Hydroxyl-Functional Acrylic Resin for Reaction with Isocyanates to Make Low VOC Acrylic Urethane Coatings

Performance Highlights:
* High gloss and image clarity
* Good durability
* Excellent solvent and chemical resistance

Formulation Advantages:
* Low VOC
* Good pigment dispersion

Application Benefits:
* Ambient cure
* Satisfactory pot life

Typical Properties:
Solids Content, %: 67
Nonvolatiles Content, %: 69
Solvent, as supplied: Methyl n-amyl ketone
Viscosity, LVF, cps: 6000-11500
Equivalent Weight: 425
Specific Gravity at 25C: 1.03
Wet Density, lbs./U.S. gal.: 8.6
Dry Density, lbs./U.S. gal.: 9.9

ROHM AND HAAS: Resins for Industrial and Maintenance Coatings (Continued):

Resin QR-1166:

New Acrylic Polyol for a Variety of Urethane Coatings with Low VOC at an Extremely Attractive Cost.

Features of Coatings Based Solely on Resin QR-1166:
* Economical Urethane Systems
* Low VOC Potential
* Excellent Dry Time
* Excellent Catalyzed Pot Life
* Good Alkali and Acid Resistance

Added Features with Reactive Modifier QM-1007 Modification:
* Lower VOC Potential
* Good Dry Time/Pot Life Balance
* Excellent Flexibility and Impact Resistance
* Good Solvent Resistance

Typical Properties:
Nonvolatiles Content, %: 73
Solvent: n-butyl acetate
Viscosity as Supplied, cps: 8000
Hydroxyl Equivalent Weight:
Solids Basis: 1000
As Supplied: 1370
Hydroxyl Number Solids Basis: 56
Dry Density, lbs/U.S gal.: 9.5
Wet Density, lbs./U.S. gal.: 8.8

Reactive Modifier QM-1007:
Reactive Diluent for Increasing Paint Solids and Improving Film Build of Acrylic and Polyester Urethane Coatings.

Performance Highlights:
(when combined with polyester or acrylic polyol)
* High gloss and depth image
* Improved durability of high solids polyester systems
* Improved flexibility (catalyzed formulations)

Formulation Advantages:
* Low VOC

Typical Properties:
Solids Content, %: 97 to 99.5
Equivalent Weight: 100
Viscosity, Brookfield cps: 100 max.
Specific Gravity at 25C: 1.06
Wet Density lbs/U.S. gal.: 8.8

ROHM AND HAAS: Resins for Industrial and Maintenance Coatings (Continued):

Resin QR-1003 and Resin QR-1004:

Performance Highlights:

* Film integrity maintained when exposed to corrosive chemicals
* Gloss and tint retention reducing the need to repaint
* Long-lasting protection because of abrasion and crack resistance

Typical Properties:

Experimental Resin QR-1003:
Solids Content, %: 57
Nonvolatiles Content, %: 60
Solvent: PM Acetate/Ethyl Benzene/Petroleum Distillates (55/25/20)
Viscosity, cps., Brookfield at 25C: 3000-6000
Wet Density lbs./U.S. gal.: 8.6
Dry Density lbs./U.S. gal.: 9.6
Hydroxyl Equivalent Wt.
(Solids): 600
(As Supplied): 1000
Bulking Values, lbs./gal.:
(Solids): .104
(As Supplied): .116

Experimental Resin QR-1004:
Solids Content, %: 57
Nonvolatiles Content, %: 60
Solvent: Xylene/PM Acetate (66/34)
Viscosity, cps., Brookfield at 25C: 3000-6000
Wet Density lbs./U.S. gal.: 8.6
Dry Density lbs./U.S. gal.: 9.6
Hydroxyl Equivalent Wt.
(Solids): 600
(As Supplied): 1000
Bulking Values, lbs./gal.
(Solids): .104
(As Supplied): .116

ROHM AND HAAS: Resins for Industrial and Maintenance Coatings (Continued):

Resin QR-1033:

Performance Highlights:
* Lacquer-like dry properties whch reduce equipment downtime
* Reduced formulation costs because of lower isocyanate demand
* Film integrity maintained when exposed to corrosive chemicals
* Gloss and tint retention, thereby reducing the need to repaint

Typical Properties:

Solids Content, %: 46.5
Nonvolatiles Content, %: 50
Solvent: PM Acetate/Ethyl Benzene/Petroleum Distillates (55/25/20)
Specific Gravity at 25C: 1.04
Viscosity, cps. at 25C: 4000-8000
Dry Density, lbs./U.S. gal.: 9.8
Wet Density, lbs./U.S. gal.: 8.5
Hydroxyl Equivalent Weight
  (Solids): 1000
  (As Supplied): 2000
Bulking Values, lbs./gal
  (Solids): .103
  (As Supplied): .118

Resin QR-1102:

Hydroxyl-Functional, Isocyanate-Reactive Acrylic Resin

Performance Highlights:
* Superior early hardness development
* Very good durability
* Low temperature cure

Typical Properties:
Solids Content, %: 52
Nonvolatiles Content, %: 55
Solvent: PM Acetate/Ethyl Benzene/NAPHTHOLITE 66/3 (55/25/20)
Specific Gravity at 25C: 1.02
Viscosity, cps at 25C: 3500-7000
Wet Density, lbs./U.S. gal.: 8.5
Dry Density, lbs./U.S. gal.: 9.5
Hydroxyl Equivalent Weight:
  (Solids): 867
  (As Supplied): 1556

ROHM AND HAAS: Resins for Industrial and Maintenance Coatings (Continued):

MAINCOTE HG-54:

Waterborne Acrylic Resin for High-Gloss, Corrosion-Resistant Paints

Performance Highlights:

* Decreased maintenance expenses and replacement costs
* Reduced need to repaint as frequently
* Protection of the paint user's valuable investment
* Excellent resistance to gloss loss, corrosion, fading, chalking, chemicals and solvents

Typical Properties:

Solids Content, %: 41.5
Viscosity, cps. Brookfield at 25C: 50-400
pH: 7.5
Bulking Value
Wet bulking value, U.S. gal./lbs.: .117
Dry bulking value, U.S. gal./lbs.: .112
Density at 25C, lbs./U.S. gal: 8.5

MAINCOTE TL-5:

Waterborne Acrylic Polymer that Crosslinks with Epoxies for use in High-Gloss, Tile-Like Coatings

Performance Highlights

* Similar performance of solvent-borne product at significant raw material savings
* Low maintenance costs
* Reduced need to paint as often
* Easy removal of difficult stains
* Long term high gloss, and resistance to chemicals often used in institutional cleaning materials
* Less need to relocate personnel during paint application
* Low odor paint
* Simplified contractor's work schedule
* Recoating within hours

Formulation Advantages:

* Low level of organic volatiles

Typical Properties:

Solids Content, %: 41.5%
pH: 7.5
Viscosity, cps., Brookfield at 25C: 200-1200
Density at 25C lbs./U.S. gal.: 8.6
Equivalent Weight (solids): 1700

ROHM AND HAAS: Resins for Industrial and Maintenance Coatings (Continued):

RHOPLEX MV-23:

Waterborne Acrylic Vehicle for Multi-purpose Primers

Performance Highlights:

* Good corrosion resistance for various metals
* Stain blocking on staining woods and for interior stains
* Alkali resistance on masonry
* Excellent adhesion to galvanized steel

Typical Properties:
Solids content, %: 43.0
Viscosity, cps., Brookfield at 25C: 300 to 1300
pH (as shipped): 9.6
Density at 25C, lbs./U.S. gal.: 8.8
Bulking Value:
Emulsion, U.S. gal./lbs.: 0.114
Emulsion solids, U.S. gal./lb.: 0.115
Particle charge: Anionic

RHOPLEX WL-51:
Economical General-Purpose Acrylic Emulsion for Air-Dry or Low Temperature Force-Dry Waterborne Coatings

Performance Highlights:
* Excellent adhesion to a variety of engineering and decorative plastics, treated and untreated metal, and wood substrates
* Good metal protection
* High gloss and image clarity
* Gasoline and solvent resistance

Formulation Advantages:
* Low VOC
* Divalent stability
* Internal viscosity control
* Blends well with other latexes and water-reducible alkyds

Application Benefits
* Outstanding spray characteristics
* Rapid hardness development
* Fast set and tack-free time

Typical Properties:
Appearance: Milky white fluid
Solids Content, wt. %/vol. %: 41.5/39.5
pH: 7.5
Density, lbs./U.S gal.: 8.6
Specific Gravity at 25C: 1.03
Brookfield Viscosity, at 25C, cps.: 200-1200

ROHM AND HAAS: Resins For Industrial and Maintenance Coatings (Continued):

RHOPLEX WL-71:

Acrylic Emulsion for Blending with Water-Reducible Alkyds

Performance Highlights of RHOPLEX WL-71/alkyd blends:

* Fast drying
* Rapid development of water resistance
* Good resistance to gasoline and solvents
* Excellent gloss and image clarity
* Very good corrosion protection
* Excellent appearance

Formulation Advantages:

* Low VOC
* Excellent compatibility with water-reducible alkyds
* Good pH and viscosity stability
* Stable with reactive pigments

Application Benefits:

* Excellent spray characteristics
* Good dip-coat and flow coat characteristics
* Reduced dry-to-touch and tack-free time
* Reduced hardness development

Typical Properties:

Solids Content, wt. %/vol. %: 41.5/39.4
pH: 7.5
Brookfield Viscosity, cps., at 25C: 300-2000
Density at 25C lbs./U.S. gal: 8.6
Minimum Film Forming Temperature, C: 52
Specific Gravity at 25C: 1.035

ROHM AND HAAS: Resins for Industrial and Maintenance Coatings (Continued):

RHOPLEX WL-81:

Acrylic Emulsion for Lacquers Offering Metal Protection and Durability

Performance Highlights:

* Very good metal protection
* Exterior durability
* Early water resistance
* Good gloss and image clarity
* Adhesion to a variety of substrates including pretreated steel, pretreated aluminum, and wood
* Print and block resistance
* Salt spray resistance

Formulation Advantages:

* Low VOC
* Internal and precise viscosity control
* Excellent modifier for water-reducible epoxy esters

Application Benefits:

* Good spray characteristics
* Rapid hardness development
* Fast set and tack-free time
* Excellent dip-coat and flow-coat characteristics
* Dip tank stability

Typical Properties:

Appearance: Milky white liquid
Solids Content, wt. %/vol. %: 41.5/39.2
Volatile Phase: Water
Brookfield Viscosity, cps., at 25C: 40-500
pH: 7.5
Minimum Film Forming Temperature, C: 57
T300, C: 60
Density at 25C., lbs/U.S. gal.: 8.6
Specific Gravity at 25C: 1.03
Freeze/Thaw Stability (5 Cycles, -15/70F): Pass
Mechanical Stability (5' Waring Blender): Pass
Heat Stability (10 Days/140F): Pass

ROHM AND HAAS: Resins for Industrial and Maintenance Coatings (Continued):

RHOPLEX WL-91:

Acrylic Emulsion for Air-Dry Lacquers with Solvent Resistance and High Gloss

Performance Highlights:

* Outstanding resistance to gasoline and solvents
* High gloss and image clarity
* Good adhesion to a variety of metal, plastic, and wood substrates
* Salt spray resistance
* Corrosion resistance
* Early water resistance
* Print and block resistance
* Excellent metal protection with reactive pigments

Formulation Advantages:

* Low VOC
* Wide formulating latitude (permits broad choice of coalescents)
* Stability with reactive pigments
* Internal and precise viscosity control
* Divalent ion stability

Application Benefits:

* Excellent spray characteristics
* Rapid hardness development
* Fast set and tack-free time

Typical Properties:

Appearance: Milky white liquid
Solids Content, wt.%/vol.%: 41.5/39.4
Volatile Phase: Water
Brookfield Viscosity at 25C: 100-1000
pH: 7.7
Minimum Film Forming Temperature, C: 52
T300, C: 58
Density at 25C, lbs./U.S. gal. 8.6
Specific Gravity at 25C: 1.03
Freeze/Thaw Stability (5 Cycles, -15/70F): Pass
Mechanical Stability (5' Waring Blender): Pass
Heat Stability (10 Days/140F): Pass

ROHM AND HAAS: Resins for Industrial and Maintenence Coatings (Continued):

RHOPLEX WL-92:

Acrylic Emulsion for Air-Dry Lacquers with Outstanding Hot-Print and Block Resistance

Performance Highlights:

* Early and higher stackability, with no sticking
* Earlier packing and shipping
* More resistance to dust and dirt damage
* Fewer rejects and returns due to coating and handling procedures
* Long-term protection
* An excellent balance of handling, appearance and resistance properties

Formulation Advantages:

* Low VOC for environmentally acceptable coatings
* The versatility to formulate for many coatings needs and substrates with only one resin, simplifying inventory management
* Compatibility with corrosion-inhibiting pigments
* Ease of formulation

Application Benefits

* Extreme versatility--application can be by spray, dip, flow or direct roll

Typical Properties:

Appearance: Light yellow opaque fluid
Solids Content, wt.%/vol.%: 42/40.2
pH: 7.5
Brookfield Viscosity, cps., at 25C: 200-600
Density at 25C, lbs./U.S. gal.: 8.6
Specific Gravity at 25C: 1.03
Minimum Film Forming Temperature, C: 23
Freeze/Thaw Stability: Protect from Freezing
Mechanical Stability: Pass
Heat Stability (10 Days/140F): Pass

ROHM AND HAAS: Resins For Industrial and Maintenance Coatings (Continued):

RHOPLEX WL-96:

Acrylic Emulsion for Economical Air-Dry Lacquers with Outstanding Adhesion to Plastics and Metals

Performance Highlights:

* Early stacking of freshly coated items, with no sticking
* Early packing and shipping
* More resistance to dust and dirt-pickup
* Fewer rejects and returns due to coating and handling procedure
* Excellent balance of handling, appearance and resistance properties

Formulation Advantages:

* Low resin and inventory costs
* Low VOC for environmentally acceptable coatings
* Reactive pigment stability for improved metal protection
* Internal and precise viscosity control--external thickeners, which detract from metal protection are not required

Application Benefits:

* Very versatile--application can be by spray, dip, flow or direct roll

Typical Properties:

Appearance: Light yellow opaque liquid
Solids Content, wt.%/vol.%: 42/39.9
pH: 7.0
Brookfield Viscosity, cps at 25: 400-1400
Density at 25C, lbs./U.S. gal.: 8.6
Specific Gravity at 25C: 1.03
MFT, C: 26
Freeze/Thaw Stability: Protect from Freezing
Mechanical Stability: Pass
Heat Stability (10 days/140F): Pass

SANNCOR INDUSTRIES, INC.: SANCURE Waterborne Urethanes:

SANCURE waterborne aliphatic and aromatic waterborne urethane polymers are fully reacted, high molecular weight, colloidal dispersions which exhibit excellent abrasion, chemical, water and impact resistance. They are easy to handle, and readily applied using all conventional techniques, and are compatible with other water-based polymers. Current applications for these polymers include:

* Adhesives and Sealants
* Concrete Coatings and Sealers
* Fabric Binders and Lamination
* Floor Finishes
* Metal Coatings
* Overprint Varnishes
* Paper Coatings
* Plastic Coatings
* Printing Inks
* Rust Treatment
* Thread Coatings
* Vacuum Metalizing Coatings
* Vinyl and Urethane Coatings
* Wood Coatings

<u>Prod. No.:</u>

Aliphatic:

815:
Solids %: 35%
Viscosity: 500
pH: 8.5-10.0
Hard film with high tensile strength and modulus values. For use in tough, UV resistant, clear and pigmented coatings on metal, concrete, wood, and plastic. Also used in gravure and flexographic inks.

822A:
Solids %: 30%
Viscosity: 1,000
pH: 9.0-11.0
Soft and extensible film for flexible substrates such as leather basecoats.

847:
Solids %: 30%
Viscosity: 100
pH: 9.0-11.0
Extremely fine particle size urethane dispersion. Excellent water resistance and exceptionally fast drying properties. For use in printing inks, concrete, wood, overprint varnishes.

SANNCOR INDUSTRIES, INC.: SANCURE Waterborne Urethanes (Continued):

Prod. No.:

849:
Solids %: 35%
Viscosity: 2,000
pH: 9.0-11.0
Very soft films with high elongation for light-stable, abrasion resistant coatings on flexible plastics, elastomerics and foams.

862:
Solids %: 35%
Viscosity: 700
pH: 9.0-11.0
Flexible films with intermediate hardness and elongation. For use as coatings on metals, plastics and elastomerics. Also for coating special substrates such as glass and ceramics.

865:
Solids %: 35%
Viscosity: 500
pH: 9.0-11.0
Flexible films which exhibits a dry surface feel. Recommended for overprint varnishes, thread coatings and other applications where heat resistance is required.

867:
Solids %: 40%
Viscosity: 700
pH: 8.0-10.0
Contains no organic co-solvent. Forms soft, flexible film. Excellent adhesion to various substrates including mylar, nylon

878:
Solids %: 38%
Viscosity: 500
pH: 8.0-10.0
Contains no organic co-solvent. Similar to SANCURE 867 but harder and drier film properties.

895:
Solids %: 35%
Viscosity: 500
pH: 9.5-10.5
Hard films. Recommended for UV resistant clear and pigmented coatings on metal, concrete, wood or plastic substrates.

SANNCOR INDUSTRIES, INC.: SANCURE Waterborne Urethanes (Continued):

Prod. No.:

896:
Solids %: 32%
Viscosity: 250
pH: 9.0-10.5
Very hard film with high tensile and modulus values. Recommended for general coatings, either clear or pigmented.

899:
Solids %: 35%
Viscosity: 500
pH: 9.0-10.5
Slightly softer than 815. Exceptionally high gloss. For use in overprint varnishes, vinyl and urethane topcoats.

1818:
Solids %: 35%
Viscosity: 500
pH: 8.0-10.0
Hard film. Exceptional gloss, durability, burnishing, removability and powdering resistance properties in high speed buffing floor polishes. For use in restorer formulations and self-polishing floor finishes.

9500:
Solids %: 40%
Viscosity: 40,000
pH: 9.0-11.0
Flexible film for fabric lamination.

Aromatic:

1507:
Solids %: 35%
Viscosity: 500
pH: 8.5-10.0
Recommended for use in pigmented coatings on metal, concrete, wood and plastic substrates. Also to reduce moisture vapor transmission.

1525:
Solids %: 60%
Viscosity: 40,000
pH: 8.5-10.0
High solids, high viscosity urethane recommended for fabric lamination.

SANNCORR INDUSTRIES, INC.: SANCURE Waterborne Urethanes (Continued):

Prod. No.

Cationic:

1200:
Solids %: 35%
Viscosity: 2,000
pH: 4.0-5.0
Aromatic polymer with excellent adhesion to metals. In combination with tannic acid transforms rust into surface ready for primer and topcoating.

1201:
Solids %: 37%
Viscosity: 1,500
pH: 3.5-5.5
Aliphatic polymer. Excellent adhesion to wide variety of substrates. Care must be taken in blending with other polymers and additives because of cationic nature.

Dull Concentrate:

857:
Solids %: 35%
Viscosity: 2,000
pH: 9.0-11.0
Recommended to impart matte finish or blend to vary gloss level.

Crosslinker:

40:
Solids %: 80%
Viscosity: 90
Recommended to improve adhesion to difficult substrates as well as enhancing chemical, water and solvent resistance. Air dry or low bake system.

XAMA-7:
Solids %: 100%
Viscosity: 1,600
Polyfunctional aziridine. To improve adhesion as well as enhancing chemical, water and solvent resistance.

87:
Solids %: 85%
Viscosity: 3,500
Recommended for use above 275F. for heat-activated applications.

SARTOMER CO.: Hydroxyl Terminated POLY BD Resins Functional Liquid Polymers:

POLY BD Resins are liquid, hydroxyl terminated polymers of butadiene (R-45HT and R-45M) with a number average molecular weight of approximately 3,000. The degree of polymerization is in the range of 50. The microstructure of both homopolymers is similar to that of polybutadiene prepared by emulsion polymerization with the exception of terminal hydroxyl groups. The predominant configuration is Trans-1,4 (60%) with approximately 20% each Cis-1,4 and Vinyl 1,2.

The terminal hydroxyl groups are primary and predominantly of the allylic type. This structural combination results in high reactivity in condensation polymerizations and the preparation of derivatives. The hydroxyl functionality of R-45HT is typically in the range of 2.4-2.6 whereas that for R-45M is somewhat lower, typically 2.2 to 2.4.

POLY BD Resins are characterized by a low volatiles content with low moisture levels. Their low viscosity allows for easy handling, both in containers and in bulk. Storage stability for all resins is excellent.

Typical Physical Properties of POLY BD Resins:

R-45HT:

Nonvolatile Material, Wt. %: 99.9
Viscosity, poise @ 30C: 50.
Hydroxyl Value, meq./gm.: 0.83
Hydroxyl Number, mg KOH/gm: 46.6
Molecular Weight, No. Average: 2800.
Moisture, Wt. %: 0.05
Specific Gravity @ 30C: 0.901
Iodine Number: 398.

R-45M:

Nonvolatile Material, Wt. %: 99.9
Viscosity, poise @ 30C: 50.
Hydroxyl Value, meq./gm.: 0.75
Hydroxyl Number, mg KOH/gm: 42.1
Molecular Weight, No. Average: 2800.
Moisture, Wt. %: 0.05
Specific Gravity @ 30C: 0.899
Iodine Number: 398.

SARTOMER CO.: Monomer/Oligomer Products:

Sartomer Product Code:

Monofunctional Monomers: Monofunctional acrylic esters are non-crosslinkable, low viscosity monomers used as reactive diluents for radiation and peroxide cure application areas.

203:

Chemical Description: Tetrahydrofurfuryl Methacrylate
Appearance: Clear Liquid
Reactive Esters, %: 98
Molecular Weight: 170
Viscosity @ 25C, cps: 2-8
Features:
* high boiling
* low viscosity
* promotes adhesion
* contains cyclic group

Applications:
* anaerobic adhesives * screen printing inks * polymer for artificial fingernails * rubber modifiers * emulsion polymerization * plastic products * wire and cable coatings

Performance Properties:
* Flexibility
* Adhesion
* Low Shrinkage
* Impact Strength

208:

Chemical Description: Cyclohexyl Methacrylate
Appearance: Clear Yellow Liquid
Reactive Esters, %: 99
Molecular Weight: 168
Viscosity @ 25C, cps: 2-10
Features:
* active hydrogen in cyclohexyl group
* high refractive index
* opticals similar to glass
* high boiling
* excellent stability

Applications:
* vinyl polymerization * rubber vulcanization * floor polishes * optical lens coating * anaerobic adhesives

Performance Properties:
* Weatherability
* Low Shrinkage
* Hardness
* Heat Resistance

SARTOMER CO.: Monomer/Oligomer Products(Continued):

Sartomer Product Code:

211:
Chemical Description: n-Hexyl Methacrylate
Appearance: Clear Liquid
Reactive Ester: 98%
Molecular Weight: 170
Viscosity @ 25C, cps: 2-10
Features:
* high solubility in aliphatic solvents
* contains hydrophobic backbone
Applications:
* adhesives * binders * finishes * viscosity index improvers * sizes for textiles, leathers, papers
Performance Properties:
* Weatherability * Flexibility
* Adhesion * Low Shrinkage
* Impact Strength * Water Resistance

220:
Chemical Description: Cyclohexyl Acrylate
Appearance: Clear Yellow Liquid
Reactive Esters, %: 96+
Molecular Weight: 154
Viscosity @ 25C, cps: 2-10
Features:
* contains cyclic group
* high refractive index
* high Tg
Applications:
* floor polishes * optical lens coating * other glass and plastic coatings
Performance Properties:
* Weatherability * Chemical Resistance
* Low Shrinkage * Hardness
* Heat Resistance * Abrasion Resistance
* Water Resistance

232:
Chemical Description: 2-Ethoxyethyl Acrylate
Appearance: Light Straw-colored Liquid
Reactive Esters, %: 91+
Molecular Weight: 144
Viscosity @ 25C, cps: 3-10
Features:
* high volatility
* low Tg
* slightly water dispersible
Applications:
* aqueous and solvent coatings * emulsion polymerization
Performance Properties:
* Flexibility * Adhesion
* Hardness * Impact Strength

SARTOMER CO.: Monomer/Oligomer Products(Continued):

<u>Sartomer Product Code:</u>

233:
Chemical Description: 2-Ethoxyethyl Methacrylate
Appearance: Light Straw-colored Liquid
Reactive Esters, %: 98
Molecular Weight: 158
Viscosity @ 25C, cps: 5-10
Features:
* Tg=15C
Applications:
* coatings and polishes * acrylic rubber vulcanization * adhesives
Performance Properties:
* Weatherability * Flexibility
* Adhesion * Hardness

242:
Chemical Description: Isodecyl Methacrylate
Appearance: Clear Liquid
Reactive Esters, %: 99
Molecular Weight: 226
Viscosity @ 25C: 5-10
Features:
* contains hydrophobic backbone
Applications:
* pressure sensitive adhesives * sealants * potting aids
Performance Properties:
* Weatherability * Flexibility
* Adhesion * Hardness
* Impact Strength * Water Resistance

244:
Chemical Description: 2-Methoxyethyl Acrylate
Appearance: Clear Liquid
Reactive Esters, %: 98
Molecular Weight: 130
Viscosity @ 25C, cps: 10-15
Features:
* slightly water dispersible
Applications:
* soft contact lenses * polyacrylate rubbers * barrier coatings * PVC impact modifiers * textile coatings
Performance Properties:
* Weatherability * Chemical Resistance
* Flexibility * Adhesion
* Low Shrinkage * Impact Strength

SARTOMER CO.: Monomer/Oligomer Products(Continued):

<u>Sartomer Product Code:</u>

256:

Chemical Description: 2(2-Ethoxyethoxy) Ethylacrylate
Appearance: Clear Yellow Liquid
Reactive Esters, %: 99
Molecular Weight: 188
Features:
* slightly water dispersible
* good reactive diluent
* low Tg

Applications:
* aqueous and solvent coatings * emulsion polymerization

Performance Properties:
* Flexibility * Adhesion
* Hardness * Impact Strength

257:

Chemical Description: Stearyl Acrylate
Appearance: Solid @ 24C
Reactive Esters, %: 99
Molecular Weight: 324
Features:
* contains long chain aliphatic hydrophobic backbone

Applications:
* slip additives * release paper coatings * pour point depressants

Performance Properties:
* Weatherability * Flexibility
* Adhesion * Hardness
* Water Resistance * Impact Strength

285:

Chemical Description: Tetrahydrofurfuryl Acrylate
Appearance: Clear Liquid
Reactive Esters, %: 98
Molecular Weight: 156
Viscosity @ 25C, cps: 2-10
Features:
* high boiling

Applications:
* pressure sensitive adhesives

Performance Properties:
* Flexibility * Adhesion
* Low Shrinkage * Hardness
* Abrasion Resistance

SARTOMER CO.: Monomer/Oligomer Products(Continued):

<u>Sartomer Product Code:</u>

313:
Chemical Description: Lauryl Methacrylate
Appearance: Clear Liquid
Reactive Esters, %: 98
Molecular Weight: 254
Viscosity @ 25C, cps: 2-10
Features:
* low toxicity
* Tg=65C
* contains long chain aliphatic hydrophobic backbone
Applications:
* adhesives * lube oil additives * caulks * sealants * plasticizers * floor wax * amphipathic stabilizers used for nonaqueous dispersions
Performance Properties:
* Weatherability
* Adhesion
* Impact Strength
* Flexibility
* Low Shrinkage
* Water Resistance

324:
Chemical Description: Stearyl Methacrylate
Appearance: Clear Liquid (Above Room Temp.)
Reactive Esters, %: 98+
Molecular Weight: 330
Viscosity @ 25C, cps.: 10 @ 26C
Features:
* Tg=38C
* melting pt. = 21.4C
* contains long chain aliphatic hydrocarbon backbone
Applications:
* pour point depressants * lubricating oil additives * paper coatings * pressure sensitive adhesives * textile treatment chemicals * varnishes * paints
Performance Properties:
* Weatherability
* Adhesion
* Impact Strength
* Flexibility
* Low Shrinkage
* Water Resistance

333:
Chemical Description: Octyl Acrylate
Appearance: Clear Liquid
Reactive Esters, %: 96+
Molecular Weight: 184
Features:
* low volatility
* contains hydrophobic backbone
Applications:
* pressure sensitive adhesives * sealants
Performance Properties
* Weatherability
* Adhesion
* Impact Strength
* Flexibility
* Low Shrinkage
* Water Resistance

SARTOMER CO.: Monomer/Oligomer Products(Continued):

Sartomer Product Code:

335:

Chemical Description: Lauryl Acrylate
Appearance: Light Colored Liquid
Reactive Esters, %: 95
Molecular Weight: 240
Viscosity @ 25C, cps: 2-10
Features:
* low volatility
* contains long chain aliphatic hydrophobic backbone
Applications:
* adhesives * coatings * lube oil additives
Performance Properties
* Flexibility * Adhesion
* Low Shrinkage

339:

Chemical Description: 2-Phenoxyethyl Acrylate
Appearance: Yellow Liquid
Reactive Esters, %: 98
Molecular Weight: 192
Viscosity @ 25C, cps: 10-15
Features:
* high boiling * low volatility
Applications:
* adhesives * inks * coatings * tile coatings
Performance Properties:
* Flexibility * Adhesion
* Low Shrinkage * Hardness
* Abrasion Resistance * Water Resistance

340:

Chemical Description: 2-Phenoxyethyl Methacrylate
Appearance: Yellow Liquid
Reactive Esters, %: 98
Molecular Weight: 206
Viscosity @ 25C, cps: 10-15
Features:
* high boiling * low volatility
* lower toxicity cyclic monofunctional
Applications:
* adhesives * inks * coatings * tile coatings
Performance Properties:
* Flexibility * Adhesion
* Low Shrinkage * Hardness
* Abrasion Resistance * Water Resistance

SARTOMER CO.: Monomer/Oligomer Products(Continued):

Sartomer Product Code:

375:
Chemical Description: Tertiary Butyl Acetate
Appearance: Clear Liquid
Reactive Esters, %: 99
Molecular Weight: 128
Viscosity @ 25C., cps: 10
Features:
* Tg=43C
* t-butyl group can disassociate easily with heat or hydrolysis
* solubility in water @ 21C=0.08 g/100g
Applications:
* electronic photoresists/masks * adhesives * coatings * co/ter polymers of acrylic acid
Performance Properties
* Low Shrinkage * Hardness
* Heat Resistance * Abrasion Resistance

378:
Chemical Description: Glycidyl Acrylate
Appearance: Clear Liquid
Reactive Esters, %: 96+
Molecular Weight: 128
Viscosity @ 25C, cps: 5-10
Features:
* contains an epoxy group
Applications:
* paper coatings * concrete polymers * coatings & inks * wood impregnators
Performance Properties:
* Chemical Resistance * Adhesion
* Low Shrinkage * Hardness
* Heat Resistance * Abrasion Resistance

379:
Chemical Description: Glycidyl Methacrylate
Appearance: Clear Liquid
Reactive Esters, %: 97+
Molecular Weight: 142
Viscosity @25C, cps: 8-10
Features:
* contains an epoxy group
Applications:
* hydrogels for contact lenses * molding & casting compounds * epoxy/acrylate hybrid systems * adhesives * powder coatings * stabilizers for non-aqueous dispersions
Performance Properties:
* Chemical Resistance * Adhesion
* Low Shrinkage * Hardness
* Heat Resistance * Abrasion Resistance

SARTOMER CO.: Monomer/Oligomer Products(Continued):

Sartomer Product Code:

395:

Chemical Description: Isodecyl Acrylate
Appearance: Clear Liquid
Reactive Esters, %: 99+
Molecular Weight: 212
Viscosity @ 25C, cps: 1-10
Features:
* low Tg * low viscosity
Applications:
* viscosity index improvers * adhesives * coatings * inks
Performance Properties:
* Flexibility * Adhesion
* Low Shrinkage * Impact Strength
* Water Resistance

423:

Chemical Description: Isobornyl Methacrylate
Appearance: Clear Yellow Liquid
Reactive Esters, %: 98
Molecular Weight: 222
Viscosity @ 25C, cps: 10-15
Features:
* cyclic group offers high Tg properties
Applications:
* acrylic powders for automatic coatings
* polymer cements
Performance Properties:
* Weatherability * Low Shrinkage
* Hardness * Heat Resistance
* Impact Strength * Water Resistance

432:

Chemical Description: Benzyl Acrylate
Appearance: Clear Liquid
Reactive Esters, %: 98+
Molecular Weight: 162
Viscosity @ 25C, cps: 5-15
Features:
* aromatic
Applications:
* inks * coatings
Performance Properties:
* Flexibility * Adhesion
* Low Shrinkage * Abrasion Resistance

SARTOMER CO.: Monomer/Oligomer Products(Continued):

Sartomer Product Code:

439:

Chemical Description: Hexyl Acrylate
Appearance: Clear Liquid
Reactive Esters, %: 99+
Molecular Weight: 156
Viscosity @ 25C, cps: 10-15
Features:
* cyclic group offers high Tg properties
Applications:
* glass coatings * polymer cements * sealants * other coatings
Performance Properties:
* Weatherability * Hardness
* Adhesion * Low Shrinkage
* Water Resistance

440:

Chemical Description: Isooctyl Acrylate
Appearance: Clear Liquid
Reactive Esters, %: 99+
Molecular Weight: 184
Viscosity @ 25C, cps. 2-10
Features:
* low viscosity * low Tg
* good reactive diluent
Applications:
* caulks & sealants * adhesives * coatings
Performance Properties:
* Weatherability * Flexibility
* Adhesion * Low Shrinkage
* Impact Strength * Water Resistance

506:

Chemical Description: Isobornyl Acrylate
Appearance: Clear Liquid
Reactive Esters, %: 96% Pure
Molecular Weight: 208
Viscosity @ 25C, cps.: 5-10
Features:
* cyclic group offers high Tg properties
* excellent reactive diluent for oligomers
Applications:
* glass & other coatings
Performance Properties:
* Weatherability * Low Shrinkage
* Hardness * Heat Resistance
* Water Resistance

SARTOMER CO.: Monomer/Oligomer Products(Continued):

Difunctional Monomers: Difunctional acrylics are considered the "workhorse" acrylic monomers. They offer a good balance of low viscosity, fast reactivity and crosslink density properties to radiation and peroxide cure systems.

Sartomer Product Code:

205:

Chemical Description: Triethylene Glycol Dimethacrylate
Appearance: Clear Liquid
Reactive Esters, %: 98
Molecular Weight: 286
Viscosity @ 25C, cps: 5-15
Features:
* low vapor pressure * low viscosity
* high boiling
Applications:
* electronic photoresists * rubber products * vinyl plastisols * cast acrylic sheets * synthetic rubber and ion exchange resins * photosensitive polymers * dental polymers * flexographic printing plates * overprint varnish coatings * "no-wax" vinyl floor coatings
Performance Properties:
* Weatherability * Flexibility
* Adhesion * Low Shrinkage
* Abrasion Resistance * Impact Strength

206:

Chemical Description: Ethylene Glycol Dimethacrylate
Appearance: Clear Liquid
Reactive Esters, %: 98+
Molecular Weight: 198
Viscosity @ 25C, cps: 5-10
Features:
* high boiling * low viscosity
* colorless
Applications:
* emulsion polymerization * encapsulants * electronic fiberglass reinforced polyesters * ion exchange resins * rubber products * acrylic sheets * paper processing additives * moisture barrier films * dental polymers * vinyl acrylic copolymers * modified acrylics
Performance Properties:
* Weatherability * Chemical Resistance
* Hardness * Heat Resistance
* Abrasion Resistance

SARTOMER CO.: Monomer/Oligomer Products(Continued):

Sartomer Product Code:

209:
Chemical Description: Tetraethylene Glycol Dimethacrylate
Appearance: Clear Liquid
Reactive Esters, %: 98+
Molecular Weight: 330
Viscosity @ 25C, cps: 5-15
Features:
* low viscosity * non-corrosive
* low vapor pressure * fast cure response
Applications:
* acrylic coatings * anaerobic adhesives * plastisols * PVC electrical insulation * coatings * rubber products
Performance Properties:
* Weatherability * Chemical Resistance
* Hardness * Heat Resistance

210:
Chemical Description: Polyethylene Glycol Dimethacrylate
Appearance: Clear Liquid
Reactive Esters, %: 98
Molecular Weight: 330
Viscosity @ 25C, cps: 5-10
Features:
* low viscosity * low vapor pressure
* fast cure response
Applications:
* castings & rubber compounds * plastisols * anaerobic adhesives * coatings for fibers, papers, and plastics
Performance Properties:
* Weatherability * Chemical Resistance
* Flexibility * Adhesion
* Hardness * Heat Resistance
* Abrasion Resistance * Impact Strength

212:
Chemical Description: 1,3 Butylene Glycol Diacrylate
Appearance: Clear Straw-colored liquid
Reactive Esters, %: 97
Molecular Weight: 198
Viscosity @ 25C, cps: 5-10
Features:
* low viscosity * high stain resistance
Applications:
* inks * photoresists * wire & cable coatings * plastic compounds
Performance Properties:
* Chemical Resistance * Abrasion Resistance
* Water Resistance

SARTOMER CO.: Monomer/Oligomer Products(Continued):

SARTOMER Product Code:

213:

Chemical Description: 1,4-Butanediol Diacrylate
Appearance: Clear Liquid
Reactive Esters, %: 97.5
Molecular Weight: 198
Viscosity @ 25C, cps: 5-15
Features:
* high solvency
Applications:
* inks * photoresists * adhesives * textile products * plastic products * rubber products
Performance Properties:
* Chemical Resistance
* Heat Resistance
* Abrasion Resistance
* Water Resistance

214:

Chemical Description: 1,4-Butanediol Dimethacrylate
Appearance: Light Colored Liquid
Reactive Esters, %: 96
Molecular Weight: 226
Viscosity @ 25C, cps: 5-15
Features:
* high solvency
Applications:
* casting compounds * rubber compounds * glass reinforced plastics * adhesives * plastisols * dental polymers * wire & cable coatings
Performance Properties:
* Chemical Resistance
* Heat Resistance
* Abrasion Resistance
* Water Resistance

230:

Chemical Description: Diethylene Glycol Diacrylate
Appearance: Clear Liquid
Reactive Esters, %: 97+
Molecular Weight: 214
Viscosity @ 25C, cps; 5-15
Features:
* low viscosity
* high boiling
Applications:
* component of toughening agent in glass reinforced plastics
Performance Properties:
* Adhesion
* Low Shrinkage
* Impact Strength
* Water Resistance

SARTOMER CO.: Monomer/Oligomer Products(Continued):

Sartomer Product Code:

231:
Chemical Description: Diethylene Glycol Dimethacrylate
Appearance: Clear Liquid
Reactive Esters, %: 98+
Molecular Weight: 242
Viscosity @ 25C, cps: 5-15
Features:
* low viscosity * high boiling
Applications:
* rubber vulcanization * photopolymer printing plates * letterpress inks * moisture barrier films * conversion coatings
Performance Properties:
* Adhesion * Low Shrinkage
* Impact Strength * Water Resistance

238:
Chemical Description: 1,6 Hexanadiol Diacrylate
Appearance: Clear Liquid
Reactive Esters, %: 97+
Molecular Weight: 226
Viscosity @ 25C, cps: 5-10
Features:
* low Tg * high boiling
* fast cure response * low volatility
Applications:
* RTV seals, caulking, gaskets * wood coatings * photoresists * other inks and coatings
Performance Properties:
* Weatherability * Flexibility
* Adhesion * Water Resistance

239:
Chemical Description: 1,6 Hexanediol Dimethacrylate
Appearance: Clear Liquid
Reactive Esters, %: 98
Molecular Weight: 254
Viscosity @ 25C, cps: 5-10
Features:
* contains hydrophobic backbone
Applications:
* casting compounds * glass fiber reinforced plastics * adhesives * textile products * rubber compounds * molding compounds
Performance Properties:
* Weatherability * Flexibility
* Adhesion * Water Resistance

SARTOMER CO.: Monomer/Oligomer Products(Continued):

Sartomer Product Code:

248:
Chemical Description: Neopentyl Glycol Dimethacrylate
Appearance: Light Yellow Liquid
Reactive Esters, %: 94.8+
Molecular Weight: 240
Viscosity @ 25C, cps: 10-30
Features:
* high boiling * low viscosity
Applications:
* castings * plastisols * coatings * rubber products * building products
Performance Properties
* Chemical Resistance * Hardness
* Heat Resistance

259:

Chemical Description: Polyethylene Glycol (200) Diacrylate
Appearance: Clear Liquid
Reactive Esters, %: 97+
Molecular Weight: ~302
Viscosity @ 25C, cps: 5-10
Features:
* high boiling * low volatility
* low Tg
Applications:
* coatings * inks * adhesives * photoresists * flexographic plates * "no wax" floor tile coatings * wire & cable products * textile products
Performance Properties:
* Flexibility * Adhesion
* Impact Strength

268:

Chemical Description: Tetraethylene Glycol Diacrylate
Appearance: Clear Liquid
Reactive Esters, %: 97.5+
Molecular Weight: 302
Viscosity @ 25C, cps: 5-15
Features:
* high boiling * low volatility
* low Tg * low toxicity
Applications:
* inks * adhesives * textiles * coatings * photoresists * graphic products * rubber products
Performance Properties:
* Flexibility * Adhesion
* Low Shrinkage * Impact Strength

SARTOMER CO.: Monomer/Oligomer Products(Continued):

<u>Sartomer Product Code:</u>

272:

Chemical Description: Triethylene Glycol Diacrylate
Appearance: Clear Liquid
Reactive Esters, %: 95
Molecular Weight: 258
Viscosity @ 25C, cps: 5-15
Features:
* low vapor pressure
* low viscosity
* high boiling

Applications:
* plastisols * coatings, fibers, papers * letterpress printing plates * chemical intermediates * building products

Performance Properties:
* Weatherability
* Chemical Resistance
* Hardness
* Heat Resistance

297:

Chemical Description: 1,3 Butylene Glycol Dimethacrylate
Appearance: Clear Straw-colored Liquid
Reactive Esters, %: 98+
Molecular Weight: 226
Viscosity @ 25C, cps: 5-10
Features:
* low viscosity

Applications:
* plastisols * cast acrylic sheets * processing aids * impregnated wood composites * glass reinforced plastics * adhesives

Performance Properties:
* Weatherability
* Chemical Resistance
* Hardness
* Heat Resistance
* Water Resistance

306:

Chemical Description: Tripropylene Glycol Diacrylate
Appearance: Clear Liquid
Reactive Esters, %: 98.5+
Molecular Weight: 300
Viscosity @ 25C, cps: 10-20
Features:
* low toxicity
* high boiling

Applications:
* floor tiles * coatings * rubber compounding * inks

Performance Properties:
* Weatherability
* Flexibility
* Adhesion
* Abrasion Resistance
* Impact Strength

SARTOMER CO.: Monomer/Oligomer Products(Continued):

Sartomer Product Code:

348:
Chemical Description: Ethoxylated Bisphenol A Dimethacrylate
Appearance: Clear Viscous Liquid
Reactive Esters, %: 98+
Molecular Weight: 452
Viscosity @ 25C, cps: 750-1000
Features:
* high boiling * fast cure response
* very low volatility
Applications:
* dental compounds * anaerobic adhesives * elastomer products * graphic art products
Performance Properties:
* Chemical Resistance * Adhesion
* Low Shrinkage * Hardness
* Heat Resistance * Abrasion Resistance

349:
Chemical Description: Ethoxylated Bisphenol A Diacrylate
Appearance: Clear Viscous Liquid
Reactive Esters, %: 98 min
Molecular Weight: 424
Viscosity @ 25C, cps: 2500
Features:
* very low volatility * high boiling
Applications:
* photoresists * coatings
Performance Properties:
* Chemical Resistance * Adhesion
* Low Shrinkage * Hardness
* Heat Resistance * Abrasion Resistance

365:
Chemical Description: Zinc Dimethacrylate
Appearance: Powdered Solid
Reactive Esters, %: 99+
Molecular Weight: 265
Viscosity @ 25C, cps: N/A
Features:
* offers low compression set properties
* slightly soluble in water
* soluble in warm acetic acid
Applications:
* rubber vulcanization * casting compounds
Performance Properties:
* Hardness * Heat Resistnace

SARTOMER CO.: Monomer/Oligomer Products (Continued):

Sartomer Product Code:

416:
Chemical Description: Zinc Diacrylate
Appearance: Powdered Solid
Reactive Esters, %: 99+
Molecular Weight: 207
Viscosity @ 25C, cps: N/A
Features:
* melting point, 109C
* slightly soluble in water; soluble in warm acetic acid
Applications:
* conductive coatings * reactive pigments
Performance Properties:
* Hardness * Heat Resistance

2000:
Chemical Description: Acrylate Terminated Monomer with Average Chain Length of C14-C15
Appearance: Clear Liquid
Reactive Esters, %: 99+
Molecular Weight: ~300
Viscosity @ 25C, cps: 10-20
Features:
* contains long chain hydrophobic backbone
* low viscosity
* low odor
Applications:
* screen print inks * coatings * adhesives
Performance Properties:
* Flexibility * Adhesion
* Low Shrinkage * Impact Strength
* Water Resistance

2100:
Chemical Description: Methacrylate Terminated Monomer with Average Chain Length of C14-C15
Appearance: Clear Liquid
Reactive Esters, %: 99+
Molecular Weight: ~300
Viscosity @ 25C, cps: 10-20
Features:
* low odor * contains hydrophobic backbone
* low viscosity
Applications:
* electrical coatings and encapsulants * paper coatings * coagent for polyvinylchlorides and polyolefins * adhesives
Performance Properties
* Flexibility * Adhesion
* Low Shrinkage * Impact Strength
* Water Resistance

SARTOMER CO.: Monomer/Oligomer Products(Continued):

Tri/Tetra/Multifunctional Monomers: Tri/tetra/multifunctional acrylates are highly reactive monomers that offer high crosslink density properties to radiation and peroxide cure systems.

Sartomer Product Code:

290:
Chemical Description: Tris(2-Hydroxy Ethyl) Isocyanurate Trimethacrylate
Appearance: White Crystalline Solid
Reactive Esters, %: 98
Molecular Weight: 437
Viscosity @ 25C, cps: 15-25M
Features:
* triazin compound * high melting point
Applications:
* electrical coatings & encapsulants * wood fillers * rubber compoundings
Performance Properties:
* Weatherability * Chemical Resistance
* Hardness * Heat Resistance

295:
Chemical Description: Pentaerythritol Tetraacrylate
Appearance: Clear Liquid
Reactive Esters, %: 95+ (65+-10 Tetra)
Molecular Weight: 352
Viscosity @ 25C, cps: 200+-50 at 38C
Features:
* melting point, 15-18C * high viscosity
* high boiling
Applications:
* photosensitive polymers * coatings
Performance Properties:
* Chemical Resistance * Adhesion
* Hardness * Heat Resistance

350:
Chemical Description: Trimethylolpropane Trimethacrylate
Appearance: Clear Liquid
Reactive Esters, %: 99
Molecular Weight: 338
Viscosity @ 25C, cps: 35-50
Features:
* high boiling * low viscosity
Applications:
* plastisols * vinyl acetate latices * moisture barrier films * elastomer coagents * polyethylene coagents
Performance Properties:
* Weatherability * Chemical Resistance
* Hardness * Heat Resistance

SARTOMER CO.: Monomer/Oligomer Products(Continued):

Sartomer Product Code:

351:
Chemical Description: Trimethylolpropane Triacrylate
Appearance: Clear Liquid
Reactive Esters, %: 97+
Molecular Weight: 296
Viscosity @ 25C, cps: 60-80
Features:
* low viscosity
* fast cure response
* low volatility
* high boiling liquid
Applications:
* vinyl acrylic latex paints * dry film photopolymer resists * flexographic, offset, and screen printing inks * exterior architectural paint bases * wood fillers * coatings
Performance Properties:
* Weatherability
* Chemical Resistance
* Hardness
* Heat Resistance

368:
Chemical Description: Tris (2-Hydroxy Ethyl) Isocyanurate Triacrylate
Appearance: White Crystalline Solid
Reactive Esters, %: 98
Molecular Weight: 423
Viscosity @ 25C, cps: 15-25M
Features:
* melting point, 52-54C
* triazin compound
Applications:
* elastomer, polyolefin and plastisol products
Performance Properties:
* Chemical Resistance
* Hardness
* Heat Resistance
* Abrasion Resistance

399:
Chemical Description: Dipentaerythritol Pentaacrylate
Appearance: Viscous Liquid
Reactive Esters, %: 97.5
Molecular Weight: 525
Viscosity @ 25C, cps: 15-25 M
Features:
* Contains pentaacrylate functionality
* rotational flexibility exists around ether moiety
Applications:
* rubber vulcanization * abrasion resistant coatings * other coatings
Performance Properties:
* Weatherability
* Chemical Resistance
* Flexibility
* Adhesion
* Hardness
* Heat Resistance
* Abrasion Resistance

SARTOMER CO.: Monomer/Oligomer Products(Continued):

Sartomer Product Code:

444:

Chemical Description: Pentaerythritol Triacrylate
Appearance: Clear Liquid
Reactive Esters, %: 97+ (50-70 Tri-ester)
Molecular Weight: ~300
Viscosity @ 25C, cps: 300-400
Features:
* high boiling
* hydroxyl pendant group
* fast cure response
Applications:
* photoresists * coatings and adhesives * wood fillers * rubber products * conversion coatings
Performance Properties:
* Chemical Resistance
* Adhesion
* Hardness
* Heat Resistance

New Property Specific Functional Acrylics: These functional acrylic monomers provide specific benefits, such as water solubility, adhesion, and lower skin irritation that enhance processing and/or finished product properties.

Water Dispersible Monomers:

252:

Chemical Description: Polyethylene Glycol (600) Dimethacrylate
Appearance: Clear Liquid
Reactive Esters, %: 99
Molecular Weight: 736
Viscosity @ 25C, cps: 65-75
Features:
* difunctional
* produces soft, flexible films
Applications:
* rubber products
Performance Properties:
* Chemical Resistance
* Flexibility
* Adhesion
* Low Shrinkage
* Hardness
* Impact Strength

SARTOMER CO.: Monomer/Oligomer Products(Continued):

Sartomer Product Code:

344:
Chemical Description: Polyethylene Glycol (400) Diacrylate
Appearance: Clear Liquid
Reactive Esters, %: 99+
Molecular Weight: 508
Viscosity @ 25C, cps: 55-65
Features:
* difunctional
* produces soft, flexible films
Applications:
* flexible coatings * overprint varnishes * offset inks * "no wax" vinyl tile coatings * dental compounds
Performance Properties:
* Chemical Resistance
* Flexibility
* Adhesion
* Low Shrinkage
* Hardness
* Impact Strength

361:
Chemical Description: Diethylaminoethyl Acrylate
Appearance: Clear Yellow Liquid
Reactive Esters, %: 99+
Molecular Weight: 171
Viscosity @ 25C, cps: 1-10
Features:
* wet adhesion promoter
* contains an amino group, a source for cationic charge density
* monofunctional
Applications:
* dye additives * coatings * resists
Performance Properties:
* Flexibility
* Adhesion
* Low Shrinkage
* Abrasion Resistance

363:
Chemical Description: Dimethylaminoethyl Acrylate
Appearance: Clear Yellow Liquid
Reactive Esters, %: 99+
Molecular Weight: ~200
Viscosity @ 25C, cps: 1-10
Features:
* wet adhesion promoter
* contains an amino group, a source for cationic charge density
* monofunctional
Applications:
* dye additives * lubricating oil additives * intermediate for silane coupling agents * intermediate for quaternary chemicals * cationic polymers * intermediate for water treatment chemical additives * conductive paper chemicals
Performance Properties:
* Flexibility
* Adhesion
* Low Shrinkage
* Abrasion Resistance

SARTOMER CO.: Monomer/Oligomer Products(Continued):

Sartomer Product Code:

607:
Chemical Description: Polypropylene Glycol Monoacrylate
Appearance: Clear Liquid
Reactive Esters, %: 99
Molecular Weight: 335-372
Viscosity @ 25C, cps.: 10-20
Features:
* monofunctional
* low skin irritation
* low volatility
Applications:
* resist applications
Performance Properties:
* Adhesion
* Low Shrinkage

604:
Chemical Description: Polypropylene Glycol Monomethacrylate
Appearance: Clear Liquid
Reactive Esters, %: 99
Molecular Weight: 350-387
Viscosity @ 25C, cps.: 10-20
Features:
* monofunctional
Applications:
* coatings * emulsion polymers * photoresists * photopolymer printing plates
Performance Properties:
* Adhesion
* Low Shrinkage

676:
Chemical Description: Butanediol Monoacrylate
Appearance: Clear Liquid
Reactive Esters, %: 98
Molecular Weight: 144
Viscosity @ 25C, cps: 10-15
Features:
* monofunctional
* contains a reactive hydroxyl moiety
Applications:
* urethane prepolymers * acrylamide cationic polymers * wet strength paper additives * homopolymers used in electrophotographic resins * copolymers and homopolymers for water treatment and mineral processing * aqueous automotive OEM finishes * wood topcoats * vinyl tile coatings * textile backcoatings * aqueous dispersions for coil coatings and wood products * flat board topcoatings
Performance Properties:
* Flexibility
* Adhesion
* Impact Strength

SARTOMER CO.: Monomer/Oligomer Products(Continued):

<u>Sartomer Product Code:</u>

9011:

Chemical Description: Trifunctional Methacrylate Ester
Appearance: Clear Liquid
Reactive Esters, %: 99+ reactive
Viscosity @25C, cps: 50-70
Features:
* same product as SARTOMER 9010 but incorporates standard inhibitor systems
* fast cure response   * trifunctional

Applications:
* plastisols

Performance Properties:
* Chemical Resistance   * Adhesion
* Low Shrinkage   * Hardness
* Heat Resistance   * Abrasion Resistance

9012:

Chemical Description: Trifunctional Acrylate Ester
Appearance: Clear Liquid
Reactive Esters, %: 98+ reactive
Viscosity @ 25C, cps: 70-90
Features:
* fast cure response   * trifunctional

Applications:
* radiation curable coatings, sealants, adhesives

Performance Properties:
* Chemical Resistance   * Adhesion
* Low Shrinkage   * Hardness
* Heat Resistance   * Abrasion Resistance

9013:

Chemical Description: Aliphatic Monofunctional Ester
Appearance: Clear Liquid
Reactive Esters, %: 98+
Viscosity @ 25C, cps: 5-15
Features:
* good reactive diluent   * monofunctional

Applications:
* inks * coatings * adhesives * photoresists

Performance Properties:
* Weatherability   * Flexibility
* Adhesion   * Low Shrinkage
* Impact Strength

SARTOMER CO.: Monomer/Oligomer Products(Continued):

Sartomer Product Code:

9014:

Chemical Description: Aliphatic Difunctional Ester
Appearance: Clear Liquid
Reactive Esters, %: 99
Viscosity @ 25C, cps: 5-20
Features:
* good reactive diluent * difunctional

Applications:
* inks * coatings * adhesives * photoresists

Performance Properties:
* Weatherability
* Adhesion
* Heat Resistance
* Chemical Resistance
* Low Shrinkage

9015:

Chemical Description: Aliphatic Monofunctional Ester
Appearance: Clear Liquid
Reactive Esters, %: 98+
Viscosity @ 25C, cps: 5-15
Features:
* good reactive diluent * monofunctional

Applications:
* inks * coatings * adhesives * photoresists

Performance Properties:
* Weatherability
* Adhesion
* Impact Strength
* Flexibility
* Low Shrinkage

9040:

Chemical Description: Alkoxylated Diacrylate Ester
Appearance: Clear Liquid
Reactive Esters, %: 99+
Viscosity @ 25C, cps: 5-20
Features:
* good reactive diluent * diacrylate

Applications:
* inks * coatings * adhesives * photoresists

Performance Properties:
* Weatherability
* Adhesion
* Heat Resistance
* Chemical Resistance
* Low Shrinkage

SARTOMER CO.: Monomer/Oligomer Products(Continued):

Low Skin Irritation Monomers:

Sartomer Product Code:

355*:
Chemical Description: Di-Trimethylpropane Tetraacrylate
Appearance: Clear Liquid
Reactive Esters, %: 99
Molecular Weight: 438
Viscosity @ 25C, cps: 752
Features:
* fast cure response * high crosslink density
Applications:
* inks * coatings
Performance Properties:
* Weatherability * Chemical Resistance
* Flexibility * Adhesion
* Hardness * Heat Resistance
* Abrasion Resistance * Water Resistance
* Not on U.S. TOSCA list. Commercially available for sale outside U.S.

454:
Chemical Description: Ethoxylated Trimethylolpropane Triacrylate
Appearance: Clear Liquid
Reactive Esters, %: 98.5
Molecular Weight: 428
Viscosity @ 25C, cps: 75+-25
Features:
* fast cure response
Applications:
* gravure and litho inks * overprint varnishes * coatings & adhesives * concrete polymers
Performance Properties:
* Chemical Resistance * Abrasion Resistance

9003**:
Chemical Description: Propoxylated Neopentyl Glycol Diacrylate
Appearance: Clear Liquid
Reactive Esters, %: 99
Molecular Weight: 328
Viscosity @ 25C, cps: 40-55
Features:
* high boiling * low viscosity
Applications:
* inks * floor tile coatings
Performance Properties:
* Adhesion * Low Shrinkage
* Abrasion Resistance * Impact Strength
**Covered by U.S. Patent #4,382,135.

SARTOMER CO.: Monomer/Oligomer Products(Continued):

Sartomer Product Code:

9020:

Chemical Description: Glyceryl Propoxy Triacrylate
Appearance: Clear Liquid
Reactive Esters, %: 98+
Molecular Weight: 428
Viscosity @ 25C, cps: 80+-10
Features:
* low odor
Applications:
* inks * coatings
Performance Properties:
* Weatherability
* Chemical Resistance
* Low Shrinkage
* Impact Strength

9021:

Chemical Description: Highly Propoxylated Glyceryl Triacrylate
Appearance: Clear Liquid
Reactive Esters, %: 99+
Viscosity @ 25C, cps: 80+-10
Features:
* low odor
Applications:
* inks * coatings
Performance Properties:
* Weatherability
* Chemical Resistance
* Flexibility
* Low Shrinkage
* Impact Strength

9041:

Chemical Description: Pentaacrylate Ester
Appearance: Viscous Liquid
Reactive Esters, %: 98
Molecular Weight: ~500
Viscosity @ 25C, cps: 20M
Features:
* fast cure response
* high crosslink density
Applications:
* coatings * inks * adhesives
Performance Properties:
* Weatherability
* Chemical Resistance
* Flexibility
* Adhesion
* Hardness
* Heat Resistance
* Abrasion Resistance
* Water Resistance

SARTOMER CO.: Monomer/Oligomer Products(Continued):

Sartomer Product Code:

9209**:

Chemical Description: Alkoxylated Aliphatic Diacrylate Ester
Appearance: Clear Liquid
Reactive Esters, %: 99.9+
Viscosity @ 25C, cps: 10-20
Features:
* fast cure response * low viscosity
* high boiling
Applications:
* inks * floor tile coatings
Performance Properties:
* Flexibility * Adhesion
* Low Shrinkage * Abrasion Resistance
* Impact Strength

** Covered by U.S. Patent #4,382,135

Fire Retardant Monomer:

640:

Chemical Description: Tetrabromo Bisphenol A Diacrylate
Appearance: Granular Solid
Reactive Esters, %: 96+
Molecular Weight: 741
Viscosity @ 25C, cps: N/A
Features:
* bromine group offers fire retardant properties
Applications:
* automotive coatings * wire and cable coatings
Performance Properties
* Adhesion * Heat Resistance
* Abrasion Resistance * Water Resistance

Viscosity Reducing Aid:

7001:

Chemical Description: Metallic Carboxylate
Appearance: Powder
Reactive Esters, %: 98+
Viscosity @ 25C, cps: N/A
Features:
* melting point <250C * soluble in warm acetic acid
Applications:
* silica filled rubber compounding products
Performance Properties:
* Adhesion * Low Shrinkage
* Abrasion Resistance

SARTOMER CO.: Monomer/Oligomer Products(Continued):

Scorch Retarding Aids:

Sartomer Product Code:

500:
Chemical Description: SARET Crosslinking Agent (Trifunctional)
Appearance: Dark Liquid
Reactive Esters, %: 95+
Viscosity @ 25C, cps: 40-60
Features:
* flash point, 212F
* compatible with alcohols, ethers, ketones
* medium functionality
Applications: rubber products
Performance Properties:
* adhesion
* low shrinkage
* abrasion resistance

515:
Chemical Description: SARET Crosslinking Agent (Trifunctional)
Appearance: Dark Liquid
Reactive Esters, %: 95+
Viscosity @ 25C, cps: 40-60
Features:
* flash point, 155F
* compatible with alcohols, ethers, ketones
* medium functionality
Applications:
* rubber products
Performance Products:
* Adhesion
* Low Shrinkage
* Abrasion Resistance

516:
Chemical Description: SARET Crosslinking Agent (Difunctional)
Appearance: Dark Liquid
Reactive Esters, %: 95+
Viscosity @ 25C, cps: 10-15
Features:
* contains non-nitroso inhibitor package
* low functionality
* low staining
* low color
Applications:
* rubber products
Performance Products:
* Chemical Resistance
* Flexibility
* Abrasion Resistance

SARTOMER CO.: Monomer/Oligomer Products(Continued):

Sartomer Product Code:

517:
Chemical Description: SARET Crosslinking Agent (Trifunctional)
Appearance: Dark Liquid
Reactive Esters %: 95+
Viscosity @ 25C, cps: 45-60
Features:
* contains non-nitroso inhibitor package
* medium functionality * low staining
* low color
Applications:
* rubber products
Performance Properties: Abrasion Resistance

518:
Chemical Description: SARET Crosslinking Agent (Tetra-functional, Average)
Appearance: Dark Liquid
Reactive Esters, %: 95+
Viscosity @ 25C, cps: 130-140
Features:
* contains non-nitroso inhibitor package
* high functionality * low staining
* low color
Applications: rubber products
Performance Properties:
* Adhesion * Low Shrinkage
* Abrasion Resistance

Pigment Dispersing Aids:

801:
Appearance: Clear Liquid
Viscosity @ 25C, cps: 360
802:
Appearance: Clear Liquid
Viscosity @25 C, cps: 280
803:
Appearance: Clear Liquid
Viscosity @ 25C, cps: 213

Features:
* viscosity reduction * pigment dispersion stability
* adhesion promotion to phosphatized steel substrates
* flow enhancement * easy pigment redispersion
* fast cure response
Applications:
* wood coatings * wood fillers
* solder masks * pigmented coatings
Performance Properties:
* Adhesion * Low Shrinkage

SARTOMER CO.: Monomer/Oligomer Products(Continued):

Other Products. These non-acrylic monomers are not typically used in high energy systems.

Sartomer Product Code:

Diol:
100:
Aliphatic Diol with Average Chain Length Approx. Equal to C14-C15
Appearance: Waxy White Solid
Features:
* soluble in H2O: <0.01g/100g at 75F
* flash point, 370F
* melting range, 38-42C
* hydroxyl#=391 mgKOH/g
Applications:
* intermediate for urethanes, polyesters, allyls, and prepolymers * low-cost replacement for 1,6 hexanediol
Performance Properties
* Weatherability * Flexibility
* Impact Strength * Water Resistance
Acetate:
322:
Chemical Description: Triethylene Glycol Diacetate
Appearance: Clear Liquid
Reactive Esters, %: 95 min.
Molecular Weight: 234
Features: plasticizer
Applications: films and coatings
Performance Properties:
* Chemical Resistance * Flexibility
* Heat Resistance
Caprolactam:
681:
Chemical Description: Vinyl Caprolactam
Appearance: Yellow Colored Solid
Reactive Esters, %: 98 Pure
Molecular Weight: 139
Viscosity @ 25C, cps: N/A
Features:
* lower toxicity
* can undergo cationic cure mechanisms
* contains polar captam group
* fast cure response
* excellent reactive diluent
Applications:
* homopolymer for potential use in water soluble coatings * vinyl caprolactam/vinyl acetate copolymers * vinyl caprolactam/methacrylate ester copolymers * coatings * lubricating oil additives * adhesives
Performance Properties:
* Adhesion * Low Shrinkage
* Hardness

SARTOMER CO.: Monomer/Oligomer Products(Continued):

Oligomers. Functional acrylic oligomers are high molecular weight, photosensitive materials which contribute major properties to finished high energy cured systems.

Epoxy:

3000:

Chemical Description: Epoxy Diacrylate
Appearance: Viscous Liquid
Reactive Esters, %: 98+
Molecular Weight: 428
Viscosity: ~1MM
Features: fast cure response
Applications:
* paper clear coatings * wood top coatings * screen inks * litho inks * polyethylene coatings * metal decorative coatings * adhesive papers * wood fillers * solder masks and photoresists
Performance Properties:
* Chemical Resistance
* Low Shrinkage
* Heat Resistance
* Water Resistance
* Adhesion
* Hardness
* Abrasion Resistance

Elastomeric:

5000:

Chemical Description: Polybutadiene Diacrylate
Appearance: Viscous Liquid
Reactive Esters, %: 99+
Molecular Weight: 3000
Viscosity @ 25C, cps: 4500-5000
Features:
* low volatility
* excellent electrical properties
* low Tg
* aids adhesion
Applications:
* optical fibers * pottants and encapsulants * conversion coatings * pressure sensitive adhesives
Performance Properties:
* Flexibility
* Low Shrinkage
* Water Resistance
* Adhesion
* Impact Strength

SARTOMER CO.: Monomer/Oligomer Products(Continued):

Urethanes:

Sartomer Product Code:

9503:

Chemical Description: Urethane Diacrylate (Soft)
Appearance: Clear Liquid
Reactive Esters, %: 99
Molecular Weight: 2000
Viscosity @ 25C, cps: 2MM @ 70F.
90M @ 100F.
13M @ 140F.

Features:
* offers good light stability and gloss retention properties
* produces soft urethanes

Applications:
* printing inks * PVC floor coatings * wood coatings * paper coatings * anaerobic adhesives * tile coatings * textile coatings * flexible overprint varnishes * outdoor applications * fiber optic coatings * automotive coatings

Performance Properties:
* Weatherability
* Flexibility
* Low Shrinkage
* Heat Resistance
* Abrasion Resistance

9504:

Chemical Description: Urethane Diacrylate (Medium)
Appearance: Clear Liquid
Reactive Esters, %: 99
Molecular Weight: 1700
Viscosity @ 25C, cps: 167M @ 70F.
10.5M @ 100F.
3250M @ 140F

Features:
* produces medium hardness urethanes
* offers good light stability and gloss retention properties

Applications: anaerobic adhesives * tile coatings * flexible overprint varnishes * outdoor applications * fiber optic coatings

Performance Properties:
* Weatherability
* Flexibility
* Low Shrinkage
* Abrasion Resistance

SARTOMER CO.: Monomer/Oligomer Products(Continued):

Sartomer Product Code:

9505:

Chemical Description: Urethane Diacrylate (Hard)
Appearance: Clear Liquid
Reactive Esters, %: 99
Molecular Weight: 1250
Viscosity @ 25C, cps: 145M @ 100F./35M @ 140F.
Features:

* offers good light stability and gloss retention properties
* produces hard urethanes

Applications:

* anaerobic adhesives * tile coatings * flexible overprint varnishes * outdoor coatings * fiber optic coatings

Performance Properties:

* Weatherability
* Flexibility
* Low Shrinkage
* Hardness
* Abrasion Resistance

MACROMER Monomers:

These are macromolecular monomers, having excellent applicability as oligomers for radiation cure systems and as grafting agents for copolymer production.

13K-PSMA:

Chemical Description: Methacrylate Terminated Polystyrene
Appearance: Granular Solid
Reactive Esters, %: 90+
Molecular Weight: 13M
Viscosity @ 25C, cps: N/A
Features: contains methacrylate functionality
Applications:

* pour point depressants * inks * coatings * solvent base sealants * glass reinforcements * UV cure heat seal adhesives * graft copolymers

Performance Properties:

* Adhesion
* Low Shrinkage
* Hardness
* Impact Strength
* Water Resistance

13K-RC:

Chemical Description: Methacrylate Terminated Polystyrene--Functional Radiation Curable Grade
Appearance: Granular Solid
Reactive Esters, %: 90+
Molecular Weight: 13M+-1500
Features: methacrylate functionality radiation cure grade
Applications:

* UV curable inks * coatings * heat seal adhesives * laminating adhesives

Performance Properties:

* Adhesion
* Low Shrinkage
* Haraness
* Impact Strength
* Water Resistance

SARTOMER CO.: Monomer/Oligomer Products(Continued):

Sartomer Product Code:

13K-PSOH:

Chemical Description: Hydroxy Terminated Polystyrene
Appearance: Granular Solid
Reactive Esters, %: 90+
Molecular Weight: 13M
Features: contains hydroxy functionality
Applications:
* UV cure heat seal adhesives
* graft copolymers
Performance Properties:
* Adhesion
* Hardness
* Water Resistance
* Low Shrinkage
* Impact Strength

MACROBASE Monomers:
Blends of MACROMER 13K-RC Monomer with SARTOMER functional acrylic monomers.

600:

Chemical Description: Blend of 13K-RC and Isooctyl Acrylate
Appearance: Liquid
Reactive Esters, %: 95+
Viscosity @ 25C, cps: 2885
Features: monofunctional, excellent adhesion
Applications:
* UV cure heat seal adhesives
* dispersants for hydrophobic pigments
* UV curable inks and coatings
Performance Properties:
* Flexibility
* Low Shrinkage
* Adhesion
* Impact Strength

610:

Chemical Description: Blend of 13K-RC and Tripropylene Glycol Diacrylate
Appearance: Liquid
Reactive Esters, %: 95+
Viscosity @ 25C, cps: 100-125M
Features: difunctional, excellent adhesion
Applications:
* UV/EB curable adhesives * dispersants for hydrophobic pigments * UV curable inks and coatings
Performance Properties:
* Flexibility
* Low Shrinkage
* Adhesion
* Impact Strength

SARTOMER CO.: Monomer/Oligomer Products(Continued):

Sartomer Product Code:

620:

Chemical Description: Blend of 13K-RC and Ethoxylated TMPTA with Tripropylene Glycol Diacrylate
Appearance: Liquid
Reactive Esters, %: 95+
Viscosity @ 25C, cps: 250M-300M
Features: high functionality, excellent adhesion
Applications:
* UV/EB curable adhesives * dispersants for hydrophobic pigments * UV curable inks and coatings
Performance Properties:
* Chemical Resistance * Hardness
* Heat Resistance

SARTOMER CO.: Monomer/Oligomer Products(Continued):

ESACURE Photoinitiators:

These materials are free radical initiators for high energy polymerizations.

Sartomer Product Code:

ESACURE KB1:

Chemical Description: Benzildimethyl Ketal
Appearance: White Powder
Molecular Weight: 256
Monomer Features:
* optimum absorption at 250-350 nm
* fast cure response
* good through cure
Applications:
* white pigmented coatings
* inks * photopolymers
* electronic photoresists
* polyester-styrene wood filler composites

ESACURE KB60:

Chemical Description: 60% Solution of Benzildimethyl Ketal
Monomer Features:
* optimum absorption at 250-350 nm
* fast cure response
* easy-to-process liquid ketal
* good through cure
Applications:
* white pigmented coatings
* inks * photopolymers
* electronic photoresists
* polyester-styrene wood filler composites

ESACURE EB3:

Chemical Description: Mixture of Benzoin Normal Butyl Ethers
Molecular Weight: 256
Monomer Features:
* easy-to-process, lower cost liquid photoinitiator
* optimum absorption at 250-350 nm
Applications:
* polyester-styrene wood filler composites

SARTOMER CO.: SMA Resins:

SMA Resins are low molecular weight copolymers of styrene and maleic anhydride in anhydride and in partial ester form. They were developed to fill a need for resins having low solution viscosities, high functionality, light color, water solubility and water insensitivity upon drying.

They are widely employed as leveling agents in floor polishes, as embrittling/anti-resoil agents in rug shampoos and as pigment dispersants in paints, inks and plastics. They are also utilized in paper sizings, coatings for metals, aqueous grinding fluids and numerous other applications,.

Typical Properties of SMA Resins:

<u>SMA</u> <u>Resin:</u>

1000:

Molecular Weight: 1600
Melting Range, C: 150-170
Acid No.: 480
15% NVM Viscosity (cps.): 17

2000:

Molecular Weight: 1700
Melting Range, C: 140-160
Acid No.: 350
15% NVM Viscosity (cps.): 26

3000:

Molecular Weight: 1900
Melting Range, C: 115-130
Acid No.: 275
15% NVM Viscosity (cps.): 52

1440:

Molecular Weight: 2500
Melting Range, C: 55-75
Acid No.: 175
15% NVM Viscosity (cps.): 27

17352:

Molecular Weight: 1700
Melting Range, C: 160-170
Acid No.: 270
15% NVM Viscosity (cps.): 24

SARTOMER CO.: SMA Resins (Continued):

SMA Resin:

2625:

Molecular Weight: 1900
Melting Range, C: 135-150
Acid No.: 220
15% NVM Viscosity (cps.): 30

3840:

Molecular Weight: 2300
Melting Range, C: 100-120
Acid No.: 105

Solution Resins:

1440 H:

Solids: 33
pH: 8.5
Viscosity (cps): 8,000
Gardner Color: 4

Vehicle #2:

Solids: 27
pH: 8.5
Viscosity (cps): 10,000
Gardner Color: <1

1538 H:

Solids: 38
pH: 8.5
Viscosity (cps): 1,000
Gardner Color: 2-3

Vehicle #3:

Solids: 32
pH: 9.2
Viscosity (cps): 3,000
Gardner Color: 3

SATURN CHEMICALS, INC.: Polymers:Acrylates:

SATURN SPS-L SPS-N Solution:
Polymerized Styrene Acrylate Copolymer:

SATURN SPS-L and SPS-N Resins are clear water white solution polymerized Saturn-Acrylic copolymers. They are thermoplastic film forming resins which dry by solvent evaporation. These resins have been developed for coatings applications which require the following properties:

Water White Clarity
Good Adhesion
Chemical Resistance
Quick Dry
Excellent Water Resistance
Low Ultra Violet Discoloration
Hard Films
Low Specific Gravity

Application Areas:

Masonry Paints	Curing Compounds
Floor Sealers	Industrial Maintenance Paints
Alkali and Acid Resistant Coatings	
Texture Coatings	Metal Primers
Clear Metal Coatings	Plaster Sealers
Inks	Barrier Coats
Gloss Coats on Paper, Luggage, Labels	
Paper and Tube Coatings	

Physical Properties:

SPS-L & SPS-N
Xylene
Solids Wt. %: 50+-1%
Solvent: Xylene
K.B.: 99
Viscosity-Gardner Holdt: T-X

SPS-L & SPS-N
Aromatic
Solids Wt. %: 50+-1%
Solvent: AR 100
K.B.: 91
Viscosity-Gardner Holdt: T-V

SATURN CHEMICALS, INC.: Polymers:Acrylates(Continued):

SATURN ALS:

Solution polymerized styrene acrylate copolymer with aliphatic solubility.

SATURN ALS is an aliphatic soluble Styrene-Acrylate Copolymer. This clear water white solution polymerized copolymer has been designed to achieve an economical aliphatic soluble polymer which exhibits an excellent balance of water, chemical, and U.V. resistance

SATURN ALS has been developed for coatings applications which require the following properties:

Exempt Mineral Spirits Solubility
Water White Clarity
Good Adhesion
Chemical Resistance
Quick Dry
Excellent Water Resistance
Chemically Neutral
Resistance to Alcohols
Low Ultra Violet Light Discoloration
Good Flexibility

Application Areas:

Masonry Paints
Curing Compounds
Floor Sealers
Industrial Maintenance Paints
Alkali and Acid Resistance Coatings
Texture Paints
Waterproofing Paints
Metal Primers
Clear Metal Coatings
Plaster Sealers
Inks
Barrier Coats on Paper, Labels
Pipe and Tube Coatings

SATURN ALS Polymer Specifications:

Solids: 50+-1%
Solvent: Aromatic Blend
Viscosity Gardner Holdt @25C: X-Z

SATURN CHEMICALS, INC.: Polymers:Hydrocarbons:

SATURN HC Resins:

SATURN HC polymers exhibit lower color, odor, and more uniformity than competitive aromatic/aliphatic polymers.

Specific Performance Properties:

- Chemical and Water Resistance
- Wide Range of Compatibility
- Quick Dry
- Soluble in Aliphatic Hydrocarbons
- Low Color
- Low Odor
- Excellent Stability
- Chemically Neutral

These important polymer properties translate into specific performance attributes in many applications such as:

- Tackify Adhesives and Rubber
- Increase Gloss of Paints & Inks
- Increase Hardness of Paints and Inks
- Increase Water & Chemical Resistance in Mastic Adhesives & Coatings
- Improve Pigment Acceptance and Filler Acceptance

These resins are the natural choice for binders or modifiers in formulations which require acid, alkali, and water resistance such as:

- Paints:
  - Leafing Paints - Bronze and Aluminum
  - Primers
  - Concrete Enamels
  - Metal Paints
- Moisture Barriers
- Caulking and Waterproofing Compounds
- Roofing Compounds
- Adhesives
- Inks
- Rubber Compounding
- Corrugated Board

Properties:

Resin # 100: Softening Point: C: 100

Resin # 110: Softening Point: C: 110

Resin # 130: Softening Point: C: 130

Resin # 140: Softening Point: C: 140

SATURN CHEMICALS, INC.: Polymers: Hydrocarbons(Continued):

SATURN HR Resins:
Heat Reactive Polymers

SATURN HR Resins are produced in continuous non catalytic computer controlled process facilities. SATURN HR polymers are lighter in color, odor, and are more uniform in physical properties throughout the resin than competitive resins produced by batch processing.

Specific Performance Properties:

Chemical and Water Resistance	Excellent Stability
Quick Dry	Low Odor
Chemically Neutral	Wide Range of Compatibility

Major Application Areas--and Noteworthy Properties

Paints - Increase Gloss Hardness and Water and Chemical Resistance
Aluminum and active Pigments - Resins are Neutral, promotes formulae stability
Adhesives - Tackify, Increase Chemical and Water Resistance
Rubber - Tackify, and Promotes Processing
Inks - Increase Gloss & Hardness
Caulking and Waterproofing Compounds - Improve Pigment and Filler Acceptance, Improves Water and Chemical Resistance
Roofing Compounds - Promotes Adhesion, Leafing and Stability

SATURN HR Resin Properties:

Resin # 100: Softening Point: C: 100

Resin # 110: Softening Point: C: 110

Resin # 130: Softening Point: C: 130

Resin # 140: Softening Point: C: 140

SATURN CHEMICALS, INC.: Polymers: Hydrocarbon(Continued):

SATURN HC Resin Solutions:
Convenient--Economical Hydrocarbon Resin Solutions

SATURN HC Resin solutions are part of a series of neutral aromatic, aliphatic thermoplastic polymers.
Polymer solutions produced directly from molten resin provide benefits for the formulator:

Key Benefits--Of Direct Molten Solutions:

1. Diminished Polymer Degradation
2. Lower Solution Viscosities
3. Increased Solution Stability
4. Increased Solubility
5. Lower Color
6. Low Odor
7. Ultra Fine Filtration

Significant Advantages of Hydrocarbon Resin Solutions in Coatings and Adhesive Systems:

Cost Reduction
Improved Water Resistance
Improved Gloss
Improved Adhesion
Improved Pigment Wetting
Increased Hardness
Decreased Hardness
Decreased Dry Time
Improved Acid & Alkali Resistance
Improved Adhesive Tack

SATURN HC Resin Properties:

Resin #: HC 100-140
Softening Point Ring and Ball degree C: 100-140 degree C
Color-Gardner: Max 50% Toluene: 7-9
Iodine Value: 90-120
Acid Value: Nil
Saponification: Nil
Ash Content Max: <0.01
Specific Gravity: 1.08-1.10
Pounds per Gallon: 9.0-9.15

SATURN CHEMICALS, INC.: Polymers: Hydrocarbons(Continued):

VEPRIN:
Hydrocarbon Resin Solution in Mineral Seal Oil Ink Formulating

VEPRIN Hydrocarbon Resins are non-heat reactive Alkylated Aromatic Polymers blended with Mineral Seal Oil specifically designed and polymerized for printing ink formulating.

VEPRIN Resins will provide the following important properties:

Low Odor: Double Processed to Assure Low Odor Characteristics.

Consistent: VEPRIN Resin Solutions are produced to Viscosity Specifications rather than Percentage of Solids.

Reducibility: VEPRIN Resins Dilute 9 to 1 with Mineral Seal Oil Without Drop Out, Tested @ 0 degree C

Consistent Color: Gardner @ 50% 12-14

Compatibility with Ink Oils and Other Resins: VEPRIN Solutions are produced from the Molten Resin yielding Minimum Polymer Degradation and best compatibility.

These resins are the natural choice for binders or modifiers in formulations which require acid, alkali, and water resistance such as:

- Paints
  - Leafing Paints--Bronze and Aluminum
  - Primers
  - Concrete Enamels
  - Metal Paints
- Curing Compounds
- Moisture Barriers
- Caulking and Waterproofing Compounds
- Roofing Compounds
- Adhesives

SATURN CHEMICALS, INC.: Polymer:Hydrocarbon(Continued):

SATURN LCH25--85% MS:
Low M.P. Adhesive Modifying Polymer

SATURN LCH25--85% M/S is a low melt point adhesive modifying polymer. This polymer is based on a heat reactive polymer, modified with naphthenic oil for final melt point control as needed. LCH 25-85% MS has the same attributes as the other LC Series polymers, namely low color odor, and a high degree of compatibility and formulation versatility. LCH25-85% MS will increase water resistance, tack and filler acceptance on typical adhesive applications.

Properties:

Ring and Ball Softening Point: 10-25C
Solids: 85% Nominal
Color-Gardner @ 50% MS: 8-10
Wt/Gal: 7.9# - 8.2# Gal.
Iodine #: 70-90
Viscosity: V-X @ 77 degree F
Acid value: Nil
Solvent: Proprietary
Type: Aromatic-Aliphatic Blend

SATURN CHEMICALS, INC.: Polymers: Hydrocarbon(Continued):

SATURN LC-70 Series:
Solution Hydrocarbon Resins

Low Color, Low Odor, Low Cost

SATURN LC-70 Series are high quality Hydrocarbon Resins. These polymers are noteworthy for their low color, odor, and excellent Aliphatic solubilities and stabilities.

Product:

CMR-70:
- Solids: 70%
- Type: Aromatic
- Gardner Color (50%): 9-10
- Special Features: Adhesive Modifier

SSR-70:
- Solids: 70%
- Type: Aromatic-Aliphatic
- Gardner Color (50%): 7-8
- Special Features: Rule 66 Solubility

DPR-70:
- Solids: 70%
- Type: Heat Reactive
- Gardner Color (50%): 8-9
- Special Features: Economical

Key Benefits:
- Aliphatic Solubility
- Low Color
- Excellent Stability
- Excellent Formulation Versatility
- Excellent Heat Aging Characteristics
- Economical
- Chemical and Water Resistant
- Chemically Neutral
- Compatibility With Many Film Formers

Physical Properties:

CMR-70:
- Softening Point Ring and Ball degree C: 90-110
- Resin Type: Aromatic

SSR-70:
- Softening Point Ring and Ball degree C: 90-110
- Resin Type: Aromatic Aliphatic

DPR-70:
- Softening Point Ring and Ball degree C: 90-110
- Resin Type: Heat Reactive

SATURN CHEMICALS, INC.: Polymers:Hydrocarbon(Continued):

SATURN (ERC45%) Emulsion Resin Concentrate:

SATURN ERC-45% is a solvent water emulsion of SATURN CMR-70 Hydrocarbon Resin. This unique emulsion which is part of Saturn's new Hydrocarbon Emulsion series will yield a clear continuous water resistant film on drying.

Key Benefits & Application Areas

1. Clear Water Deposited Film
2. Passes ASTMC-309-81
3. Water & Chemical Resistant Film
4. Can Be Used As A Concrete Sealer
5. Excellent Compatibility With Other Film Formers
6. Adds Water & Chemical Resistance to Other Emulsions Paint and Coating System

Composition:

SATURN DPR Resin:	45.0%
Mineral Spirits:	20.0%
Water:	32.0%
Emulsifiers & Other Additives:	3.0%

Typical Physical Properties:

Solids: 45.0%
Viscosity Brookfield: 4-8000 cps
Weight/gallon: 8.26-8.32#/gal.
California VOC: $\frac{293 \text{ grams}}{\text{liter.}}$

SATURN CHEMICALS, INC.: Polymers:Hydrocarbon(Continued):

SATURN HCRE Series Emulsions are solvent--water emulsions of hydrocarbon resins with excellent compatibilities with most adhesive binder systems, such as styrene butadiene, acrylic, and PVA, emulsions. These emulsions provide increased emulsion stability, water resistance and adhesion to in situ emulsions.

Key Benefits to Adnesive Formulations:

Excellent water resistance
Increased tack
Increased wet strength
Increased adhesion to metal
Emulsion stability

H.C. Resin Emulsion:

General Composition	
Percent SATURN CMR-70	49.0
Percent Mineral Spirits	21.5
Percent Water	27.0
Percent Additives	2.5

Typical Physical Properties:	
Solids:	49-51%
pH:	9-10
Viscosity, Brookfield	1400-2000 cps
Weight/gallon	8.35-8.45#/gallon

HC Resin Emulsion:
Oil Modified:

General Composition:	
Percent SATURN CMR-70	34.7
Percent Mineral Spirits	14.9
Percent Process Oil	20.9
Percent Water	27.2
Percent Additives	2.3

Typical Physical Properties:	
Solids:	54-55%
pH:	9-10
Viscosity, Brookfield:	250-500 cps
Weight/gallon:	8.25-8.35#/gal.

SATURN CHEMICALS, INC.: Polymers: Hydrocarbon(Continued):

SATURN SSA-70 Acid Functional Hydrocarbon Polymer:

SATURN SSA-70 is an acid functional hydrocarbon polymer. This light colored functional polymer is designed for applacations in the adhesive and coatings industry. Its acid functionality facilitates the in situ manufacture of hydrocarbon emulsions for the adhesives industry. Emulsions based on SSA-70 exhibit improved stability, binder wetting and adhesion.

In the paint and coatings industry, the acid functional SSA-70 exhibits improved compatibility with many film formers, increased pigment wetting properties, and improved substrate adhesion.

Key Benefits:

Readily Emulsifiable
Improves Emulsion Stability
Aliphatic Soluble
Improves Pigment Wetting
Improves Compatibility
Low Color
Excellent Heat Aging Charaacteristics
Improves Adhesion
Improves Chemical & Water Resistance
Economical

Solids: 70+-1% MS
Softening Point R&B degree C: 110-125
Color Gardner as Supplied 70%: 9-13
@ 50%: 8-12
Iodine #: 110-140
Viscosity Brookfield: 5-8000 cps

Wt/Gallon: 8.3-8.4

Acid #: 5-10
Sap #: 0
Ash Content: <.01%

Resin Type: Aromatic Aliphatic

SATURN CHEMICALS, INC.: Polymers: Hydrocarbon(Continued):

SATURN DPR:
Heat Reactive Flake Resins:

SATURN DPR Resin Series is a low cost premium polymer. SATURN DPR polymers are a cost effective series that has excellent color, stability and uniform properties.

Applications Areas and Noteworthy Properties:

Paints - Increase Gloss Hardness and Water and Chemical Resistance
Aluminum and active Pigments - Resins are Neutral, promotes formulae stability
Adhesives - Tackify, Increase Chemical and Water Resistance
Rubber - Tackify, and Promotes Processing
Inks - Increase Gloss & Hardness
Caulking and Waterprooofing Compounds - Improve Pigment and Filler Acceptance, Improve Water and Chemical Resistance
Roofing Compounds - Promotes Adhesion, Leafing and Stability

DPR 100 Properties:

DPR-100:

Softening Point C: 100
Color-Gardner Max in 50% Toluene: 10-12
Iodine Value: 160-190
Acid Value: Nil
Saponification Value: Nil
Ash Content Max: <0.01
Specific Gravity: 1.08-1.10
Pound per Gallon: 9.0-9.15

DPR-140:

Softening Point C: 140
Color-Gardner Max in 50% Toluene: 11-13
Iodine Value: 160-190
Acid Value: Nil
Saponification Value: Nil
Ash Content Max.: <0.01
Specific Gravity: 1.08-1.10

SATURN CHEMICALS, INC.: Polymers:Hydrocarbon(Continued):

SATURN HWPE-70%:
Emulsifiable Hydrocarbon Resin Solution

SATURN HWPE-70% is a 70% hydrocarbon resin solution specially blended with emulsifiers and wetting agents. This polymer is noteworthy for its low color, odor, excellent aliphatic solubilities and ease of emulsification.

Primary Uses:

* Preparation of Stable Emulsions
* Emulsion Curing Compounds
* Emulsion White Pigmented Curing Compounds
* Emulsion Adhesive Modifier

Key Benefits

* Aliphatic Solubility
* Excellent Stability
* Low Color
* Chemical & Water Resistance

SATURN HWPE-70% Physical Properties:

Solids: 70%
Softening Point Ring and Ball degree C: 80-90
Viscosity Brookfield: 6,860+-2,000
Resin Type: Aromatic
Aliphatic

SATURN CMR Resins:
For Low Cost Curing Compounds

Modern concrete curing is accomplished by the forming of a moisture retentive membrane over the freshly poured concrete. These membranes provide low moisture vapor transmission rates yielding controlled water loss and the higher physical properties of concrete.

SATURN CMR Hydrocarbon Resins have been successfuly used as the film former for efficient low cost curing membranes. The excellent chemical resistance helps prevent efflorescence, dusting and spalling of the coated concrete. Curing compounds, based on SATURN CMR Resins, provide excellent quick drying membranes which pass A.S.T.M. C309-84 Type 1.

SATURN CHEMICALS CO., INC.: Polymers: Hydrocarbon(Continued):

SATURN SPS and ALS Resins for Premium Curing Compounds:

Modern concrete curing is accomplished by the forming of a moisture retentive membrane over the freshly poured concrete. These membranes provide low moisture vapour transmission rates yielding controlled water loss and the highest physical properties of the concrete.

Premium concrete curing compounds based on SATURN SPS and ALS continued performing throughout the job, since penetration of these coatings into the concrete yields a continuous membrane able to resist water and dirt pick up. Mortar and concrete drippings, dirt and oil, are easily removed from concrete cured by membranes based on SATURN SPS and ALS.

Properties of SPS and ALS Curing Compounds:

- Cure-Harden-Seal
- Low Moisture Transmission Rates
- Water White Solutions
- Quick Drying
- Relative Non Yellowing
- Prevents Efflorescence, Dusting, and Spalling
- Reduce Clean-Up and Construction Stains
- Exempt Solvents
- Excellent Toughness and Chemical Resistance
- Good Base for Adhesives

SATURN CHEMICALS, INC.: Polymers:Styrenes:

SATURN SPC:
Solution Polymerized Styrene Acrylate Polymer

SATURN SPC is a solution polymerized Styrene Acrylate polymer at 50% solids in Xylene or AROMATIC 100 solvents. SATURN SPC has been designed as an economical polymer for applications which require improved water resistance and hardness characteristics. Since SPC is a solution polymerized polymer it does not contain residual emulsification agents which cause water sensitivity, yellowing and turbid films and solutions.

Coatings formulated with SATURN SPC will have good non-yellowing characteristics and excellent chemical and water resistance properties.

Formulating Properties:

- Water White Clarity
- Good Adhesion
- Chemical Resistance
- Quick Dry
- Excellent Water Resistance
- Resistance to Low KB Solvents & Alcohol
- Superior Ultra Violet Light Properties vs Styrene Butadiene and Chlorinated Rubber
- Hard Films
- Low Specific Gravity

Application Areas:

- Curing Compounds
- Floor Sealers
- Alkali and Acid Resistance Coatings
- Masonry Paints
- Clear Metal Coatings
- Primers
- Barrier Coatings
- Pipe and Tube Coatings

Physical Properties SATURN SPC:
In Xylene:
Solids Weight %: 50+-1%
Solvent: Xylene
K.B.: 99
Viscosity-Gardner Holdt: Z-Z1

In Aromatic:
Solids Weight %: 50+-1%
Solvent: AROMATIC 100 Solvent
K.B.: 91
Viscosity-Gardner Holdt: Z-Z1

SATURN CHEMICALS, INC.: Polymers:Styrene(Continued):

SATURN SAE:
Styrene Acrylate Chemical Resistant Emulsion 45% Solids

SATURN SAE is an emulsion of a solvent base styrene acrylic polymer designed to provide solvent base performance from a water emulsion system. Films of SATURN SAE when reduced with water as per the reduction formula listed below--are crystal clear water white polymer films which are resistant to water, chemicals, and exterior weathering. These products can be used as coatings for concrete, floor sealers, and waterprooofing, texture, and masonry paints. Coatings and paints formulated from SATURN SAE emulsions will meet the requirements of California VOC regulations.

COMPOSITION:

SATURN Styrene Acrylate Polymer:	45.00%
Solvent (AR-100):	24.23%
Water:	26.08%
Emulsifiers and Other Additives:	4.69%
	100.00%

Typical Physical Properties:

Solids: 45.0+1.0%
Viscosity Brookfield: 4-8000 cps

SATURN CHEMICALS, INC.: Polymers:Styrene(Continued):

SATURN C.S.P. 65% AR-100/M.S.:
Clear Sealant Polymer Base

SATURN C.S.P. 65% is a water white crystal clear Styrene Acrylate Polymer specially polymerized for use as a base system for the production of clear sealants and caulks.

Sealants formulated with SATURN C.S.P. 65 exhibit useful properties such as listed below:

Excellent Adhesion to glass-metal, etc.
Good Flexibility throughout standard temperature ranges.
Low Shrinkage
Good Aging Properties
Good Water Resistance
Toughness
Self Healing Cracks
Non Red Label
Good Package Stability
Good Flow and Gunning Characteristic
Clarity
Pigment Acceptability (Do not use with active pigments such as ZnO.)

Physical Properties:

Solid Weight %: 65%
Solvent Blend: 35% of a 50-50 blend of AROMATIC 100 and Mineral Spirits
Flash Point: Greater than 105 degree F.
Weight/Gallon: 7.94
Specific Gravity Resin: 1.04
Color-Gardner: Below 1
Clarity: Crystal Clear
Flexibility: Passes 3/16" mandrel
Brookfield Viscosity: 40-60,000 C.P.S.

SYNRAY CORP.: Alkyd Resins:

Long Oil Alkyds:

Product:

SYNRAY 1180-98:
Solids+-1%: 98
Viscosity Gardner-Holt: X-Z1
Type Oil: Soya
Phthalic Anhydride %: Isophthalic 11
Application: Conforms to TT-P-102. For exterior oil base house paint.

SYNRAY 2355-60:
Solids+-1%: 60
Solvent: Exempt MS
Viscosity Gardner-Holt: T-W
Type Oil: TOFA
Phthalic Anhydride %: 24
Application: Conforms to TT-R-266 Type 5 (@ 60% N.V.). High quality architectural enamels.

SYNRAY 23550-60:
Solids+-1%: 60
Solvent: Exempt MS
Viscosity Gardner-Holt: U-X
Type Oil: Mixed Oils
Phthalic Anhydride %: 23
Application: "No-frills" economical finishes. Meets performance requirements for TT-R-266.

SYNRAY 2365-60:
Solids+-1%: 60
Solvent: Exempt MS
Viscosity Gardner-Holt: T-W
Type Oil: Soya
Phthalic Anhydride %: 24
Application: Conforms to TT-R-266 Type I Class B. High quality architectural enamels.

SYNRAY 2464-60:
Solids+-1%: 60
Solvent: Exempt MS
Viscosity Gardner-Holt: X-Z
Type Oil: Soya
Phthalic Anhydride %: 24
Application: Heavier viscosity version of 2365-60.

SYNRAY 2464-70X:
Solids+-1%: 70
Solvent: Xylol
Viscosity Gardner-Holt: O-S
Type Oil: Soya
Phthalic Anhydride %: 24
Application: Conforms to TT-R-266 Type I Class C. Used with vinyl resins for heavy duty maintenance and marine coatings.

SYNRAY CORP.: Alkyd Resins(Continued):

Medium Oil Alkyds:

Product:

SYNRAY 2550-55:

Solids+-1%: 55
Solvent: Exempt MS
Viscosity Gardner-Holt: X-Z
Acid No. Max.: 10
Type Oil: Soya
Phthalic Anhydride %: 25
Application: Interior & exterior enamels & semi-gloss machinery and floor paints, with water extendability to 30% N.V.

SYNRAY 3152-60:

Solids+-1%: 60
Solvent: Toluol
Viscosity Gardner-Holt: E-F
Acid No. Max.: 10
Type Oil: Soya
Phthalic Anhydride %: 31
Application: Conforms to TT-P-115 Type III. Fast dry traffic paint-compatible with chlorinated rubber.

SYNRAY 3153-50:

Solids+-1%: 50
Solvent: Exempt MS
Viscosity Gardner-Holt: U-Z
Acid No. Max.: 12
Type Oil: Soya/Linseed
Phthalic Anhydride %: 31
Application: Conforms to TT-R-266 Type III. Exterior primers, enamels and machinery finishes.

SYNRAY 3154-50:

Solids+-1%: 50
Solvent: Exempt MS
Viscosity Gardner-Holt: U-Z
Acid No. Max.: 12
Type Oil: TOFA
Phthalic Anhydride %: 31
Application: Conforms to TT-R-266 Type VI. Metal primers enamels and machinery finishes.

SYNRAY CORP.: Alkyd Resins(Continued):

Medium Oil Alkyds(Continued):

Product:

SYNRAY 3155-50V:
Solids+-1%: 50
Solvent: Exempt VM&P
Viscosity Gardner-Holt: X-Z
Acid No. Max.: 12
Type Oil: Soya
Phthalic Anhydride %: 31
Application: Conforms to TT-P-115E Type I. Traffic and runway marking paint.

SYNRAY 31540-50:
Solids+-1%: 50
Solvent: Exempt MS
Viscosity Gardner-Holt: U-Z
Acid No. Max.: 12
Type Oil: Mixed Oils
Phthalic Anhydride %: 31
Application: "No-frills" economical finishes. Meets performance requirements for TT-R-266.

SYNRAY 3155-50:
Solids+-1%: 50
Solvent: Exempt MS
Viscosity Gardner-Holt: U-Z
Acid No. Max.: 12
Type Oil: Soya
Phthalic Anhydride %: 31
Application: Conforms to TT-R-266 Type IV. Metal primers and machinery finishes.

SYNRAY 3156-50:
Solids+-1%: 50
Solvent: Exempt MS
Viscosity Gardner-Holt: U-Z
Acid No. Max.: 12
Type Oil: Soya
Phthalic Anhydride %: Dicarboxylic Acid 31
Application: "Tuf-stuff" extender resin for urethane floor varnishes.

SYNRAY 3255-70:
Solids+-1%: 70
Solvent: Xylol
Viscosity Gardner-Holt: Z2-Z4
Acid No. Max.: 8
Type Oil: DCO
Phthalic Anhydride %: 32
Application: Fast air dry with excellent color & gloss with resistance to taping.

SYNRAY CORP.: Alkyd Resins(Continued):

Medium Oil Alkyds(Continued):

Product:

SYNRAY 3549-60:

Solids+-1%: 60
Solvent: Exempt VM&P
Viscosity Gardner-Holt: Z1-Z3
Acid No. Max.: 8
Type Oil: Soya
Phthalic Anhydride %: 35
Application: Conforms to Texas & N.J. Type IV specifications. Very fast dry traffic paint. Compatible with chlorinated rubber.

SYNRAY 3550-60:

Solids+-1%: 60
Solvent: Exempt VM&P
Viscosity Gardner-Holt: Z5-Z7
Acid No. Max.: 12
Type Oil: Soya
Phthalic Anhydride %: 35
Application: Traffic vehicle conforming to North Carolina state spec.

SYNRAY 3746-50:

Solids+-1%: 50
Solvent: Exempt MS
Viscosity Gardner-Holt: Z5-Z6
Acid No. Max.: 10
Phthalic Anhydride %: 37
Application: High drink resin for low cost paints.

Short Oil Alkyds:

AQUA-DRY 2833-75:

Solids+-1%: 75
Solvent: n-Butanol
Viscosity Gardner-Holt: Z3-Z5
Acid No. Max.: 45-50
Type Oil: TOFA
Phthalic Anhydride %: 28
Application: Water reducible alkyd for air dry and baking enamels, with good impact resistance, flexibility and gloss retention.

SYNRAY CORP.: Alkyd Resins(Continued):

Short Oil Alkyds(Continued):

Product:

SYNRAY 3539-50:
Solids+-1%: 50
Solvent: Xylol
Viscosity Gardner-Holt: Z-Z2
Acid No. Max.: 10
Type Oil: TOFA
Phthalic Anhydride %: 35
Application: Chain-stop alkyd. Very fast drying primers & finishes.

SYNRAY 4021-50:
Solids+-1%: 50
Solvent: Exempt Aromatic
Viscosity Gardner-Holt: Z1-Z3
Acid No. Max.: 8
Type Oil: TOFA
Phthalic Anhydride %: 40
Application: Chain-stop alkyd. Very fast drying primers & finishes.

SYNRAY 4031-60:
Solids+-1%: 60
Solvent: Toluol
Viscosity Gardner-Holt: Z4-Z6
Acid No. Max.: 10
Type Oil: Soya
Phthalic Anhydride %: 40
Application: Traffic vehicle conforming to Tenn. state spec.

SYNRAY 4034-50:
Solids+-1%: 50
Solvent: Xylol/VM&P
Viscosity Gardner-Holt: Z5-Z6+
Acid No. Max.: 10
Type Oil: TOFA
Phthalic Anhydride %: 40
Application: High drink alkyd. Cross links with melamines. Used in high gloss bake enamels.

SYNRAY 4040-60:
Solids+-1%: 60
Solvent: Exempt Aromatic
Viscosity Gardner-Holt: X-Z
Acid No. Max.: 20
Type Oil: Linseed
Phthalic Anhydride %: 40
Application: Conforms to requirements of TT-P-1757.

SYNRAY CORP.: Alkyd Resins(Continued):

Short Oil Alkyds(Continued):

Product:

SYNRAY 4239-60:
Solids+-1%: 60
Solvent: Toluol/VM&P
Viscosity Gardner-Holt: Z5-Z6+
Acid No. Max.: 12
Type Oil: Soya
Phthalic Anhydride %: 42
Applications: High heat rapid set traffic marking paint for states requiring "no closure" to traffic. Meets TT-P-85 spec.

SYNRAY 4747-60:
Solids+-1%: 60
Solvent: Exempt MS
Viscosity Gardner-Holt: U-W
Acid No. Max.: 6
Type Oil: Safflower
Phthalic Anhydride %: Chlorendic Anhydride 47
Application: Conforms to requirements of DOD-R-21417. Interior fire retardant paints.

Flat Alkyds:

BURNOK IB-32L-40:
Solids+-1%: 40
Solvent: Low Odor MS
Viscosity: Gel
Acid No. Max.: 10
Type Oil: Fatty Acids
Phthalic Anhydride %: Present
Modification: Polyamide
Application: Low viscosity, thixotropic alkyd for flat & gloss.

BURNOK IB-32-40:
Solids+-1%: 40
Solvent: Low Odor MS
Viscosity: Gel
Acid No. Max.: 10
Type Oil: Fatty Acids
Phthalic Anhydride %: Present
Modification: Polyamide
Application: High viscosity, thixotropic alkyd for flat & gloss.

SYNRAY CORP.: Alkyd Resins(Continued):

Flat Alkyds(Continued):

Product:

SYNFLAT 3743-30:
Solids+-1%: 30
Solvent: Low Odor MS
Viscosity Gardner-Holt: X-Z
Acid No. Max.: 10
Type Oil: Soya
Phthalic Anhydride %: 37
Application: Low odor alkyd for interior flat paints.

Copolymerized Alkyds:

SYNRENE 1725-50:
Solids+-1%: 50
Solvent: Exempt VM&P
Viscosity Gardner-Holt: Z-Z2
Acid No. Max.: 10
Type Oil: Fatty Acids
Phthalic Anhydride %: Present
Modification: VT
Applications: Vinyl-toluene alkyd for fast drying industrial finish.

SYNRENE 2030-60:
Solids+-1%: 60
Solvent: Exempt VM&P
Viscosity Gardner-Holt: Z-Z2
Acid No. Max.: 10
Type Oil: Fatty Acids
Phthalic Anhydride %: Present
Modification: VT
Applications: Vinyl-toluene alkyd for fast drying industrial finishes.

SYNRENE 2031-50:
Solids+-1%: 50
Solvent: Exempt Aromatic
Viscosity Gardner-Holt: T-V
Type Oil: Fatty Acids/DCO
Phthalic Anhydride %: Present
Modification: Styrene
Application: Conforms to TTE-515, TT-E-516 fast drying primer & equipment finishes.

SYNRAY CORP: Alkyd Resins(Continued):

Copolymerized Alkyds(Continued):

Product:

SYNCONE 28E-50:

Solids+-1%: 50
Solvent: Exempt Aromatic
Viscosity Gardner-Holt: U-X
Acid No. Max.: 15
Type Oil: Soya/Fatty Acids
Phthalic Anhydride %: 30
Modification: Silicone
Application: Conforms to TT-P-28. High heat resistant aluminum paints.

SYNCONE 490-60:

Solids+-1%: 60
Solvent: Exempt MS
Viscosity Gardner-Holt: W-Z
Acid No. Max.: 12
Type Oil: Soya
Phthalic Anhydride %: 16
Modification: Silicone
Applications: Conforms to TT-E-490. High gloss retention for exterior enamels.

Non-Drying Alkyds:

SYNPLAST 4233-60:

Solids %: 60
Solvent: Xylol
Viscosity Gardner-Holt: X-Z
Acid No. Max.: 6
Type Oil: Coconut
Phthalic Anhydride %: 42
Application: Lacquer plasticizer for white baking enamels.

SYNPLAST 7070-65:

Solids %: 65
Solvent: Xylol
Viscosity Gardner-Holt: Z4-Z5
Acid No. Max. 25
Type Oil: Castor
Phthalic Anhydride %: 42
Application: Non-lifting plasticizer and lacquer primer.

SYNRAY CORP.: Alkyd Resins(Continued):

Polyester & Oil Free Alkyds:

Product:

SYNPLEX 5100-68:
Solids+-1%: 68
Solvent: Super Hi-Flash Naphtha
Viscosity Gardner Holt: Z4-Z6
Acid No. Max.: 22
Phthalic Anhydride %: Present
Application: OH=100 roller and coil coating resin

SYNPLEX 5150-70:
Solids+-1%: 70
Solvent: Ethyl Glycol Ac
Viscosity Gardner-Holt: Z4-Z6
Acid No. Max.: 8
Phthalic Anhydride %: Present
Applications: Gloss & semigloss finishes with good "skydrol" resistance. Conforms to MIL-C-83286.

SYNPLEX 5200-100:
Solids+-1%: 100
Viscosity Gardner-Holt: X-Y
Acid No. Max.: 5
Phthalic Anhydride %: Present
Applications: High solids baking alkyd. Low temp cure.

SYNPLEX 5265-60:
Solids+-1%: 60
Viscosity Gardner-Holt: J-M
Solvent: Ethyl Glycol Ac
Acid No. Max.: 13
Phthalic Anhydride %: Present
Application: OH=265. Gloss & semigloss 2 component aircraft finish where "skydrol" resistance is needed. Conforms to MIL-C-46168B.

Rosin & Phenolic Modified Alkyds:

SYNRAY 2250-55:
Solids+-1%: 55
Solvent: Exempt MS
Viscosity Gardner-Holt: X-Z
Acid No. Max.: 10
Type Oil: Linseed/Fatty Acids
Phthalic Anhydride %: 22
Modification: Rosin
Application: Industrial maintenance paints.

SYNRAY CORP.: Alkyd Resins, Etc.(Continued):

Rosin & Phenolic Modified Alkyds(Continued):

Product:

SYNRAY 2829-60:

Solids+-1%: 60
Solvent: Xylol/VM&P
Viscosity Gardner-Holt: Z2-Z4
Acid No. Max.: 12
Type Oil: Linseed/Fatty Acids
Phthalic Anhydride %: 27
Modification: Rosin Phenolic
Application: Fast dry industrial enamels.

SYNRAY 2934-45:

Solids+-1%: 45
Solvent: Exempt VM&P
Viscosity Gardner-Holt: Z-Z2
Acid No. Max.: 10
Type Oil: Mixed Fatty Acids
Phthalic Anhydride %: 29
Modification: Rosin
Application: Low cost drum enamels and primers.

SYNRAY 3937-50:

Solids+-1%: 50
Solvent: Exempt Aromatic
Viscosity Gardner-Holt: X-Z
Acid No. Max.: 10
Type Oil: Soya
Phthalic Anhydride %: 39
Modification: Phenolic
Application: Conforms to TT-E-485 Type IV. Industrial and non-lifting primers.

Phenolic Varnishes:

SYNVAR PR-65:

Solids+-1%: 65
Solvent: Exempt VM&P
Viscosity Gardner-Holt: B-D
Type Oil: Tung/Linseed
Modification: Phenolic
Application: Abrasion resistant traffic paint.

SYNRAY CORP.: Alkyd Resins, Etc.(Continued):

Phenolic Varnishes(Continued):

Product:

SYNVAR 119-58:

Solids+-1%: 58
Solvent: Exempt MS
Butanol
Viscosity: Gardner-Holt: C-F:
Type Oil: Tung/Linseed
Modification: Phenolic
Applications: Conforms to TT-V-119 spar varnish

SYNVAR 15189-55:

Solids+-1%: 55
Solvent: Exempt MS
Viscosity: Gardner-Holt: B-D
Type Oil: Tung/Linseed
Modification: Phenolic
Application: Conforms to AASHO M-69-70 Type II & TT-P-38 aluminum mixing varnish.

SYNVAR 15218-I-60:

Solids+-1%: 60
Solvent: Xylol/MS
Viscosity: Gardner-Holt: L-N
Type Oil: Tung/Linseed
Modification: Phenolic
Application: Conforms to DOD-V-15218 Type I marine mixing varnish.

SYNVAR 15218-II-60:

Solids+-1%: 60
Solvent: Exempt MS
Viscosity Gardner-Holt: L-N
Type Oil: Tung/Linseed
Modification: Phenolic
Application: Conforms to DOD-V-15128 Type II marine mixing varnish.

SYNRAY CORP.: Alkyd Resins, etc.(Continued):

Urethane & Epoxy Modified Alkyds:

Product:

SYNTHANE 2168-60:

Solids+-1%: 60
Solvent: Exempt MS
Viscosity Gardner-Holt: Z1-Z3
Acid No. Max.: 2
Type Oil: Sunflower
Modification: TDI
Application: Urethane modified alkyd for high quality varnishes.

SYNTHANE 2169-60:

Solids+-1%: 60
Solvent: Exempt MS
Viscosity Gardner-Holt: Z1-Z3
Acid No. Max.: 2
Type Oil: Linseed
Modification: TDI
Application: Urethane modified alkyd for high quality varnishes.

SYNPOXY:

Solids+-1%: 60
Solvent: Exempt MS
Viscosity Gardner-Holt: Z1-Z3
Acid No. Max.: 10
Type Oil: TOFA
Modification: Epoxy
Apearance: Tough abrasion and chemical resistant coatings. Concrete and deck paints, primer and maintenance paints.

SYNRAY CORP.: High Solids and Water Reducible Resins:

High Solids Alkyd - Air Dry:

Product:

SYNRAY 2830-75:
Solids+-1%: 75
Solvent: Toluene/n-Butyl Ac.
Viscosity Gardner-Holt: Z3-Z5
Acid No. Max.: 12
Type Oil: Fatty Acids
Phthalic Anhydride %: 25
Modification: Rosin
Application: Drum enamels & primers.

SYNRAY 2835-75:
Solids+-1%: 75
Solvent: Toluene/n-Butyl Ac.
Viscosity Gardner-Holt: Z3-Z5
Acid No. Max.: 12
Type Oil: Fatty Acids
Phthalic Anhydride %: 22
Modification: Rosin/Phenolic
Application: Quick dry enamels & primer

SYNRAY 3234-75:
Solids+-1%: 75
Solvent: n-Butyl Ac.
Viscosity Gardner-Holt: Z3-Z5
Acid No. Max.: 10
Type Oil: Soya
Phthalic Anhydride %: 31
Application: High solids chain-stopped alkyd for fast dry industrial coatings.

SYNRAY 3430-75:
Solids+-1%: 75
Solvent: MEK/Toluene
Viscosity Gardner-Holt: Z3-Z5
Acid No. Max.: 10
Type Oil: Soya
Phthalic Anhydride %: 30
Application: Very fast chain-stopped alkyd.

SYNRAY 3927-75:
Solids+-1%: 75
Solvent: MPK/Toluene
Viscosity Gardner-Holt: Z2-Z4
Acid No. Max.: 15
Type Oil: Fatty Acids
Phthalic Anhydride %: 39
Applications: High solids, fast dry alkyds for industrial finishes.

SYNRAY CORP.: High Solids and Water Reducible Resins(Continued):

High Solids Alkyd - Air Dry(Continued):

Product:

SYNRAY 5412-86:

Solids+-1%: 85
Solvent: PM Acetate
Viscosity Gardner-Holt: Z-Z2
Acid No. Max.: 10
Type Oil: Soya
Phthalic Anhydride %: Present
Application: High solids, medium oil air dry alkyd, as possible replacement for TTR-266 Type III.

SYNRAY 5438-80:

Solids+-1%: 80
Solvent: MIBK/Toluene
Viscosity Gardner-Holt: Z5-Z6+
Acid No. Max.: 30
Type Oil: Linseed/Tung
Phthalic Anhydride %: 38
Modification: Rosin/Phenolic
Application: Low VOC lift resistant primers and interior enamels.

SYNRAY 5430-75:

Solids+-1%: 75
Solvent: MIBK
Viscosity Gardner-Holt: Z3-Z5
Acid No. Max.: 8
Type Oil: Soya
Phthalic Anhydride %: 30
Application: Very fast air-dry chain-stopped alkyd with very good exterior durability.

SYNRAY 6024-80:

Solids+-1%: 80
Solvent: PM Acetate
Viscosity Gardner-Holt: Z1-Z3
Acid No. Max.: 5
Type Oil: Linseed
Phthalic Anhydride %: 23
Application: High solids, fast dry long oil alkyd for interior and exterior enamels.

SYNRAY CORP.: High Solids and Water Reducible Resins(Continued):

High Solids Polyester - Bake & Urethane:

Product:

SYNPLEX 5150-70:
Solids+-1%: 70
Solvent: PM Acetate
Viscosity Gardner-Holt: Z-Z2
Acid No. Max.: 3
Phthalic Anhydride %: Present
Application: Gloss & semigloss 2 component urethane finishes OH=150

SYNPLEX 5100-100:
Solids+-1%: 100
Viscosity Gardner-Holt: Z4-Z5
Acid No. Max.: 3
Phthalic Anhydride %: Present
Application: High solids baking alkyd. Low temperature cure.

SYNPLEX 5210-100:
Solids+-1%: 100
Viscosity Gardner-Holt: Z2-Z4
Acid No. Max.: 3
Phthalic Anhydride %: Present
Application: Low molecular weight oil-free polyester for low temperature curing high solids industrial finishes.

SYNPLEX 5265-76:
Solids+-1%: 75
Solvent: MPK
Viscosity Gardner-Holt: Z3-Z5
Acid No. Max.: 6
Phthalic Anhydride %: Present
Application: High solids gloss & semigloss 2 component urethane finishes OH=200

SYNPLEX 5310-75:
Solids+-1%: 75
Solvent: EGMBE/MAK
Viscosity Gardner-Holt: Z4-Z6
Acid No. Max.: 2
Modification: Acrylic
Application: Acrylic thermosetting resin with low-temp-cure response. Good impact resistance and exterior durability.

SYNRAY CORP.: High Solids and Water Reducible Resins(Continued):

High Solids Alkyd - Bake & Urethane:

SYNRAY 5423-90:

Solids+-1%: 90
Solvent: EEP
Viscosity Gardner-Holt: Z5-Z6
Acid No. Max.: 12
Type Oil: Fatty Acids
Phthalic Anhydride %: 23
Application: Low VOC industrial baking enamels, with good chemical resistance.

SYNRAY 5429-90:

Solids+-1%: 90
Solvent: PM Acetate
Viscosity Gardner-Holt: Z4-Z6
Acid No. Max.: 10
Type Oil: Fatty Acids
Phthalic Anhydride %: 29
Application: High solids alkyd for industrial bake finishes.

SYNRAY 5444-80:

Solids+-1%: 80
Solvent: MIBK/Isopropyl Alcohol
Viscosity Gardner-Holt: Z4-Z6
Acid No. Max.: 10
Type Oil: Coconut
Phthalic Anhydride %: 44
Application: High quality, non yellowing finishes & lacquer plasticizer.

SYNRAY 5447-90:

Solids+-1%: 90
Solvent: EEP/Toluene
Viscosity Gardner-Holt: Z2-Z4
Acid No. Max.: 10
Type Oil: Fatty Acids
Phthalic Anhydride %: 47
Application: High solids, economical, industrial baking enamels.

SYNRAY CORP.: High Solids and Water Reducible Resins(Continued):

Water Reducible Alkyd:

Product:

AQUA-STAIN 1585-45:
Solids+-1%: 45
Solvent: EGMBE/Water
Viscosity Gardner-Holt U-W
Type Oil: Linseed
Application: Interior & exterior water reducible stains.

SYN-AQUA 2833-75:
Solids+-1%: 75
Solvent: sec-Butyl Alcohol
Viscosity Gardner-Holt: Z3-Z5
Type Oil: TOFA
Phthalic Anhydride %: Present
Modification: Resin
Application: Fast dry industrial primers & enamels with good corrosion resistance & stability.

SYN-AQUA 5815-75:
Solids+-1%: 75
Solvent: PROPASOL P/sec Butyl Alcohol
Viscosity Gardner-Holt: Z3-Z5
Type Oil: Fatty Acids
Phthalic Anhydride %: Isophthalic
Modification: MMA/VT
Application: Air dry or bake primers & enamels with good hardness and flexibility.

SYN-AQUA 5821-70:
Solids+-1%: 70
Solvent: EGMBE
Viscosity Gardner-Holt: Z4-Z6
Type Oil: Fatty Acids
Phthalic Anhydride %: Present
Modification: Styrene
Application: Fast dry industrial primers and topcoats.

SYN-AQUA 5919-75:
Solids+-1%: 75
Solvent: PROPASOL P/EGMBE
Viscosity Gardner-Holt: Z5-Z6
Type Oil: Fatty Acids
Phthalic Anhydride %: Present
Application: Trade sale interior & exterior coatings. Good stability.

SYNRAY CORP.: High Solids and Water Reducible Resins(Continued):

Water Reducible Alkyd(Continued):

Product:

SYN-AQUA 5932-75:

Solids+-1%: 75
Solvent: PROPASOL P/sec Butyl Alcohol
Viscosity Gardner-Holt: Z4-Z6
Acid No. Max.: 45-50
Type Oil: Fatty Acids
Phthalic Anhydride %: Present
Application: General purpose coatings with fast air dry and good corrosion resistance.

SYN-AQUA 5933-75:

Solids+-1%: 75
Solvent: PROPASOL P/sec Butyl Alcohol
Viscosity Gardner-Holt: Z6-Z7
Acid No. Max.: 39-43
Type Oil: Fatty Acids
Phthalic Anhydride %: Present
Application: Very fast air dry with very good corrosion resistance.

SYN-AQUA 5935-75:

Solids+-1%: 75
Solvent: EGMBE/sec Butyl Alcohol
Viscosity Gardner-Holt: Z6-Z7
Acid No. Max.: 37-42
Type Oil: Fatty Acids
Phthalic Anhydride %: Present
Application: Fast dry water reducible resin with excellent corrosion resistance.

TEGO CHEMIE SERVICE USA: Resins:

SILIKOPHEN P50/X:

Binder for high temperature anticorrosion paints resistant up to 650C/1,200F. Coatings applied to industrial equipment, chimneys, exhaust tubes, etc.

Chemical Composition: Phenyl methyl polysiloxane resin

Physical Data:
* Solids content: 50+-1.5%
* Viscosity at 25C: 30-50 (cSt)
* Specific gravity at 25C: 1.02-1.03 g/cm3
* Colour: clear to slightly cloudy
* Flash Point: 26C/80F
* Solvent: xylene

Areas of application:
High temperature anticorrosive coating for:
* industrial equipment
* chimneys
* exhaust tubes

SILIKOPHEN P50/300:

Binder for high temperature anticorrosion paints resistant up to 650C/1,200F. Coatings applied to industrial equipment, chimneys, exhaust tubes, etc.

Chemical Composition: Phenyl methyl polysiloxane resin

Physical data:
Solids content: 50+-2%
Viscosity at 25C: 200-400 (cSt)
Specific gravity at 25C: 1.02-1.03 g/cm3
Colour: clear to slightly cloudy
Flash point: 26C/80F
Solvent: Xylene

Areas of application:
High temperature anticorrosion coating for:
* industrial equipment
* chimneys
* exhaust tubes

TEGO CHEMIE SERVICE USA: Resins(Continued):

SILIKOPHEN P80/20:

Binder for high temperature corrosion resistant coatings up to 650C/1200F. Typical coating applications include industrial equipment, boiler and furnace stacks, engine parts, woodstoves and high temperature electric/gas heaters. SILIKOPHEN P80/20 offers excellent air dry characteristics thus improving the adhesion, hardness and corrosion resistance of the air dried paint film.

Chemical Composition: Phenyl methyl polysiloxane resin

Physical data:

% Solids content: 70%+-2.0%
Viscosity @ 25C: 700+-200 (cSt)
Specific gravity @ 25C: 1.10 g/cm3
Color: clear to slightly cloudy
Flash point: 28C/82F
Solvent: xylene

Areas of application:
* Industrial Equipment
* Boiler and Furnace Stacks
* Engine Parts
* Woodstoves
* High Temperature Electric/Gas Heaters

SILIKOPHEN P80/X:

Binder for high temperature anticorrosion paints resistant up to 650C/1,200F. Coatings applied to industrial equipment, chimneys, exhaust pipes, etc.

Chemical Composition: Phenyl methyl polysiloxane resin

Physical data:

* Solids content: 80+-2%
* Viscosity at 25C: 2,600+-300 (cSt)
* Specific Gravity at 25C: approx. 1.14g/cm3
* Flash point: approx. 30C/86F
* Solvent: xylene/methanol
* Colour: Clear to slightly cloudy

Areas of application:
High temperature anticorrosion coating for:
* industrial equipment
* chimneys
* exhaust tubes

TEGO CHEMIE SERVICE USA: Resins(Continued):

SILIKOPHEN P40/W:

Binder for water based high temperature anticorrosion paints resistant up to 650C/1200F. Coatings applied to industrial equipment chimneys, exhaust tubes, etc.

Chemical Composition: Emulsion of phenyl methyl polysiloxane resin

Physical data:

* Solids content: 40+-1.5%
* Viscosity @25C: 20-30 mm2s-1 (cSt)
* Specific gravity @ 25C: approx. 1.04 g/cm3
* Flash point: 38C/100F
* Solvent: Water/Xylene (less than 10%)

Areas of application:

High temperature anticorrosive coating for:
* industrial equipment
* chimneys
* exhaust tubes

SILIKOFTAL HTL 2:

Binder for decorative high-quality exterior coatings of cookware and electrical domestic appliances, and heaters.

Chemical composition: Silicone-modified polyester resin.

Physical data:

* Solids content: 58.5+-1.5%
* Viscosity at 25C: approx. 1,000 (cSt)
* Density at 25C: 1.10 g/cm3
* Flash point: Approx. 40C/105F
* Solvent: methoxy butyl acetate

Areas of application:

Decorative coatings of:
* cookware, ie: pots, pans
* bakeware
* electrical domestic appliances, ie: toasters, grills
* decorative coatings for high temperature applications up to 250C/482F.

TEGO CHEMIE SERVICE USA: Resins(Continued):

SILIKOFTAL HTS:

Binder for decorative high-quality exterior coatings of cookware and electrical domestic appliances.

Chemical Composition: Silicone-modified polyester resin

Physical data:
* Solids content: 58.5%+-1.5%
* Viscosity at 25C: approx. 300 (cSt)
* Density at 25C: 1.1 g/cm3
* Flash point approx. 50C/122F
* Solvent: methoxy butyl acetate

Areas of application:
Decorative coatings of:
* cookware, ie: pots, pans
* bakeware
* electrical domestic appliances, ie: toasters, grills
* decorative coatings for high temperature applications up to 300C/572F

THIBAUT AND WALKER: Alkyd Resins:

SUPER ALKYD Short Oil Semi-Oxidizing:

Item:

501-50XV:
% NV: 50
Solvent: Xylol
VMP
Viscosity (G-H): Z5-Z7
Type Oil: TOFA
Description: High drink, color retentive baking vehicle.

533-50XV:
% NV: 50
Solvent: Xylol
VMP
Viscosity (G-H): Z5-Z6+
Type Oil: TOFA
Description: Standard quality baking finishes.

537-50X:
% NV: 50
Solvent: Xylol
Viscosity (G-H): Z-Z2
Type Oil: TOFA
Description: Medium viscosity baking alkyd, all xylol version of 501-50XV.

543-45G:
% NV: 45
Solvent: Sol G
Viscosity (G-H): Y-Z
Type Oil: Soya
Description: A short oil vehicle in slow solvent especially adaptable for roller coating or similar applications.

545-50X:
% NV: 50
Solvent: Xylol
Viscosity (G-H): N-P
Type Oil: TOFA
Description: A universal grinding vehicle with excellent low temperature baking properties. Low viscosity for ease of handling.

547-50X:
% NV: 50
Solvent: Xylol
Viscosity (G-H): W-Y
Type Oil: TOFA
Description: Medium viscosity for low temperature baking and/or electrostatic spray

THIBAUT AND WALKER: Alkyd Resins(Continued):

SUPER ALKYD Short Oil Semi-Oxidizing:

Item:

564-60T:
% NV: 60
Solvent: Toluol
Viscosity (G-H): Z5-Z6
Type Oil: Soya
Description: Very fast air dry, designed for hot line traffic paints.

564-50T:
% NV: 50
Solvent: Toluol
Viscosity (G-H): X-Z
Type Oil: Soya
Description: Lower solids version of 564-60T for ease of handling.

616-60T:
% NV: 60
Solvent: Toluol
Viscosity (G-H): Z5-Z6
Type Oil: TOFA
Description: TOFA version of 564-60T.

620-60T:
% NV: 60
Solvent: Toluol
Viscosity (G-H): Z2-Z4
Type Oil: TOFA
Description: Lower viscosity version of 616-60T with equal air dry & cure properties.

SUPER ALKYD Medium Oil:

400-43LO:
% NV: 43
Solvent: Low odor MS
Viscosity (G-H): X-Y
Type Oil: Soya
Description: A high drink medium oil alkyd in low odor Mineral Spirits with low solids for ease of handling.

400-50M:
% NV: 50
Solvent: MS
Viscosity (G-H): Z4-Z5
Type Oil: Soya
Description: High viscosity medium oil alkyd for improved performance and economy.

THIBAUT AND WALKER: Alkyd Resins(Continued):

SUPER ALKYD Medium Oil(Continued):

Item:

402-50M:
% NV: 50
Solvent: MS
Viscosity (G-H): X-Z
Type Oil: Soya/Linseed
Description: An all purpose linseed/soya, medium alkyd. Meets requirements for TT-R-266D, Type III.

407-55M:
% NV: 55
Solvent: MS
Viscosity (G-H): Y-Z1
Type Oil: Soya
Description: An all purpose and marine paint alkyd.

415-50MO:
% NV: 50
Solvent: OMS
Viscosity (G-H): Z1-Z3
Type Oil: Soya
Description: General purpose odorless medium oil alkyd.

440-50M:
% NV: 50
Solvent: MS
Viscosity (G-H): X-Z
Type Oil: Soya
Description: All soya version of 402-50M. Meets requirements for TT-R-266D, Type IV.

449-50V:
% NV: 50
Solvent: VMP
Viscosity (G-H): N-P
Type Oil: Soya
Description: Traffic paints and drum enamels.

449-60V:
% NV: 60
Solvent: VMP
Viscosity (G-H): Z3-Z5
Type Oil: Soya
Description: Traffic paints and drum enamels. Meets NJ DOT Type 4 spec.

THIBAUT AND WALKER: Alkyd Resins(Continued):

Alkyd Resins:

SUPER ALKYD Medium Oil(Continued):

Item:

451-50M:
% NV: 50
Solvent: MS
Viscosity (G-H): X-Z
Type Oil: TOFA
Description: TOFA version of 440-50M. Meets requirements of TT-R-266D, Type VI.

465-50M:
% NV: 50
Solvent: MS/
Xylol
Viscosity (G-H): Z4-Z5
Type Oil: TOFA
Description: High drink, general purpose TOFA medium alkyd. TOFA version of 489-50M.

473-50M:
% NV: 50
Solvent: MS
Viscosity (G-H): Z1-Z3
Type Oil: Soya
Description: Higher viscosity version of 440-50M.

477-50M:
% NV: 50
Solvent: MS
Viscosity (G-H): Z4-Z5
Type Oil: TOFA
Description: TOFA version of 400-50M.

478-50MO:
% NV: 50
Solvent: OMS
Viscosity (G-H): Z1-Z3
Type Oil: TOFA
Description: TOFA version of 415-50MO.

489-50M:
% NV: 50
Solvent: MS/Xylol
Viscosity (G-H): Z4-Z6
Type Oil: Soya
Description: Higher viscosity of 400-50M

THIBAUT AND WALKER: Alkyd Resins(Continued):

SUPER ALKYD Medium Oil(Continued):

Item:

492-45M:
% NV: 45
Solvent: MS
Viscosity (G-H): C-F
Type Oil: Soya
Description: Medium oil alkyd that with driers added meets requirements for TT-V-109C, Interior Alkyd Resin Varnish.

493-50V:
% NV: 50
Solvent: VMP
Viscosity (G-H): X-Z
Type Oil: Soya/Linseed
Description: Traffic vehicle, fast drying version of T & W 402-50M.

494-55M:
% NV: 55
Solvent: Blend
Viscosity (G-H): Z3-Z5
Type Oil: Mixed Drying
Description: Medium-long oil designed to accept large amounts of water, up to 20% or more by volume in paints.

496-43M:
% NV: 43
Solvent: MS
Viscosity (G-H): U-X
Type Oil: Soya
Description: Similar to 400-50M but with greater compatibility and at lower solids for ease of handling.

604-50V:
% NV: 50
Solvent: VMP
Viscosity (G-H): Z1-Z3
Type Oil: Soya
Description: High viscosity, medium oil for use in traffic paints and drum enamels

605-60T:
% NV: 60
Solvent: Toluol
Viscosity (G-H): D-H
Type Oil: Soya
Description: Fast dry traffic paints, Meets NJ DOT Type 3A

THIBAUT AND WALKER: Alkyd Resins(Continued):

SUPER ALKYD Medium Oil(Continued):

Item:

615-45V:
% NV: 45
Solvent: VMP
Viscosity (G-H): Z1-Z3
Type Oil: Soya
Description: Fast dry traffic paints, primers, enamels and structural steel coatings.

618-50M:
% NV: 50
Solvent: MS
Viscosity (G-H): W-Y
Type Oil: Mixed Drying
Description: A mixed drying oil, containing IPA for improved flexibility.

SUPER ALKYD Long Oil:

Item:

300-60M:
% NV: 60
Solvent: MS
Viscosity (G-H): U-W
Type Oil: Soya
Description: Long oil soya. Meets requirements for TT-R-266D, Type I, Class B.

300-70M:
% NV: 70
Solvent: MS
Viscosity (G-H): Z1-Z3
Type Oil: Soya
Description: Long oil soya. Meets requirements for TT-R-266D, Type I, Class A.

303-50MD:
% NV: 50
Solvent: MS
Viscosity (G-H): H-J
Type Oil: Soya
Description: Alkyd varnish suitable for all interior and exterior trim work as well as floor and decks. Contains driers.

THIBAUT AND WALKER: Alkyd Resins(Continued):

SUPER ALKYD Long Oil(Continued):

Item:

305-70M:
% NV: 70
Solvent: MS
Viscosity (G-H): Y-Z1
Type Oil: Soya
Description: Vinyl compatible long oil soya. Meets requirements for TT-R-266D, Type I, Class C, Solvent M.

305-70X:
% NV: 70
Solvent: Xylol
Viscosity (G-H): O-S
Type Oil: Soya
Description: Vinyl compatible long oil soya. Meets requirements for TT-R-266D, Type I, Class C, Solvent X.

308-100:
% NV: 100
Viscosity (G-H): W-Y
Type Oil: TOFA
Description: Latex modifier. Exterior house paints.

310-50M:
% NV: 50
Solvent: MS
Viscosity (G-H): T-V
Type Oil: Soya
Description: Quality non-yellowing and non-penetrating enamels.

317-60M:
% NV: 60
Solvent: MS
Viscosity (G-H): U-W
Type Oil: Soya
Description: Chlorinated alkyd.

328-90M:
% NV: 90
Solvent: MS
Viscosity (G-H): R-V
Type Oil: Linseed
Description: Flexible long oil for exterior paints.

THIBAUT AND WALKER: Alkyd Resins(Continued):

SUPER ALKYD Long Oil(Continued):

Item:

329-60M:
% NV: 60
Solvent: MS
Viscosity (G-H): T-W
Type Oil: TOFA
Description: TOFA version of 300-60M SUPER ALKYD. Lower solids version of 329-70M.

329-70M:
% NV: 70
Solvent: MS
Viscosity (G-H): Z4-Z6
Type Oil: TOFA
Description: Long oil TOFA. Meets requirements for TT-R-266D, Type V. TOFA version of 300-70M.

332-60M:
% NV: 60
Solvent: MS
Viscosity (G-H): X-Z
Type Oil: Soya
Description: Higher viscosity version of 300-60M for architectural enamels.

345-100:
% NV: 100
Viscosity (G-H): W-Y
Type Oil: Soya
Description: Latex modifier. Exterior house paint.

354-55M:
% NV: 55
Viscosity (G-H): T-W
Type Oil: Soya
Description: Lower solids version of 332-60M for ease of handling.

362-60MO:
% NV: 60
Solvent: OMS
Viscosity (G-H): V-W
Type Oil: Soya
Description: General purpose odorless long oil alkyd.

THIBAUT AND WALKER: Alkyd Resins(Continued):

SUPER ALKYD Long Oil(Continued):

Item:

363-90M:
% NV: 90
Solvent: MS
Viscosity (G-H): W-Y
Type Oil: Mixed Drying
Description: Flexible long oil for exterior coatings and a latex modifier.

366-60M:
% NV: 60
Solvent: MS
Viscosity (G-H): T-V
Type Oil: Linseed
Description: Long oil linseed. Meets requirements for TT-R-266D, Type II, Class B.

366-70M:
% NV: 70
Solvent: MS
Viscosity (G-H): Z-Z2
Type Oil: Linseed
Description: Long oil linseed. Meets requirements for TT-R-266D, Type II, Class A.

367-55MO:
% NV: 55
Solvent: OMS
Viscosity (G-H): X-Z
Type Oil: Soya
Description: Odorless version of 332-60M with lower solids for ease of handling.

368-55M:
% NV: 55
Solvent: MS
Viscosity (G-H): W-X
Type Oil: TOFA
Description: TOFA version of 354-55M.

377-60M:
% NV: 60
Solvent: MS
Viscosity (G-H): W-Y
Type Oil: TOFA
Description: Higher viscosity version of 329-60M. TOFA version of 332-60M.

THIBAUT AND WALKER: Alkyd Resins(Continued):

SUPER ALKYD Long Oil(Continued):

Item:

483-60M:
% NV: 60
Solvent: MS
Viscosity (G-H): Z-Z2
Type Oil: Soya
Description: Higher viscosity version of 332-60M.

601-70M:
% NV: 70
Solvent: MS
Viscosity (G-H): R-V
Type Oil: Mixed Drying
Description: Oxidizing binder designed for deflocculating and wetting of pigments.

SUPER ALKYD Flat Alkyds:

486-30P:
% NV: 30
Solvent: Sol 140
Viscosity: T-W
Type Oil: Soya
Description: Low odor flat designed for maximum non-penetration.

499-30M:
% NV: 30
Solvent: MS
Viscosity (G-H): X-Z
Type Oil: Soya
Description: Higher viscosity and more economical (M. sp.) version of 486-30P.

608-30M:
% NV: 30
Solvent: MS
Viscosity (G-H): U-V
Type Oil: TOFA
Description: A non-penetrating alkyd that allows high pigment loading.

THIBAUT AND WALKER: Alkyd Resins(Concentrated):

SUPER ALKYD Flat Alkyds(Continued):

Item:

610-35M:
% NV: 35
Solvent: MS
Viscosity (G-H): W-Y
Type Oil: TOFA
Description: A medium viscosity version of 608-30M with excellent hold-out properties.

619-30M:
% NV: 30
Solvent: MS
Viscosity (G-H): X-Z
Type Oil: Soya
Description: Maximum non-penetrating flat with best excellent economy due to its high reduced viscosity.

701-35MO:
% NV: 35
Solvent: OMS
Viscosity (G-H): Z4-Z6
Type Oil: TOFA
Description: A high viscosity odorless flat with excellent non-penetration.

703-35MO:
% NV: 35
Solvent: OMS
Viscosity (G-H): V-X
Type Oil: TOFA
Description: Odorless flat vehicle with good non-penetration and excellent compatibility.

707-30MO:
% NV: 30
Solvent: OMS
Viscosity (G-H): Z-Z2
Type Oil: TOFA
Description: Lower viscosity version of 701-35MO for ease of handling.

708-40M:
% NV: 40
Solvent: MS
Viscosity (G-H): Z-Z2
Type Oil: TOFA
Description: Interior flats and Shake and Shingle paints.

709-40M:
% NV: 40
Solvent: MS
Viscosity (G-H): X-Z
Type Oil: TOFA
Description: Non-penetrating flats, and Shake and Shingle paints.

THIBAUT AND WALKER: Concentrated Varnish Stains:

<u>Color:</u>

Dark Mahogany:
- % NV: 33
- Solvent: M.sp./Aromatic
- Wt/Gal.: 7.20
- Modifier: Phenolic Resin

Light Mahogany:
- % NV: 33
- Solvent: M.sp./Aromatic
- Wt./Gal.: 7.20
- Modifier: Phenolic Resin

Light Oak:
- % NV: 28
- Solvent: M.sp./Aromatic
- Wt./Gal.: 7.20
- Modifier: Phenolic Resin

Dark Oak:
- % NV: 34
- Solvent: M.sp./Aromatic
- Wt./Gal.: 7.20
- Modifier: Phenolic Resin

Walnut:
- % NV: 30
- Solvent: M.sp./Aromatic
- Wt./Gal.: 7.20
- Modifier: Phenolic Resin

Maple:
- % NV: 36
- Solvent: M.sp./Aromatic
- Wt./Gal.: 7.20
- Modifier: Phenolic Resin

Rosewood:
- % NV: 30
- Solvent: M.sp./Aromatic
- Wt./Gal.:7.20
- Modifier: Phenolic Resin

Dark Brown Pine:
- % NV: 30
- Solvent: M.sp./Aromatic
- Wt./Gal.: 72.0
- Modifier: Phenolic Resin

Description: Produces oil stain finishes with brilliance of color and clarity of tone that bring out the natural grain and beauty of woods. Widely compatible with solvent-based polymers to create a variety of desirable characteristics.

Special colors available upon request.

THIBAUT & WALKER: High Solids Resins:

SUPER ALKYD Air Dry Alkyds:

Item:

574-75TK:
% NV: 75
Solvent: Toluol/MPK
Viscosity (G-H): Z1-Z3
Type Oil: Soya
Description: Very fast set to touch air dry short oil used in drum enamels and other fast dry air dry applications. Low temperature bakes.

800-80TK:
% NV: 80
Solvent: Toluol/MPK
Viscosity (G-H): Z1-Z3
Type Oil: Soya
Description: High solids medium for quick drying enamels, primers and structural steel coatings and traffic paints.

801-75TK:
% NV: 75
Solvent: Toluol/MPK
Viscosity (G-H): Z1-Z3
Type Oil: TOFA
Description: Rosin ester modified version of 574-75TK for faster tack free air dry with good flexibility. Comparable to rosin/phenol modified high solids alkyds.

802-75:
% NV: 75
Solvent: Toluol/n-Butanol
Viscosity (G-H): Z3-Z4
Type Oil: TOFA
Description: A rosin ester modified air or forced air dry alkyd for dip or spray applications.

802-70:
% NV: 70
Solvent: Toluol/n-Butanol
Viscosity (G-H): X-Z
Type Oil: TOFA
Description: Lower solids version of T & W 802-75.

803-75:
% NV: 75
Solvent: Xylol/n-Butanol
Viscosity (G-H): Z3-Z4
Type Oil: TOFA
Description: Fast air dry metal coating with excellent chemical resistance

THIBAUT & WALKER: High Solids Resins(Continued):

SUPER ALKYD Air Dry Alkyds(Continued):

Item:

805-80:
% NV: 80
Solvent: ARO 100
Viscosity (G-H): Z6
Type Oil: Soya
Description: High solids medium oil for standard air dry dip applications.

806-80:
% NV: 80
Solvent: ARO 100
Viscosity (G-H): Z6
Type Oil: TOFA
Description: Rosin ester modified medium oil for fast dry dip applications.

810-80:
% NV: 80
Solvent: ARO 100/MAK
Viscosity (G-H): Z3-Z5
Type Oil: Soya
Description: An air dry medium oil similar to T&W 800-80TK but with a Flash Point >100F.

SUPER ALKYD Baking Alkyds:

840-90:
% NV: 90
Solvent: PMAc
Viscosity (G-H): Z4-Z5
Type Oil: TOFA
Description: Good performance high solids bake or low bake alkyd.

845-90:
% NV: 90
Solvent: PMAc
Viscosity (G-H): Z3-Z4
Type Oil: TOFA
Description: Economical high solids baking alkyd with superior performance.

846-90:
% NV: 90
Solvent: Xylol
Viscosity (G-H): Z5-Z6+
Type Oil: TOFA
Description: Economical baking alkyd with good performance.

THIBAUT & WALKER: High Solids Resins(Continued):

SUPER POLYESTER Baking Polyesters:

Item:

880-90:
% NV: 90
Solvent: PMAc
Viscosity (G-H): Z4-Z5
Description: Standard performance, economical oil free polyester.

890-85:
% NV: 85
Solvent: PMAc
Viscosity (G-H): Z4-Z5
Description: High performance oil free polyester.

THIBAUT & WALKER: SUPER ALKYD Modified-Oxidizing Alkyds:

Item:

346-60M:
% NV: 60
Solvent: MS
Viscosity (G-H): V-Y approx.
Type Oil: Soya
Modifier: Silicone
Description: Meets TT-E-490C and TT-E-1593B.

420-50M:
% NV: 50
Solvent: MS
Viscosity (G-H): W-Y
Type Oil: Mixed Drying
Modifier: Rosin/Phenol
Description: Phenolated medium oil for abrasion resistance.

439-50M:
% NV: 50
Solvent: MS
Viscosity (G-H): T-W
Type Oil: Soya/Linseed
Modifier: Rosin
Description: General purpose vehicle to enhance gloss and dry.

441-50M:
% NV: 50
Solvent: MS
Viscosity (G-H): I-K
Type Oil: Soya
Modifier: Epoxy
Description: Epoxy ester for fast dry toughness, adhesion and chemical resistance.

442-50M:
% NV: 50
Solvent: MS
Viscosity (G-H): L-N
Type Oil: Linseed
Modifier: Epoxy
Description: Linseed version of 441-50M, with equal properties.

475-50M:
% NV: 50
Solvent: MS
Viscosity (G-H): T-V
Type Oil: TOFA
Modifier: Rosin
Description: Gloss, dry and hardness enhancer.

THIBAUT AND WALKER: SUPER ALKYD Modified-Oxidizing Alkyds (Continued):

Item:

549-38S:
% NV: 38
Solvent: MS
Viscosity (G-H): Q-T
Type Oil: TOFA
Modifier: Rosin
Description: For use in spackling compound

566-45V:
% NV: 45
Solvent: VMP
Viscosity (G-H): Z1-Z3
Type Oil: TOFA
Modifier: Rosin-Ester
Description: Fast dry drum and toy enamel, Tannin stain blockers, floor and wood sealers.

567-50VT:
% NV: 50
Solvent: VMP
Toluol
(Exempt Blend)
Viscosity (G-H): X-Y
Type Oil: TOFA
Modifier: Rosin-Ester
Description: Economical alternative to fast dry phenolated alkyds while retaining similar properties. Drum & toy enamels, Tannin stain blocker.

614-45M:
% NV: 45
Solvent: MS
Viscosity (G-H): Z1-Z4
Type Oil: Mixed Drying
Modifier: Rosin
Description: Medium-short oil with good flexibility. Gloss and dry enhancer.

650-60M:
% NV: 60
Solvent: MS
Viscosity (G-H): Z1-Z3
Type Oil: TOFA
Modifier: Epoxy
Description: Higher solids TOFA version of 441 or 442-50M.

LV3045-45MD:
% NV: 45
Solvent: MS
Viscosity (G-H): A5-A
Type Oil: Drying
Modifier: Hydrocarbon Resin
Description: Aluminum vehicle for interior and exterior use.

THIBAUT AND WALKER: SUPER THANE Oil Modified Urethanes:

Item:

331-60M:
% NV: 60
Solvent: MS
Viscosity (G-H): Z-Z2
Type Oil: Linseed
Description: Fast dry, abrasion resistant coating. Gym Floor finishes.

340-50M:
% NV: 50
Solvent: MS
Viscosity (G-H): Z-Z3
Type Oil: Linseed
Description: Economical finishes, either glossy or satin. Fast hardness development.

370-60X:
% NV: 60
Solvent: Xylol
Viscosity (G-H): R-U
Type Oil: Linseed
Description: Fast drying, good abrasion resistance. Industrial enamels. Aerosols.

371-45M:
% NV: 45
Solvent: MS
Viscosity (G-H): T-V
Type Oil: Soya
Description: Designed for maximum compatibility with alkyds, including mediums. Especially designed to be used in conjunction with T & W 496-45M. Excellent for flats.

385-55M:
% NV: 55
Solvent: MS
Viscosity (G-H): W-X
Type Oil: Soya
Description: Excellent for satins, designed to maintain original flatted gloss during storage of varnish.

390-50M:
% NV: 50
Solvent: MS
Viscosity (G-H): Z1-Z3
Type Oil: Soya
Description: Similar to T & W 385-55M but at higher viscosity and reduced and lower solids. Ideal for economical flats and stains.

THIBAUT & WALKER: SUPER THANE Oil Modified Urethanes(Continued):

Item:

429-50M:
% NV: 50
Solvent: MS
Viscosity (G-H): Z1-Z3
Type Oil: Linseed
Description: Excellent hardness, fast dry, affords good economics.

432-50M:
% NV: 50
Solvent: MS
Viscosity (G-H): Y-Z2
Type Oil: Soya
Description: Excellent for water, detergent and chemical resistance. Best resin for satins or flats.

458-60M:
% NV: 60
Solvent: MS
Viscosity (G-H): Z-Z2
Type Oil: Soya
Description: For color retentive enamels and floor finishes.

469-50X:
% NV: 50
Solvent: Xylol
Viscosity (G-H): Z1-Z3
Type Oil: Linseed
Description: Fastest drying, hardest and most abrasive resistant. Aerosols, traffic markers.

472-47M:
% NV: 50
Solvent: MS
Viscosity (G-H): O-R
Type Oil: Soya
Description: Polyurethane containing ASA.

495-50M:
% NV: 50
Solvent: MS
Viscosity (G-H): Z1-Z3
Type Oil: Soya
Description: Excellent water, detergent and chemical resistance. Best for flats and satins.

LV3019-45MD:
% NV: 45
Solvent: MS
Viscosity (G-H): B-D
Type Oil: Soya
Description: Urethane varnish blend ready for use. Contains driers.

THIBAUT AND WALKER: Water Reducible Resins:

SUPER ALKYD Alkyds:

Item:

900-75:
% NV: 75
Solvent: PGMBE
Viscosity (G-H): Z2-Z4
Type Oil: Mixed
Description: Water reducible medium-short oil alkyd for air, forced air, and baking applications.

901-75:
% NV: 75
Solvent: PGMBE/sec-Butanol
Viscosity (G-H): Z3-Z5
Type Oil: Mixed
Description: Water reducible medium alkyd for air, forced air and bake applications with good gloss and moisture resistance.

902-75:
% NV: 75
Solvent: PGMBE
Viscosity (G-H): Z3-Z5
Type Oil: Soya
Description: Water reducible air dry alkyd for early moisture resistance.

SUPER POLYESTER Oil Free Polyesters:

950-75:
% NV: 75
Solvent: PGMBE
Viscosity (G-H): Z4-Z6
Description: General purpose, water reducible oil free polyester for industrial baking coatings.

960-75:
% NV: 75
Solvent: PGMBE
Viscosity (G-H): Z4-Z6
Description: A high performance water reducible polyester that offers excellent color and gloss retention, outstanding weatherability and exceptional hardness and impact resistance. T & W 950-75 is recommended for coatings requiring excellent exterior durability.

THIBAUT AND WALKER: Water Reducible Resins(Continued):

SUPER THANE Oil Modified Urethanes:

Item:

975-70:
% NV: 70
Solvent: PGMBE/sec-Butanol
Viscosity (G-H): Z4-Z6
Type Oil: TOFA
Description: Water reducible oil modified urethane for general purpose varnish applications.

980-70:
% NV: 70
Solvent: PGMBE/sec-Butanol
Viscosity (G-H): Z4-Z6
Type Oil: Soya
Description: Soya version of T & W 975-70

UNION CARBIDE CORP.: Base Resins:

Base Resins:

Resin System:

Vinyl:
UCAR Solution Vinyls:
VYES, VP-200, VMCA ***
VAGH, VMCH, VYHH, VERR, VROH, etc. ****
AW-850, -150, etc. **

Epoxy:
Cycloaliphatic Epoxides:
ERL-4221, -4299 ***
CYRACURE Resins ***

Epoxidized Vinyls:
UCAR Solution VERR ****

Customized & Specialty Epoxides: ****

Phenolic:

UCAR Phenolic Baking Resins:
BKS-2600, BKR-2620 ****
BKS-2900, BKS-2910 ***
BK-5918 **** *****
BKUA-2370, BKUA-2392 **

UCAR Phenolic Oil-Soluble Resins:
CK-2400 ****
CK-2500, CK-1634 ***

Ethyl Silicates:
UCAR Ethyl Silicates:
ES-40 ****
TEOS ****
ESP ****
ESP-E ****
ES Condensed ****

Acrylics:
UCAR Functional Acrylics:
882, 883, and UCARLNK Crosslinker 888 ****
UCARLNK Crosslinker XL-20 ****

Phenoxy:
UCAR Phenoxy Resins:
PKHA, PKHC, PKHH, PKHJ, PXKH-6956 ****
PXKS-6978 **
Other:
BAKELITE Butyrals (PVB):
XYHL, XYSG ****

UNION CARBIDE CORP.: Intermediates/Additives/Reactants:

Cycloaliphatic Epoxides:
ERL-4221(Acid Scavenger) ****

Plasticizers:
FLEXOL EP-8, EPO, LOE ****

Cycloaliphatic Epoxides:
CYRACURE Diluents, Flexibilizers, Photoinitiators

Polyols:
TONE Diols, Triols **** ***

UCAR Solution Vinyls:
VP-200, VYES **** ***

Crosslinkers:
UCARLNK XL-20 **

Monomers:
E-Caprolactone **** ***
TONE M-100 **** ***

UCAR Butyl Phenol Monomer (PTBP): ** *** ****

Monomers:
E-Caprolactone

Monomers:
TONE M-100 **** ***
E-Caprolactone **** ***

Diluents:
TONE Polyols ***

Crosslinkers:
UCAR Phenolic BRL-6891, -1100

UNION CARBIDE CORP.: Modifying Resins:

Key:
*: High Solids
**: Waterborne
***: Radiation Cure
****: Solvent
*****: Powder

Polyols:
TONE Diols, Triols ****

UCAR Phenoxy Resins ****

UCAR Phenolic Resin Hardeners:
BRWE Novolac Products **** *** *****
BKUA Dispersions **
Resoles **** *** *****

UCAR Solution Vinyls:
VYES ***

Polyols:
TONE Diols, Triols *** ****

UCAR Solution Vinyls:
VAGH, VAGD, VROH ****
VYES, VP-200 ***
AW-850, -150 **

UCAR Phenoxy Resins ****

Polyols:
TONE Diols, Triols *** ****

UCAR Solution Vinyls:
VMCH, VMCA, VYHH, VYLF ****
AW-850, -150, etc. ****
VYES ***

UCAR Phenolic Resins:
BKUA Dispersions ****
BK-5918, BRWE-5300 **** ***

UCAR Solution Vinyls:
AW-850, -150, etc. **
VMCH, VAGH, etc. ****

UCAR Phenolic Resins:
BKUA-2370, -2392 Dispersions **

Polyols:
TONE Diols, Triols ****

Alkyd & Oleoresinous Modifiers:
UCAR Phenolic Resins **** ** ***
UCAR Solution Vinyls **** ** ***
UCAR Phenoxy Resins ****

Nitrocellulose Modifiers:
UCAR Solution Vinyl VYNC ****

UNION CARBIDE CORP.: UCAR Coating Resins: End Uses/Products:

Product Finishes:
- Wood Furniture & Fixtures: UCAR Phenolic Resins
  - UCAR Solution Vinyls
  - FLEXOL Epoxy Plasticizers
  - UCARLNK Crosslinkers
- Wood Flat Stock: UCAR Phenolic Resins
  - UCAR Solution Vinyls
  - FLEXOL Epoxy Plasticizers
- Metal Furniture & Fixtures: UCAR Phenolic Resins
  - UCAR Solution Vinyls
  - Cycloaliphatic Epoxies
  - CYRACURE Resins
  - FLEXOL Epoxy Plasticizers
  - TONE Caprolactone Polyols & Monomers
- Containers & Closures: UCAR Phenolic Resins
  - UCAR Solution Vinyls
  - UCAR Phenoxy Resins
  - Cycolaliphatic Epoxides
  - CYRACURE Resins
  - FLEXOL Epoxy Plasticizers
  - TONE Caprolactone Polyols & Monomers
- Sheet, Strip & Coil: UCAR Phenolic Resins
  - UCAR Solution Vinyls
  - BAKELITE PVB
  - UCAR Phenoxy Resins
  - Cycloaliphatic Epoxides
  - CYRACURE Resins
  - FLEXOL Epoxy Plasticizers
  - TONE Caprolactone Polyols & Monomers
- Major Appliances: UCAR Phenolic Resins
  - UCAR Solution Vinyls
  - UCAR Phenoxy Resins
  - Cycloaliphatic Epoxides
  - CYRACURE Resins
  - FLEXOL Epoxy Plasticizers
  - TONE Caprolactone Polyols & Monomers
- Other Appliances: UCAR Phenolic Resins
  - UCAR Solution Vinyls
  - Cycloaliphatic Epoxides
  - CYRACURE Resins
  - FLEXOL Epoxy Plasticizers
  - TONE Caprolactone Polyols & Monomers
- Automotive:
  - Top Coat: Cyclolaliphatic Epoxides
    - TONE Caprolactone Polyols & Monomers
  - Primer: UCAR Phenoxy Resins
    - Cycloaliphatic Epoxides
    - TONE Caprolactone Polyols & Monomers

UNION CARBIDE CORP.: UCAR Coating Resins: End Uses/Products (Continued):

Product Finishes(Continued):

Automotive:
Underbody & Parts: UCAR Phenolic Resins
UCAR Solution Vinyls
UCAR Phenoxy Resins
Cycloaliphatic Epoxides
TONE Caprolactone Polyols & Monomers
Trucks & Buses: UCAR Solution Vinyls
Cycloaliphatic Epoxides
FLEXOL Epoxy Plasticizers
TONE Caprolactone Polyols & Monomers
Railroad: UCAR Phenolic Resins
UCAR Solution Vinyls
Cycloaliphatic Epoxides
FLEXOL Epoxy Plasticizers
TONE Caprolactone Polyols & Monomers
Other Transportation: UCAR Phenolic Resins
UCAR Solution Vinyls
UCAR Phenoxy Resins
Cycloaliphatic Epoxides
FLEXOL Epoxy Plasticizers
TONE Caprolactone Polyols & Monomers
UCARLNK Crosslinkers
Machinery & Equipment: UCAR Phenolic Resins
UCAR Solution Vinyls
UCAR Phenoxy Resins
Cyclolaliphatic Epoxides
FLEXOL Epoxy Plasticizers
TONE Caprolactone Polyols & Monomers
Electrical Insulation: UCAR Phenolic Resins
UCAR Phenoxy Resins
Cylcoaliphatic Epoxides
CYRACURE Resins
FLEXOL Epoxy Plasticizers
TONE Caprolactone Polyols & Monomers
Marine: UCAR Phenolic Resins
UCAR Solution Vinyls
UCAR Ethyl Silicates
BAKELITE PVB
Cyclolaliphatic Epoxides
TONE Caprolactone Polyols & Monomers

UNION CARBIDE CORP.: UCAR Coating Resins: End Uses/Products (Continued):

Paper, Foil & Film: UCAR Phenolic Resins
UCAR Solution Vinyls
UCAR Phenoxy Resins
Cycloaliphatic Epoxides
CYRACURE Resins
FLEXOL Epoxy Plasticizers
TONE Caprolactone Polyols & Monomers

Special Purpose Coatings:

Maintenance: UCAR Phenolic Resins
UCAR Solution Vinyls
UCAR Ethyl Silicates
BAKELITE PVB
UCAR Phenoxy Resins
Cycloaliphatic Epoxides
FLEXOL Epoxy Plasticizers
TONE Caprolactone Polyols & Monomers
UCARLNK Crosslinkers
Auto Refinishing: Cycloaliphatic Epoxides
TONE Caprolactone Polyols & Monomers
UCARLNK Crosslinkers
Traffic Paints: UCAR Phenolic Resins
UCAR Solution Vinyls
TONE Caprolactone Polyols & Monomers
Roof and Deck: TONE Caprolactone Polyols & Monomers
Aerosol: TONE Caprolactone Polyols & Monomers

Architectural:

Interior Water-Based: UCARLNK Crosslinkers

UNOCAL CHEMICALS DIVISION: 76 Polymers For Architectural Coatings (Continued):

76 RES Acrylics:

Product:

76 RES 6510:
Solids %: 65
pH: 9.0
Visc. cps: 800
Comments: Fine particle size and high molecular weight provide excellent grain crack resistance and tint retention. Higher solids allow greater formulating latitude.

76 RES Styrene Butadiene:

76 RES 4470:
Solids %: 50
pH: 9.0
Visc. cps: 350
Comments: Excellent for interior dry wall primers and coatings which require high moisture resistance properties and excellent cold temperature touch-up.

UNOCAL CHEMICALS DIVISION: 76 Polymers For Architectural Coatings:

76 RES Vinyl Acrylics:

Product:

76 RES 661:
Solids %: 55
pH: 4.5
Visc. cps: 1000
Comments: Promotes freeze-thaw and mechanical stability in compounded paints. Excellent in flats and semi-gloss systems.

76 RES 3011:
Solids %: 55
pH: 3.0
Visc. cps: 2000
Comments: General purpose binder for interior and exterior paints. Provides good flow and leveling properties.

76 RES 3016:
Solids %: 55
pH: 3.0
Visc. cps: 2100
Comments: Effective binder for interior flats and semi-gloss paints where a slightly harder finish is desired. Useful in contractor grade semi-gloss systems.

76 RES 3077:
Solids %: 55
pH: 4.5
Visc. cps: 2150
Comments: High performance vehicle for interior, exterior and semi-gloss paints. Excellent scrub resistance coupled with good flow and leveling properies. Excellent adhesion over chalky surfaces.

76 RES 3078:
Solids %: 65
pH: 5.0
Visc. cps: 2500
Comments: General purpose high solids vehicle with good tint retention and enamel holdout properties.

76 RES 3079:
Solids %: 65
pH: 4.5
Visc. cps: 1950
Comments: Low viscosity version of 76 RES 3078.

UNOCAL CHEMICALS DIVISION: 76 Polymers For Architectural Coatings (Continued):

76 RES Vinyl Acrylics:

Product:

76 RES 3082:
Solids %: 55
pH: 4.5
Visc. cps: 400
Comments: Exceptional vehicle for premium flat, semi-gloss and eggshell finishes. Provides superior wet adhesion, excellent scrub resistance, and stain removal properties.

76 RES Styrene Acrylics:

76 RES 1018:
Solids %: 50
pH: 8.5
Visc. cps: 2000
Comments: Highly effective in water-resistant coatings such as clears for masonry surfaces, vertical wall and textured finishes. Provides good gloss and clarity. Exhibits high pigment binding capacity.

76 RES 1019:
Solids %: 50
pH: 8.5
Visc. cps: 3900
Comments: Provides flexibility, alkali resistance and water resistance in high performance roof coatings and mastics. Exhibits high pigment binding capacity and adhesion to a variety of substrates.

76 RES Vinyl Acetates:

76 RES 3001:
Solids %: 58
pH: 4.5
Visc. cps: 1600
Comments: Extremely high pigment binding properties produce economical flats with good leveling. Excellent for ready-mixed tape joint cements.

76 RES 3001P:
Solids %: 60
pH: 4.5
Visc. cps: 2000
Comments: Plasticized homopolymer for contractor-grade flats and ready-mixed tape joint cements.

Section 10
Rheological/Viscosity Control Agents

ALCO CHEMICAL CORP.: ALCOGUM Thickeners and Rheology Modifiers:

<u>Product Name:</u>

ALCOGUM VEP-I
Viscoelastic Polymer

ALCOGUM VEP-I is a CAD designed sodium polyacrylate thickener. Its unique manufacturing process ensures a clean product.

ALCOGUM VEP-I is a high viscosity aqueous solution of sodium polyacrylate. It is ready to use as received.

ALCOGUM VEP-I may be used to thicken styrene butadiene, neoprene, acrylic, vinyl acetate, and ethylene vinyl acetate latex compounds. In general, it is a satisfactory thickener for latexes of alkaline pHs.

Some suggested applications for ALCOGUM VEP-I are:

* Carpet Backing
* Adhesives
* Textile Coatings
* Dispersions
* Latex Foam
* Emulsions
* Nonwovens
* Protective Colloid

Some of the outstanding properties of this water-soluble synthetic resin are:

Latex Thickening Action: Immediate and Extremely Effective
High Viscosities at Low Concentrations
Ready to Use Liquid
Resistance to Mildew or Bacteria
Exceptional for Low Penetration Coatings
Exceptional Heat Stability

ALCOGUM VEP-I is a high viscosity aqueous solution of sodium polyacrylate. It is ready to use as received.

Typical Properties:
Total Solids: 12.5%
pH: 9.0
Viscosity, 25C, cps: Typically 20,000
Appearance: Clear to slightly opaque light colored liquid

ALCO CHEMICAL CORP.: ALCOGUM Thickeners and Rheology Modifiers: (Continued):

ALCOGUM VEP-II
Viscoelastic Polymer

ALCOGUM VEP-II is a CAD designed sodium polyacrylate thickener. Its unique manufacturing properties ensures a clean product.

ALCOGUM VEP-II is a high viscosity aqueous solution of sodium polyacrylate. It is ready to use as received.

ALCOGUM VEP-II may be used to thicken styrene vinyl, neoprene, acrylic, vinyl acetate, and ethylene vinyl acetate latex compounds. In general, it is a satisfactory thickener for latexes of alkaline pHs.

Suggested applications for ALCOGUM VEP-II are:

* Carpet Backing
* Adhesives
* Textile Coatings
* Dispersions
* Latex Foam
* Emulsions
* Nonwovens
* Protective Colloid

Some of the outstanding properties of this water-soluble synthetic resin are:

Latex Thickening Action: Immediate and Extremely Effective
High Viscosities at Low Concentrations
Ready to Use Liquid
Resistance to Mildew or Bacteria
Excellent Heat Stability

ALCOGUM VEP-II is a high viscosity aqueous solution of sodium polyacrylate. It is ready to use as received.

Typical Properties:
Total Solids: 14.0%
pH: 9.0
Appearance: Clear to slightly opaque light colored liquid
Viscosity, 20 rpm, RV, 25C, cps: Typically 10,000

ALCO CHEMICAL CORP.: ALCOGUM Thickeners and Rheology Modifiers: (Continued):

ALCOGUM 9710

ALCOGUM 9710 is a sodium polyacrylate thickener which can be used in the thickening of natural and synthetic latexes for a wide variety of applications.

ALCOGUM 9710 may be used to thicken styrene-butadiene, neoprene, acrylic, vinyl acetate and ethylene-vinyl acetate latex compounds. In general, it is a satisfactory thickener for latex which has an alkaline reaction, although some success has been achieved in neutral or slightly acid systems.

ALCOGUM 9710 sets criteria of CFR, Title 21, Sections:

175.300
175.105
176.170 (limited to 2% solids on coating solids)
176.180

In addition, this product has also been disclosed on USDA as a thickener in coatings and adhesives in contact with fresh foods.

Typical Properties:

Total Solids: 12.5%
Appearance: Clear to slightly opaque light colored liquid
pH: 12.0
Viscosity, 20 rpm, RV, 25C, cps: 30,000

Meets requirements of TSCA.

ALCO CHEMICAL CORP.: ALCOGUM Thickeners and Rheology Modifiers (Continued):

ALCOGUM 296-W
Acrylic Thickener and Protective Colloid

ALCOGUM 296-W is recommended primarily as a stabilizer, protective colloid, and thickener for water based systems, such as natural and synthetic rubber latices, and polymeric dispersions, such as polyvinyl acetate and polyacrylate emulsions. It possesses unique adhesive, suspending, emulsifying, and film-forming characteristics, making it ideal for any other industrial use.

ALCOGUM 296-W is a high viscosity aqueous solution of sodium polyacrylate. It is an anionic colloid furnished as a whitish colored homogenous solution at 15% solids, ready to use.

Some of the outstanding properties of this water-soluble synthetic resin are:

Latex Thickening Action: Immediate and Extremely Effective
Non-Yellowing, Non-Oxidizing
High Viscosities at Low Concentrations
Resistance to Mildew or Bacteria
Liquid Ready to Use: No cooking Required

Some suggested applications for ALCOGUM 296-W are as a: Protective Colloid; Adhesive Additive; Paint Additive; Latex Thickener; and Cement Additive.

ALCOGUM 296-W is a high viscosity aqueous solution of sodium polyacrylate. It is an anionic colloid furnished as a whitish colored--homogeneous solution at 15% solids, ready to use.

Typical Properties:
Total Solids: 14.7-17.2
Appearance: Opaque Liquid
Viscosity, 20 rpm, RV, 25C, cps: 13,000-26,000
pH: 8.3-10.8

ALCO CHEMICAL CORP.: ALCOGUM Thickeners and Rheology Modifiers (Continued):

ALCOGUM AN-10
ALCOGUM AN-8
ALCOGUM AN-6

ALCOGUM AN-10 is a sodium polyacrylate thickener which finds extensive use in the thickening of natural and synthetic latexes for a wide variety of applications.

ALCOGUM AN-10 and its dilutions, ALCOGUM AN-8 and ALCOGUM AN-6, may be used to thicken styrene-butadiene, acrylonitrile butadiene, neoprene, acrylic, vinyl acetate and ethylene-vinyl acetate latex compounds. In general they are most satisfactory thickeners for latexes which have an alkaline reaction although some success has been achieved in neutral or slightly acid systems.

ALCOGUM AN-10 and its dilutions are used in a variety of applications including:

* Carpet Backing
* Latex Foam
* Adhesives
* Emulsions
* Textile Coatings
* Nonwovens
* Dispersions

ALCOGUM AN-10, and its dilutions, ALCOGUM AN-8 and ALCOGUM AN-6, are sodium polyacrylate thickeners which find extensive use in the thickening of natural and synthetic latexes for a wide variety of applications.

Typical Properties:

AN-10:
Total Solids: 10.0%
pH: 9.0-11.0
Viscosity, 20 rpm, RV, 25C, cps: 70,000

AN-8:
Total Solids: 8.0%
pH: 9.0-11.0
Viscosity, 20 rpm, RV, 25C, cps: 60,000

AN-6:
Total Solids: 6.0%
pH: 9.0-11.0
Viscosity, 20 rpm, RV, 25C, cps: 40,000

Color: Clear to slightly opaque light colored liquid

ALCO CHEMICAL CORP.: ALCOGUM Thickeners and Rheology Modifiers (Continued):

ALCOGUM S7H

ALCOGUM S7H is a polyacrylic acid with the following typical properties:

Appearance: clear, colorless liquid
Molecular Weight: 1,000,000 by GPC
pH: 2.1+-3.0
Total Solids: 15.0+-1.0%

ALCOGUM S7H polyacrylic acid is supplied in the form of a clear, stable, aqueous solution. ALCOGUM AR7H is suitable in a number of applications as a replacement for natural gums. Unlike the natural gums, ALCOGUM S7H is stable to hydrolysis and is not susceptible to bacterial degradation.

The ammonium, potassium, and sodium salts of ALCOGUM S7H can be used as thickeners in aqueous solutions at pH 7.0 and over, while also providing dispersion and suspension properties.

CAS No. 9003-01-4

ALCOGUM S20

ALCOGUM S20 is a polyacrylic acid with the following typical properties:

Appearance: clear, colorless liquid
Molecular Weight: 2,000,000 AMW
pH: 2.1-3.0
Total Solids: 12.0+-0.5%

ALCOGUM S20 polyacrylic acid is supplied in the form of a clear, stable, aqueous solution. ALCOGUM S20 is suitable in a number of applications as a replacement for natural gums. Unlike the natural gums, ALCOGUM S20 is stable to hydrolysis and is not susceptible to bacterial degradation.

The ammonium, potassium, and sodium salts of ALCOGUM S20 can be used as thickeners in aqueous systems at pH 7.0 and over, while also providing dispersion and suspension properties.

The rate of resolubilization of films of ALCOGUM S20 can be controlled by immersion of the dried film in a 10% solution of various mineral acids or multivalent cations.

CAS No. 9003-01-4

ALCO CHEMICAL CORP.: ALCOGUM Thickeners and Rheology Modifiers (Continued):

ALCOGUM L-11

ALCOGUM L-11 is a reactive, acidic thickening agent of the type commonly referred to as an in-situ thickener. It is supplied as a stable latex emulsion. The addition of any of a variety of alkaline agents results in the neutralization of the acidic groups and the formation of a clear, viscous solution. This solution is an effective thickener for many systems, including various synthetic latexes, synthetic latex compounds and synthetic latex coatings.

Typical Properties:
- Total Solids: 28.0+/-1.0%
- pH: 2.0-4.0
- Viscosity: 14.0 cps
- Polymer Type: Cross-Linked
- Emulsion Type: Anionic
- Equivalent Weight: 236
- Specific Gravity: 1.050
- Appearance: Milky White Liquid
- Flow Characteristic: Short

ALCOGUM L-12

ALCOGUM L-12 is a reactive, acidic thickening agent of the type commonly referred to as an in-situ thickener. It is supplied as a stable latex emulsion. The addition of any of a variety of alkaline agents results in the neutralization of the acidic groups and the formation of a clear, viscous solution. This solution is an effective thickener for many systems, including various synthetic latexes, synthetic latex compounds and synthetic latex coatings.

Typical Properties:
- Total Solids: 28.5+/-1.0%
- pH: 2.0-3.5
- Viscosity: 20.0 cps
- Polymer Type: Cross-Linked
- Emulsion Type: Anionic
- Equivalent Weight: 230
- Specific Gravity: 1.050
- Appearance: Milky White Liquid
- Flow Characteristic: Short

ALCO CHEMICAL CORP.: ALCOGUM Thickeners and Rheology Modifiers (Continued):

ALCOGUM L-15

ALCOGUM L-15 is a reactive, acidic thickening agent of the type commonly referred to as an in-situ thickener. It is supplied as a stable latex emulsion. The addition of any of a variety of alkaline agents results in the neutralization of the acidic groups and the formation of a clear, viscous solution. This solution is an effective thickener for many systems, including various synthetic latexes, synthetic latex compounds and synthetic latex coatings.

Typical Properties:

Total Solids: 30.0+/1.0%
pH: 2.2-3.5
Viscosity: 15.0 cps
Polymer Type: Cross-Linked
Emulsion Type: Anionic
Equivalent Weight: 236
Specific Gravity: 1.050
Appearance: Milky White Liquid
Flow Characteristic: Short

ALCOGUM L-31

ALCOGUM L-31 is a reactive, acidic thickening agent of the type commonly referred to an as in-situ thickener. It is supplied as a stable latex emulsion. The addition of any of a variety of alkaline agents results in the neutralization of the acidic groups and the formation of a clear, viscous solution. This solution is an effective thickener for many systems, including various synthetic latexes, synthetic latex compounds and synthetic latex coatings.

Typical Properties:

Total Solids: 40.0+/-1.0%
pH: 2.2-3.4
Viscosity: 20.0 cps
Emulsion Type: Anionic
Equivalent Weight: 263
Specific Gravity: 1.050
Appearance: Milky White Liquid
Flow Characteristic: Short

ALCO CHEMICAL CORP.: ALCOGUM Thickeners and Rheology Modifiers (Continued):

ALCOGUM L-35

ALCOGUM L-35 is a reactive, acidic thickening agent of the type commonly referred to as an in-situ thickener. It is supplied as a stable latex emulsion. The addition of any of a variety of alkaline agents results in the neutralization of the acidic groups and the formation of a clear, viscous solution. This solution is an effective thickener for many systems, including various synthetic latexes, synthetic latex compounds and synthetic latex coatings.

Typical Properties:
- Total Solids: 20.0+/-1.0%
- pH: 2.2-4.0
- Viscosity: 50.0 cps
- Emulsion Type: Anionic
- Specific Gravity: 1.055
- Appearance: Milky White Liquid

ALCOGUM SL-70

ALCOGUM SL-70 is an alkali activated rheology modifier. ALCOGUM SL-70 is an acrylic associative polymer formulated to give structure to filled compoounds for use in a wide range of applications.

Typical Emulsion Properties:
- Color: milk white emulsion
- Solids: 30%
- Viscosity: 100 cps
- pH: 2.5-4.0

ALCOGUM SL-70 can be utilized by addition into systems before neutralization to pH 8-9.

When added to higher pH systems, pre-dilution will allow better mixing.

The water solutions of associative acrylic polymers have been shown to undergo an increase in viscosity when used in conjunction with nonionic surfactants.

ALCO CHEMICAL CORP.: ALCOGUM Thickeners and Rheology Modifiers (Continued):

ALCOGUM SL-76

ALCOGUM SL-76 is an alkali activated rheology modifier. ALCOGUM SL-76 is an acrylic associative polymer formulated to give structure to filled compounds in paint applications.

Typical Emulsion Properties:
Color: Milk white emuslion
Solids: 30%
Viscosity: 100 cps
pH: 2.5-4.0

ALCOGUM SL-76 should be utilized by addition into systems before neutralization to pH 8-9.

When added to higher pH systems, pre-dilution will allow better mixing.

The water solutions of associative acrylic polymers have been shown to undergo an increase in viscosity when used in conjunction with nonionic surfactants.

ALCOGUM SL-98

ALCOGUM SL-98 is an alkali activated rheology modifier. ALCOGUM SL-98 is an acrylic associative polymer formulated to give structure to filled compounds for use in a wide range of applications.

Typical Emulsion Properties:
Color: milk white emulsion
Solids: 30%
Viscosity: 100 cps
pH: 2.5-4.0

ALCOGUM SL-98 can be utilized by addition into systems before neutralization to pH 8-9.

When added to higher pH systems, pre-dilution will allow better mixing.

The water solutions of associative acrylic polymers have been shown to undergo an increase in viscosity when used in conjunction with nonionic surfactants.

ALCO CHEMICAL CORP.: ALCOGUM Thickeners and Rheology Modifiers (Continued):

ALCOGUM 382
High Efficiency Thickener for
Adhesive and Coating Applications

ALCOGUM 382 is a high molecular weight polymer supplied as a water-in-oil emulsion. This product can be easily added to any aqueous formulation to produce rapid thickening. ALCOGUM 382 is compatible with most synthetic and natural latices.

Some of the key features of this polymer are:
- Useful Over a Wide pH Range
- Easy to Handle, Mix and Use
- No Alkali Adjustment is Required
- Resistant to Bacterial Contamination

ALCOGUM 382 represents a new approach to thickeners for adhesives and coatings. The product should be used as is and can replace cellulosic and alkali type thickeners to add viscosity and modify rheology of most aqueous formulations.

Typical Properties:
- Solids Content: 31%
- pH: 7.0
- Specific Gravity: 1.05
- Flash Point (open cup): 108 C
- Viscosity, 20 rpm, RV, 25C, cps: 100
- Appearance: Clear liquid
- Rheology: Very pseudo plastic

ALLIED COLLOIDS: VISCALEX/RHEOVIS CR Thickening Agents:

The VISCALEX Range:

VISCALEX HV30:

A highly thixotropic alkali activated thickening agent, supplied as a 30% active low viscosity acrylic copolymer emulsion. VISCALEX HV30 is particularly recommended for "textured" emulsion paints for interior and exterior use. Application in high P.V.C. matt finishes is also possible, but its use in medium to low P.V.C. finishes is not recommended.

VISCALEX 'V' series:

Alkali activated acrylic emulsion copolymers developed for three specific areas in emulsion paint manufacture:

VISCALEX VM:

for high to medium P.V.C. formulations i.e. 95-50% P.V.C.

VISCALEX VS:

for predominantly semi-gloss finish formulations in the range 25-35% P.V.C.

VISCALEX VG:

for low P.V.C. systems, i.e. gloss emulsion paint formulations 15-20% P.V.C.

VISCALEX VG2 and VG3 are available in this series.

Higher viscosity paints are achieved using VISCALEX VG3.

RHEOVIS CR:

A new improved alkali activated thickener for matt emulsion paints.

RHEOVIS CR is an acrylic copolymer emulsion in water designed to replace conventional cellulosic thickeners in high p.v.c. emulsion paints.

Typical Properties--VISCALEX Range/RHEOVIS CR:

HV30/'V' Series:

Nature: Acrylic copolymer emulsion
Physical Form: Low viscosity liquid
Solid Content: 30+-1%
pH: 3.0-5.0 2.0-3.5
Viscosity (as supplied): approx 5 cps (Brookfield 20 rpm)
Specific Gravity: 1.05

CR:

Nature: Acrylic copolymer emulsion
Physical Form: Low viscosity liquid
Solid Content: 30+-1%
pH: 2-3.5
Viscosity (as supplied): approx 40 cps (Brookfield 20 rpm)
Specific Gravity: 1.10

CABOT CORP.: CAB-O-SIL Fumed Silica:

CAB-O-SIL fumed silica is fumed silicon dioxide, a material which is generally regarded as unique in industry because of its unusual particle characteristics. CAB-O-SIL fumed silica's extremely small particle size, its enormous surface area, its high purity and its chain-forming tendencies set it apart in a class of its own.

Foremost among CAB-O-SIL fumed silica's several contributions to product performance is its ability to modify a product's flow properties by building a three-dimensional network which alternately forms and disrupts in response to the degree of shear forces present--thereby controlling flow characteristics in accordance with the formulator's exact requirements.

Grades of CAB-O-SIL Fumed Silica:

CAB-O-SIL fumed silica is available in treated and untreated grades. The untreated grades vary in surface area, bulk density and thickening efficiency. These are available in aqueous dispersions, also. The aqueous dispersions are available under the name CAB-O-SPERSE dispersions.

Grade:

Standard: L-90
Densed: L-90D
Surface Area (m2/g): 100+-15
Nominal Particle Diameter (Microns): .027
Thickening Efficiency (Standard Grade): Low

Standard: LM-130
Densed: LM-130D
Surface Area (m2/g): 130+-15
Nominal Particle Diameter (Microns): .021
Thickening Efficiency (Standard Grade): Moderate

Standard: LM-5
Densed: LM-7
Surface Area (m2/g): 160+-15
Nominal Particle Diameter (Microns): .017
Thickening Efficiency (Standard Grade): Moderate

Standard: M-5
Densed: MS-7
Surface Area (m2/g): 200+-25
Nominal Particle Diameter (Microns): .014
Thickening Efficiency (Standard Grade): High

Standard: PTG
Surface Area (m2/g): 200+-20
Nominal Particle Diameter (Microns): .014
Thickening Efficiency (Standard Grade): Very High

CABOT CORP.: CAB-O-SIL Fumed Silica(Continued):

Grade:

Standard: MS-55
Densed: MS-75
Surface Area (m2/g): 255+-15
Nominal Particle Diameter (Microns): .011
Thickening Efficiency (Standard Grade): High

Standard: HS-5
Densed: HS-7
Surface Area (m2/g): 325+-25
Nominal Particle Diameter (Microns): .008
Thickening Efficiency (Standard Grade): High

Standard: EH-5
Densed: S-17
Surface Area (m2/g): 380+-30
Nominal Particle Diameter (Microns): .007
Thickening Efficiency (Standard Grade): Very High

Property:
pH (4% Aqueous Slurry): 3.7-4.3
325 Mesh Residue (%): 0.02 Max
Specific Gravity: 2.2
Refractive Index: 1.46
X-Ray Form: Amorphous
Assay (% SiO2): >99.8

CAB-O-SIL TS-720 Hydrophobic Fumed Silica:

Viscosity Control: effective thickener*good temperature stability*low moisture pick up

Antisag: excellent stability in epoxies*heat resistant

Antisettling: imparts good suspension*any pigment settling is readily redispersable

Reinforcement: good reinforcement*no crepe hardening on aging in silicone rubber

Water resistance: improves hydrophobicity*low moisture content*low moisture pick up*keeps hygroscopic powders free flowing

DANIEL PRODUCTS CO.: DAPRO 7:

Viscosity Stabilizer and Anti-Gelling Agent for Solvent Thinned Coatings

DAPRO 7 is a chemical inhibitor which prevents drastic viscosity increases, gelling, or livering where these problems are caused by either vehicle polymerization prior to application or by gradual changes in the vehicle adsorption on pigment surfaces. Premature vehicle polymerization occurs when a reactive system is subjected to excessive energy. Reactivity of the system can result from the presence of a highly reactive vehicle such as a vinyl toluene treated alkyd or short oil alkyd, the use of an insufficiently blocked catalyst, or the presence of a pigment that contains available metallic ions to act as a catalyst or drier (iron blue, lead chromate or natural umber).

Specifications:

Active Ingredients: 66.0%
Solvent: 34.0% (Xylol)
Wt./Gal. (Sp.Gr.): 7.5 lbs. (0.90)
Viscosity (Gardner-Holdt): B-D
Flash Point: 81F/27.2C
Appearance: Amber Liquid

DANIEL PRODUCTS CO.: SUSPEND-AYD Modified Bentone Gel:

For Pigment Suspension and Flow Control of Solvent Thinned Coatings

SUSPEND-AYD Gels are dispersions of modified BENTONES which are up to 50% more effective than conventional BENTONE slurries. These dispersions give immediate full thixotropic effects and require no further milling. They exhibit no after-bodying or viscosity drift on aging, even at elevated temperatures.

SUSPEND-AYD 1 and 2 are produced by carefully controlled swelling of BENTONE 34 and 38, respectively, to achieve a balance of efficiency and ease of use. Batch-to-batch uniformity assures reproducible and reliable paint performance.

SUSPEND-AYD Gels can be used in a wide range of trade sales and industrial coatings to:

* Improve pigment suspension
* Improve sag resistance
* Reduce penetration
* Reduce brush drag
* Reduce floating and flooding

DANIEL PRODUCTS CO.: VISC-AYD 812:

Bodying Agent for Solvent Thinned Coatings

VISC-AYD 812 with Activator 814 is a broadly effective bodying agent which increases Newtonian viscosity with no affect on thixotropy. It will not adversely affect brushing, leveling, gloss, color, hiding, color acceptance, color retention or fineness of grind. It is effective in most major types of solvent thinned trade sales paints, including house paints, enamels and alkyd flats.

VISC-AYD 812 with Activator 814 can be incorporated into a mill base or added by simple stirring to a finished paint. In either case, the viscosity increase is usually complete within 24 hours and will remain constant, provide the paint itself is stable.

VISC-AYD 812 Base
(Crosslinking Dispersion)

% Pigment: 30.0
% Vehicle Solids: 32.0
Solvent: Exempt Mineral Spirits
Wt./Gal. (Sp.Gr.): 9.2 (1.10)
Flash Point: 104F/40.0C

VISC-AYD 814 Activator:

% Vehicle Solids: 70.0
Solvent: Exempt Mineral Spirits
Wt./Gal. (Sp.Gr.): 7.5 (0.90)
Flash Point: 104F/40.0C

DEGUSSA CORP.: AEROSIL for Lacquers and Paints:

AEROSIL is a highly dispersed, amorphous silica which is derived through flame hydrolysis of silicon tetrachloride.

Types of AEROSIL for the Coatings Industry:

The following grades are useful in particular:
AEROSIL 200
AEROSIL 300
AEROSIL 380
AEROSIL 380
AEROSIL TT600
AEROSIL COK84
AEROSIL R972
AEROSIL R974

Which advantages does AEROSIL offer in lacquers and paints?
* Control of viscosity and thixotropy
* Improvement in the suspension behaviour of pigments
* Increase in the resistance to water
* Prevention of gassing with zinc dust paints
* Favourable effect on adhesion and flexibility
* Positive influence on the dispersibility
* Improvement in the free-flow characteristics of powdered coatings
* Flatting of special coating systems

Physico-chemical Data of AEROSIL:

AEROSIL 200:
Appearance: fluffy white powder
BET-surface area: m2/g: 200+-25
Average primary particle size nm: 12
SiO2 %: >99.8

AEROSIL 300:
Appearance: fluffy white powder
BET-surface area: m2/g: 300+-30
Average primary particle size nm: 7
SiO2 %: >99.8

AEROSIL 380:
Appearance: fluffy white powder
BET-surface area: m2/g: 380+-30
Average primary particle size nm: 7
SiO2 %: >99.8

DEGUSSA CORP.: AEROSIL for Lacquers and Paints(Continued):

Physico-chemical Data of AEROSIL(Continued):

AEROSIL TT600:
Appearance: fluffy white powder
BET-surface area m2/g: 200+-50
Average primary particle size: 40
SiO2 %: >99.8

AEROSIL COK84:
Appearance: fluffy white powder
BET-surface area m2/g: 170+-30
SiO2 %: 82-86

AEROSIL R972:
Appearance: fluffy white powder
BET-surface area m2/g: 110+-20
Average primary particle size: 16
SiO2 %: >99.8

AEROSIL R974:
Appearance: fluffy white powder
BET-surface area m2/g: 170+-20
Average primary particle size: 12
SiO2 %: >99.8

DUPONT CO.: KEVLAR Aramid Pulp:

A better alternative for viscosity control and reinforcement in...

* Adhesives
* Sealants
* Coatings
* Caulkings
* Mastics

When reinforcing sealants and adhesives with KEVLAR pulp you get better thixotropy. A very small amount added to resins provides excellent "sag" or "slump" resistance, so sealants and adhesives stay where you want them to stay.

Even more remarkable is the shear thinning effect provided with KEVLAR.

KEVLAR also offers a unique combination of high tensile strength, low density, high temperature stability, good chemical resistance and high surface area.

You get the ideal properties you need:

* Better thixotropy
* High temperature stability
* Good chemical resistance
* Increased tensile strength and modulus
* Increased tear strength
* Easily processed
* Environmentally safe
* Cost effective

ENGELHARD CORP.: Attapulgite Specialty Thickeners and Sorbents:

The ATTAGEL thickeners effectively form thixotropic colloidal gels in ionic and non-ionic aqueous solutions and most organic liquids to develop thickening and/or suspension properties. They are used in paints and coatings, offering rapid dispersibility and effective thixotropic thickening. In adhesives, caulks, sealants and wallboard compounds, they provide sag resistance, viscosity control, spreadability and excellent feathering at low cost. When used in light-weight auto-body filler, ATTAGEL thickeners control "float-out" and settling. In asphalt coatings, they build excellent thixotropic characteristics in asphalt cutbacks and emulsions. Other important applications for ATTAGEL products include cosmetics, polishes, ceramic binders, molecular sieve binders, wax emulsions, and lubricants.

ATTASORB, ATTACOTE, and the LVM and RVM granular sorbents are highly absorptive, and, preferentially adsorptive for polar molecules in solutions. The powdered grades are used as free-flow aids in laundry powders, chemicals, and fire extinguisher powders where they help to absorb moisture which can cause caking.

Mineral Thickeners:

Dry Grades:

ATTAGEL 50:
Average particle size, um: 0.1
Hegman: 6.0-7.0
Oil Absorption: 115
Color: Light Cream
Specific Gravity: 2.36

ATTAGEL 40:
Average particle size, um: 0.1
Hegman: 5.0-6.0
Oil Absorption: 110
Color: Light Cream
Specific Gravity: 2.36

Liquid Products:

ATTAFLOW FL Liquid Thickener:
Average particle size, um: 0.1
Color: Grey
Specific Gravity: 1.2

ENGELHARD CORP.: Attapulgite Specialty Thickeners and Sorbents (Continued):

Thermally Activated Sorptive Products:

Powdered Grades:

ATTASORB LVM:
Average particle size, um: 2
Oil absorption: 100
Color: Cream

ATTASORB RVM:
Average particle size, um: 2
Oil absorption: 105
Color: Light Cream

ATTACOTE:
Average particle size, um: 18
Oil absorption: 65
Color: Cream

ATTACOTE RVM:
Average particle size, um: 18
Oil absorption: 65
Color: Light Cream

Granular Grades:

LVM*:
8/16, 16/30, 25/50, 50/80, 100/200, 100/Up, 200/Up Mesh Sizes:
Color: Cream

* LVM=low volatile matter

RVM**:
8/16, 16/30, 25/50, 50/80, 100/200, 100/Up, 200/Up Mesh Sizes:
Color: Light Cream

** RVM=regular volatile matter

Typical Chemical Analysis of Engelhard Attapulgite Products: (Volatile-Free Basis)

Silicon (SiO_2)	65.9%
Aluminum (Al_2O_3)	12.2
Magnesium (MgO)	11.5
Iron (Fe_2O_3)	3.6
Calcium (CaO)	4.3
Phosphorous (P_2O_5)	1.1
Potassium (K_2O)	0.8
Titanium (TiO_2)	0.5
Trace Elements	0.1
	100.0%

INTERSTAB CHEMICALS: Viscosity Control Agents:

It is necessary to control the viscosity and viscosity dependent properties of all coatings. INTERSTAB offers two products useful in controlling the viscosity of solvent based paints. Both products, PUFFING AGENT and PUFFING AGENT #2, are liquids and as such can be added at any time during manufacture.

PUFFING AGENT:

PUFFING AGENT is a chemically modified liquid organic compound in mineral spirits which is effective in increasing the viscosity of coating formulations. PUFFING AGENT is particularly effective in oil paints which contain reactive pigments such as zinc oxide. When no reactive pigments are present the addition of Noury Dry Calcium or Noury Dry Zinc (0.2% Metal) is useful in obtaining maximum viscosity.

PUFFING AGENT #2:

PUFFING AGENT #2 is a synthetic organic liquid compound, designed to increase the viscosity of solvent thinned coatings. It is particularly effective in medium and long oil alkyds, oil modified polyurethanes, styrenated alkyds, and epoxy esters.

INTERLITE VCA:

INTERLITE VCA is a viscosity control additive designed for use in vinyl solution coatings and organosols containing an active resin solvent. INTERLITE VCA will not adversely affect the heat and light stability of the coating during processing or after application.

Physical Properties:

PUFFING AGENT:

Viscosity (G/H-Max.): Z
Form: Clear Liquid

PUFFING AGENT #2:

Viscosity (G/H-Max.): Z
Form: Clear Liquid

INTERLITE VCA:

Viscosity (G/H-Max): D
Form: Clear Liquid

KELCO DIVISION: KELZAN Xanthan Gum:

KELZAN industrial grade xanthan gum is a natural high-molecular weight branched polysaccharide. It functions as a hydrophilic colloid to thicken, suspend, and stabilize water-based systems.

Product Features and Benefits:

* Viscosity which shows practically no change with temperature variations.
* Solubility and stability in acid solutions.
* Stability in alkaline solutions.
* Suspending properties for hard to suspend solids.
* Compatibility and stability in solutions containing high concentrations of various salts.
* Heat stability, even at very high temperatures.
* Solubility in hot or cold water.
* Extreme pseudoplasticity (high viscosity under low shear, but low viscosity under high shear).
* Freeze-thaw stability.
* High viscosity at low concentrations.
* Good lubricating value and pleasing skin feel in aqueous solutions.
* Highly effective emulsion stabilization.

Grades:

KELZAN is available in one viscosity grade. Recognizing the special requirements of certain industrial applications, Kelco has developed a number of modified KELZAN products with special functionality:

KELZAN D: for increased effectiveness in systems containing other gums
KELZAN M: for use in systems which require a lower salt content
KELZAN S: for systems requiring dispersibility

Typical Chemical and Physical Properties:

Composition: xanthan gum
Color and form: free-flowing cream-colored powder
Particle size (approx): 28 mesh, Tyler standard screen (595 micron)
pH: 7.0+-1.5

NL CHEMICALS, INC.: NL Rheological Additives:

Solvent Based Coatings:
Anti-Settling Additives:

M-P-A MS:
Chemical Description: An organic, high solids paste dispersed in mineral spirits.
Principal Uses: For trade sales paints where a minimum processing temperature of about 140F (60C) is reached.
Density: 0.83
N.V. (%): 40
Volatile: Rule 66 MS
Rheology: Imparts non-settling with only small viscosity increase. Also controls sag.
Levels of Use: 0.5% to 1.0%

M-P-A X:
Chemical Description: An organic, high solids paste dispersed in xylene.
Principal Uses: For industrial finishes where a minimum processing temperature of about 110F (43C) is reached.
Density: 0.88
N.V. (%): 40
Volatile: Xylene
Rheology: Imparts non-settling with only small viscosity increase.
Levels of Use: 0.5-1.0%

M-P-A 60 MS:
Chemical Description: A lower solids version of M-P-A MS--for easier dispersion.
Principal Uses: Same as M-P-A MS
Density: 0.82
N.V. (%): 24
Volatile: Rule 66 MS
Rheology: Same as M-P-A X.
Levels of Use: 0.8% to 1.7%

M-P-A 60X:
Chemical Description: A lower solids version of M-P-A X--for easier dispersion.
Principal Uses: Same as M-P-A X.
Density: 0.87
N.V (%): 24
Volatile: Xylene
Rheology: Same as M-P-A X
Levels of Use: 0.8% to 1.7%

NL CHEMICALS, INC.: NL Rheological Additives(Continued):

Solvent Based Coatings:
Anti-Settling Additives(Continued):

M-P-A 60T:
Chemical Description: A low solids, organic paste dispersed in toluene--for easy dispersion.
Principal Uses: Same as M-P-A X.
Density: 0.87
N.V. (%): 24
Volatile: Toluene
Rheology: Same as M-P-A X.
Levels of Use: 0.8 to 1.7%

M-P-A 1078X:
Chemical Description: A high solids, organic paste dispersed in xylene--specially processed for increased efficiency and easier dispersion.
Principal Uses: Same as M-P-A X. Except minimum processing temperature is 100F (38C.)
Density: 0.89
N.V. (%): 40
Volatile: Xylene
Rheology: Same as M-P-A X, but more efficient and easier to use
Levels of Use: 0.3% to 0.8%

M-P-A 2000X:
M-P-A 2000T:
Chemical Description: A pourable organic dispersion of xylene or toluene--developed for high efficiency. Easy incorporation and very easy dispersion.
Principal Uses: Same as M-P-A 1078 X.
Density: 0.88
N.V. (%): 20
Volatile: Xylene or Toluene
Rheology: Same as M-P-A 1078 X, but pourable.
Levels of Use: 0.6% to 1.6%

BENTONE 14:
Chemical Description: Organic derivative of a special smectite clay.
Principal Uses: For trade sales and industrials--no minimum or maximum processing temperature.
Density: 2.30
N.V. (%): 100
Rheology: Imparts non-settling with only a small viscosity increase.
Levels Of Use: 0.5% to 1.0%

NL CHEMICALS, INC: NL Rheological Additives(Continued):

Solvent Based Coatings:
Thixotropic Additives:

THIXCIN R:
Chemical Description: An organic derivative of castor oil.
Principal Uses: For aliphatic solvent systems, especially trade sales paints where processing temperatures fall within 95F to 130F (35C to 54C).
Density: 1.02
N.V. (%): 100
Rheology: Excellent thixotropy--controls sag and leveling, increases viscosity and imparts non-settling.
Levels of Use: 0.2% to 0.8%

THIXCIN GR:
Chemical Description: Inorganically modified derivative of castor oil. Easy dispersing version of THIXCIN R.
Principal Uses: Same as THIXCIN R.
Density: 1.47
N.V. (%): 100
Rheology: Same as THIXCIN R.
Levels of Use: 0.3 to 1.2%

THIXATROL ST:
Chemical Description: An organic derivative of castor oil.
Principal Uses: For aliphatic trade sales systems where processing temperatures fall within 135F to 165F (57C to 74C). For aromatic industrial systems where processing temperatures fall within 90F to 120F (33C to 49C).
Density: 1.02
N.V. (%): 100
Rheology: Excellent thixotropy--controls sag and leveling, increases viscosity and imparts non-settling.
Levels of Use: 0.2% to 0.8%

THIXATROL GST:
Chemical Description: Inorganically modified derivative of castor oil. Easy dispersing version of THIXATROL ST.
Principal Uses: Same as THIXATROL ST.
Density: 1.46
N.V. (%): 100
Rheology: Same as THIXATROL ST.
Levels of Use: 0.3 to 0.8%

NL CHEMICALS, INC.: NL Rheological Additives(Continued):

Solvent Based Coatings:
Thixotropic Additives(Continued):

RHEOX 1:
Chemical Description: An organic paste dispersed in mineral spirits.
Principal Uses: For architectural trade sales finishes. High temperature processing 140F to 190F (60C to 88C).
Density: 0.81
N.V. (%): 45
Volatile: Rule 66 MS
Rheology: Less thixotropy than THIXATROL ST. Controls sag and leveling. Increases viscosity and imparts non-settling.
Levels of Use: 0.4 to 2.0%

RHEOX 53 MS:
Chemical Description: An organic paste dispersed in mineral spirits.
Principal Uses: For architectural trade sales finishes where processing temperatures fall within 130F to 180F (54C to 82C).
Density: 0.85
N.V. (%): 40
Volatile: Rule 66 MS
Rheology: Develops a highly thixotropic buttery consistency-controls sag & leveling, increases viscosity & imparts non-settling. Greater thixotropy than RHEOX-1.
Levels of Use: 0.4 to 2.0%

RHEOX 53X:
Chemical Description: An organic paste dispersed in xylene.
Principal Uses: For aromatic industrial finishes where processing temperatures fall within 100F to 140F (38C to 60C).
Density: 0.89
N.V. (%): 40
Volatile: Xylene
Rheology: Excellent thixotropy--controls sag and leveling, increases viscosity and imparts non-settling.
Levels of Use: 0.6% to 3.0%

POST-4:
Chemical Description: Complex derivative of castor oil.
Principal Uses: Post additive for trade sales and industrial finishes to control sag and pigment settling. No shear required.
Density: 1.03
N.V. (%): 71
Rheology: Contributes to non-settling. Increases sag resistance of trade sales paints.
Levels of Use: 0.3% to 1.0%

NL CHEMICALS, INC.: NL Rheological Additives(Continued):

Solvent Based Coatings:
Thixotropic Additives(Continued):

BENTONE 27:
Chemical Description: Organic derivative of a special smectite clay.
Principal Uses: For intermediate to high polarity organic systems, especially industrial finishes. Normally requires a polar activator, but has no processing temperature requirements.
Density: 1.80
N.V. (%): 100
Rheology: Very good thixotropy--controls sag and leveling, increases viscosity and imparts non-settling.
Levels Of Use: 0.2% to 0.9%

BENTONE 34:
Chemical Description: Organic derivative of a special smectite clay.
Principal Uses: For low to intermediate polarity organic systems, including trade sales and industrial finishes. Requires polar activity for maximum efficiency but has no processing temperature requirements.
Density: 1.70
N.V. (%): 100
Rheology: Same as BENTONE 27
Levels of Use: 0.2% to 0.9%

BENTONE 38:
Chemical Description: Organic derivative of a special smectite clay.
Principal Uses: Same as BENTONE 34, plus has an edge on efficiency in systems at both extremely low and high intermediate ends of polarity range.
Density: 1.70
N.V. (%): 100
Rheology: Same as BENTONE 27
Levels of Use: 0.2% to 0.9%

BENTONE SA-38:
Chemical Description: Organic derivative of a special smectite clay.
Principal Uses: Same as BENTONE 38, but chemically self-activating. (Requires no polar additive but a minimum processing temperature of 110F (43C) is recommended.)
Density: 1.50
N.V. (%): 100
Rheology: Same as BENTONE 27
Levels of Use: 0.2% to 0.9%

NL CHEMICALS, INC.: NL Rheological Additives(Continued):

Solvent Based Coatings:
Thixotropic Additives(Continued):

BENTONE GEL OMS:

Chemical Description: An organo-clay dispersed in mineral spirits--for easier dispersion.

Principal Uses: Completely dispersed and gelled form of BENTONE. May be used as a post additive.

Density: 0.85
N.V. (%): 15
Volatile: Rule 66 MS
Rheology: Same as BENTONE 27
Levels of Use: 1.5% to 4.0%

THIXSEAL 1084:

Chemical Description: Inorganically modified derivative of castor oil.

Principal Uses: Sag and slump control agent for caulks, sealants and mastics where processing temperatures fall within 140F to 160F (60C to 71C).

Density: 1.36
N.V. (%): 100
Rheology: Very good thixotropy in coatings. Controls sag and leveling, increases viscosity and imparts non-settling.
Levels of Use: 0.2% to 3.0%

NL CHEMICALS, INC.: NL Rheological Additives(Continued):

Water Based Coatings:
Anti-Settling Additive:

M-P-A 1075:
Chemical Description: A 100% organic, high solids paste dispersed in butanol--for easy dispersion
Principal Uses: For both trade sales and industrial finishes. May also be used in organic solvent systems.
Density: 0.83
N.V. (%): 45
Volatile: Butanol
Rheology: Imparts non-settling with only small viscosity increase.
Levels of Use: 0.5% to 1.0%

Thickeners:

BENTONE LT:
Chemical Description: Refined organo-clay mineral-easy dispersing
Principal Uses: For all latex paint systems. Excellent leveling and flow control.
Density: 1.8
N.V. (%): 100
Levels of Use: 0.1% to 0.7%

BENTONE EW:
Chemical Description: Highly beneficiated smectite clay--easy dispersing.
Principal Uses: For water reducible industrial finishes and inks.
Density: 2.5
N.V.: 100
Levels of Use: 0.3% to 1.0%

Rheology: Excellent rheology. Controls sag and leveling, increases viscosity and imparts non-settling.

RHEOLATE 1:
Chemical Description: A low viscosity, ready to use polymeric additive.
Principal Uses: For water borne systems.
Density: 1.07
N.V.: 30
Volatile: Water
Rheology: Excellent balance of rheological properties. pH must be adjusted to 8.5 to 10.0.
Levels of Use: 0.3% to 1.7%

STANCHEM INC.: Polymers and Polymeric Specialties: Thickeners and Additives:

STANCHEM Product:

6006:

Effective "in situ" thickener for a variety of emulsion systems. Can be used as pigment dispersant.

Polymer Type: Acrylic copolymer
Viscosity (cps): <50
Solids % (min.): 27.5
Average Particle Size, M: 0.05-0.1
Tg, C (Theoretical): 14
pH @ 25C: 3.0-4.0
Weight lb/gal: 8.7
Ionic Charge: Anionic

6070:

Superior additive for water-base thickening applications. Acrylic emulsion with good mechanical shear resistance.

Polymer Type: Acrylic copolymer
Viscosity (cps): <100
Solids % (min.): 29.5
Average Particle Size, M: 0.05-0.1
Tg, C (Theoretical): 14
pH @ 25C: 2.0-3.5
Weight lb/gal: 8.7
Ionic Charge: Anionic

TROY CHEMICAL CORP.: Rheology Modifiers:

TROYTHIX Rheology Agents

If Your Coating Needs...
* Thixotropy
* Sag Control
* Pigment Suspension
* Hold Out
* Leveling
* Brushability
* Viscosity Increase

Troythix Products:

XYZ:
- Physical Form: Powder
- Solvent System: Aliphatic
- Rheological Properties:
  - Type of Flow: Thixotropic
  - Apparent Viscosity Increase
  - Antisettling
  - Hold Out
  - Brushability

XD:
- Physical Form: Powder
- Solvent System: Aliphatic
- Rheological Properties:
  - Type of Flow: Thixotropic
  - Apparent Viscosity Increase
  - Antisettling
  - Hold Out
  - Brushability

A:
- Physical Form: Powder
- Solvent System: Aliphatic Aromatic Oxygenated
- Type of Flow: Thixotropic
- Apparent Viscosity Increase
- Antisettling
- Hold Out
- Brushability

TROY CHEMICAL CORP.: Rheology Modifiers(Continued):

TROYTHIX Products:

PSA 1:
Physical Form: Powder
Solvent System: Aliphatic Aromatic Oxygenated
Rheological Properties:
Type of Flow: Thixotropic
Antisettling
Hold Out

ANTISETTLE:
Physical Form: Powder
Solvent System: Aliphatic Aromatic Oxygenated
Rheological Properties:
Type of Flow: Newtonian
Antisettling

ANTISETTLE S:
Physical Form: Powder
Solvent System: Aliphatic Aromatic
Type of Flow: Thixotropic
Rheological Properties:
Type Of Flow: Thixotropic
Apparent Viscosity Increase
Antisettling
Hold Out
Brushability

ANTI SAG 4:
Physical Form: Liquid
Solvent System: Aliphatic Aromatic Oxygenated
Rheological Properties
Type of Flow: Thixotropic
Antisettling
Hold Out
Brushability

ANTI-SAG S:
Physical Form: Liquid
Solvent: Aliphatic
Rheological Properties:
Type of Flow: Thixotropic
Antisettling
Hold Out
Brushability

TROY CHEMICAL CORP.: Rheology Modifiers(Continued):

TROYTHIX Products:

21BA:
- Physical Form: Liquid
- Solvent Used: Aliphatic
- Rheological Properties:
  - Type of Flow: Thixotropic
  - Apparent Viscosity Increase
  - Antisettling
  - Hold Out
  - Brushability

42BA:
- Physical Form: Liquid
- Solvent System: Aliphatic Aromatic Oxygenated
- Rheological Properties:
  - Type Of Flow: Newtonian
  - Apparent Viscosity Increase

Water Base Products:

LLBA:
- Physical Form: Liquid
- Solvent System: Water
- Rheological Properties:
  - Type of Flow: Nonionic
  - Apparent Viscosity Increase
  - Antisettling
  - Hold Out
  - Brushability

R.T. VANDERBILT CO., INC.: Paint/Paper Products:

VAN GEL
Thixotrope for Aqueous Systems

VAN GEL is a magnesium aluminum silicate that is specially formulated for easy incorporation and consistent performance in aqueous systems. Aqueous dispersions of VAN GEL are highly thixotropic gels at low solids and are resistant to bacterial and enzymatic degradation.

The use of VAN GEL in a coating formula produces a product with no separation and with uniform thixotropic consistency that requires no stirring while promoting good flow and brushing characteristics. The result is "dripless" application and leveling with no sag.

Composition: Magnesium aluminum silicate
Physical form: Flakes
Color: Off-white
Odor: None
Density: 2.6+-0.05 Mg/m3
Moisture: 12% Maximum
Viscosity: 800+-400 cps.
Hegman Fineness of 4% dispersion: 7 Minimum
pH of 4% dispersion: 9.5+-0.5

Section 11
Silicone Additives

DOW CORNING CORP.: Silicone Additives:

Spectrum One:

Product:

DOW CORNING 3 Additive:
Typical Benefits: To prevent pigment separation and flotation, improve leveling, flow-out and gloss.
Functionality: Silanol
Primary Polymer Systems: Most solvent-based polymers including alkyd, acrylic, epoxy, urethane, vinyl, polyester and silicone.

DOW CORNING 7 Additive:
Typical Benefits: A defoaming agent for solvent-based resins, paints and inks.
Functionality: None
Primary Polymer Systems: Same as DOW CORNING 3 additive.

DOW CORNING 11 Additive:
Typical Benefits: To increase the mar resistance of a wide variety of solvent-based films by reducing surface friction. Improves gloss, flow-out and wetting. Also used to prevent pigment separation and flotation.
Functionality: Carbonol
Primary Polymer Systems: Same as DOW CORNING 3 additive.

DOW CORNING 54 Additive:
Typical Benefits: Designed for use in solvent-based and high-solids systems. May work in some water-borne and powder coating systems. In addition, may improve abrasion resistance, water repellency and flow-out in urethane systems.
Functionality: Carbonol
Primary Polymer Systems: Same as DOW CORNING 3 additive plus water-borne systems.
Comments: DOW CORNING 54 additive, DOW CORNING 56 additive and DOW CORNING 57 additive are especially good for use with higher solids and solventless coatings because they contain no additional solvent. Care should be taken that they are added in low enough concentrations to prevent contamination.
DOW CORNING 54 additive and DOW CORNING 57 additive may be effective in powder coatings to improve wetting, leveling and mar resistance and may also decrease pigment separation.

DOW CORNING 56 Additive:
Typical Benefits: Recommended for improving leveling, flow-out, gloss and pigment dispersion in solvent-based systems. May also work as a defoamer for some systems.
Functionality: None
Primary Polymer Systems: Same as DOW CORNING 3 additive.
Comments: See comments on DOW CORNING 54 additive. Typically used to improve the appearance of solvent-based paints applied by curtain coating or spraying. May also be used to stabilize and orient aluminum flake in metallic finishes.

DOW CORNING CORP.: Silicone Additives(Continued):

Spectrum One(Continued):

DOW CORNING Z-6070 Silane:

Typical Benefits: Treatment of inorganic pigments such as talc, clay and silica to render the surfaces hydrophobic. Aids in reduction of grind times by improving pigment dispersion. May improve adhesion of polymers to substrate when used as a primer.

Functionality: Methoxy silane

Primary Polymer Systems: Improves adhesion when used as a primer with acrylics, polyesters and urethanes.

Comments: Reacts with water to form methanol. Designated as a methyltrimethoxysilane.

Spectrum Two:

DOW CORNING 14 Additive:

Typical Benefits: For water-based coatings to improve mar resistance, leveling, flow-out and gloss. Can be used to eliminate pigment floating. Can also be used in solvent systems.

Functionality: Carbonol

Primary Polymer Systems: Polyesters, alkyds and acrylic emulsions.

DOW CORNING 29 Additive:

Typical Benefits: Used primarily for leveling and flow-out. Imparts mar resistance and antiblocking. Prevents pigment floating.

Functionality: Carbonol

Primary Polymer Systems: Acrylics, epoxies, polyesters, and urethanes.

Comments: Can be used as an additive to varnish or lacquer overprint varnishes to improve adhesion, mar resistance, leveling and gloss. Prevents blocking.

DOW CORNING 51 Additive:

Typical Benefits: Imparts mar resistance and slip to water-based systems. May also be effective with certain solvent-based systems--in particular, when an alcohol or polar solvent is present.

Functionality: None

Primary Polymer Systems: Water-based polymer systems such as acrylic emulsions, and some alcohol-containing solvent systems.

Comments: A water dilutable, ultra-high molecular weight silicone dispersion. It can be added to water-based inks and latex paints to impart mar resistance and slip.

DOW CORNING CORP.: Silicone Additives(Continued):

Spectrum Two:

DOW CORNING 57 Additive:
Typical Benefits: To improve flow-out, mar resistance and gloss in water-based, solvent-based, high-solids, solventless and powder coatings.
Functionality: Carbonol
Primary Polymer Systems: Same as DOW CORNING 3 additive.
Comments: See comments on DOW CORNING 54 additive.

DOW CORNING 59 Additive:
Typical Benefits: A defoaming agent for water-based systems.
Functionality: None
Primary Polymer Systems: Acrylics and other water-based emulsions.
Comments: Especially effective in acrylic emulsion paints and inks.
Can be used in the production of water-based vehicles.

DOW CORNING 64 Additive:
Typical Benefits: Imparts water resistance to cured films. Can provide improved mar resistance and reduced pigment floating.
Functionality: Amine
Primary Polymer Systems: Urethanes, acrylics and polyesters.

Spectrum Three:

DOW CORNING 21 Additive:
Typical Benefits: To improve adhesion of a variety of polymer systems to glass and metal substrates. Acts as a room temperature curing agent for silicone resin-based coatings. Improves adhesion and water resistance of epoxy adhesives.
Functionality: Amine-methoxy
Primary Polymer Systems: As an additive: acrylic, alkyd, polyester, urethane and epoxy.
As a primer: Epoxy
Comments: Particularly effective with epoxy systems when used as a primer or additive. With epoxy systems, blend and package with epoxy curing agents only. Package stability should be checked when used as an additive. Designated as an aminomethoxysilane.

DOW CORNING 25 Additive:
Typical Benefits: Particularly effective in promoting adhesion of a wide variety of polymer systems to glass, aluminum and steel when used either as an additive or a primer.
Functionality: Epoxy-methoxy
Primary Polymer Systems: Acrylic, alkyd, polyester, urethane and epoxy.
Comments: Of the current adhesion promoters, DOW CORNING 25 additive has generally been found to be most effective. An epoxy-methoxy compound.

DOW CORNING CORP.: Silicone Additives(Continued):

Spectrum Three(Continued):

DOW CORNING Z-6030 Silane:

Typical Benefits: Used as an additive, promotes adhesion of a variety of polymer systems to aluminum.
Functionality: Methacrylate-methoxy
Primary Polymer Systems: Acrylic, alkyd, polyester, urethane and epoxy.
Comments: Has not been found to be effective as a primer. Designated as a methacryloxypropyltrimethoxysilane.

DOW CORNING Z-6032 Silane:

Typical Benefits: Used as an additive, promotes adhesion of a variety of polymer systems to aluminum and glass.
Functionality: Vinylbenzyl-amine-methoxy
Primary Polymer Systems: Acrylic, alkyd, polyester, urethane, epoxy and PVAC.
Comments: Has not been found to be effective as a primer. Designated as an N-Z-(vinylbenzylamino)-ethyl-3-aminopizopyltrimethoxysilane monohydrogen chloride.

DOW CORNING Z-6040 Silane:

Typical Benefits: Particularly effective in promoting adhesion of a wide variety of polymer systems to glass, aluminum and steel when used either as an additive or a primer. Improves water resistance of some urethane adhesives.
Functionality: Epoxy-methoxy
Primary Polymer Systems: Acrylic, alkyd, polyester, urethane and epoxy.
Comments: Over-all, an excellent adhesion promoter. Designated as a glycidoxypropltrimethoxysilane.

Other Available Additives:

DOW CORNING 190 Surfactant:

Typical Benefits: Improves mar resistance, leveling, flowout and gloss in all system types. Aids in reduction of pigment floating.
Functionality: Carbonol
Primary Polymer Systems: Acrylic emulsions, alkyds and polyesters.

DOW CORNING CORP.: Silicone Additives(Continued):

Other Available Additives:

200 Fluid 10-100 cSt:

Typical Benefits: A versatile additive for improving film defects such as pigment separation and wetting. Increases mar resistance and slip. For use in solvent solventless systems.
Functionality: None
Primary Polymer Systems: Same as DOW CORNING 3 additive.
Comments: Recommended lower viscosity products to improve compatibility and wetting.

DOW CORNING 290 Paintable Release:

Typical Benefits: Designed to improve recoatability of aqueous coatings. May improve mar resistance, flow-out and antiblocking characteristics in aqueous systems.
Functionality: None
Primary Polymer Systems: Acrylic, alkyd

DOW CORNING HV-490 Emulsion:

Typical Benefits: Imparts slip and release properties to water-based films.
Functionality: None
Primary Polymer Systems: Water-based polymer systems including acrylics, alkyds and urethanes.
Comments: Recommend lower viscosity products to improve compatibility and wetting.

DOW CORNING 3-5024 Silicone Water-Based Emulsion:

Typical Benefits: Particularly effective in increasing the water and mar resistance of water-based systems.
Functionality: None
Primary Polymer Systems: Water-based polymer systems including acrylics, alkyds and urethanes.
Comments: 1-component, anionically stabilized, water-based silicone elastomer. Low viscosity, unfilled, opaque emulsion.

DOW CORNING 3-5025 Silicone Water-Based Emulsion:

Typical Benefits: Same benefits as DOW CORNING 3-5024 silicone water-based emulsion.
Functionality: None
Primary Polymer Systems: Water-based polymer systems including acrylics, alkyds and urethanes.
Comments: 1-component, anionically stabilized, water-based silicone elastomer. Supplied as a medium viscosity, filled, opaque emulsion.

DOW CORNING CORP.: Silicone Additives-Typical Properties:

<u>Product:</u>

Spectrum One:

DOW CORNING 3 Additive:
Color: Clear
Percent Solids: 10
Solvent: Toluene
Viscosity at 77F (25C): 1.0 cSt

DOW CORNING 7 Additive:
Color: Clear
Percent Solids: 5
Solvent: Methyl-isobutyl ketone
Viscosity at 77F (25C): 0.8 cSt

DOW CORNING 11 Additive:
Color: Clear
Percent Solids: 10
Solvent: Toluene
Viscosity at 77F (25C): 1.5 cSt

DOW CORNING 54 Additive:
Color: Clear
Percent Solids: 100
Solvent: None
Viscosity at 77F (25C): 170 cSt

DOW CORNING 56 Additive:
Color: Clear
Percent Solids: 100
Solvent: None
Viscosity at 77F (25C): 1400 cSt

DOW CORNING Z-6070 Silane:
Color: Clear
Percent Solids: 98
Solvent: None
Viscosity at 77F (25C): 2 cSt

Spectrum Two:

DOW CORNING 14 Additive:
Color: Clear
Percent Solids: 10
Solvent: Isopropanol
Viscosity at 77F (25C): 4.0 cSt

DOW CORNING CORP.: Silicone Additives-Typical Properties (Continued):

Spectrum Two(Continued):

DOW CORNING 29 Additive:
Color: Light amber
Percent Solids: 100
Solvent: None
Viscosity at 77F (25C): 300 cSt at 40C

DOW CORNING 51 Additive:
Color: Hazy
Percent Solids: 80
Solvent: Water
Viscosity at 77F (25C): 200-500 cSt

DOW CORNING 57 Additive:
Color: Light amber
Percent Solids: 100
Solvent: None
Viscosity at 77F (25C): 300 cSt

DOW CORNING 59 Additive:
Color: White
Percent Solids: 10
Solvent: Water
Viscosity at 77F (25C): 2000 cSt

DOW CORNING 64 Additive:
Color: Clear
Percent Solids: 50
Solvent: Aliphatic solvents and isopropyl alcohol
Viscosity at 77F (25C): 150 cSt

Spectrum Three:

DOW CORNING 21 Additive:
Color: Light amber
Percent Solids: 50
Solvent: n-butanol
Viscosity at 77F (25C): 6.0 cSt

DOW CORNING 25 Additive:
Color: Light amber
Percent Solids: 100
Solvent: None
Viscosity at 77F (25C): 850 cSt

DOW CORNING Z-6030 Silane:
Color: Clear
Percent Solids: 100
Solvent: None
Viscosity at 77F (25C): 2 cSt

DOW CORNING CORP.: Silicone Additives-Typical Properties (Continued):

Spectrum Three(Continued):

DOW CORNING Z-6032 Silane:
Color: Golden brown
Percent Solids: 40
Solvent: Methanol
Viscosity at 77F (25C): 2 cSt

DOW CORNING Z-6040 Silane:
Color: Clear
Percent Solids: 100
Solvent: None
Viscosity at 77F (25C): 3 cSt

Other Available Additives:

DOW CORNING 190 Surfactant:
Color: Light amber
Percent Solids: 100
Solvent: None
Viscosity at 77F (25C): 1500 cSt

200 Fluid 10-100 cSt:
Color: Clear
Percent Solids: 100
Solvent: None
Viscosity at 77F(25C): 10 to 100 cSt

DOW CORNING 290 Paintable Release:
Color: White
Percent Solids: 50
Solvent: Water
Viscosity at 77F (25C): 80 cSt

DOW CORNING HV-490 Emulsion:
Color: White
Percent Solids: 36
Solvent: Water
Viscosity at 77F (25C): 100 cSt

DOW CORNING 3-5024 Silicone Water-Based Emulsion:
Color: White
Percent Solids: 40
Solvent: Water
Viscosity at 77F (25C): 550 cSt

DOW CORNING 3-5025 Silicone Water-Based Emulsion:
Color: White
Percent Solids: 40
Solvent: Water
Viscosity at 77F (25C): 20,000-70,000 St

GENERAL ELECTRIC CO.: Silicone Additives:

Product:

SF 69:
Silicone Content: 100%
Internal Solvent: None
Reducing Solvent: Aliphatic & aromatic hydrocarbons
Viscosity @ 25C (77F) cps: 4-40
Surface Tension @ 25C (77F): Approx dynes/cm: 20.0
Specific Gravity @ 25C (77F): 0.96

SF 96 (350 M):
Silicone Content: 100%
Internal Solvent: None
Reducing Solvent: Aliphatic & aromatic hydrocarbons
Viscosity @ 25C (77F) cps: 350
Surface Tension @ 25C (77F): Approx dynes/cm: 21.1
Specific Gravity @ 25C (77F): 0.97

VISCASIL (60M):
Silicone Content: 100%
Internal Solvent: None
Reducing Solvent: Aliphatic & aromatic hydrocarbons
Viscosity @ 25C (77F) cps: 60,000
Surface Tension @ 25C (77F): Approx. dynes/cm: 21.3
Specific Gravity @ 25C (77F): 0.98

SF 1023:
Silicone Content: 100%
Internal Solvent: None
Reducing Solvent: Aliphatic & aromatic hydrocarbons
Viscosity @ 25C (77F) cps: 45-65
Surface Tension @ 25C (77F): Approx. dynes/cm: 24.0
Specific Gravity @ 25C (77F): 1.03

SF 1188:
Silicone Content: 100%
Internal Solvent: None
Reducing Solvent: Aliphatic & aromatic hydrocarbons & H2O
Viscosity @ 25C (77F) cps: 800-1200
Surface Tension @ 25C (77F): 25.5
Specific Gravity @ 25C (77F): 1.04

SR 107:
Silicone Content: 60%
Internal Solvent: AROMATIC 150
Reducing Solvent: Petroleum Spirits
Viscosity @ 25C (77F) cps: 8-40
Surface Tension @ 25C (77F): 24.6
Specific Gravity @ 25C (77F): 1.04

GENERAL ELECTRIC CO.: Silicone Additives(Continued):

<u>Product:</u>

SR 882:
Silicone Content: 80%
Internal Solvent: Toluene
Reducing Solvent: Aromatic hydrocarbons & blends with aliphatic solvents
Viscosity @ 25C (77F): 200-1000
Surface Tension @ 25C (77F): Approx. dynes/cm: 200-1000
Specific Gravity @ 25C (77F): 1.14

AF 60:
Silicone Content: 30%
Internal Solvent: Water
Reducing Solvent: Water
Viscosity @ 25C (77F): 750
Surface Tension @ 25C (77F): Approx. dynes/cm.: 32.2
Specific Gravity @ 25C (77F): 1.01

AF 72:
Silicone Content: 30%
Internal Solvent: Water
Reducing Solvent: Water
Viscosity @ 25C (77F): 1,000
Surface Tension @ 25C (77F): Approx. dynes/cm.: 1,000
Specific Gravity @ 25C (77F): 29.1

AF 66:
Silicone Content: 100%
Internal Solvent: None
Reducing Solvent: Aliphatic & aromatic hydrocarbons
Viscosity @ 25C (77F): 500
Surface Tension @ 25C (77F): Approx. dynes/cm.: 500
Specific Gravity @ 25C (77F): 21.1

AF 70:
Silicone Content: 100%
Internal Solvent: None
Reducing Solvent: Aliphatic & aromatic hydrocarbons
Viscosity @ 25C (77F): 1,500
Surface Tension @ 25C (77F): Approx. dynes/cm.: 21.1
Specific Gravity @ 25C (77F): 1.01

AF 75:
Silicone Content: 10%
Internal Solvent: Water
Reducing Solvent: Water
Viscosity @ 25C (77F): Approx. dynes/cm.: 2,500
Surface Tension @ 25C (77F): Approx. dynes/cm.: 29.1
Specific Gravity @ 25C (77F): 1.02

GENERAL ELECTRIC CO.: Silicone Additives(Continued):

Product:

AF 9020:
- Silicone Content: 20%
- Internal Solvent: Water
- Reducing Solvent: Water
- Viscosity @ 25C (77F) cps: 2,500
- Surface Tension @ 25C (77F): Approx. dynes/cm: 31.0
- Specific Gravity @ 25C (77F): 1.01

Application:
- Flow Control
- Pigment Dispersion
- Improved Gloss
- Improved Slip
- Improved Washability
- Reduced "Orange Peel"
- "Wetter" Spray Coating
- Bubble Control
- Water Repellency
- Water Spot Resistance
- Foam Reduction
- Decreased Pigment Float
- Improved Brushing
- Weather Durability
- Silking Control
- Mar Resistance

GENESEE POLYMERS CORP.: Epoxy Functional Silicone Solutions:

EXP-36-X20 Epoxy Functional Silicone Solution:

A 20% active epoxy functional silicone fluid in xylene.

Typical Properties:
- Viscosity: 5 Cstks.
- Wt./Gallon: 7.5 lbs.
- Specific Gravity: 0.90
- % Active: 20%
- Equivalent Weight per Epoxy (EEW) (Solution): 2847 gms.
- Appearance: Thin, Clear Liquid
- Solvent: Xylene
- Flash Point: 79F

Special Features:
- Forms Highly Adhesive, Flexible Coatings
- Extremely Durable
- Rust Inhibition
- Clear, Cured Film
- Abrasion Resistant
- Excellent Bonding
- Release Properties
- Excellent Weathering
- Chemical Resistant
- Curable at Ambient or Elevated Temperatures

EXP-38-X20 Epoxy Functional Silicone Solution:

A 20% Active Epoxy Functional Silicone Fluid in Xylene.

Typical Properties:
- Viscosity: 162 Cstks.
- Wt./Gallon: 7.5 lbs.
- Specific Gravity: 0.91
- % Active: 20%
- Equivalent Weight per Epoxy (EEW) (Solution): 5180 gms.
- Appearance: Light straw, Hazy liquid
- Solvent: Xylene
- Flash Point (TCC): 79F

Special Features:
- Abrasion Resistant
- Durable
- Rust Inhibition
- Clear
- Non-cured film removable with solvent
- Flexibility
- Excellent Bonding
- Release Properties
- Excellent Weathering

GENESEE POLYMERS CORP.: Silicone Emulsion/Wax Emulsion:

GP-4-E Silicone Emulsion:

A 50% silicone solids emulsion of a film forming polyamino-functional silicone polymer.

Typical Properties:
Appearance: White, opaque liquid
Odor: Mild
Carrier: Water
Organic Solvents: None
Type of Emulsifiers: Non-ionic
pH: 10.0
Silicone Solids: 50%
Specific Gravity: 0.96
Wt./Gallon: 8.0 lbs.
Silicone Functionality: Amine-alkyl
Special Features:
Stable emulsion
Dilutable with hard or soft water
No washoff of silicone polymer
High amine functionality
Excellent detergent resistance
Forms a cured film

GP-256 Wax Emulsion:

A 37% solids emulsion of oxidized polyethylene wax in water carrier.

Typical Properties:
Appearance: Light tan, opaque liquid
Active Ingredient: 30%
Carrier: Water
Specific Gravity: 0.986
Wt./Gallon: 8.2 lbs.
Emulsifier Type (Primary): Non-ionic, Tridecylalcohol ethoxylate
Polymer Properties-Hardness: 5.5 mm
Melt Point: 104C
pH: 8.5
Special Features:
Extremely fine particle size emulsion
Dilutable with hard or soft water
Excellent long term emulsion stability
Imparts slip resistance in floor polish films

GENESEE POLYMERS CORP.: Silicone Fluids:

GP-4 Silicone Fluid:

An amine functional silicone polymer.

Typical Properties:
- Appearance: Light straw, clear liquid
- Viscosity: 65 cstks.
- Flash Point: >200F (Closed Cup)
- Wt./Gallon: 8.0 lbs.
- Specific Gravity: 0.97
- % Active: 100%
- Amine No.: 75

Special Features:
- High amine-alkyl content
- Stable indefinitely in closed containers
- Good heat stability

GP-5 Emulsifiable Paintable Silicone:

Water-less concentrate containing a methylalkylarylpolysiloxane fluid and non-ionic emulsifiers.

Typical Properties:
- Silicone Content: 80%
- Non-ionic Emulsifiers: 14%
- Aliphatic Hydrocarbon (Naphtha): 6%
- Wt./Gallon: 8.0 lbs.
- Specific Gravity: 0.96
- Appearance: Clear to hazy liquid
- Odor: Mild

Special Features:
- Indefinite Stability
- Ease of Emulsification
- Lower Shipping Costs
- Will not freeze

EXP-28 Silicone Fluid(Paint Additive):

A 100% active silicone product.

Typical Properties:
- Silicone Content: 100%
- Viscosity @ 25C: 4-30 Cstks.
- Surface Tension @ 25C: 20 Dynes/cm
- Specific Gravity: 0.96
- Lbs./Gravity: 8.0 lbs.
- Flash Point: 170F (closed cup)
- Appearance: Clear, water white liquid

Special Features:
- High purity
- Dilutable in aliphatic hydrocarbons
- Stable solution
- Effective at low levels
- High surface activity

GENESEE POLYMERS CORP.: Silicone Fluids(Continued):

GP-70 Paintable Silicone Fluid:

A 100% active methyl/alkylaryl silicone.

Typical Properties:
Appearance: Light straw clear liquid
Viscosity: 500 cstks.
Flash Point: >200F (Closed Cup)
Wt./Gallon: 7.3 lbs.
Specific Gravity: 0.87
% Active: 100%
Special Features:
Ease of Cleaning
Lubricity
Organo-compatibility

GP-70-S Paintable Silicone Fluid:

A 100% active methyl/alkylaryl silicone.

Typical Properties:
Appearance: Light straw clear liquid
Viscosity: 1200 cstks
Flash Point: >300F (Closed Cup)
Wt./Gallon: 7.3 lbs.
Specific Gravity: 0.87
% Active: 100%
Special Features:
Ease of Cleaning
Lubricity
Organo-compatibility

GP-71 Mercapto Modified Silicone Fluid:

A dimethyl silicone copolymer fluid.

Typical Properties:
Appearance: Water-white, light straw liquid
Odor: Sulfur Odor
Specific Gravity: 0.97
Flash Point: >450F (PM-CC)
Wt./Gallon: 8.0 lbs.
% Active: 100%
% Solvents: None
Viscosity 77F: 150 cstks
Special Features:
Reactive mercapto functionality
High affinity for metal surfaces
Corrosion inhibiting properties
Better release agent than conventional silicone fluids in many applications
Greater high temperature and oxidative stability than conventional silicones.

GENESEE POLYMERS CORP.: Silicone Fluids (Continued):

GP-74 Paintable Silicone Fluid:

100% active methyl/alkylaryl silicone.

Typical Properties:
- Appearance: Light Straw Clear Liquid
- Viscosity: 800 cstks.
- Flash Point: >200F (closed cup)
- Wt./Gallon: 8.0 pounds
- Specific Gravity: 0.97
- % Active: 100%

Special Features:
- Good wetting properties
- Good lubrication
- Ease of cleaning
- Good heat stability
- Chemically inert

GP-7200 Silicone Fluid:

A mercapto functional methyl alkyl ('paintable') silicone polymer.

Typical Properties:
- Appearance: Yellow, clear liquid
- Odor: Sulfur odor
- Specific Gravity: 0.97
- Wt./Gallon: 8.0 lbs.
- % Active: 100%
- Flash Point: >300F (PMCC)
- Viscosity, 25C: Moderately viscous

Special Features:
- Ease of Cleaning
- Good Lubrication
- Good High Temperature Stability
- Reactive with Sulfur & Peroxide Cured Elastomers

GENESEE POLYMERS CORP.: Silicones:

GP-RA-158 Silicone Polish Additive:

A high gloss, detergent resistant polish additive.

Typical Properties:
Appearance: Clear, moderately viscous liquid
Odor: Solvent
Active Ingredient: 50%
Solvent: Mineral Spirits
Viscosity: 2000 cstks. (approx.)
Specific Gravity: 0.86
Flash Point: 99F (closed cup)
Wt./Gallon: 7 lbs.
Special Features:
Stable solvent solution
Contains amino and alkoxy functionality
Excellent detergent resistance

GP-RA-159 Silicone Polish Additive:

A high gloss, detergent resistant polish additive in an alcohol/mineral spirits solvent blend.

Typical Properties:
Appearance: Clear, water-white liquid
Odor: Isopropanol
Active Ingredient: 50% (Silicone)
Viscosity (typical): 200 cstks.
Specific Gravity: 0.86
Flash Point (P-M CC): 52F
Solvent Composition: Mineral Spirits and Isopropanol
Wt./Gallon: 7 lbs.
Special Features:
Stable silicone/solvent solution
Contains amino and alkoxy functionality
Fast drying solvent blend
Excellent detergent resistance

GENESEE POLYMERS CORP.: Silicones(Continued):

GP-86-AE Silicone Emulsion:

A 50% silicone solids emulsion of a film forming amino-functional silicone polymer.

Typical Properties:
- Appearance: White, opaque fluid
- Odor: Mild
- Carrier: Water
- Organic Solvents: None
- Type Emulsifiers: Non-ionic
- pH: 9.5
- Silicone Solids: 50%
- Specific Gravity: 0.96
- Wt./Gallon: 8.0 lbs.

Special Features:
- Stable emulsion
- Dilutable with hard or soft water
- No washoff of silicone polymer
- High amino-functionality
- Excellent detergent resistance
- High silicone content (50%)

Section 12
Titanium Dioxides

DUPONT CO.: TI-PURE Titanium Dioxide:

DUPONT white pigments for Coatings

Rutile Titanium Dioxide:

<u>Grade:</u>

R-100:
Sp. Gr.: 4.2
% Min. TiO2: 97
Meets ASTM D-476 Spec: Type II
Principal Applications: Interior gloss finishes--trade sales and industrial

R-900:
Sp. Gr.: 4.0
% Min. TiO2: 94
Meets ASTM D-476 Spec: Type II
Principal Applications: General purpose--especially high gloss finishes

R-901:
Sp. Gr.: 3.8
% Min. TiO2: 85
Meets ASTM D-476 Spec: Type III
Principal Applications: General purpose trade sales--especially emulsion flats

R-902:
Sp. Gr.: 4.0
% Min. TiO2: 91
Meets ASTM D-476 Spec: Type IV
Principal Applications: General purpose for interior and exterior finishes

R-931:
Sp. Gr.: 3.6
% Min. TiO2: 80
Meets ASTM D-476 Spec: Type III
Principal Applications: High PVC interior and exterior flat emulsion and alkyd paints

R-960:
Sp. Gr.: 3.9
% Min. TiO2: 89
Meets ASTM D-476 Spec: Type IV
Principal Applications: Exterior finishes requiring outstanding gloss, tint retention and chalk resistance.

DUPONT CO.: TI-PURE Titanium Dioxide(Continued):

DuPont white pigments for Coatings(Continued)

Rutile-Slurries:

Grade:

R-940:
Slurry Density: Lb/Gal: 19.5
% Solids: 76.5
Brookfield Viscosity cp: <1000
Principal Applications: Aqueous coatings--gloss and semi-gloss. Slurry counterpart of R-900.

R-941:
Slurry Density: Lb/Gal: 15.8
% Solids: 64.5
Brookfield Viscosity cp: <500
Principal Applications: Aqueous coatings--high PVC flats. Slurry counterpart of R-931

R-942:
Slurry Density: Lb/Gal: 19.4
% Solids: 76.5
Brookfield Viscosity cp: <1000
Principal Applications: Exterior aqueous coatings. Slurry counterpart of R-902.

DUPONT CO.: TI-PURE Titanium Dioxide(Continued):

Product Description:

TI-PURE R-100 is a very low binder demand, easy dispersing pigment of interest for interior industrial and trade sales applications.

TI-PURE R-900 (R-940 Slurry) offers particular advantage in high gloss and semi-gloss finishes because of its high hiding power at low pigment volume concentrations plus excellent brightness, ease of dispersion, gloss and cleanness of tint. Its chalk resistance is satisfactory for many exterior trade sales paints. Accordingly, it can be considered as general purpose pigment.

TI-PURE R-901 has outstanding hiding efficiency in all types of flat paints and finds wide acceptance in emulsion paints.

TI-PURE R-902 (R-942 Slurry) has an excellent balance of chalk resistance and optical values in a broad range of exterior finishes. It is recommended for its high quality performance in automotive refinish and exterior trade sales paint. High level of chalk resistance in addition to its optical properties make it a multipurpose pigment for both interior and exterior coatings.

TI-PURE R-931 (R-941 Slurry) provide a good balance between hiding power and film integrity in high PVC flat pigments. They are most efficient in paints containing 2.0 lb. or more TiO_2 per gallon with the PVC ranging between 45% and 60%.

TI-PURE R-960 provides outstanding chalk resistance, tint retention and gloss retention along with excellent low-shear dispersion, color acceptance (resists flocculation) and hiding power. Because of its excellent exterior durability, it is the pigment of choice for automotive OEM topcoats and coil coatings for residential and industrial siding.

TI-PURE Rutile Slurries retain the performance properties of their respective dry TI-PURE counterparts. However, they offer significant advantages in ease of handling, reduced waste, lower raw materials and dispersion costs.

KERR-MCGEE CHEMICAL CORP.: TRONOX Titanium Dioxides:

TRONOX Chloride Process Titanium Dioxide CR-800:

TRONOX CR-800 is a low oil absorption treated rutile pigment produced by the chloride process. It has high gloss producing properties for interior trade sales paints and industrial coatings. It may be used in exterior applications where the maximum chalk resistance of CR-822 is not required. It exhibits high brightness and excellent tint strength in gloss and semi-gloss enamels.

CR-800 is also recommended as high gloss, low abrasivity pigment for letter press, gravure, polyamid, and other inks.

Typical Properties:

TiO2 content, minimum	95%
Density	
Specific gravity	4.1
As pounds per gallon	34.4 lb/gal
One pound bulks-gallons	0.0291 gal/lb
Oil absorption	17 lb oil/100 lb pigment
Water demand	18 lb water/100 lb pigment
pH	7.0
Specific resistance	11,000 ohms
Average particle diameter	0.23 um
Chalk resistance	medium
Specification	ASTM D 476-73, type II, III

TRONOX Chloride Process Titanium Dioxide CR-800PG:

TRONOX CR-800PG is of great advantage when used in those plastics applications requiring a blue white, excellent dispersing TiO2. In addition to its high tinting strength it also exhibits comparatively low abrasivity.

CR-800PG has found wide acceptance in powder coatings, polyethylene film and vinyl sheet goods. It is also used in most other common plastics applications. It meets the Plastic Pipe Institute's guide-lines for formulation substitution and is authorized by the National Sanitation Foundation for general interchangeability in NSF-listed formulations.

Typical Properties:

TiO2, content, minimum	95%
Solid density	4.1 g/ml
Bulk density	40-55 lb/cu ft
Oil absorption	17 lb oil/100 lb pigment
Average particle size	0.19 um
pH	6.0-8.0
Specific resistance	7600 ohms
Grit on 325 mesh, maximum	0.004%

KERR-MCGEE CHEMICAL CORP.: TRONOX Titanium Dioxides(Continued):

TRONOX Chloride Process Titanium Dioxide CR-800S:

TRONOX CR-800S is the slurry counterpart of TRONOX CR-800. The high gloss producing and strength properties of CR-800 have been preserved in its slurry form.

TRONOX CR-800S is recommended for interior gloss and semi-gloss water-borne coatings, as well as exterior water-borne coatings where the maximum durability of CR-822 is not required.

TRONOX CR-800S may also be used in flat coatings when it is supplemented with functional extenders to obtain the required dry hiding properties.

Specifications:

Solids	76.5+-1%
Volatile	water
Density	19.3-19.9 lb/gal
Viscosity @ 100 RPM	300-500 cp
pH	8.5-9.5

Typical Viscosity Profile:

RPM	Viscosity, cp
100	400
50	650
20	1300
10	2300

TRONOX Chloride Process Titanium Dioxide CR-801:

TRONOX CR-801 is a medium oil absorption treated rutile pigment produced by the chloride process and is designed as a general purpose trade sales pigment. It is characterized by high brightness and high strength at all pigments volume concentrations such as in gloss enamels, semi-gloss enamels, satin and flat latex paints, and flat solvent thinned paints. It may be used in exterior coatings where the chalk resistance of CR-821 or CR-822 is not required.

Typical Properties:

TiO2 content, minimum	93%
Density	
Specific gravity	4.0
Scattering coefficient per pound TiO2	
In enamels	4.2
In flat latex	3.7
Oil absorption	26 lb oil/100 lb pigment
Water demand	30 lb water/100 lb pigment
pH	7.5
Specific resistance	6000 ohms
Average particle diameter	0.24 um
Chalk resistance	medium
Specification	ASTM D476-73 Type II, III

KERR-MCGEE CHEMICAL CORP.: TRONOX Titanium Dioxides(Continued):

TRONOX Chloride Process Titanium Dioxide CR-813:

TRONOX CR-813 is a high oil absorption treated rutile pigment which has been designed specifically for interior flat architectural paints, both latex and solvent based, where it exhibits outstanding brightness, highest strength and an exceptionally clean tint tone. A change to CR-813 from lower dry hiding pigments in such finishes permits formulators to achieve equal opacity at lower cost, with no loss of film integrity.

TRONOX CR-813 is also recommended as the highest hiding pigment for low gloss water based flexographic inks.

Typical Properties:

TiO2 content, minimum	87%
Density	
Specific Gravity	3.85
Scattering coefficient per pound TiO2	
In enamels	3.9
In flat latex	4.4
Oil absorption	36.5 lb oil/100 lb pigment
Water demand	44 lb water/100 lb pigment
pH	7.5
Specific resistance	6000 ohms
Average particle diameter	0.25 um
Chalk resistance	medium
Specification	ASTM 476-73 Type III

TRONOX Chloride Process Titanium Dioxide CR-813S:

TRONOX CR-813S is the slurry counterpart of TRONOX CR-813. The excellent high dry hiding properties of CR-813 have been preserved in its slurry form.

TRONOX CR-813S is recommended for interior flat water-borne coatings, such as flat latex wall paints and ceiling tile and paper coatings.

Specifications:

Solids	64.5+-1%
Volatile	water
Density	15.6-16.2 lb/gal
Viscosity @ 100 RPM	125-250 cp
pH	8.5-9.5

Typical Viscosity Profile

RPM	Viscosity, cp
100	170
50	270
20	490
10	800

KERR-MCGEE CHEMICAL CORP.: TRONOX Titanium Dioxides(Continued):

TRONOX Chloride Process Titanium Dioxide CR-821:

TRONOX CR-821 is a treated rutile pigment produced by the chloride process. It has been designed to be a universal high gloss pigment for both interior and exterior film application. It has a high initial gloss, ease of dispersion, opacifying power and good exterior weathering properties.

CR-821 is recommended for all classes of coatings: architectural, industrial and maintenance; solvent thinned or water thinned; interior or exterior.

Typical Properties:

TiO_2 content, minimum	92%
Density	
Specific gravity	4.0
Scattering coefficient per pound TiO_2	
In enamels	4.1
In flat latex	3.0
Oil absorption	18.6 lb oil/100 lb pigment
Water demand	21.0 lb water/100 lb pigment
pH	7.8
Specific resistance	6,500 ohms
Average particle diameter	0.22 um
Chalk resistance	high
Specification	ASTM D 476-84 Type III,IV

TRONOX CR-821S

TRONOX CR-821S titanium dioxide pigment slurry is a water dispersion of TRONOX CR-821. It is recommended for all classes of interior or exterior water thinned coatings, inks, and adhesives.

TRONOX CR-821's major application will be in architectural latex paints. Its high gloss and strength properties make it especially suitable for gloss and semigloss finishes. With selected functional extenders it gives high hiding in eggshell and flat finishes also. Because of its good weather resistance properties it is recommended for exterior house paints of all types as well.

Typical Properties:

Solids, %	76.5+-1
Volatile	water
Density, lbs./gals.	19.2-19.9
Viscosity @ 60 RPM, cp	200-300
pH	8.5-9.5

KERR-MCGEE CHEMICAL CORP.: TRONOX Titanium Dioxides(Continued):

TRONOX Chloride Process Titanium Dioxide CR-822:

TRONOX CR-822 is a treated rutile pigment produced by the chloride process. It has been designed to have the highest degree of gloss and tint retention properties in exterior film applications. It is characterized by an optimum balance of high initial gloss, ease of dispersion, opacifying power and exterior weathering properties.

CR-822 is recommended for all interior and exterior enamels and lacquers in architectural, industrial and maintenance finishes. It is also recommended for exterior plastics applications, particularly tinted vinyl siding.

Typical Properties:

TiO_2 content, minimum	90%
Density	
Specific Gravity	4.01
Scattering coefficient per pound TiO_2	
In enamels	4.1
In flat latex	3.1
Oil absorption	19 lb oil/100 lb pigment
Water demand	21 lb water/100 lb pigment
pH	7.7
Specific resistance	7,500 ohms
Average particle diameter	0.22 um
Chalk resistance	highest
Specification	ASTM D476-73 Type III,IV

TRONOX Chloride Process Titanium Dioxide CR-822S:

TRONOX CR-822S is the slurry counterpart of TRONOX CR-822. The excellent durability, strength, and gloss producing properties of CR-822 have been preserved in its slurry form.

TRONOX CR-822S is recommended primarily for maximum durability in gloss and semi-gloss exterior water-borne coatings. However, it may also be used in interior coatings advantageously and may even be used in flat coatings when it is supplemented with functional extenders to obtain the required dry hiding properties.

Specifications:

Solids	76.5+-1%
Volatile	water
Density	19.1-19.7 lb/gal
Viscosity @ 100 RPM	175-300 cp
pH	8.5-9.5

Typical Viscosity Profile:

RPM	Viscosity, cp
100	225
50	350
20	650
10	1100

KERR-MCGEE CHEMICAL CORP.: TRONOX Titanium Dioxides(Continued):

TRONOX Chloride Process Titanium Dioxide CR-834:

TRONOX CR-834 is specifically designed for plastics and plastisol applications. While used with most all resins it is widely accepted in rigid vinyl pipe and siding formulations, ABS and polystyrene.

It meets the Plastic Pipe Institute guidelines for formulation substitution and is authorized by the National Sanitation Foundation for general interchangeability in NSF listed formulations.

It is characterized by:

* good dispersibility
* high tinting strength and opacity
* blue tint tone
* excellent color retention after processing with problem resins such as polyvinyl chloride and ABS

Typical Values

TiO_2 content, minimum	97%
Density, solid	4.2 g/ml
Oil absorption	15 lb oil/100 lb pigment
Average particle size	0.19 um
Lightfastness	excellent
Rutile content of contained TiO_2	100%
pH	6 to 7
Specific resistance	7,600 ohms
Resistance to anti-oxidant discoloration	excellent
Bulk density	40-55 lbs/ft 3

NL CHEMICALS, INC.: TITANOX Titanium Dioxide Pigment:

Anatase pigmentary:

Physical Properties:

1000 Paper:
Specific Gravity: 3.9
Percent TiO2, Min.: 98.5
Oil Absorption: 25.0
Percent Retained on 325 Mesh: 0.05
Chalk Resistance: None
pH: 7-9
A.S.T.M. D476-73 Type: I
General Product Applications: Paint, Exterior, Flat/Paint, Traffic. Paper: Filler, Coating. White Sidewall Tires.

1070 General Purpose:
Specific Gravity: 3.8
Percent TiO2, Min.: 95.0
Remainder: Al
Oil Absorption: 20.5
Percent Retained on 325 Mesh: 0.01
Chalk Resistance: None
pH: 7-8
A.S.T.M. D476-73 Type: I
General Product Applications: Paint: Exterior, Flat/Paint: Traffic. Paper: Coating. Ceramic Mill Add. Synthetic Fibers.

1072 Fibers:
Specific Gravity: 3.9
Percent TiO2: 98.5
Oil Absorption: 19.0
Percent Retained on 325 Mesh: 0.01
Chalk Resistance: None
pH: 7-8
A.S.T.M. D476-73 Type: I
General Product Applications: Synthetic Fibers

Rutile Pigmentary:

2020 General Purpose:
Specific Gravity: 4.1
Percent TiO2, Min.: 94.0
Remainder: Al
Oil Absorption: 18.5
Percent Retained on 325 Mesh: 0.01
Chalk Resistance: Mod.
pH: 7-8
A.S.T.M. D476 Type: II, III
General Product Applications: Paint, Interior: Gloss, Semi-Gloss, Flat/Exterior: Gloss, Semi-Gloss, Flat/Other: Masonry, Product Finishes. Plastics & Rubber: Polyvinyl Chloride, Plastisol. Printing Inks: Flexographic, Metal Decorating.

NL CHEMICALS, INC.: TITANOX Titanium Dioxide Pigment(Continued):

Rutile Pigmentary(Continued):

2071 Plastics:
Specific Gravity: 4.1
Percent TiO2, Min.: 96.5
Remainder: Al
Oil Absorption: 16.5
Percent Retained on 325 Mesh: 0.01
Chalk Resistance: None
pH: 6-8
A.S.T.M. D476-73 Type: II
General Product Applications: Plastics & Rubber: Polyvinyl Chloride, Polyethylene, Polystyrene

2073 Plastics:
Specific Gravity: 4.2
Percent TiO2, Min.: 97.5
Remainder: Al
Oil Absorption: 17.0
Percent Retained on 325 Mesh: 0.01
Chalk Resistance: None
pH: 7-8
A.S.T.M. D476-73 Type: II
General Product Applications: Plastics & Rubber: Polyvinyl Chloride, Polyethylene, Polystyrene

2081 Laminates:
Specific Gravity: 4.0
Percent TiO2, Min.: 91.0
Remainder: Al, Si
Oil Absorption: 28.0
Chalk Resistance: High
pH: 7-8
A.S.T.M. D476-73 Type: III, IV
General Product Applications: Melamine Laminate

2090 Printing Inks:
Specific Gravity: 4.1
Percent TiO2, Min.: 94.0
Remainder: Al
Oil Absorption: 19.5
Percent Retained on 325 Mesh: 0.01
Chalk Resistance: Mod.
pH: 7-8
A.S.T.M. D476-73 Type: II, III
General Product Applications: Printing Inks: Flexographic, Gravure, Metal Decorating.

NL CHEMICALS: TITANOX Titanium Dioxide Pigment(Continued):

Rutile, Pigmentary(Continued):

2101 Multi-Purpose:
Specific Gravity: 4.0
Percent TiO2, Min.: 92.0
Remainder: Al, Si
Oil Absorption: 21.0
Percent Retained on 325 Mesh: 0.01
Chalk Resistance: High
pH: 7-8
A.S.T.M. D476-73 Type: II, III, IV
General Product Applications: Paint: Interior, Gloss, Semi-Gloss, Flat/Paint: Exterior, Gloss, Semi-Gloss, Flat/Traffic, Masonry, Product Finishes, Automotive

2131 Multi-Purpose:
Specific Gravity: 3.7
Percent TiO2, Min.: 80.0
Remainder: Al, Si
Oil Absorption: 44.0
Percent Retained on 325 Mesh: 0.01
Chalk Resistance: Mod.
pH: 7-9
A.S.T.M. D476-73 Type: III
General Product Applications: Paint: Interior, Semi-Gloss, Flat/Exterior: Flat/Traffic, Masonry, Product Finishes/Paper: Filler, Coating

2160 Chalk Resistance:
Specific Gravity: 4.0
Percent TiO2, Min.: 90.0
Remainder: Al, Si
Oil Absorption: 18.0
Percent Retained on 325 Mesh: 0.01
Chalk Resistance: V. High
pH: 7-8
A.S.T.M. D476-73 Type: III, IV
General Product Applications: Paint: Interior, Gloss, Semi-Gloss, Flat/Exterior, Gloss, Semi-Gloss, Flat, Masonry, Product Finishes, Automotive/Plastics & Rubber: Polyvinyl Chloride, Plastisol.

Non-Pigmentary:

3020:
Percent TiO2, Min.: 99.5
Percent Retained on 325 Mesh: 80.0
General Product Applications: Glass

Pre-Dispersed Anatase: Slurry Properties:

1050 Slurry:
Solids: 73.0 | Viscosity: 600 cps
pH: 9-10 | TiO2#/g: 13.3
Counterpart: 1000
General Product Applications: Paper: Filler, Coating

Section 13
Miscellaneous

ACETO CORP.: CATAFOR Additives:

CATAFOR CA-80

CATAFOR CA-80 is an electrostatic spray paint additive that can substantially alter the conductivity/resistivity properties of coatings intended for electrostatic spray application. Additions of small amounts of CATAFOR CA-80 to coatings adjust the electrical properties to the correct degree for good sprayability.

CATAFOR CA-100

CATAFOR CA-100 is a newly developed highly effective additive that imparts exceptional electro-conductive properties to many polymers. It is especially effective in polyurethanes, polystyrene and PVC. CATAFOR CA-100 has minimal effect on polymer properties.

Typical Properties:

Activity:	100% Active
Composition:	High molecular weight quaternary ammonium ethosulfate
Appearance:	Waxy solid
Melting Range:	75-80C
Flash Point:	None, non-explosive

CATAFOR PP-50

CATAFOR PP-50 is a new liquid version of CATAFOR CA-100 anti-static additive which has been developed for use in polyurethane bunstock formulations. It is a fluid blend of a non-corrosive, high molecular weight quaternary ammononium ethosulfate in trichloropropyl phosphate (TCPP) flame retardant plasticizer. This product combines the following advantages:

1. Easy to handle and use.
2. Outstanding anti-static properties.
3. The TCPP solvent is a familiar, widely used component of PU foam

CATAFOR PP-50 is also recommended for use in vinyl formulations and other polymer systems where the plasticizer properties of TCPP can be used.

Typical Properties:

Composition:	50% bw CATAFOR CA-100 in TCPP
Appearance:	Clear liquid
Viscosity:	360 cps @ 25C
Specific Gravity:	1.131 @ 25C
% Water:	0.5 Max.
Flash Point:	100C, C.O.C.
Pour Point:	5C (41F)

ACETO CORP.: CATAFOR Additives(Continued):

CATAFOR PU

CATAFOR PU is a highly effective liquid additive designed to impart exceptional electro-conductive properties to polyurethanes. It is a fluid blend of high molecular weight quaternary ammonium ethosulfate as an 80% solution in 20% 1,4 Butanediol. This product combines the following advantages:

1. Easy to handle and use.
2. Outstanding anti-static properties.
3. Optimum compatibility characteristics with the polyol component.

Typical Properties:
Composition: 80% CATAFOR CA-100--20% 1,4 Butanediol
Appearance: Slightly viscous liquid
Viscosity: 300 cps at 25C
Freezing Point: <0C
Specific Gravity, 25C: 0.995
Flash Point: >100C, C.O.C.

ACETO CORP.: CELACOL COURLOSE Water-Soluble Cellulose Ethers:

They are available in many grades and types to enable them to be used to the best advantage as film formers, colloids, thickeners, binders, adhesives and water-controlling agents.

The principal water-soluble cellulose ethers:

COURLOSE:

SCMC: sodium carboxymethyl cellulose-technical and purified

CELACOL:

HPMC: hydroxypropyl methyl cellulose

MC: methyl cellulose

Standard Viscosity Grades of CELACOL and COURLOSE:

CELACOL:

Grade	Viscosity cps (2% aqueous solution 20C)
M20	17-23
M100	85-115
M450	350-550
M1000	850-1150
M2500	2200-2800
M5000	4000-6000
M10000	8000-12000
MM10	9-11
MM100	85-115
MMPR1P	*
HPM450	350-550
HPM2500	2200-2800
HPM5000	4000-6000
HPM15000	13000-17000
HPM30000	25000-30000 approx *

* Viscosity of standard MMPR 1P measured at 1.5% conc. w/v in 80:20 v/v methylene chloride: methanol mix--200-250 cps.

COURLOSE Technical A&N:

Grade	Viscosity cps (1% aqueous 25C)	Active % (dry basis)
Alkaline		
A200	3-6	60-64
A590	5-10	59-61
A600	7-12	60-64
A610	6-10	68-72
A650	12-20	60-65
A500HDS	3-5	59-61
A600HDS	5-10	58-60
A650HDS	10-20	59-61
A800HDS	50-100	59-61
Neutral		
N600	12-25	62-64
N602	12-25	62-64

ACETO CORP.: Chlorinated Rubbers:

CR-5

Chlorine Content, Weight Percent	65 Minimum (Average 68)
Appearance	White, Granular Powder
Viscosity Range, 20% Weight Solution in Toluene @ 25C. in cps.	4-7
Moisture & Volatiles, Weight Percent	1.5 Maximum (Average 0.9)
Insoluble In Toluol, Weight Percent	0.3 Maximum (Average 0.15)
Effect of Dry Heat on Film	Stable up to and at 130C.

CR-10

Chlorine Content, Weight Percent	65 Minimum (Average 68)
Appearance	White, Granular Powder
Viscosity Range, 20% Weight Solution in Toluene @ 25C. in cps.	8-13
Moisture & Volatiles, Weight Percent	1.5 Maximum (Average 0.9)
Insoluble in Toluol, Weight Percent	0.3 Maximum (Average 0.15)
Effect of Dry Heat on Film	Stable up to and at 130C.

CR-20

Chlorine Content, Weight Percent	65 Minimum (Average 68)
Appearance	White, Granular Powder
Viscosity Range, 20% Weight Solution in Toluene @ 25C. in cps.	15-25
Moisture & Volatiles, Weight Percent	1.5 Maximum (Average 0.9)
Insoluble in Toluol, Weight Percent	0.3 Maximum (Average 0.15)
Effect of Dry Heat on Film	Stable up to and at 130C.

ACETO CORP.: Products:

Additive C

Description:	Non-volatile, silicone free stabilizer
Appearance:	Clear to yellowish liquid
Viscosity:	Medium viscose
Assay:	100% active substance
Compatibility:	Completely soluble in almost all resins and multifunctional monomers
pH:	4
Flash Point:	>100C

A-S-A T-300-20AK

A-S-A T-300-20AK is a special polyamide dispersion, which build up a strong structure in almost all organic media such as paint, inks, sealants, etc.

1) Advantage
   1. Smaller incorporating amount than other additives exhibit enough thixotropy to prevent sagging and settling.
   2. Both prior milling addition and post-addition can be employed for this additive.
   3. Trouble free from gloss reduction.
   4. Independence of phase compositions.
2) Properties
   Appearance: Light yellow paste
   Nonvolatile: 20%
   Volatile: Xylene
   Alcohol

Butyraldoxime

Appearance:	Colorless to light yellow liquid
Purity:	98.5% min.
Color:	Max. 30 Apha
Viscosity:	Max. Gardner A3
Specific Gravity:	0.908
Solubility:	Completely miscible with aliphatic and aromatic hydrocarbons, ethanol and ether
Flashpoint A.P.c.c.:	58C

ACETO CORP.: Products (Continued):

Methyl Ethyl Ketoxime

Specifications:	
Assay	99.0 minimum
Color	APHA 5
Appearance	Clear, colorless liquid
Physical Properties:	
Specific Gravity @ 25C	0.922
Boiling Point	152C
Flash Point (Cleveland open cup)	69C

Photoinitiator
Ethyl para Dimethyl Amino Benzoate

Appearance:	Pinkish cream powder with faint odor
Melting point:	58C.

Use: in pigmented coatings as a photoactivator in combination with 2-isopropyl thioxanthone

Photoinitiator
2-Isopropyl Thioxanthone

Appearance:	White to pale cream powder with almost no odor.
Melting Point:	55-70C.

Use: in pigmented coatings at levels of 0.2-1.5% in combination with Ethyl para Dimethyl Amino Benzoate at 2-5%

Stabilizer 5

Description:	Polymeric stabilizer, free of phenolic components. 50% Solution in Trimethylolpropane Triacrylate
Appearance:	Yellow, clear solution
Viscosity:	Medium
Flash point:	>100C
Compatibility:	Completely soluble in almost all resins and multifunctional monomers.
pH:	7.5
Application:	In-can stabilizer for acrylated UV Systems. Use at levels of 1-3%.

ACETO CORP.: Products(Continued):

TAZ-O

Ethylene Imine derivative of Pentaerythritol Triacrylate

Appearance	Clear, amber liquid
Solids Content	100%
Aziridine Content meq/g	6.76
Aziridine Functionality	2.76
Density g/cc at 25C	1.185
Viscosity, cps at 25C	1092
Freezing Point	-17C
Vapor Pressure	<0.1 mm @ 25C
Refractive Index n 20D	1.4816

Zinc Oxide Red Seal

Physical:	
Mean particle size (micron)	0.25
Specific surface area (BET) $m^2 g^{-1}$	4
Bulk density	0.5 gcm^{-3}
% retained on 240 BS sieve	0.1
% moisture (105C.)	0.15
Chemical:	
ZnO	99.7%
Pb	0.15%
Cd	0.03%
As	2 ppm
Cu	3 ppm
Mn	2 ppm
Fe	10 ppm
SO_3	200 ppm
Cl	150 ppm
Water solubles	0.1%
Acid insolubles	0.01%
Loss of ignition	0.15%

ALLIED COLLOIDS, INC.: GLASWAX E1 High Melting Point Polyethylene Wax Emulsion:

Typical Properties:

Chemical Type: Polyethylene wax
Physical form: Opaque emulsion
Solids: 30%
Solvent: Water
pH: 9.0
Viscosity: 25 cps 30%
Initial softening point: >137C
Specific gravity: 0.995

GLASWAX E1 is a hard, high melting point polyethylene wax supplied as a 30% emulsion in water. It has been developed to give good slip and abrasion resistance properties in water-based inks and lacquers. The high melting point makes it ideal for applications where heat resistance is a priority, such as pre-print liner inks or heat seal resistant overprint varnishes.

AO POLYMER CORP.: AOCURE Catalysts:

Typical Properties:

AOCURE MC:

Equivalent Weight/NCO: 700
Viscosity, cps: 800
Specific gravity @ 25C: 1.15
Wt./gallon, lbs: 9.1
Solids, %: 83
Color: Amber

AOCURE MCF:

Equivalent Weight/NCO: 700
Viscosity, cps: 1000
Specific gravity @ 25C: 1.18
Wt./gallon, lbs: 9.5
Solids, %: 96
Color: Amber

Product Description:

AOCURE MC and AOCURE MCF are aromatic amine curing agents, recommended for use with AOTHANE prepolymers to formulate two-component elastomeric urethane coatings. Pigments, fillers and additives can be readily dispersed in these curing agents with a high speed disperser. AOCURE MCF was developed as replacement for AOCURE MC in fire retardant formulations.

AOCURE H

AOCURE H is a hydroxy-terminated curing agent for use with AOTHANE prepolymers to formulate a wide range of low modules, two component sealants.

Typical Properties:

Equivalent Weight: 1850
Specific Gravity @ 25C: 1.03
Weight per Gallon lbs.: 8.45
Viscosity, cps: 1000-1200
Solids, %: 100

Suggested Uses:

AOCURE H can be used with AOTHANES 2002, 2003, 2004, 3002-5, and 3075 to formulate low Shore A durometer sealants.

AO POLYMER CORP.: Products:

ESTELONE C-250
(Epoxy Curing Agent)

ESTELONE epoxy curing agents have been designed for use in powder coating systems where a wide latitude of cure schedules and excellent latency are required. It may be employed alone as the sole binder or utilized to accelerate "dicy", anhydrides, amines, and acids. ESTELONE C-250 is supplied as a free flowing non-moisture sensitive powder which does not require grinding prior to use.

Physical Properties:

Form: Free flow powder
Color (MgO=100): 95
Particle Size (average): 40 microns

AMORWAX - 1300P (Powder)
AMORWAX - 1300M (Micronized)
AMORWAX - 1300SM (Super micronized)

AMORWAX-1300 is a high melting synthetic bisamide wax recommended for resins and plastics application.

It functions as a polymer modifier, internal and external lubricant, anti-blocking, anti-tack, anti-static, mold release, mar resistance, and corrosion resistance agent.

Properties:

Melting Point: 143-147C
Flash Point: Approx. 290C
Acid Value: 7
Color-Gardner: 5
Specific Gravity at 25C: 0.97-0.99
Ash: 0.5%

Particle Size:

AMORWAX-1300P -- Approximately 70 microns
AMORWAX-1300M -- Approximately 5 microns
AMORWAX-1300SM -- Approximately 2-4 microns (80% is 2 microns)

Applications:

AMORWAX 1300 is suitable for hot melt adhesives, coatings, plastics, rubber, foundry products, powdered metals and others. Recommended use levels are between 0.2 and 3.5%.

R. E. CARROLL, INC.: ASARCO Zinc Oxide:

Zinc Oxide is the commercially most important compound of zinc with applications in the rubber, paint, chemical, ceramics, glass, paper and agricultural industries.

ASARCO has the flexibility to provide French process zinc oxide in a wide selection of grades. Special grades are also available.

Grade:

AZO 44:
Use: Ceramics, Glass
Particle Shape, Size: Nodular, Extra large
Min % ZnO: 99.60

AZO 66:
Use: Rubber, Chemical
Particle Shape, Size: Nodular, Medium

AZO 66LP:
Use: Coatings
Particle Shape, Size: Nodular, Large
Min. ZnO: 99.60

AZO 66TT:
Use: Rubber, Chemical
Particle Shape, Size: Surface Treated, Nodular, Medium
Min % ZnO: 99.60

AZO 66USP:
Use: Pharmaceutical
Particle Shape, Size: Nodular, Medium
Min % ZnO: 99.60

AZO 77:
Use: Rubber, Chemical
Particle Shape, Size: Nodular, Fine
Min % ZnO: 99.60

AZO 77TT:
Use: Rubber, Chemical
Particle Shape, Size: Surface Treated, Nodular, Fine
Min % ZnO: 99.60

AZO 77S:
Use: Coatings, Chemical
Particle Shape, Size: Nodular, Fine
Min % ZnO: 99.60

AZO 77USP:
Use: Pharmaceutical
Particle Shape, Size: Nodular, Fine
Min % ZnO: 99.60

R. E. CARROLL, INC.: FEDERATED GENCO Zinc Dust:

No. 1 (660 and L670)
and
No. 50:

Standard Grades:
Federated Genco No. 1 Zinc Dust normally is used in inoganic and organic zinc coatings, in thread compounds and as a reducing agent in chemical reactions.
Federated Genco No. 50 Zinc Dust is the same dust as the No. 1 grade but approximately 0.5% unslaked lime (CaO) has been added.

Typical Analysis:

No. 1:

Chemical Properties:
Total Zinc %: 99.0
Metallic Zinc %: 96.5
Physical Properties:
Particle Size:
Average Dia. um: 5.5-7.5
Median Dia. um: 6.0-8.5
Specific Gravity: 7.1
Bulking Value gal/lb: 0.0169
Sieve Analysis
Through 100 mesh %: 100
Through 325 mesh %: 96.0

No. 50:

Chemical Properties:
Total Zinc %: 98.5
Metallic Zinc %: 96.0
CaO %: 0.5
Physical Properties:
Particle Size:
Average Dia. um: 5.0-7.0
Median Dia. um: 6.0-8.5
Specific Gravity: 7.1
Bulking value gal./lb: 0.0170
Sieve Analysis
Through 100 Mesh %: 100
Through 325 Mesh %: 96.0

R.E.CARROLL, INC.: FEDERATED GENCO Zinc Dust(Continued):

L-15, L-10 and L-350:

Low Micron Grades:

These three low micron zinc dusts are produced with a variety of physical and chemical properties to fulfill the many requirements of low micron dusts. The L-15, frequently referred to as superfine, has high metallic zinc content and a narrow particle size distribution with essentially no particle greater than 25 micron.

The L-10 grade, ultrafine, has an even narrower distribution with essentially no particle greater than 15 micron.

The L-350 grade is used where slower settling characteristics or more controlled reaction rates are required.

Federated Genco low micron grades are used in thin film shop primers in aerosol spray formulations as reducing agents and to recover precious metals.

L-15:
- Total Zinc %: 99.0
- Metallic Zinc %: 96.5
- Particle Size:
  - Average Dia. um: 5.0
  - Median Dia. um: 5.5
- Specific Gravity: 7.1
- Bulking Value gal/lb.: 0.0170
- Less than 15 um, %: 98.5
- Less Than 10 um, %: 94.0

L-10:
- Total Zinc %: 98.9
- Metallic Zinc %: 95.5
- Particle Size
  - Average Dia. um: 4.0
  - Mediam Dia. um: 4.5
- Specific Gravity: 7.1
- Bulking Value gal./lb.: 0.0170
- Less than 15 um, %: 99.8
- Less than 10 um, %: 97.5

L-350:
- Total Zinc %: 98.7
- Metallic Zinc %: 94.0
- Particle Size:
  - Average Dia. um: 4.0
  - Median Dia. um: 4.5
- Specific Gravity: 7.0
- Bulking Value gal./lb.: 0.0170
- Screen Analysis
  - Through 100 Mesh %: 100
  - Through 325 Mesh %: 99.7

COOK PAINT AND VARNISH CO.: COROC Resins and Additives:
Acrylic:

A-72-A260
Acrylic Flow Control Agent

COROC Resin A-72-A260 is a non-silicone flow and leveling control agent. Its effect on any solvent system is based upon its surface tension modification. It does not affect any other property, especially initial or recoatable adhesion. It can be incorporated into your formulation by direct addition either prior to or after pigment dispersing.

Physical Characteristics:
- Nonvolatile: 60+-1
- Solvent:
  - xylene: 100%
- Viscosity: B-C

Suggested Uses:
The compatibility of A-72-A260 has been determined and is effective with the following resin systems:

Alkyd	Acrylic
Epoxy	Polyester
Vinyl	Varnish

A-620-A2
Acrylic Flow Control Agent

COROC Resin A-620-A2 is non-silicone flow and leveling control agent. Its effect on any solvent system is based upon its surface tension modification. It does not affect any other property, especially initial or recoatable adhesion. It can be incorporated into your formulation by direct addition either prior to or after pigment dispersing.

Physical Characteristics:
- Nonvolatile: 60+-1
- Solvent:
  - xylene: 100%
- Viscosity: A-C

Suggested Uses:
The compatibility of A-620-A2 has been determined and is effective with the following resin systems:

Alkyd	Acrylic
Epoxy	Polyester
Vinyl	Varnish

COOK PAINT AND VARNISH CO.: COROC Resins and Additives:
Acrylic(Continued):

A-2201-M
Acrylic Flow Control Agent

A-2201-M is a surface tension modifier for nonaqueous coating systems. A-2201-M is compatible with alkyds, oil free polyesters, acrylics, and certain epoxy modified resins. Because A-2201-M contains no silicone, it should not affect recoatability.

When A-2201-M is incorporated at levels of 0.1-2%, the following film properties will be significantly affected:
- Reduction of fisheyes, craters, and orange peel
- Improve flow and leveling
- Improved gloss
- Eliminate air bubbles
- Improve wetting of the substrate

Physical Characteristics:
- Nonvolatile: 70+-1
- Solvent:
  - xylene: 56%
  - butyl alcohol: 44%
- Viscosity: D-E

A-2678-M
Acrylic Flow Agent for Water-Borne Systems

COROC A-2678-M is a non-silicone flow and leveling control agent for water reducible systems. Its ability to promote leveling and flow is based upon its ability to modify surface tensions. A-2678-M has no effect upon other properties of the film, especially adhesion and recoatability. It can be incorporated into a formulation by direct addition either prior to or after pigment dispersion. A-2678-M is already neutralized and reduced in water.

Physical Characteristics:
- Nonvolatile: 50+-1
- Solvent:
  - water: 55.0
  - dipropylene glycol monomethyl ether: 23.5
  - propylene glycol monomethyl ether: 2.6
  - dimethylamino ethanol: 18.9
- Viscosity: W-Y

Suggested Uses:

A-2678-M has been found to be compatible with most water reducible resins: acrylics, alkyds, epoxy esters, and polyesters.

COOK PAINT AND VARNISH CO.: COROC Resins and Additives:
Acrylic(Continued):

XA-2678-M1
Acrylic Flow Agent

COROC XA-2678-M1 is a non-silicone flow and leveling control agent. It is an unneutralized version of A-2678-M and is supplied at 80% N.V. in a 90/10 blend of Di-Propylene Glycol Monomethyl Ether and Propylene Glycol Monomethyl Ether. Its ability to promote leveling and flow is based upon its modification of a system's surface tension. Data to date indicates it does not affect adhesion or recoat properties. XA-2678-M1 can be incorporated by direct addition either prior to or after pigment dispersion. It should be neutralized with amine when utilized in water reducible systems.

Typical Physical Characteristics:
- Nonvolatile: 80+-1
- Solvent:
  - Di-PGMME: 90%
  - PGMME: 10%
- Viscosity: Z6+

Suggested Uses:
XA-2678-M1 has been found to be compatible with acrylics, alkyds, epoxy esters and polyesters.

A-3661-E3
Acrylic Wetting and Dispersing Agent

A-3661-E3 is an acrylic grinding agent designed to aid the formulator in pigment wetting and dispersion. A-3661-E3 may be used in acrylic and alkyd formulation, but is especially useful in oil free polyesters where pigment dispersion can be a challenge. Optimum results are obtained by grinding the pigment in the primary resin with A-3661-E3. Poorer quality grinds (concerning viscosity and film gloss) are obtained when the melamine is incorporated in the grind.

Physical Characteristics:
- Nonvolatile: 48+-1
- Solvent:
  - propylene glycol mono-methyl ether acetate: 100%
- Viscosity: Z-Z2

COOK PAINT AND VARNISH CO.: COROC Resins and Additives:
Acrylic(Continued):

XA-4031-A2
Experimental Acrylic Air Release Agent

COROC XA-4301-A2 is a silicone free surface tension modifier for nonaqueous coating systems. As well as being an excellent air release agent, XA-4031-A2 significantly improves flow and leveling in a variety of systems. A stepwise addition is recommended to determine the optimum level for a particular system. Experience has shown concentrations varying from 0.1% -2.0% to be most effective.

Typical Physical Characteristics:
- Nonvolatile: 60+-1
- Solvent:
  - Xylene: 100%
- Viscosity: Alt.-A

COSAN CHEMICAL CORP.: Products:

C-FLOC

The C-FLOC line of flocculants is designed for use in primary treatment of water-based paint waste. The C-FLOC products are water soluble cationic polymers which flocculate and coagulate electrically-charged particles. These flocculants function by bridging opposite-charged particles, thereby creating large agglomerates which separate and settle to the bottom.

C-FLOC 9:
- Appearance: Clear amber liquids
- Viscosity: 300-500 cps
- Freeze-Thaw: Three cycles stable
- Wt. Per Gallon: 8.8 lbs.

C-FLOC 18:
- Appearance: Clear amber liquids
- Viscosity: 7,500-10,000 cps
- Freeze-Thaw: Three cycles stable
- Wt. Per Gallon: 9 lbs.

C-FLOC 36:
- Appearance: Clear amber liquids
- Viscosity: 55-150 cps
- Freeze-Thaw: Three cycles stable
- Wt. Per Gallon: 8.6 lbs.

COSAN JTA-20
E.P.A. Reg. No. 8489-3

COSAN JTA-20 is a highly functional, soluble, phenylmercury acetate which has been deposited on an inert, synthetic, hydrous calcium silicate to produce a free-flowing, relativly non-dusting powder which is simple to incorporate into dry formulations. COSAN JTA-20 contains 20% phenylmercury acetate as the active ingredient equivalent to 12% mercury expressed as metal.

Specifications:
- Phenylmercury Acetate, minimum: 20%
- Mercury Metal Content, minimum: 12%
- Particle Size (average): 2.1 microns
- Bulk Density: 2.8 lbs per gal.
- Appearance: Fine, free-flowing, white powder

COSAN CHEMICAL CORP.: Products(Continued):

COSPERSE 70
Liquid Dispersing and Grinding Aid

COSPERSE 70 is a liquid dispersing and grinding aid which can be incorporated in a variety of trade sales and industrial formulations. COSPERSE 70 meets Rule 66 and Regulation 3 requirements.

COSPERSE 70 will provide the following the following advantages to the formulator:

1. Greater color development of pigment.
2. Retardation of pigment settling.
3. Improved pigment dispersion eliminates excessive grinding time. Result: increased production.
4. Prevention of flocculation.
5. As aid in converting formulations based in ball or roller mills to high speed mills or dispersers.

Specifications:
- Viscosity: Gardner L-O
- Solids: 70% Plus or Minus 1%
- Acid: 26-32

COTIN 234
E.P.A. Reg. No. 8489-8

COTIN 234 is a formulated solution of a triorganotin salt which has a high degree of antifungal activity.

COTIN 234 exhibits excellent mildew resistance/preservative activity in a great variety of materials and formulations and due to its unique formulation offers excellent compatibility in latex and other water emulsion formulations.

COTIN 234 can be used where mercury containing preservatives have been omitted either due to the general toxicoligical picture, or for elimination of sulfide staining characteristics.

Specifications:
- Tributyltin salicylate: 48%
  - (Tin as Metal: 13%)
- Specific gravity: 1.09-1.10

COTIN 300
E.P.A. Reg. No. 2749-119-8589

COTIN 300 is a colorless to light yellow liquid which is highly effective against gram positive bacteria and fungi. Broad spectrum antimicrobial activity may be achieved by combination with quaternary ammonium compounds, organomercurials, etc.

Specifications:
- Bis (tributyltin) oxide: 97.4% minimum
- Tin content: 39-40%

COSAN CHEMICAL CORP.: Products (Continued):

COSKIN B

COSKIN B is an anti-skinning agent which is soluble in all proportions with oil and oleoresinous compositions. The use of COSKIN B controls skinning in a paint formulation without affecting other physical characteristics of the paint.

COSKIN B, Butyraldoxime, proves to be an effective anti-skinning agent in most paint formulations for these reasons:

Effectiveness: COSKIN B provides anti-skinning benefits with minimum effect on drying time.

Volatility: COSKIN B is completely volatile and will not cause yellowing or other objectionable film properties.

Chemically: COSKIN B is neutral. It will not cause unpredictable reactions in the can or on paint film properties.

Specifications:
- Appearance: Clear, Colorless Liquid
- Specific Gravity at 80F: 0.902-0.912

COSKIN M

The use of COSKIN M controls skinning in a paint formulation without effecting other physical characteristics of the paint. COSKIN M--methyl ethyl ketoxime--proves to be an effective anti-skinning agent in most paint formulations for these reasons:

1. Effectiveness — COSKIN M provide anti-skinning benefits with minimum effect on drying time.
2. Odor — COSKIN M is for all intents and purposes practically odorless
3. Color — COSKIN M is water white in color. It will not discolor pale tints or whites.
4. Volatility — COSKIN M is completely volatile and will not cause yellowing or other objectionable film properties.
5. Chemically — COSKIN M is neutral. It will not cause unpredictable reactions in the can or in paint film.

Specifications:
- Specific Gravity @ 75F: 0.910-0.930
- Color (Gardner): 1 max.
- Lbs./gal.: 7.7
- Appearance: Clear bright
- Tag Closed Cup Flash Point: 146F

DANIEL PRODUCTS CO.: DAPRO Interfacial Tension Modifiers:

For Minimizing Crawling, Fish Eyes, and Cratering

W-77:
for water systems only

U-99:
for water and solvent systems

S-65:
for solvent systems only

DAPRO Interfacial Tension Modifier are compounds which can reduce interfacial tension by forming an absorption layer of intermediate surface tension between the liquid/liquid or the liquid/solid phases and, thereby can:

* Eliminate or diminish film defects such as crawling, fish-eyes, and some forms of cratering.
* Promote spreading and uniform film formation on hard-to-wet or contaminated surfaces

All DAPRO Interfacial Tension Modifiers are silicone-free blends of surface-active materials. DAPRO U-99 and W-77 differ only in their solvent composition. DAPRO U-99 contains a high percentage of 2-Butoxyethanol for compatibility with both water and solvent based coatings and inks. DAPRO W-77, containing a blend of 2-Butoxyethanol and water, is effective in water based systems only. DAPRO S-65, being a blend of 2-Butoxyethanol and water, is effective in water based systems only. DAPRO S-65, being a blend of hydrophobic materials in n-Butanol and high flash aromatics, is for use in solvent thinned systems only.

DAPRO W-77:
Suggested Level of Use: 0.25-1.0%
Total Solids: 50%
Wt./Gal. (Sp. Gr.): 8.5 Lbs. (1.02)
Viscosity (Gardner-Holdt): A3
FDA Cleared Components: Yes

DAPRO U-99:
Suggested Level of Use: 0.25-1.0%
Total Solids: 50%
Wt./Gal. (Sp.Gr.): 8.4 Lbs (1.01)
Viscosity (Gardner-Holdt): A3
FDA Cleared Components: Yes

DAPRO S-65:
Suggested Level Of Use: 0.1-0.5%
Total Solids: 75%
Wt./Gal. (Sp.Gr.): 8.2 Lbs. (0.98)
Viscosity (Gardner-Holdt): Z-Z3
FDA Cleared Components: No

DANIEL PRODUCTS CO.: SLIP-AYD Surface Conditioners:

SLIP-AYD Surface Conditioners are dispersions of different types of low molecular weight polyethylene or polymeric wax, each available in a variety of solvents or solvent/vehicle blends.

Particle Size: Any wax-type additive may have a tendency to reduce gloss in high gloss coatings.
Ease of Incorporation: SLIP-AYD Surface Conditioners have been thoroughly dispersed to mix easily into a coating at any point of manufacture, including the let down stage.
Effectiveness: The effect of a wax can be altered by changing the solvent or vehicle in which the wax is dispersed.
Trouble-Free Usefulness: SLIP-AYD Surface Conditioners are stable dispersions that will not reagglomerate or seed out on storage or in use.
Solvent Selection: Dispersions based on the same wax can have dramatic differences in performance depending on the solvents used in the dispersion.

Polymeric Wax:

Softening Point: 195-200F/90.6-93.3C
Hardness(ASTM D-5): 3.5-4.0
Typical Properties: Excellent slip; excellent mar resistance; little effect on gloss.
Typical applications: Exterior can coatings, appliance finishes, high gloss coil coatings, paper and foil coatings.

<u>SLIP-AYD:</u>

SL-92:
Miscible with: Solvent or Water
Wax: 19.0%
Dispersed in: 81.0% Isopropanol
Flash Point: 63F/17.2C
Wt./Gal. (Sp.Gr.): 6.7 Lbs. (0.80)

SL-133*:
Miscible with: Solvent only
Wax: 20.0%
Dispersed in: 60.0% High Flash Aromatics
20.0% Melamine
Flash Point: 113F/45.0C.
Wt./Gal. (Sp.Gr.): 7.5 Lbs. (0.90)

SL-140:
Miscible with: Water only
Wax: 24.0%
Dispersed in: 55.0% Glycol
21.0% Water
Flash Point: 225F/107.2C
Wt./Gal. (Sp.Gr.): 8.5 Lbs. (1.02)

* Does not have FDA clearance

DANIEL PRODUCTS CO.: SLIP-AYD Surface Conditioners(Continued):

Polymeric Wax:

Softening Point: 200-210F/93.3-98.9C
Hardness (ASTM D-5): 2.0-2.5
Typical Properties: Excellent slip; good blocking resistance; very little effect on gloss; finer particle size; FDA clearance.
Typical applications: Interior can coatings; high gloss coil coatings; flexible food packaging.
Typical use levels: 0.25% to 1.0% SLIP-AYD based on total coating weight.

SLIP-AYD:

SL-404:
Miscible with: Solvent or Water
Wax: 18.0%
Dispersed in: 82.0% 2-Butoxyethanol
Flash Point: 157F/69.4C
Wt./Gal. (Sp.Gr.): 7.5 Lbs. (0.90)

SL-417:
Miscible with: Solvent or Water
Wax: 20.0%
Dispersed in: 80.0% Isopropanol
Flash Point: 63F/17.2C
Wt./Gal. (Sp. Gr.): 6.7 Lbs. (0.80)

SL-425:
Miscible with: Solvent only
Wax: 20.0%
Dispersed in: 75.0% Xylol
5.0% Polymeric Plasticizer
Flash Point: 80F/26.7C
Wt./Gal. (Sp. Gr.): 7.3 Lbs. (0.88)

SL-512*:
Miscible with: Water only
Wax: 35.0%
Dispersed in: 53.0% Water
10.0% Propylene Glycol
2.0% Surfactant
Flash Point: 225F/107.2C.
Wt./Gal. (Sp. Gr.): 8.3 Lbs. (1.00)

SL-523:
Miscible with: Solvent or Water
Wax: 18.0%
Dispersed in: 82.0% Isopropanol
Flash Point: 63F/17.2C
Wt. Gal. (Sp. Gr.): 6.7 Lbs. (0.80)

* Does not have FDA clearance

DANIEL PRODUCTS CO.: SLIP-AYD Surface Conditioners(Continued):

Polyethylene

Softening Point: 220-225F/104.4-107.2C
Hardness (ASTM D-5): 3.0-3.5
Typical Properties: Good abrasion resistance; good blocking resistance; excellent versatility; FDA clearance.
Typical applications: Interior can coatings; appliance and general industrial finishes; paper coatings; coil coatings and synthetic lacquers.
Typical use levels: 0.5-3.0% SLIP-AYD based on total coating weight.

SLIP-AYD:

SL-18:
Miscible with: Solvent or Water
Polyethylene: 21.0%
Dispersed in: 79.0% 2-Butoxyethanol
Flash Point: 157F/69.4C
Wt./Gal. (Sp. Gr.): 7.5 Lbs. (0.90)

SL-31:
Miscible with: Solvent only
Polyethylene: 18.0%
Dispersed in: 82.0% Xylol
Flash Point: 80F/26.7C
Wt./Gal. (Sp. Gr.): 7.2 Lbs. (0.86)

SL-50:
Miscible with: Solvent only
Polyethylene: 22.5%
Dispersed in: 61.5% Xylol
16.0% Polymeric Plasticizer
Flash Point: 80F/26.7C
Wt./Gal. (Sp. Gr.): 7.3 Lbs. (0.88)

SL-78:
Miscible with: Solvent only
Polyethylene: 22.0%
Dispersed in: 71.0% Xylol
7.0% Short Oil Coconut Oil Alkyd
Flash Point: 80F/26.7C
Wt./Gal. (Sp. Gr.): 7.4 Lbs. (0.89)

SL-530:
Miscible with: Solvent or Water
Polyethylene: 18.0%
Dispersed in: 82.0% 2-Butoxyethanol
Flash Point: 157F/69.4C
Wt./Gal. (Sp. Gr.): 7.5 Lbs. (0.90)

DANIEL PRODUCTS CO.: SLIP-AYD Surface Conditioners(Continued):

Polyethylene

Softening Point: 280-290F/137.8-143.3C
Hardness (ASTM D-5): <0.5
Typical Properties: Excellent abrasion resistance; excellent blocking resistance; may cause noticeable reduction in gloss.
Typical Applications: Supermarket shelving; low gloss coil coatings; prefinished paneling; epoxy systems; transfer coatings; label varnishes and metal decorating.
Typical use level: 1.0% to 4.0% SLIP-AYD based on total coating weight.

SLIP-AYD:

SL-177:
Miscible with: Solvent only
Polyethylene: 24.0%
Dispersed in: 76.0% Xylol
Flash Point: 80F/26.7C
Wt./Gal. (Sp. Gr.): 7.2 Lbs. (0.86)

SL-280:
Miscible with: Solvent only
Polyethylene: 20.0%
Dispersed in: 80.0% n-Butanol
Flash Point: 96F/35.6C
Wt./Gal. (Sp.Gr.): 6.9 Lbs. (0.83)

SL-295A:
Miscible with: Water only
Polyethylene: 21.0%
Dispersed in: 42.0% 2-Butoxyethanol
37.0% Water
Flash Point: 157F/69.4C
Wt./Gal. (Sp.Gr.): 7.9 Lbs.(0.95)

SL-300:
Miscible with: Water only
Polyethylene: 30.0%
Dispersed in: 59.0% Water
10.0% Glycol
1.0% Surfactant
Flash Point: 225F/107.2C
Wt./Gal. (Sp.Gr.): 8.4 Lbs. (1.01)

SL-551:
Miscible with: Solvent only
Polyethylene: 18.5%
Dispersed in: 61.5% High Flash Aromatics
20.0% n-Butanol
Flash Point: 96F/35.6C
Wt./Gal. (Sp. Gr.): 7.4 Lbs.(0.89)

DANIEL PRODUCTS CO.: SLIP-AYD SL 526 Ultra-fine Surface Conditioner:

For Improving Surface Properties of High Solids Polyester Coatings

SLIP-AYD SL 526 Ultra-Fine Surface Conditioner is a dispersion of a polymeric wax in an acrylic vehicle modified for improved wetting. It has been specifically designed to optimize a variety of surface properties of coatings based on high solids polyester vehicles. SL 526 will

* Reduce coefficient of friction
* Improve mar and scuff resistance
* Increase substrate wetting
* Increase flow and leveling
* Increase sharpness of reflected image
* Eliminate surface irregularities
* Reduce foaming

Specifications:

Composition:

Polymeric Wax	15.0%
Vehicle Solids	25.0%
PM Acetate	51.7%
n-Butanol	1.6%
AROMATIC 100	6.7%
	100.0%

Wt./Gal. (Sp.Gr.): 8.0 lbs. (0.96)
Flash Point: 108F (42C)
FDA Clearance: No

Suggested Level of Use: 0.5% to 2.0% of Total Formula

Wax Specifications:
Softening Range: 200-210F (93-99C)
Hardness (ASTM D-5): 2.0-2.5

DOCK RESINS CORP.: DORESCO Plasticizers:

Product:

DORESCO AKN3-15:

DORESCO AKN3-15 is an excellent quality azaleic acid castor oil modified plasticizer.

Typical physical constants:

Solids: 80+-1%
Solvent: Toluol
Viscosity: Z4 approx.
Acid No.: 15 max.
Color: 10 max.
Lbs./Gal.: 8.0

DORESCO M34-13:

DORESCO M34-13 is a polymeric plasticizer designed primarily for the textile field.

DORESCO M34-13 is a polyester with a low acid number and an appreciable amount of unreacted hydroxyl groups.

Typical physical constants:

Solids: 100%
Color (Gardner): 4 max.
Acid Number: 3.0 max.
Visc. (Gardner): Y-Z1

DORESCO PMP1-3U:

DORESCO PMP1-3U is a saturated polyester designed to produce premium grade greases.

Typical physical constants:

Viscosity (Gardner): R-U+
Acid No.: 5 Max.
Color (Gardner): 2-3
% Active: 100

DORESCO PMP7-1:

DORESCO PMP7-1 is a primary polymeric plasticizer recommended for polyvinyl chloride processors, solution grade vinyls, and lacquers.

Typical physical constants:

Viscosity (Gardner): Z5
Acid No.: 1.1
Color: 2-3
SPI Volatility: .6%

DREW INDUSTRIAL DIVISION: Flocculants for Paint Waste Treatment:

AMERFLOC 485:
Coagulant

AMERFLOC 485 coagulant is a medium molecular weight, highly charged cationic liquid polymer. Its use has been approved by the EPA for treatment of potable water.

AMERFLOC 485 is a highly effective primary coagulant for water and wastewater clarification.

Typical Properties:
Appearance: Light amber liquid
Specific Gravity (25C): 1.16
Viscosity (25C): 650 cps

AMERFLOC 490:
Coagulant

AMERFLOC 490 coagulant is a medium to high molecular weight, highly charged cationic liquid polymer. It has been approved by the EPA for treatment of potable water.

AMERFLOC 490 is a highly effective primary coagulant for water and wastewater clarification.

Typical Properties:
Appearance: Amber liquid
Specific Gravity (25C): 1.16
Viscosity (25C): 5000 cps

DREWFLOC 2270 Anionic Flocculant:

DREWFLOC 2270 is a very high molecular weight, highly anionic, single component, emulsion polymer.

DREWFLOC 2270 is highly effective in dewatering industrial slurries as well as in water clarification applications.

Typical Properties:
Appearance: Pale gray, opaque liquid
Specific Gravity (25C): 1.00+-0.02
Product Viscosity (25C): 500-2500 cps

DREW INDUSTRIAL DIVISION: Flocculants for Paint Waste Treatment (Continued):

DREWFLOC 270:
flocculant

DREWFLOC 270 is a high molecular weight, highly charged anionic, granular polymer.
DREWFLOC 270 is highly effective in dewatering industrial slurries as well as in water and wastewater clarification applications.

Typical Properties:
Appearance: White free-flowing powder
pH of 0.3% solution: 7.5
Ionic Charge: High anionic

AMERFLOC PLUS 5485:
Fluid Flocculant

AMERFLOC PLUS 5485 fluid flocculant is a very high molecular weight, highly charged cationic polymer.
AMERFLOC PLUS 5485 is highly effective in dewatering industrial and municipal sludges as well as in wastewater clarification applications.

Typical Properties:
Appearance: White viscous liquid
Specific Gravity (25C): 1.05
Product Viscosity (25C): 1500 cps

AMERFLOC PLUS 5270:
Fluid Flocculant

AMERFLOC PLUS 5270 fluid flocculant is a very high molecular weight, highly charged anionic polymer.
AMERFLOC PLUS 5270 is highly effective in dewatering industrial sludges as well as in wastewater and kraft mill clarification applications.

Typical Properties:
Appearance: White viscous liquid
Specific Gravity (25C): 1.10
Product Viscosity (25C): 1500 cps

DUPONT CO.: DuPont DBE Solvent:

For your formulation and reformulation solvent needs in bake finish coatings and industrial clean-up applications, DuPont DBE Solvent is the clear solution. When blended with other low toxicity, low cost solvents, DBE (dibasic ester) Solvent is compatible with virtually all coil and sheet coatings resin systems.

For industrial cleanup, DuPont DBE provides a high performance alternative to methylene chloride.

Look into the many advantages of using DBE Solvent:

* has low toxicity
* is low cost
* blends easily with other low toxicity, low cost solvents
* improves adhesion and flexibility
* reduces solvent popping, keeps film open
* improves leveling and flow out
* reduces orange peel
* provides a high performance alternative to methylene chloride for industrial cleanup

Choose DBE Solvent for visible improvement in coatings quality

* As a coalescing aid in waterborne coatings
* As an anti-flocculant
* For increased "color development" in tinted waterbase systems

A clear case for DBE solvent in industrial cleaning applications:

* for cleanout of polyurethane application equipment
* for cleanout of paint application systems
* for cleanup of unsaturated polyester resins
* for cleanup of uncured automotive paint

EASTMAN CHEMICAL PRODUCTS, INC.: Chlorinated Polyolefins--Adhesion Promoter--Plastics and Nonplastics:

Type:

CP-153-2:
- Form: 25 wt % in xylene
- Specific Gravity at 25/25C: 0.970
- Viscosity at 25C, cP: 500-800
- Flash Point, TCC F(C): 81 (27)

CP-343-1:
- Form: 25 wt % in xylene
- Specific Gravity at 25/25C: 0.898
- Viscosity at 25C, cP: 100
- Flash Point, TCC F (C): 81 (27)

CP-343-1:
- Form: 25 wt % in xylene
- Specific Gravity at 25/25C: 0.902
- Viscosity at 25C, cP: 28
- Flash Point, TCC F (C): 42 (6)

CP-343-1:
- Form: 40 wt % in xylene
- Specific Gravity at 25/25C: 0.930

CP-343-1:
- Form: 50 wt % in xylene
- Specific Gravity at 25/25C: 0.935
- Viscosity at 25C, cP: >100,000
- Flash Point, TCC F (C): 83 (28)

CP-343-1:
- Form: 100% powder
- Specific Gravity at 25/25C: 1.03
- Flash Point, TCC F (C): 525 (274)

CP-343-3:
- Form: 25 wt % in xylene
- Specific Gravity at 25/25C: 1.108
- Viscosity at 25C, cP: 42
- Flash Point, TCC F (C): 74 (23)

CP-343-3:
- Form: 25 wt % in toluene
- Specific Gravity at 25/25C: 1.108
- Viscosity at 25C, cP: 25
- Flash Point, TCC F (C): 42 (6)

CP-343-3:
- Form: 50 wt % in xylene
- Specific Gravity at 25/25C: 1.108
- Viscosity at 25C, cP: 1030
- Flash Point, TCC F (C): 80 (27)

CP-515-2:
- Form: 40 wt % in xylene
- Specific Gravity at 25/25C: 0.955
- Viscosity at 25C, cP: 80
- Flash Point, TCC F (C): 82 (28)

CP-515-2:
- Form: 40 wt % in AROMATIC 100
- Viscosity at 25C, cP: 220
- Specific Gravity at 25C/25C: 0.944
- Flash Point, TCC F (C): 107

EASTMAN CHEMICAL PRODUCTS, INC.: EASTMAN Cellulose Esters:

Cellulose Acetate Butyrate

<u>Type:</u>

CAB-171-15S:
  Viscosity: Seconds: 15.0
             Poises: 57.0
  Acyl Content: Acetyl, %: 29.5
                Butyryl, %: 17
  Hydroxyl Content, %: 1.5
  Melting Range, C: 230-240
  Tg, C: 161

CAB-381-0.1:
  Viscosity: Seconds: 0.1
             Poises: 0.38
  Acyl Content: Acetyl, %: 13.0
                Butyryl, %: 37
  Hydroxyl Content, %: 1.5
  Melting Range, C: 155-165
  Tg, C: 123

CAB-381-0.5:
  Viscosity: Seconds: 0.5
             Poises: 1.9
  Acyl Content: Acetyl, %: 13.0
                Butyryl, %: 37
  Hydroxyl Content, %: 1.5
  Melting Range, C: 155-165
  Tg, C: 130

CAB-381-2:
  Viscosity: Seconds: 2.0
             Poises: 7.6
  Acyl Content: Acetyl, %: 13.0
                Butyral, %: 37
  Hydroxyl Content, %: 1.5
  Melting Range, C: 171-184
  Tg, C: 133

CAB-381-20:
  Viscosity: Seconds: 20.0
             Poises: 76.0
  Acyl Content: Acetyl, %: 13.0
                Butyral, %: 37
  Hydroxyl Content, %: 1.8
  Melting Range, C: 195-205
  Tg, C: 141

EASTMAN CHEMICAL PRODUCTS, INC.: EASTMAN Cellulose Esters (Continued):

Cellulose Acetate Butyrate(Continued):

Type:

CAB-500-1:
Viscosity: Seconds: 1.0
Poises: 3.8
Acyl Content: Acetyl, %: 5.0
Butyral, %: 49
Hydroxyl Content, %: 0.5
Melting Range, C: 165-175
Tg, C: 104

CAB-500-5:
Viscosity: Seconds: 5.0
Poises: 19.0
Acyl Content: Acetyl, %: 5.0
Butyral, %: 49
Hydroxyl Content, %: 0.8
Melting Range, C: 165-175
Tg, C: 96

CAB-531-1:
Viscosity: Seconds: 1.0
Poises: 3.8
Acyl Content: Acetyl, %: 2.8
Butyral, %: 50
Hydroxyl Content, %: 1.7
Melting Range, C: 135-150
Tg, C: 115

CAB-551-0.01:
Viscosity: Seconds: 0.01
Poises: 0.038
Acyl Content: Acetyl, %: 2.0
Butyral, %: 53
Hydroxyl Content, %: 1.5
Melting Range, C: 110-125
Tg, C: 85

CAB-551-0.1:
Viscosity: Seconds: 0.1
Poises: 0.038
Acyl Content: Acetyl, %: 2.0
Butyral, %: 52
Hydroxyl Content, %: 1.8
Melting Range, C: 130-140
Tg, C: 101

EASTMAN CHEMICAL PRODUCTS, INC.: EASTMAN Cellulose Esters (Continued):

Cellulose Acetate Butyrate:

Type:

CAB-551-0.2:
Viscosity: Seconds: 0.2
Poises: 0.76
Acyl Content: Acetyl, %: 2.0
Butyryl, %: 52
Hydroxyl Content, %: 1.8
Melting Range, C: 130-140
Tg, C: 101

CAB-553-0.4:
Viscosity: Seconds: 0.4
Poises: 1.5
Acyl Content: Acetyl, %: 2.0
Butyral, %: 47
Hydroxyl Content, %: 4.3
Melting Range, C: 150-160
Tg, C: 136

Cellulose Acetate Butyrate Carboxylate:

EASTACEL 1412:
Viscosity: Seconds: 0.1
Poises: 0.38
Acid Number: 12.5
Acyl Content: Acetyl, %: 13.5
Butyryl, %: 32.5
Hydroxyl Content, %: 1.8
Carboxyl Content, %: 1.0
Active Oxygen: 0.26
Melting Range, C: 141-150
Tg, C: 114

EASTACEL 1420:
Viscosity: Seconds: 0.03
Poises: 0.11
Acid Number: 20
Acyl Content: Acetyl, %: 13.5
Butyryl, %: 31.5
Hydroxyl Content, %: 1.8
Carboxyl Content, %: 1.6
Active Oxygen: 0.42
Melting Range, C: 134-142
Tg, C: 104

EASTMAN CHEMICAL PRODUCTS, INC.: EASTMAN Cellulose Esters (Continued):

Cellulose Acetate Propionate:

Type:

CAP-482-0.5:
Viscosity: Seconds: 0.5
Poises: 1.9
Acetyl Content %: 2.5
Combined Propionyl Content, %: 45
Hydroxyl Content, %: 2.8
Melting Range, C: 188-210
Tg, C: 142

CAP-482-20:
Viscosity: Seconds: 20.0
Poises: 76.0
Acetyl Content %: 2.5
Combined Propionyl Content, %: 46
Hydroxyl Content, %: 2.0
Melting Range, C: 188-210
Tg, C: 147

CAP-504-0.2:
Viscosity: Seconds: 0.2
Poises: 0.76
Acetyl Content %: 2.5
Combined Propionyl Content, %: 40
Hydroxyl Content, %: 5.0
Melting Range, C: 188-210
Tg, C: 159

Cellulose Acetate:

CA-320S:
Viscosity: Seconds: 0.05
Poises: 0.2
Acetyl Content %: 32.0
Combined Acetic Acid, %: 44.7
Hydroxyl Content, %: 8.4
Melting Range, C: 190-269

CA-394-60S:
Viscosity: Seconds: 60
Poises: 228.0
Acetyl Content %: 39.4
Combined Acetic Acid, %: 55.0
Hydroxyl Content, %: 4.0
Melting Range, C: 240-260
Tg, C: 186

EASTMAN CHEMICAL PRODUCTS, INC.: EASTMAN Cellulose Esters (Continued):

Cellulose Acetate(Continued):

Type:

CA-398-3:
Viscosity: Seconds: 3
Poises: 11.4
Acetyl Content %: 39.8
Combined Acetic Acid, %: 55.5
Hydroxyl Content, %: 3.5
Melting Range, %: 230-250
Tg, C: 180

CA-398-6:
Viscosity: Seconds: 6
Poises: 22.8
Acetyl Content %: 39.8
Combined Acetic Acid, %: 55.5
Hydroxyl Content, %: 3.5
Melting Range, %: 230-250
Tg, C: 182

CA-398-10:
Viscosity: Seconds: 10
Poises: 38.0
Acetyl Content %: 39.8
Combined Acetic Acid, %: 55.5
Hydroxyl Content, %: 3.5
Melting Range, %: 230-250
Tg, C: 185

CA-398-30:
Viscosity: Seconds: 30
Poises: 114.0
Acetyl Content %: 39.8
Combined Acetic Acid, %: 55.5
Hydroxyl Content, %: 3.5
Melting Range, C: 230-250
Tg, C: 189

CA-435-75S:
Viscosity: Seconds: 75
Poises: 285
Acetyl Content %: 43.5
Combined Acetic Acid, %: 60.7
Hydroxyl Content, %: 0.94
Melting Range, C: 265-299

CA-436-80S:
Viscosity: Seconds: 80
Poises: 304
Acetyl Content %: 43.7
Combined Acetic Acid, %: 61.0
Hydroxyl Content, %: 0.82
Melting Range, C: 269-300

FLORIDIN CO.: MIN-U-GEL 400 Gelling Agent:

Gelling Water and Organic Liquids with MIN-U-GEL 400

MIN-U-GEL 400 is a specially processed fine-particle-size, colloidal gelling agent prepared from the clay mineral attapulgite. When dispersed and reflocculated, MIN-U-GEL 400 can be used to thicken a variety of water and organic liquids such as tap water, ionically contaminated waters (brines, sea water, etc.), aliphatic and aromatic organics, esters, alcohols, oils, liquid fats and others.

Gelling Properties:

Gel characteristics obtained with MIN-U-GEL 400 can be described as yield-pseudoplastic and thixotropic. Such gelled liquids exhibit (1) gel strength, (2) high apparent viscosities at low shear rates, (3) low apparent viscosities when subjected to high shear rates, and (4) fairly rapid recoverability of original properties when shear is halted.

MIN-U-GEL 400 Typical Product Data:

Color	light tan-gray
Free Moisture, %	10.6
Dry Particle Size	
-25 um	99.3
-44 um (325 mesh)	99.9
Residue on 325 Mesh screen	0.017
pH	9.6
Particle Density, g/ml	2.4
Bulking Value,	
lbs/gal	20
gals/lb	0.05
Average Dispersed Particle Size, um	0.12
Fineness of Grind (Hegman Gauge)	7.5
Packed Bulk Density, lbs/ft 3	26
Brookfield Viscosity, cP (7% clay dispersion in water)	2200

Chemical Analysis
(volatile free basis, 1200F)

	%
Si as SiO2	66.21
Al as Al2O3	11.71
Fe as Fe2O3	4.02
Ti as TiO2	0.55
P as P2O5	0.99
Ca as CaO	2.92
Mg as MgO	9.70
K as K2O	1.07
C as CO2	2.57
S as SO4	0.25
Other	0.01

HARWICK CHEMICAL CORP.: AQUAMIX Aqueous Dispersions and Emulsions:

AQUAMIX	Active Material	% Activity	Pounds/Gallons
101	AGERITE STALITE Emulsion	50	8.4
102	AGERITE WHITE Dispersion	50	9.3
103	Ammonium Caseinate	10	8.5
104	Antimony Oxide Dispersion	60	17.2
105	Antioxidant 2246 Dispersion	40	8.9
106	Antioxidant 425 Dispersion	40	8.6
107	BHT Dispersion	50	8.7
108	Chlorinated Paraffin Dispersion	65	11.3
109	DOP Emulsion	65	8.2
110	FLECTOL Dispersion	50	8.6
111	MBTS Dispersion	50	10.1
112	NAUGAWHITE Emulsion	60	8.1
114	SANTOVAR A Dispersion	50	8.5
115	S-W Crystals Dispersion	40	8.7
116	S-W Powder Dispersion	40	8.5
118	Sodium Silicofluoride Dispersion	50	12.5
119	Sodium Silicofluoride Dispersion	25	10.2
120	Rubbermaker's Sulfur Dispersion	68	12.9
121	TMTD Dispersion	55	10.0
122	DPTT Dispersion	50	10.1
123	Titanium Dioxide Dispersion	50	13.4
124	Titanium Dioxide Dispersion	60	15.2
125	WINGSTAY L Dispersion	50	8.8
126	WINGSTAY 29 Emulsion	50	8.6
127	WINGSTAY L Emulsion	50	8.4
128	ZMBT Dispersion	50	10.5
129	Zinc Dibutyl Dithiocarbamate Dispersion	50	9.3
130	Zinc Diethyl Dithiocarbamate Dispersion	50	10.1
131	Zinc Dimethyl Dithiocarbamate Dispersion	50	10.7
132	Zinc Oxide Dispersion	60	16.6
137	Special WINGSTAY Emulsion	50	8.18
142	Zinc Oxide Dispersion	50	14.36
156	Antioxidant SP Emulsion	50	8.68
160	STANGARD 500 Dispersion	50	9.52
162	Accelerator Thiocarbanilide Dispersion	50	9.52
165	DPG Dispersion	50	9.19
167	Zinc Dibenzyldithiocarbamate Dispersion	40	9.5
168	WINGSTAY 100 Dispersion	40	8.93
170	Chlorinated Paraffin/Antimony Oxide Disp.	65	12.53

HARWICK CHEMICAL CORP.: AQUAMIX Aqueous Dispersions and Emulsions (Continued):

AQUAMIX	Active Material	% Activity	Pounds/Gallon
179	TMTM Dispersion	50	9.85
180	Cure System for Latex Foam Dispersion	60	10.86
181	TETD Dispersion	50	9.6
186	Water Washed Clay Dispersion	65	19.79
189	AMINOX Dispersion	50	9.02
192	PERMANEX WSP Dispersion	40	8.93
199	Clay Based Slab Dip	43	11.36
215	STAN PLAS 1200 Emulsion	60	8.0
228	TOPANOL CA Dispersion	40	8.6
223	Zinc Carbonate Dispersion (Transparent Zinc Oxide)	45	12.94
239	PICCOPALE 100 Emulsion	48	8.10
242	FIREBRAKE ZB Dispersion	50	12.27
245	POLYPALE ESTER #10 Emulsion	48	8.26
246	STABELITE ESTER 10	48	8.26
247	PENTALYN A Emulsion	48	8.26
257	WINGSTAY L/SYNERGIST Emulsion	50	8.43
275	IRGANOX MD-1024 Dispersion	40	8.76
280	PVI Dispersion	50	9.51
289	WINGSTAY L/SYNERGIST Emulsion	50	8.43
290	WINGSTAY L/SYNERGIST	50	8.35
302	PENTALYN H Emulsion	48	8.26
303	IRGANOX 1077 Emulsion	40	8.43
304	SULFASAN R Dispersion	50	9.68
312	WINGSTAY 29/SYNERGIST Emulsion	50	8.51
318	PENTAL A Emulsion	48	8.26
321	WINGTACK 95 Emulsion	48	7.93
323	CDMC (Accelerator) Dispersion	50	10.6
325	IRGANOX 565 Dispersion	25	8.51
327	IRGANOX 1010 Dispersion	40	9.43
329	Zinc Stearate Dispersion	45	8.76
330	Potassium Oleate	20	8.43
331	TEFLON TL115A Dispersion	60	12.52
339	TINUVIN 144 Dispersion	40	8.6
343	DETU Dispersion	50	8.93
344	Deca Bromo DiPhenyl Oxide/Antimony Oxide	65	16.03
345	ZIPX (Accelerator) Dispersion	55	10.43

HARWICK CHEMICAL CORP.: AQUAMIX Aqueous Dispersions and Emulsions(Continued):

AQUAMIX	Active Material	% Activity	Pounds/Gallon
346	IRGANOX L57 Emulsion	50	8.18
352	TINUVIN 765 Emulsion	50	8.35
354	PLASTICHLOR 48-260 Emulsion	65	9.51
376	Ozone Resistant Wax Emulsion	50	7.93
377	EASTOTAC H-100 Resin Emulsion	45	8.26
378	NEVTAC 100 Resin Emulsion	45	8.01
385	Polybutene Emulsion	65	7.76
386	SANTOVAR A	40	8.67
387	DOA Emulsion	65	7.93
388	PERMANEX WSP	50	9.18
391	DIDP Emulsion	65	8.09
394	Zinc Stearate	25	8.60
396	WINGSTAY 29/SYNERGIST Emulsion	50	8.43
398	FORAL 85 Emulsion	48	8.26
399	Litharge	70	22.53
404	IRGANOX 1076/SYNERGIST	40	8.26
405	TCP Emulsion	65	9.18
406	1,3, Butylene Glycol Dimetharylate Emul.	50	8.35
414	WINGSTAY K Emulsion	60	7.59

INTERSTAB CHEMICALS: Anti-Skinning Agents:

Skinning is a problem common to all, oxidizable coatings. Interstab offers two products effective in controlling skinning of coatings. Both products are ketoximes and as such function predominantly by complexing the active drier in the paint system. Since they are not anti-oxidants they do not adversely affect dry time at effective use levels.

ASKA:

ASKA, Anti-Skinning Agent, is 100% methylethyl-ketoxime and is effective in preventing skinning of all oxidizable vehicles with minimal or no effect on dry time. Excellent results are obtained in alkyds, oil, oil modified urethane, epoxy esters and oleo-resinous based paints. ANTI-SKINNING AGENT has also been found effective in improving stability of water-reducible alkyd paints.

ASKA 319:

ASKA 319 is 100% butyraldoxime and is effective in preventing skinning of all oxidizable paints. It is particularly effective in oleoresinous and oil based paints.

Physical Properties:

Anti-skinning Agent
Viscosity (G/H-Max.): A
Form: Clear Liquid

ASKA 319:
Viscosity (G/H-Max.): A
Form: Clear Liquid

INTERSTAB CHEMICALS: Loss Of Dry Additives:

INTERDRY Cobalt 21%:

INTERDRY Cobalt 21 is a lead free, smooth pourable paste designed for use as a feeder drier to counteract the loss-of-dry effect that usually results when pigments with high drier absorptive capacity are used in coating formulations.

INTERDRY DS:

INTERDRY DS is a liquid, lead-free, loss of dry additive effective in minimizing or eliminating loss-of-dry that may develop in coatings during storage. It is specifically designed to stabilize and maintain the original drying characteristics of paints formulated with pigments having high drier absorption capacity.

INTERSTAB CHEMICALS: Specialty Control Products:

PREVENTOL:

PREVENTOL is a white, fine particle size extender pigment. It controls or eliminates flooding and floating by improving pigment wetting. PREVENTOL is compatible with all solvent based paints and should be added to the paint during the pigment dispersion phase of manufacture.

ANTI-SETTLING AGENT:

ANTI-SETTLING AGENT is a chemically processed clay that prevents pigment settling and caking and aids in wetting and proper dispersion. Unlike other anti-settling agents it will not appreciably change viscosities of paints.

STABILISAL C-1:

A specially prepared organic compound designed to stabilize lacquers against greening, and as an anti-gelation agent for nitrocellulose-based bronzing lacquers. STABILISAL C-1 is specially recommended for clear metal lacquers to:

* Prevent greening of dipping lacquers in tank
* Restore original color of lacquers already turned green
* Prevent gelation of nitrocellulose-based bronzing lacquers

PLASTOFLEX 520:

A phthalate ester type plasticizer offered primarily for its utility in the lacquer coating field. PLASTOFLEX 520 is of exceptional interest to the lacquer formulator because it is economical to use and possesses solvent characteristics midway between the solvent and non-solvent types of plasticizers for nitrocellulose.

ANTI-SAGGING AGENT:

ANTI-SAGGING AGENT is a 100% active organic compound which imparts thixotropy to coatings vehicles to prevent sagging. ANTI-SAGGING AGENT does not adversely affect film properties or package stability.

INTERWAX P:

INTERWAX P contains rubbery butenes dispersed in paraffin wax and is useful in reducing the moisture permeability of wax coatings. Because the butenes are thoroughly dispersed in the wax, INTERWAX P can be dissolved easily without prolonged heating and stirring. The use of INTERWAX P results in improved low temperature flexibility, decreased moisture vapor permeability, and better longevity versus wax coatings alone. The melting point of INTERWAX P is 133-135F and it is considered safe for use in chewing gum. (21CFR 121.1059).

INTERSTAB CHEMICALS: Specialty Control Products(Continued):

CALCIUM PULP:
CALCIUM PULP is a specially processed soap of linoleic acid developed for use in solvent based systems. CALCIUM PULP is a useful ingredient in putty formulations to prevent oil pigment separation. In paint it aids suspension and controls penetration. It also improves package stability and reduces sagging.

VISTAC A:
VISTAC A combines pronounced rubber softening action with tack characteristics. It imparts adhesion without significantly reducing cohesive strength of the finished product.

ANTI-LIVERING AGENT:
ANTI-LIVERING AGENT is a fatty ester glycol ether widely used in problems involving reactive pigments in paints. It is unique in that it is relatively low in acid number and usually requires only a small dosage. ANTI-LIVERING AGENT has been widely used with Iron Blue and Earth Colors (i.e., umbers and red oxides) in oleoresinous vehicles and lacquers. It is recommended that efficacy of ANTI-LIVERING AGENT be determined in each specific formulation

Physical Properties:

PREVENTOL:
Viscosity (G/H-Max.): N/A
Form: Powder
ANTI-SETTLING AGENT:
Viscosity (G/H-Max.): N/A
Form: Powder
STABILISAL C-1
Viscosity (G/H-Max.): D
Form: Clear Liquid
PLASTOFLEX 520:
Viscosity (G/H-Max.): A
Form: Clear Liquid
ANTI-SAGGING AGENT:
Viscosity (G/H-Max.): D
Form: Clear Liquid
INTERWAX P:
Viscosity: 3800 cps @ 120C
Form: Solid Wax
CALCIUM PULP:
Viscosity: Paste
Form: Creamy Pulp
VISTAC A:
Viscosity (G/H-Max.): Z6
Form: Clear Liquid
ANTI-LIVERING AGENT:
Viscosity (G/H-Max.): A
Form: Clear Liquid

KING INDUSTRIES: DISLON Products:

Product:

Anti-Sag & Anti-Settle Agents(Viscosity Increasing):

DISLON 4110:

Form: Powder
Chemical Type: Complex castor oil derivative
Applications:
- Solventless coatings, sealants, caulks
- Aromatic solvent coatings
- Aliphatic solvent coatings

Features:
- Good sag & slump control
- Good flow & leveling

DISLON 6500:

Form: Powder
Chemical Type: Polyamide
Applications:
- Solventless coatings, sealants, caulks
- Aromatic solvent coatings
- Aliphatic solvent coatings

Features:
- Prevent settling of heavy pigments.
- Non-seeding.
- Good heat stability.
- Excellent stability on aging.

DISLON 6900-20X:

Form: Paste
Chemical Type: Polyamide in xylene
Applications:
- All non-aqueous systems, including gel coats and high gloss topcoats.
- Anti-sag for high build coatings.
- Anti-settle for heavy pigments.

Features:
- No grinding required.
- Excellent anti-sag & settle.
- Age stability.
- Non-seeding.
- Master batch can be added to letdown.

KING INDUSTRIES: DISLON Products(Continued):

Anti-Settle & Anti-Sag Agents(Low Viscosity Increasing):

<u>Product:</u>

DISLON TP-202:
Form: Powder
Chemical Type: Complex synthetic wax
Applications:
Recommended for hot melt finishes, esp. traffic paints.
Features:
Anti-settle for pigments & glass beads.
Prevents slumping.
Improves wear resistance.

DISLON 4200:
Form: Paste
Chemical Type: Polyolefin in xylene
Applications:
All non-aqueous pigment dispersed systems
Features:
Stabilizes pigment dispersion.
Non-seeding.
May be synergistic with other thixotropes.

DISLON 4200-20:
Form: Soft paste
Chemical Type: Polyolefin in xylene
Applications:
All non-aqueous pigment dispersed systems
Features:
Easier dispersing version of 4200.

DISLON 4200-10:
Form: Liquid
Chemical Type: Polyolefin in xylene
Applications:
All non-aqueous pigment dispersed systems.
Features:
Post addition.

DISLON NS-30:
Form: Paste
Chemical Type: Complex polyolefin & polyamide in xylene
Applications:
All non-aqueous pigment dispersed systems.
Anti-sag for industrial maintenance coatings.
Anti-settle for heavy pigments.
Features:
Excellent pigment dispersion.
Non-seeding.
Stability on aging.
More effective than 4200 series.

KING INDUSTRIES: DISLON Products(Continued):

Anti-Flooding, Anti-Floating, Anti-Silking:

Product:

DISLON 201SP:
Form: Powder
Chemical Type: Nonionic surfactant on inorganic pigment
Applications:
Most non-aqueous pigment dispersed systems.
Recommended for air-dry alkyds and acrylic paints, acrylic baking enamels.
Features:
Eliminate floating in spray applications.
Prevent cratering & pinholes.
Good recoatability.

DISLON 230:
Form: Liquid
Chemical Type: High MW nonionic surfactant in xylene
Applications:
All non-aqueous coatings & inks.
Anti-float in baking enamels and lacquers. Defoamer for epoxy, urethane, & CI-rubber finishes.
Features:
Non silicone
Good gloss & leveling
Defoaming, deairating.
Recoatability.

DISLON 2200:
Form: Liquid
Chemical Type: Silicone modified aliphatic polycarboxylic acid in xylene
Applications:
Recommended for solvent systems with high polarity such as lacquers & baking enamels.
Features:
Stabilizes pigment dispersion
Reduces grind time
Improves hiding.

DISLON KS-273N:
Form: Liquid
Chemical Type: High MW anionic surfactant in xylene
Applications:
All non-aqueous systems.
Particularly effective in baking enamels & epoxy and urethane floor coatings.
Features:
Stabilizes pigment dispersion.
Improves adhesion.
Non silicone--excellent recoatability.

KING INDUSTRIES: DISLON Products(Continued):

Leveling:

Product:

DISLON L-1980:

Form: Liquid
Chemical Type: Polyacrylate
Applications:
Most solvent-based coatings, particularly high solids.
Urethanes.
Powder coatings.
Features:
Excellent flow & leveling.
Promotes pigment dispersion.
Good recoatability.
Good gloss & leveling without pinholes or craters.

DISLON L-1982:

Form: Liquid
Chemical Type: Polyacrylate
Applications:
Same as L-1980
Features:
Very good flow & leveling
May be more cost effective.

DISLON PL-430:

Form: Powder
Chemical Type: 100% Organic polymers
Applications:
Powder coatings, particularly isocyanate-cure polyesters.
Features:
Improves flow, surface smoothness, gloss, image reflectance.
Reduces cratering.

DISLON PL-620:

Form: Powder
Chemical Type: 100% Organic polymers
Applications:
Powder coatings, especially if pinholing is a problem.
Features:
Same, plus excellent anti-pinholing.

KING INDUSTRIES: DISLON Products(Continued):

Wetting, Dispersing, Stabilizing:

DISLON 2150:

Form: Liquid
Chemical Type: Aliphatic carboxylic acid in xylene
Applications:
Recommended for systems with polar vehicles--lacquers, baking enamels.
Systems with high levels of carbonate pigments or TiO_2
Features:
Improves wetting.
Prevents floc & settling in storage. Synergistic with anti-floating agents.

DISLON 7004:

Form: Liquid
Chemical Type: Polyether ester anionic surfactant in xylene
Applications:
All non-aqueous pigmented systems.
Recommended for carbon black & phthalocyanine blue.
Features:
Reduces grind time.
Reduces oil absorption & permits higher pigment concentrations.
Improves gloss & leveling

DISLON KS-873N:

Form: Liquid
Chemical Type: High MW surfactant in xylene
Applications:
All non-aqueous pigmented systems, particularly alkyd & acrylic baking enamels.
Features:
Prevents flocculation.
Improves color development.
Improves leveling & gloss.
Improves adhesion.

KING INDUSTRIES: DISLON Products(Continued):

Defoaming & Leveling (Non-silicone):

Product:

DISLON 1925:

Form: Liquid
Chemical Type: Vinyl polymer in xylene & esters
Applications:
Unsaturated polyesters, urethanes, epoxies.
Particularly effective on wood substrates.
Features:
Minimizes craters, brush marks.
Enhances recoatability.

DISLON 1950:

Form: Liquid
Chemical Type: Vinyl polymer in mineral spirits
Applications:
Most air dry solvent-based paints--architectural, marine, heavy duty. Epoxies. Acid-cured alkyds. Chlorinated rubber.
Features:
Prevents foaming in production and air spray, brush & roller applications.

DISLON 1970:

Form: Liquid
Chemical Type: High MW surfactant in xylene
Applications:
Baking enamels--coil, can, auto coatings.
High build urethanes & epoxies.
Features:
Effective defoaming with leveling, gloss, adhesion, recoatability.

DISLON OX-70:

Form: Liquid
Chemical Type: High MW surfactant in mineral spirits/ toluene
Applications:
High viscosity paints & mastics.
Urethane & epoxy floor coatings.
Baking enamels. Coil coatings.
Sanding sealers.
Features:
Strongest defoamer.
Recoatable.

KING INDUSTRIES: DISLON Products(Continued):

Defoaming & Leveling(Non-silicone)(Continued):

Product:

DISLON OX-880:

Form: Liquid
Chemical Type: High MW surfactant in toluene
Applications:
Most solvent-based systems.
Ambient-cure & multi-component paints.
Baking enamels. Mastics.
Features:
Recoatable
Releases air entrapment from fumed silica.

Mar Resistance:

DISLON S-5X03-E:

Form: Paste
Chemical Type: Complex polyolefin in xylene
Applications:
Wood coatings. Baking enamels, particularly coil & can coatings.
Features:
Anti-blocking.
Improves slip & wear resistance.
Increases stain resistance.
Recoatable.

KING INDUSTRIES: NACURE and K-CURE Acid Catalysts:

Product:

NACURE 155:

Acid Type: DNNDSA
% Active: 55
Volatile: Isobutanol
Acid No.: 112-116
Application:

Appliance	Auto
Can	Coil
Electrocoat	Electrostatic
Gen. Industrial	Wood
Ink	

Recommended Uses/Applications: General purpose catalyst. Excellent water, detergent and salt spray resistance. Recommended for high solids and waterborne metal and wood finishes, electrostatic spraying.

NACURE 1051:

Acid Type: DNNSA
% Active: 50
Volatile: Ethylene glycol monobutyl ether
Acid No.: 60-64
Application:

Appliance	Can
Coil	Electrocoat
Electrostatic	Gen. Industrial
Ink	

Recommended Uses/Applications: Slower curing. Best water and corrosion resistance. Recommended for high temperature applications on metal. Excellent overbake resistance.

NACURE 5076:

Acid Type: DDBSA
% Active: 70
Volatile: Isopropanol
Acid No.: 130-140
Application:

Auto	Can
Gen. Industrial	Paper
Ink	

Recommended Uses/Applications: Cost effective. Complies with FDA CFR 21, Section 175--300 (b)(3)xiii(a) and (b)

KING INDUSTRIES: NACURE & K-CURE Acid Catalysts(Continued):

Product:

NACURE 4054:
Acid Type: Acid Phosphate
% Active: 50
Volatile: Isobutanol
Acid No.: 155-165
Application:
Appliance — Auto
Gen. Industrial — Wood
Recommended Uses/Applications: Weak acid catalyst for high-NH/polymeric melamines and phenolic crosslinkers. Excellent solubility with good package stability, salt spray, adhesion and water resistance.

K-CURE 1040:
Acid Type: p-TSA
% Active: 40
Volatile: Isopropanol
Acid No.: 130-140
Application:
Appliance — Auto
Can — Coil
Gen. Industrial — Paper
Wood — Ink
Recommended Uses/Applications: General purpose, high quality grade acid. For applications where weathering and exterior durability are needed. Highest gloss. Fast cure.

K-CURE 1040W:
Acid Type: p-TSA
% Active: 40
Volatile: Water
Acid No. 130-140
Application:
Paper — Wood
Ink
Recommended Uses/Applications: Same as above, but non-flammable for WB applications.

K-CURE 129B:
Acid Type: Mixed acids
% Active: 50
Volatile: Methanol/n-Butanol
Acid No. 200-210
Application:
Paper — Wood
Ink
Recommended Uses/Applications: Very fast cure. Recommended for wood and paper coatings

KING INDUSTRIES: NACURE & K-CURE Acid Catalysts(Continued):

Product:

K-CURE 129W:
Acid Type: Mixed Acids
% Active: 50
Volatile: Water
Acid No.: 200-210
Application:
Paper   Wood
Ink
Recommended Uses/Applications: Water based inks and wood coatings (Non-flammable).

KING INDUSTRIES: NACURE Blocked Catalysts:

NACURE X49-110:
Acid Type: DNNDSA
% Active: 25
Volatile: Isobutanol/Isopropanol
pH: 6.5-7.5
Application:
Appliance   Auto
Can   Coil
Electrocoat   Electrostatic
Gen. Industrial
Recommended Uses/Applications: Best overall properties. Water and corrosion resistance, adhesion. Excellent for electrostatic spraying and waterborne coatings.

NACURE 3525:
Acid Type: DNNDSA
% Active: 25
Volatile: Isobutanol/Isopropanol
pH: 7.0-8.0
Application:
Appliance   Auto
Can   Coil
Electrocoat   Electrostatic
Gen. Industrial
Recommended Uses/Applications: Better solubility than NX49-110 but slower curing. Metaldeco, coil coating. Good adhesion and salt spray resistance.

NACURE 1323:
Acid Type: DNNSA
% Active: 21
Volatile: Xylene
Acid No. 6.8-7.5
Application:
Appliance   Can
Electrostatic
Recommended Uses/Applications: Recommended for high temperature applications. Excellent solubility in aromatic and aliphatic solvents.

KING INDUSTRIES: NACURE Blocked Catalysts(Continued):

Product:

NACURE 1419:

Acid Type: DNNSA
% Active: 30
Volatile: Xylene/MIBK
Application:
Can
Coil
Electrocoat
Electrostatic
Gen. Industrial

Recommended Uses/Applications: Excellent for electrostatic spraying. High bake applications for maximum water, detergent and salt spray resistance.

NACURE 5225:

Acid Type: DDBSA
% Active: 25
Volatile: Isopropanol
Acid No.: 6.0-7.0
Application:
Auto
Can
Electrostatic
Gen. Industrial

Recommended Uses/Applications: Best stability in high solids enamels. Good solubility in aliphatic solvents.

NACURE 5925:

Acid Type: DDBSA
% Active: 25
Volatile: Isopropanol
pH: 7.0-7.5
Application:
Auto
Can
Gen. Industrial

Recommended Uses/Applications: Good stability. Complies with FDA CFR 21, Section 175-300(b)(3)xii(a) and (b).

NACURE 2500:

Acid Type: p-TSA
% Active: 25
Volatile: Isopropanol
Acid No. 6.0-7.0
Application:
Appliance
Auto
Coil
Gen. Industrial

Recommended Uses/Applications: Low cure. Excellent stability in waterborne and high solids systems.

KING INDUSTRIES: NACURE Blocked Catalysts(Continued):

Product:

NACURE 2501:
Acid Type: TSA
% Active: 25
Volatile: Methanol/Isopropanol
pH: 6.0-7.2
Application:
Auto — Coil
Electrostatic — Gen. Industrial
Recommended Uses/Applications: Slightly higher electrical resistance than N-2500. Better ketone solubility.

NACURE 2530:
Acid Type: p-TSA
% Active: 25
Volatile: Methanol/Isopropanol
pH: 5.75-6.5
Application:
Auto
Gen. Industrial
Wood
Recommended Uses/Applications: Low cure. Low tendency to yellow or wrinkle.

NACURE 2522:
Acid Type: p-TSA
% Active: 25
Volatile: Methanol/Isopropanol
pH: 3.0-5.0
Application:
Auto — Coil
Gen. Industrial — Paper
Wood
Recommended Uses/Applications: Low cure. For fast cure schedule with short dwell times.

NACURE 2558:
Acid Type: p-TSA
% Active: 25
Volatile: Ethylene Glycol
Acid No.: 3.5-4.5
Application:
Auto — Coil
Gen. Industrial
Recommended Uses/Applications: Excellent for controlling popping and blistering in HS and WB coatings.

NACURE 4167:
Acid Type: Acid Phosphate
% Active: 25
Volatile: Isopropanol/Isobutanol
pH: 6.8-7.3
Application: Gen. Industrial
Recommended Uses/Applications: Blocked phosphate for high-NH/polymeric melamines. Good solubility.

KING INDUSTRIES: Specialty Products:

Polyester Polyols:

K-FLEX:

188:
% Solids: 100
Viscosity (cps.): 10,000
Hydroxyl No.: 230
Hydroxyl Equiv. Weight: 244

148:
% Solids: 100
Viscosity (cps.): 4000
Hydroxyl No.: 235
Hydroxyl Equiv. Weight: 239

Applications: Modifier in Acrylic, Alkyd, Epoxy & Polyester formulations. Improves flexibility, adhesion, film hardness and VOC.

XP 171-90:
% Solids: 88
Viscosity (cps.): 5000
Hydroxyl No.: 130
Hydroxyl Equiv. Weight: 460
Applications: Modifier in Urethane, Acrylic, Alkyd, Epoxy & Polyester formulations. Improves flexibility, adhesion, storage stability and VOC.

Acrylic Oligomers:

K-Cryl:

200-6/MAK:
% Solids: 75
Solvent: MAK
Viscosity (cps.): 3000
Acid No. (Solids): 25
Hydroxyl No. (Solids): 92

200-6/MIBK:
% Solids: 75
Solvent: MIBK
Viscosity (cps.): 3500
Acid No. (Solids): 28
Hydroxyl No. (Solids): 94

Applications: Alkyd, Polyester and Urethane modifier. Pigment grinding vehicle.

KING INDUSTRIES: Specialty Products(Continued):

Acrylic Oligomers(Continued):

K-CRYL:

23-80:

% Solids: 80
Solvent: MIBK
Viscosity (cps.): 6000
Acid No. (Solids): 26
Hydroxyl No. (Solids): 64
Applications: Same as K-CRYL 200-6 with greater film flexibility

Rust and Corrosion Inhibitors:

NACORR:

1151:

% Metal: 6.5% Ba
% Active (DNNSA): 50
Volatile: Mineral Spirits
Weiqht/Gallon (lbs.): 8.02

1351:

% Metal: 2.1% Ca
% Active (DNNSA): 50
Volatile: Mineral Spririts
Weight/Gallon (lbs.): 7.73

1552:

% Metal: 3.3% Zn
% Active (DNNSA): 50
Volatile: Ethylene glycol monobutyl ether
Weight/Gallon (lbs.): 8.29

Applications: Coatings demanding strict rust and corrosion resistance properties. Primers, maintenance coatings.

LONZA INC.: ACRAWAX C:

ACRAWAX C is a unique grade of N,N' ethylenebisstearamide designed to provide the ultimate performance in a multitude of industrial products' applications.

Offering an exceptional balance of chemical and physical properties, ACRAWAX C affords broad flexibility in formulating performance products for demanding end uses.

ACRAWAX C is manufactured to an exceedingly high level of purity. This combined with its unique structure contributes to overall ease of handling and quality end products.

A choice of five physical forms of ACRAWAX C is available to fit most processing schemes

* Prilled
* Beaded
* Powdered
* Atomized
* Dispersion

Typical Physical Properties:

Color/Form: Cream/Hard Waxy Solid
Gardner Color, 1963: 5 max
Melting Point, C: 140-145
Acid Value: 8 max
Flash Point, C: 285
Specific Gravity: 0.97
Ash, %: 0.05

Applications:

Adhesives:

ACRAWAX C is utilized as an anti-block and process aid in thermoplastic heat sealing and pressure sensitive adhesives to reduce cold flow and melt viscosity.

Wire Drawing

Coatings:

ACRAWAX C is the preferred slip agent for food and beverage container resinous coatings.

In paints, stains, and lacquers, ACRAWAX C improves weatherability and insures homogeneous leveling.

For printing inks and pigmented vehicles, ACRAWAX C is an efficient grinding and dispersing agent.

ACRAWAX C will not effect substrate adhesion or optical properties.

Asphalt Specialties
Powder Metallurgy
Paper Defoamers
Foundry Resins
Electrical Products
Wax and Resinous Blends

FDA Status:

ACRAWAX C is produced with raw materials and under conditions which are approved for use in the manufacture of a variety of industrial products intended for food contact applications.

MILLIKEN CHEMICALS: SYN FAC Polyols:

Formulating for High Solids Coatings

The use of reactive diluents in paint and coating recipes is a relatively simple way of reducing volatile organic content (VOC).

However, many products employed as reactive diluents have certain drawbacks, i.e., they are too volatile and/or too water sensitive or cause other deficiencies in the coating.

Milliken Chemical's SYN FAC line of reactive diluents is based on ethoxylated and/or propoxylated Bisphenol-A. These polyols offer many advantages to the paint and coating industry.

Benefits:

* Liquid at room temperature.
* Non-volatile at normal bake temperatures.
* Difunctional, fully reactive chain linkers.
* These polyols function as completely slow solvents which react into the resin matrix.
* Non-water sensitive.

In addition, SYN FAC polyols do not adversely affect important film properties. In fact certain properties are enhanced by the SYN FACs. These include impact resistance, flex, adhesion and flow properties in certain coatings.

Milliken Chemical produces a wide range of Bisphenol-A based polyols of various molecular weights and viscosities. Two products that have found wide use in the paint and coatings industry.

SYN FAC 8009:

Primarily used in water based systems to upgrade film properties.

SYN FAC 8031:

Designed for solvent based systems to reduce VOC and upgrade film properties.

SYN FAC 8009:

SYN FAC 8009 is an ethylene oxide adduct of Bisphenol A. Its primary hydroxyl difunctionality in combination with the double phenolic rings gives this polyol property enhancement value in many paints and coatings.

SYN FAC 8031:

SYN FAC 8031 effectively improves solvent borne coatings by imparting greater flexibility, impact strength and higher solids. SYN FAC 8031 is an ethylene oxide adduct of Bisphenol A with primary hydroxyl difunctionality. It has a low degree of water sensitivity.

MOONEY CHEMICALS, INC.: Additives:

DRI-RX Drier Accelerator:
Specific Gravity 25/25C.: 0.987
% Active: 30
Viscosity, Gardner: A-5
Color: Amber
Flash Point: 90F.

VISCOTROL A Rheology Modifier:
Specific Gravity: 0.980
Form: White Powder
Melting Point: 86C.

SKINO #1 (Anti-Skinning Agent):
Specific Gravity, 25/25C: 0.907
NVM: nil
Viscosity, Gardner: A
Color: Water White
Weight: 7.55 pounds per gallon
Flash Point: 160F.

SKINO #2 (Anti-Skinning Agent):
Specific Gravity: 0.92
NVM: Nil
Color: Water white to pale straw
Viscosity: A-Gardner
Weight: 7.6 pounds per gallon

21% Cobalt TEN-CEM "Feeder" Drier:
% Cobalt: 21.0 minimum
Viscosity: Fluid
Grind, Hegman: 5H minimum
Lbs/Gal: 9.9
% NVM: 71.0
Color: Royal Blue

SPURSO (Dispersing Agent):
Specific Gravity: 0.961
% Solids: 70.0
Color: Amber
Flash Point: 104F. (minimum)
Lbs/Gal: 8.00
Viscosity: Fluid

HEX-CEM LFD Lead Free "Feeder" Drier:
Color: Bluish White
Weight: 10.9 pound per gallon
Specific Gravity, 25/25C.: 78.0
Viscosity, Seconds, 77F.: 6
Fineness of Grind, Hegman: 5
Flash Point, Minimum: 104F.
% NVM: 78

NL CHEMICALS, INC.: Anticorrosive Pigments:

NALZIN 2
Anticorrosive Pigment

Cost Effective Corrosion Protection for Both Solvent and Aqueous Systems

NALZIN 2 is an improved, second generation zinc phospho oxide, replacing NALZIN SC-1.

Advantages:

* High corrosion inhibition activity
* Very cost effective
* Non-toxic
* Low water solubility; gives both long term protection and stability
* Small particle size allows fine grind and high gloss
* Easy deagglomeration for rapid and clean dispersion
* Low oil absorption increases formulating flexibility
* Low water and dispersant demands
* White color with low tint strength permits formulation of any color
* Minimizes tendency of latex paints to flash rust
* Effective in solvent and water based paints.

ONCOR M50
Anticorrosive Pigment

With an ONCOR M50 paint system, you can repaint up to three bridges for the cost you would expect to pay for repainting one.

For what it costs to protect a bridge with a poor-wetting inorganic primer paint system, you can protect three comparable bridges with an ONCOR M50 Defense-in-Depth paint system.

PFIZER PIGMENTS INC.: PFERRITAN Heat Stable Pigments:

New line of inorganic synthetic pigments with:

* Clean, rich tan colors
* Exceptional heat stability to well above 600F
* Reliability and consistency

Typical properties:
Color Index: Pigment Yellow 119
Description: Tan Heat Stable Pigments
General Formula: ZnO Fe2O3
General Composition:
Fe2O3: 66.0% min.
ZnO: 33.0% min.
Specific Gravity: 5.20
+325 Mesh: >0.20%
Oil Absorption: 25
Water Soluble Salts: <0.20%
Loss on Ignition: 0.25%
pH: 5.0
Moisture: <1.0%

End Uses:
Coil Coatings
Plastics
Roofing Granules
Other Applications:
* Ceramics
* Rubber
* High Bake Coatings Systems

POLYVINYL CHEMICALS INC.: NEOCRYL Special Purpose Polymer Additives:

AP-2650:
Type: Silicone
Solids: 25%
Viscosity: 10 cps
Wt. Per Gallon: 7.6 lbs.
Solvent System: Toluene
Features/Applications: Silicone-type defoaming and de-aerating agent for water-borne coatings and solvent-based coatings. Particularly recommended for use in difficult to control water-borne curtain coating applications.

CX-100:
Type: Polyaziridine
Solids: 100%
Viscosity: 200 cps
Wt. per Gallon: 8.9 lbs.
Features/Applications: A trifunctional, carboxyl reactive crosslinker. Incorporating 0.7 equivalenets of CX-100 with carboxyl containing water-borne vehicles greatly enhances chemical, water, and solvent resistance. Adheison to difficult to wet substrates is also improved in selected systems.

RAYBO CHEMICAL CO.: Specialized Additives:

Product:

RAYBO 3-ANTISILK:
Features: Floating and Silking Control
System: Solvent
Comments: Highly Effective

RAYBO 6-ANTISAG
Features: Controls Sagging and Settling
System: Solvent
Comments: No Puffing

RAYBO 15-FLOINDUCER:
Features: Eliminates Brushmarks
System: Solvent
Comments: Flow, Flexibility

RAYBO 26-SLIKIT:
Features: Scratch Resistance Additive (Air Dry)
System: Solvent
Comments: Free of Silicone

RAYBO 38-NILFLO:
Features: Controls Sagging and Settling
System: Solvent
Comments: Highly Compatible

RAYBO 41-SPANGLE:
Features: Improves Stability and Brightness in Aluminum Paints
System: Solvent
Comments: Very Effective

RAYBO 43-MARHARD:
Features: Improves Mar Resistance (Baked)
System: Solvent
Comments: Acts on Surface

RAYBO 56-MAGNAMIX:
Features: Improves Wetting, Dispersion, Anti-Settling
System: Solvent
Comments: Improves Film

RAYBO 57-OPTISPERSE HS:
Features: Improves Dispersion and Gloss
System: Solvent, Water Reducible
Special Comments: Cost-effective; 100% Solids

RAYBO CHEMICAL CO.: Specialized Additives(Continued):

Product:

RAYBO 60-NORUST:
Features: Eliminates Can Corrosion and Nailhead Rust
System: Water
Comments: Versatile Rust Inhibitor
RAYBO 61-AQUAWET:
Features: Eliminates Fisheyes, Craters, Crawling
System: Water, Water/Reducible
Comments: Cationic
RAYBO 62-HYDROFLO:
Features: Flow Control Agent
System: Water Reducible
Comments: No Intercoat Problems
RAYBO 63-DISPERSE:
Features: Unique Dispersing Agent
System: Solvent, Water
Comments: Low Cost
RAYBO 68-HYDROSLIP:
Features: Improves Mar Resistance and Slip
System: Water, Water/Reducible
Comments: Works on Top Surface
RAYBO 72-VISCON:
Features: Increases Viscosity in Alkyds
System: Solvent, Water/Reducible
Comments: Not Thixotropic
RAYBO 74-RELEASE-AIR:
Features: Eliminates Air Entrainmnt-High Solids
System: Solvent
Comments: For Primers and Topcoats
RAYBO 77-ALGEL:
Features: Aluminum Octoate Gels Without Heat
System: Solvent
Comments: Efficient and Effective
RAYBO 82-ANTISTAT:
Features: Eliminates Static Charge
System: Solvent
Comments: Highly Compatible
RAYBO 85-RUSTIB:
Features: Stops Rusting Under Paints
System: Solvent
Comments: Cationic Wetting
RAYBO 94-SUSPERSE B:
Features: Controls Sagging and Settling
System: Solvent
Comments: Very Economical

REYNOLDS METALS CO.: Aluminum Atomized Powder:

Fine Atomized Powder

Grade:

40:
Average Particle Size Fisher (u): 28
Typical Surface Area cm2/g: 500

120:
Average Particle Size Fisher (u): 18
Typical Surface Area cm2/g: 800

200:
Average Particle Size Fisher (u): 16
Typical Surface Area cm2/g: 1000

1-511:
Average Particle Size Fisher (u): 15
Typical Surface Area cm2/g: 1200

1-231:
Average Particle Size Fisher (u): 10
Typical Surface Area cm2/g: 1800

1-131:
Average Particle Size Fisher (u): 7
Typical Surface Area cm2/g: 2600

400:
Average Particle Size Fisher (u): 5
Typical Surface Area cm2/g: 3500

S-792:
Average Particle Size Fisher (u): 21
Typical Surface Area cm2/g: 850

SANDOZ CHEMICALS CORP.: Additives for Plastics, Coatings:

These plastic additives are designed for incorporation in the mass for injection, and blow molding, extruded articles and for films, tapes, fibers.

Additives for coatings are designed for incorporation in paints and lacquers.

Product Name:

LEUCOPHOR EGM:
Chemical Class: Coumarine Derivative
Form: Powder
Scope of Application: LD-PE, HD-PE, PP, PVC, PP Fibers, PES Fibers

LEUCOPHOR B-302:
Chemical Class: Stilbene Derivative
Form: Liquid
Scope of Application: Ink Aqueous, Coatings Aqueous

LEUCOPHOR B-432:
Chemical Class: Stilbene Derivative
Form: Liquid
Scope of Application: Ink Aqueous, Coatings Aqueous

LEUCOPHOR B-635:
Chemical Class: Stilbene Derivative
Form: Liquid
Scope of Application: Ink Aqueous, Coatings Aqueous

LEUCOPHOR C-7002:
Chemical Class: Stilbene Derivative
Form: Liquid
Scope of Application: Ink Aqueous, Coatings Aqueous

LEUCOPHOR C-6708:
Chemical Class: Stilbene Derivative
Form: Liquid
Scope of Application: Ink Aqueous, Coatings Aqueous

LEUCOPHOR P-306:
Chemical Class: Stilbene Derivative
Form: Liquid
Scope of Application: Ink Aqueous, Coatings Aqueous

LEUCOPHOR P-631:
Chemical Class: Stilbene Derivative
Form: Liquid
Scope of Application: Ink Aqueous, Coatings Aqueous

SANDOZ CHEMICALS CORP.: Additives for Plastics, Coatings (Continued):

Product Name:

LEUCOPHOR P-539:
Chemical Class: Stilbene Derivative
Form: Liquid
Scope of Application: Ink Aqueous, Coatings Aqueous
LEUCOPHOR C-6902:
Chemical Class: Stilbene Derivative
Form: Liquid
Scope of Application: Ink Aqueous, Coatings Aqueous
LEUCOPHOR L:
Chemical Class: Stilbene Derivative
Form: Liquid
Scope of Application: Ink Aqueous, Coatings Aqueous
LEUCOPHOR 909:
Chemical Class: Stilbene Derivative
Form: Liquid
Scope of Application: Ink Aqueous, Coatings Aqueous
LEUCOPHOR BCR:
Chemical Class: Stilbene Derivative
Form: Liquid
Scope of Application: Ink Aqueous, Coatings Aqueous
LEUCOPHOR B-150:
Chemical Class: Stilbene Derivative
Form: Liquid
Scope of Application: Ink Aqueous, Coatings Aqueous
LEUCOPHOR P-300:
Chemical Class: Stilbene Derivative
Form: Liquid
Scope of Application: Ink Aqueous, Coatings Aqueous
LEUCOPHOR B-310:
Chemical Class: Stilbene Derivative
Form: Liquid
Scope of Application: Ink Aqueous, Coatings Aqueous
ACLARAT 8678:
Chemical Class: Coumarine Derivative
Form: Granule
Scope of Application: Ink Aqueous, Coatings Aqueous
VP SANDOFLAM 5086:
Chemical Class: Aliphatic bromine containing phosphate
Form: Powder
Scope of Application: Coatings Aqueous, PUR, PMMA, Other Polymers, UP
SANDOFLAM 5087:
Chemical Class: Polyphosphonate
Scope of Application: Coatings Aqueous, PUR, PMMA, UP
Form: Liquid

SANDOZ CHEMICALS CORP.: SANDUVOR Ultra Violet Light Absorbers:

Ultra Violet Absorbers and Light Stabilizers for Coatings

SANDUVOR 3206:
A liquid, oxalanilide UV absorber which offers low volatility, ease of handling, low viscosity, low color, and high compatibility. SANDUVOR 3206 is recommended for automotive OEM and refinish paints. It can also be used very effectively in high grade industrial coatings. SANDUVOR 3206 contains 20% xylene and 80% active UV absorbers.

SANDUVOR VSU Powder:
A highly effective UV absorber of the oxanilide class. The commercial form is a white powder which is readily soluble in most coating solvents. It is recommended for use in a wide variety of binder systems, such as acrylates, acrylic melamines, alkyd melamines and two-component polyurethanes. SANDUVOR VSU is recommended for polymer protection in plastics, particularly PVC and transparent plastics.

SANDUVOR 3050:
A hindered amine light stabilizer developed primarily for base coat, clear coat automotive coatings but has been found to perform very well in one coat refinish and industrial applications. SANDUVOR 3050 is a highly viscous waxy liquid, which is restored to a free flowing liquid by slight warming to 120F. SANDUVOR 3050 shows tremendous improvements in weathering of many types of automotive and industrial paints with respect to gloss retention, crack prevention and polymer protection.

<u>New Development Products:</u>

VP SANDUVOR 3040:
A UV Absorber of the 2 hydroxybenzophenone class currently offered on an experimental basis. The product's form is slightly yellowish, odorless, free-flowing powder. VP SANDUVOR 3040 can be applied in systems such as lacquers based on polyurethanes, phenol-based alkyd lacquers, amine or amide cured epoxies, etc. It is especially recommended for wood top coat enamels where it is particularly applicable for preventing the darkening of wood.

SANDUVOR 3046:
A light stabilizer based on a nickel quencher which is offered on an experimental basis. In addition, the quencher contains a sterically hindered amine (HALS) as a ligand. The product is very suitable for the stabilization of various binder systems and is especially suited for automotive top coats where excellent results have been noted with respect to crack prevention, gloss retention, and polymer protection without adverse affects to film cure. The commercial form is a 50% solution of active material in xylene. SANDUVOR 3046 is non-volatile at the usual baking conditions of automotive paints. SANDUVOR 3046 is also applicable in industrial finishes.

SANDOZ CHEMICALS CORP.: SANDUVOR Ultra Violet Light Absorbers (Continued):

VP SANDUVOR 3052:

A light stabilizer of the sterically hindered amine class which is also offered on a research and development basis. The application areas are high quality enamels of all kinds. Under the usual application and baking conditions, SANDUVOR 3052 is non-volatile, and is a 100% active low viscosity liquid whose recommneded application amounts are 1.0-2.0% based on binder solids. SANDUVOR 3052 is especially suitable for automotive refinish and industrial coatings.

VP SANDUVOR 3220:

Currently a research and development light stabilizer system. The product contains a nickel HALS quencher and a UV absorber. The form of the product is a low viscosity, greenish-yellow oil containing approximately 20% xylene. Under normal baking conditions, similar to those used with automotive enamels, VP SANDUVOR 3220 is essentially non-volatile. VP SANDUVOR 3220 performs commendably in automotive enamels, as well as industrial finishes.

VP SANDUVOR 3212:

An experimental product incorporating a UV light stabilizing system which has been developed primarily for automotive applications and is non-volatile. VP SANDUVOR 3212 has produced significant results as a stabilization system for automotive and industrial paints.

VP SANDUVOR 3041 Dispersion:

A provisional, 30% active, aqueous dispersion of a 2-hydroxybenzophenone type UV absorber, it contains the active substance in a very fine particle size. The product can be added to aqueous dispersions or emulsions of binders of all kinds. Because of its maximum UV absorption in the range of 270-340 nm, VP SANDUVOR 3041 Dispersion is suitable for the UV stabilization of binders and pigments in virtually any aqueous coating system, but is especially useful in wood enamels where it prevents the darkening of wood. SANDUVOR 3041 Dispersion prevents color changes and reduces gloss losses and crack formation due to the influence of light and weather.

VP SANDUVOR VSU Dispersion:

A 30% dispersion of an oxalanilide type UV absorber which contains the active substance dispersed to a very fine particle size. As a research product it is recommended for addition to aqueous dispersions or emulsions of binders of all kinds. It will prevent color changes, reduce gloss losses and prevent crack formation due to the influence of light and weather.

VP SANDUVOR 3051 Dispersion:

An experimental aqueous dispersion of a sterically hindered amine light stabilizer (HALS) which contains the 30% active substance. It can be added to aqueous dispersions or emulsions of binders all kinds. Significantly improves the weatherability of aqueous industrial coatings with respect to gloss retention, crack prevention and polymer protection.

SHAMROCK CHEMICALS CORP.: SHAMROCK Micronized Powders:

SHAMROCK NEPTUNE I

Typical Properties:
Appearance: White free-flowing powder
Grades: SP5 N1
Hegman: 6 7+
Microns: 18 5
Specific Gravity: 0.95
Penetration: <1
Initial Melt/Softening Point F: 194
Final Melt Point F: 235
Flash Point F: >250

Description:
SHAMROCK NEPTUNE I has been developed for use with water reducible coatings. A high melt polyethylene wax, NEPTUNE I has been specifically processed to produce a water-loving stir-in powder. Offered in two particle size grades, the material will rapidly disperse in aqueous systems with a minimum amount of mixing.

TSCA Registered
Meets requirements for FDA 21 CFR 175.105, 175.300, 175.320, 176.170, 176.180, 176.200, USDA.

SHAMROCK SST-2

Typical Properties
Appearance: White free-flowing powder
Grade: N5
Hegman: 7
Microns: 12.5
Specific Gravity: 2.15
Melt Point F: 610
Flash Point F: >600

Description:
SHAMROCK SST-2 is a micronized PTFE with a grind of 7.0 on the Hegman Grind Gauge. PTFE powders may be used in combination with other waxes to maximize abrasion resistance and slip (lubricity). SST-2 may be used alone when high-bake temperatures are required.
SST-2P is the gray version PTFE offered at a reduced cost.
SST-2K, a near black PTFE, is the most economical.
SST-2P and SST-2K do not carry FDA status.

TSCA Registered
Meets requirements for FDA 21 CFR 175.105, 175.300, 175.320, USDA

SHAMROCK CHEMICALS CORP.: SHAMROCK Micronized Powders (Continued):

SHAMROCK SST-3

Typical Properties:
- Appearance: White free-flowing powder
- Grade: N1
- Hegman: 7+
- Microns: 5
- Specific Gravity: 2.15
- Melt Point F: 610
- Flash Point F: >600

Description:

SHAMROCK SST-3 is a micronized PTFE with a grind of 7+ on the Hegman Grind Gauge. PTFE powders may be used in combination with other waxes to maximize abrasion resistance and slip (lubricity). SST-3 may be used alone when high-bake temperatures are required.

SST-3P is the gray version PTFE offered at a reduced cost.

SST-3K, a near black PTFE, is the most economical.

SST-P and SST-3K do not carry FDA status.

TSCA Registered.

Meets requirements for FDA 21 CFR 175.105, 175.300, 175.320, USDA.

SHAMROCK SST-3H

Typical Properties:
- Average Microns: 5
- Specific Gravity: 2.15
- Melt Point F: 610
- Flash Point F: >600

Description:

SST-3 is the Shamrock designation for fine particle size, stir-in PTFE powder. SST-3H is an especially fine particle size version, prepared for use in aqueous systems, notably water based can coatings.

SST-3 is a pure PTFE, free of surfactants or wetting agents.

TSCA Registered.

Meets requirements for FDA 21 CFR 175.105, 175.300, 175.320.

SHAMROCK CHEMICALS CORP.: SHAMROCK Micronized Powders (Continued):

SHAMROCK S-232

Typical Properties:
Appearance: White free-flowing powder
Grade: N1
Hegman: 7+
Microns: 5
Specific Gravity: 1.1
Melt Point F: 235
Initial Melt/Softening Point F: 165
Flash Point F: >500

Description:
SHAMROCK S-232, an alloy of waxes creating a synergistic effect, offers the user increased lubricity, hardness, anti-blocking characteristics, and abrasion and scuff resistance, without sacrificing intercoat adhesion and with minimal effect on gloss and clarity.

TSCA Registered
Meets requirements for FDA 21 CFR 175.105, 175.300, 175.320, 176.170, 176.180, 176.200, USDA.

SHAMROCK S-363

Typical Properties:
Appearance: White free-flowing powder
Hegman: 7+
Microns: 5
Specific Gravity: 0.94
Melt Point: 288F/142C

Description:
SHAMROCK S-363 is an alloy of polypropylene and a modifying polyethylene. This formulation produces unique performance properties in both ink and general industrial formulations.

The process technique employed offers the formulator the use of an ultra-fine polypropylene to obtain scuff resistance previously not available.

S-363 can be dispersed easily into water or solvent based formulations.

TSCA Registered.

SHAMROCK CHEMICALS CORP.: SHAMROCK Micronized Powders (Continued):

SHAMROCK S-368
Water Dispersible Wax
NEPTUNE Series

Typical Properties:
Appearance: White free-flowing powder
Grade: N5
Hegman: 7
Microns: 12.5
Specific Gravity: 0.93
Melt Point F: 208
Softening Point F: 158
Flash Point F: >250

Description:
SHAMROCK NEPTUNE S-368 is a stable water dispersible grade of polyethylene wax. This hydrophilic wax forms a true dispersion in water without addition of co-solvents.
SHAMROCK NEPTUNE S-368 imparts lubricity, abrasion resistance, and anti-blocking properties to all aqueous based systems. One non-coating application employs NEPTUNE S-368 as an additive in laminatng adhesives to improve its die-cutting capability.

TSCA Registered.
Meets requirements for FDA 21 CFR 175.105, 175.300, 175.320, 176.170, 176.180, USDA.

SHAMROCK S-379

Typical Properties:
Appearance: White free-flowing powder

Grades:	H	N3	N
Hegman:	7	7+	7+
Microns:	12.5	8.0	5.0

Specific Gravity: 0.93
Melt Point F: 208
Initial Melt/Softening Point F: 158

Description:
SHAMROCK S-379 is a synthetic polyethylene wax with the typical straight chain molecular structure. It differs from branched chain microcrystalline wax in that it is considerably less soluble in ordinary solvents.
S-379H is the most economical stir-in wax. It provides good rub and slip in most coatings. S-379 N3 and S-379 N are only slightly higher priced, for use where greater fineness of grind is required.

TSCA Registered.
Meets requirements for FDA 21 CFR 175.105, 175.300, 175.320, 176.170, 176.180, USDA

SHAMROCK CHEMICALS CORP.: SHAMROCK Micronized Powders (Continued):

SHAMROCK S-381

Typical Properties:
Appearance: White free-flowing powder
Grade: N1
Hegman: 7+
Microns: 4
Specific Gravity: 0.93
Melt Point F: 206
Initial Melt/Softening Point F: 158
Flash Point F: >250

Description:

SHAMROCK S-381 is classified as a modified micronized blend of waxes having the performance of polypropylene, without the typical poor U.V. resistance of polypropylene. S-381 can be stirred in to produce a semi-gloss finish or added with amorphous silicas to produce satin or flat finishes with excellent mar resistance and reduction of coefficient of friction. Ratios of silica to S-381 should be 1:1 for satin finishes and 2:1 for flat finishes.

S-381 is an efficient flatting agent.

S-381 wets out in aqueous systems much more readily than most waxes.

The addition of S-381 grants the formulator an opportunity to moderate the gloss of the system while retaining surface hardness characteristics of the system.

TSCA Registered
Meets requirements for FDA 21 CFR 175.105, 175.300, 175.320, 176.170, 176.180, USDA

SHAMROCK S-394

Typical Properties:
Appearance: White free-flowing powder

Grade:	N5	N1
Hegman:	7	7+
Microns:	12.5	5

Specific Gravity: 0.95
Melt Point F: 235
Initial Melt/Softening Point F: 194
Flash Point: >250

SHAMROCK CHEMICALS CORP.: SHAMROCK Micronized Powders (Continued):

SHAMROCK S-394(Continued)

Description:
SHAMROCK S-394, a polyethylene wax, offers excellent recoatability, hardness and anti-blocking characteristics for the formulator. Supplied in two standard grinds, N5 and N1, S-394 has a fairly high melt point, with a softening point only a few degrees lower. The narrow temperature range reduces the solubility of the wax.

TSCA Registered.
Meets requirements for FDA 21 CFR 174.105, 175.300, 175.320, 176.170, 176.180, 176.200, USDA.

SHAMROCK S-395

Typical Properties:
Appearance: White free-flowing powder

Grade:	SP5	N5
Hegman:	6+	7
Microns:	18	12.5

Specific Gravity: 0.95
Melt Point F: 258
Initial Melt/Softening Point F: 212
Flash Point F: >250

Description:
SHAMROCK S-395, a crystalline grade of polyethylene, offers increased hardness, recoatability, abrasion resistance, and anti-blocking characteristics to the formulator. Supplied in two standard grades, SP5 and N5, S-395 has a high melt point, with a softening point only slightly lower in temperature. The narrow temperature range reduces the solubility of the wax.

S-395 is an excellent product for improving taber abrasion resistance, for can slide resistance for shelving, and for other abrasive type properties. A particle size close to the dry film thickness of the coating will optimize the abrasion resistance.

TSCA Registered
Meets requirements for FDA 21 CFR 175.105, 175.300, 175.320, 176.170, 176.180, 176.200, USDA.

SHAMROCK CHEMICALS CORP.: SHAMROCK Micronized Powders (Continued):

SHAMROCK S-400

Typical Properties:

Appearance: Free-Flowing Powder

Grades:	N1	N5	SP5
Microns:	5.0	12.5	18.0
Hegman:	7+	7	6+

Specific Gravity: 0.98
Melting Point: 285F
Flash Point: >450F

Description:

S-400 is a synthetic wax which is easily dispersed under high speed mixing equipment. It is used in coatings to reduce the coefficient of friction or to increase slip with a minimal effect on gloss. S-400 is used in coatings to reduce coin marking on high gloss whites, and is especially useful in preventing finger-nail marring.

S-400 can also be used in conjunction with other types of waxes to obtain a combination of good abrasion resistance while maintaining slip. S-400 comes in three different particle sizes. The formulator can choose the particle size based on the dry film thickness.

TSCA Registered
Meets requirements for FDA 21 CFR 175.105, 175.300, 176.170, 176.180, USDA.

SHAMROCK CHEMICALS CORP.: SHAMROCK Micronized Powders (Continued):

SHAMROCK S-NAUBA 5021

Typical Properties:
- Appearance: Off-white powder
- Grade: N5
- Hegman: 7+
- Microns: 12.5
- Specific Gravity: 0.995
- Melt Point F: 185
- Initial Melt/Softening Point F: 174
- Flash Point F: >250

Description:
SHAMROCK S-NAUBA 5021 is a finely ground prime yellow Carnauba wax, enabling ease of dispersion with high-speed mixers into either solvent or water based systems. Because of the low temperature melt flow characteristics, S-NAUBA 5021 is especially suited for bake systems achieving surface temperature greater than 200F. Levels in the 2% range will result in improved surface hardness.

TSCA Registered
Meets requirements for FDA 21 CFR 175.105, 175.300, 175.320, 176.170, 176.180, 176.200, USDA.

SHAMROCK VERSAFLOW 266 and VERSAFLOW 263 Liquid Polyethylene for Solvent Systems

Typical Properties:

VERSAFLOW 266:
- Appearance: Clear Water-White
- Concentration: 67%
- Type: Polymer Lubricant
- Solvent: AROMATIC 100
- Color: Clear
- Specific Gravity: 0.87
- Flash Point F: 101
- Boiling Point F: 320
- Viscosity: 30 to 80 cps at 77F

VERSAFLOW 263:
- Appearance: Clear water-white solution
- Concentration: 67%
- Type: Polymer Lubricant
- Solvent: Toluene
- Color: Clear
- Specific Gravity: 0.87
- Flash Point F: 45
- Boiling Point F: 232
- Viscosity: 30 to 80 cps at 77F

SHAMROCK CHEMICALS CORP.: SHAMROCK Micronized Powders (Continued):

SHAMROCK VERSAFLOW 266 and VERSAFLOW 263(Continued)

VERSAFLOW 266 is a high concentration solution of a polymerized low molecular weight lubricating polymer that offers the formulator an alternative to silicone. It is ideal for use in ambient or high temperature baked enamel coatings. VERSAFLOW 266 will not affect gloss or clarity. It will increase slip (lower the COF) and improve mar resistance.

Since VERSAFLOW 266 is a surface active material, it will allow the user to obtain many of the beneficial features of silicones without their inherent problems. VERSAFLOW 266 may be added to all paint and coating systems that are compatible with aromatic or aliphatic solvents.

VERSAFLOW 263 is available to formulators who prefer the use of this polymer in toluene solvent. VERSAFLOW BASE, which is solvent free, is also available. This surface active polymer is also available in an emulsifiable form for water reducible systems.

TSCA Registered.

VERSAFLOW BASE

Typical Properties:
- Appearance: Pale Amber
- Concentration: 100% polyethylene
- Specific Gravity: 0.86
- Weight per Gallon: 7.16
- Flash Point: >250F
- Boiling Point: >600F
- Viscosity: 900+-50 cps @ 70F

Description:

VERSAFLOW BASE is offered to those customers for which shipments of flammable or combustible products may be a problem, i.e. international.

Regardless of the solvent carrier, performance properties of this surface-active polyethylene will remain constant.

TSCA Registered.

SHAMROCK CHEMICALS CORP.: SHAMROCK Micronized Powders (Continued):

SHAMROCK Texture Series

Texture:

Ultra-Fine:
Average Range (Microns): 18
Melt Point F: 330
Specific Gravity: 0.90

5378:
Average Range (Microns): 25
Melt Point F: 330
Specific Gravity: 0.90

5380:
Average Range (Microns): 40
Melt Point F: 330
Specific Gravity: 0.90

5382:
Average Range (Microns): 55
Melt Point F: 330
Specific Gravity: 0.90

5384:
Average Range (Microns): 75
Melt Point F: 330
Specific Gravity: 0.90

5386:
Average Range (Microns): 90
Melt Point F: 330
Specific Gravity: 0.90

Appearance: Free-Flowing Powder

Description:
The TEXTURE Series are crystalline polypropylenes with a broad range of particle sizes. The TEXTURE Series, stir-in powders that can be easily dispersed under high-speed mixing equipment, are primarily used as texturing agents.
The high melt viscosity of TEXTURE makes it useful for bake systems as well as air dry systems. Because of this high melt viscosity, TEXTURE will not lose its particle definition at temperatures over 330F.
After the average particle size is carefully controlled, the material is screened to eliminate oversize particles. This process results in a "suede-like" feel.
The solvent resistance of TEXTURE enables it to be compatible with all resin systems, both solvent and water base.

TSCA registered. Meets requirements for FDA 21CFR 177.1520.

THE SHERWIN-WILLIAMS CO.: MOLY-WHITE:

Environmentally Safe, White, Corrosion-Inhibiting Pigments for Solvent and Water-Based Paint Systems

MOLY-WHITE MZAP:
For Solvent or Water-Based Coatings

MZAP is a new, low cost white corrosion inhibitor. MZAP is a composite of calcium zinc molybdate and zinc phosphate pigment.

MOLY-WHITE 212:
For Water-Based Coatings

212 is a basic calcium-zinc molybdate designed for water-dispersible and latex systems.

MOLY-WHITE 101:
For Solvent-Based Coatings

101 is a basic zinc molybdate used with solvent-based paint vehicles including oils, alkyds, modified alkyds, epoxy esters and polyesters.

MOLY-WHITE 331:
For Solvent-Based Coatings

331 is a lower cost version of the 101.

MOLY-WHITE ZNP:
For Solvent-Based Coatings

ZNP is a new version of 331 and provides additional cost savings. It contains an ingredient which provides adhesion over rusted steel and is recommended for solvent-based vehicles.

SONNEBORN DIVISION: SACI Rust Preventives:

SACI-100 & SACI-100A:
Corrosion preventive concentrates (solvent-oil dispersions)

SACI-100 and 100A corrosion inhibitor concentrates are thixotropic, polar compounds which are designed to be used as bases in the formulation of corrosion preventive coatings. SACI-100 and 100A concentrates will yield films which are soft, oily, self-healing and have excellent fingerprint removal and acid neutralization properties.

SACI-100:
Viscosity, cps: 12,000
Non Volatile by Weight: 60%
Color: Tan

SACI-100A:
Viscosity, cps: 17,000
Non Volatile by Weight: 60%
Color: Lt. Brown

SACI-200 & SACI-200A
Corrosion preventive concentrates (oil dispersions)

SACI-200 oil based corrosion inhibitor concentrates are thixotropic, polar compounds which are designed to be used as bases in the formulation of corrosion preventive coatings. SACI-200 corrosion preventive concentrates will yield films which are soft, oily, self-healing and have excellent fingerprint removal and acid neutralization properties.

SACI-200:
Viscosity, cps: 90,000
Non Volatile by Weight: 96% min.
Color: Lt. Brown

SACI-200A:
Viscosity, cps: 95,000
Non Volatile by Weight: 96% min.
Color: Brown

SONNEBORN DIVISION: SACI Rust Preventives(Continued):

SACI-300 & SACI-300A
Corrosion preventive concentrates (solvent-oil-wax dispersions)

SACI-300 and 300A corrosion inhibitor concentrates are thixotropic, polar compounds which are designed to be used as bases in the formulation of corrosion preventive coatings. SACI-300 and 300A will yield films which are moderately firm and waxy.

SACI-300:
Viscosity, cps: 65,000
Non Volatile by Weight: 60%
Color: Tan

SACI-300A:
Viscosity, cps: 45,000
Non Volatile by Weight: 60%
Color: Lt. Brown

SACI-450W, 445A & 445M
Corrosion preventive concentrates (waterborne emulsions)

SACI-450W, 445A & 445M waterborne corrosion preventives feature improved corrosion protection, superior stability and higher solids content than previous waterborne products offered.

SACI-450W:
Viscosity, cps: 30,000
Non Volatiles by Weight: 48% min.
Color: White

SACI-445A:
Viscosity, cps: 4,000
Non Volatiles by Weight: 45% min.
Color: White

SACI-445M:
Viscosity, cps: 20,000
Non Volatiles by Weight: 45% min.
Color: White

SACI-500A
Corrosion preventive concentrate (solvent-wax dispersion)

A solvent-based concentrate for use in formulating protective coatings of metal surfaces. SACI-500A diluted with additional solvent will meet the requirements of Government Specification MIl-C-83933A (MR).

Viscosity, cps: 55,000
Nonvolatile by wt, %: 70.0
Color: Tan

SONNEBORN DIVISION: SACI Rust Preventives(Continued):

SACI 552-A Rust Preventive Concentrate
Solvent-Wax Dispersion

SACI 552-A is a solvent based rust preventative widely used in the preparation of coatings for metal surfaces protection, and which meets the requirements of Government Specification MIL-C-83933A(MR).

Viscosity, cps: 5,000
Nonvolatile by wt, %: 52
Color: Tan

SACI-700
Corrosion preventive formulating concentrate

Formulated products based upon SACI-700 can be used to protect iron, steel, galvanized steel, aluminum and other metallic and non-metallic surfaces.

Viscosity, cps: 60,000
Non Volatile by wt, %: 50%
Color: Tan

SACI-760
Corrosion preventive formulating concentrate

SACI-760 corrosion inhibitor concentrate is designed to allow the formulation of higher solids, lower volatile organic compound (VOC) finished products

Viscosity, cps: 60,000
Non-Volatile by Weight: 60%
Color: Tan

SACI-2400
Corrosion preventive concentrate (solvent-wax dispersion)

Can be used to formulate protective coatings for virtually all metal surfaces. This product will typically yield films which are moderately firm and waxy.

Viscosity: 50,000 cps
Non Volatiles by Weight: 65%
Color: Light Tan

SONNEBORN DIVISION: SACI Rust Preventives(Continued):

SACI-2452
Corrosion preventive concentrate(solvent-wax dispersion)

A solvent based, wax-modified dispersion which can be used to formulate protective coatings for virtually all metal surfaces. This product will typically yield films which are moderately firm and waxy.

Viscosity, Brookfield: 5,000 cps
Non Volatiles by Weight: 52% min.
Color: Light Tan

SACI-8100
Corrosion preventive concentrate

A high solids base designed to allow the formulation of corrosion inhibitors yielding tough abrasion and solvent resistant films. It is especially useful where clear, thin, hard film coatings are desired.

Viscosity, Brookfield, cps: 10,000
Non Volatile by Weight: 70%
Color: Light Tan

SACI-8200
Corrosion preventive concentrate

SACI-8200 high solids corrosion inhibitor concentrate is designed to allow the formulation of low VOC, automotive or industrial coatings. Coatings based on 8200 will yield firm, glossy, abrasion resistant films which are impervious to solvents such as mineral spirits, heptane and gasoline.

Viscosity, Brookfield, cps: 12,000
Non Volatile by Weight: 80%
Color: Dark Tan

SYNTHRON, INC.: Powder Coatings Industry:

Flow Modifiers for Powder Coatings

ACRYLRON MFP Powder:
Sixty-seven percent active powdered acrylate copolymer flow modifier. ACRYLRON MFP Powder may be incorporated into specific FDA regulated coatings.

ACRYLON MFB Powder:
Sixty-seven percent active powdered acrylate flow modifier.

ACRYLRON MFP-F:
Acrylate copolymer designed to meet the requirements of CFR-21, Section 175.300. This liquid additive is attractive for resin manufacturers that make a flow modifier containing masterbatches for powder coatings.

Silicone Surface Active Additives

MODAREZ SLB:
Seventy percent active silicone powder. Active ingredient is SILWET S-7605 (Formerly SILWET L-5310). MODAREZ SLB is easily dispersed in powder coatings and much easier to clean up in production.

MODAREZ SPA:
Seventy percent active silicone powder. MODAREZ SPA is more easily dispersed in powder coatings than silicone fluids.

Antistat

ASHSTAT 50-P-2:
Fifty percent active polymeric additive to control build-up of static charges in powder coatings.

SYNTHRON, INC.: Specialty Additives for the Coatings, Ink, and Adhesive Industry:

Aqueous Phase Dispersants, Dispersing Aids, and Thickeners:

ACRYLRON CA:
Polyacrylic acid, can be modified to produce thickening and dispersant properties.

ACRYMUL AM-162S:
Polyacrylate thickener for water based system.

MASQUOL AW6 & AW7:
Polyacrylate dispersants for deposit control boiler water service.

MODAREZ COU:
Thermocoagulant, anti-migrant for water based polymer systems.

MODAREZ PF 162:
Dispersing and leveling aid for water based polymer systems.

MODAREZ PF 756:
Pigment dispersant and leveling aid for water based polymer systems. PF 756 enhances optical brightener effects.

PROTESOL DSL-40:
Water soluble organic dispersant for pigments and fillers.

PROTESOL DSL-200:
Water soluble organic dispersant for pigments and fillers. Very low free sulfate content.

Non-Aqueous Phase Leveling, Flow, and Deaerating Additives:

ACRYLRON MFP-F:
Flow, leveling, and hydrophobic wetting agent. Acrylate copolymer(CFR-21; Section 175.300)

ACRYLRON MFP-B:
Flow, leveling, and hydrophobic wetting agent. Acrylate polymer.

ACRYLRON MFP-P40:
Antifoam agent for non-aqueous hydraulic fluids.

MODAREZ TMF-50:
Low viscosity solvent solution for leveling, flow, and pigment wetting. (CFR-21; 175.300)

MODAREZ TMB-50:
Low viscosity solvent solution for leveling, flow, and pigment wetting.

SYNTHRON, INC.: Specialty Additives for the Coatings, Ink, and Adhesive Industry(Continued):

Antifoams:

MOUSSEX 561 PL:
Non-silicone, durable, high temperature stable, antifoam.

MOUSSEX 910 SE:
Modified silicone, dilution stable, antifoam.

MOUSSEX 941 PL:
Non-silicone, durable, high temperature stable antifoam. Very easy to disperse.

MOUSSEX 954 HL:
Modified oil based antifoam. Does not contain silicone.

Cationic Additives:

GLYTAC A-100:
Epoxy quaternary product used for making cationic additives.

MULTIFLOC VP 103:
Cationic polymer, dispersant and fixative for the paper industry. Effective in neutral and alkaline size systems.

Surfactants:

NOVEPOX TAN 117:
Diepoxide surfactant for making water dispersible epoxies.

NOVEPOX TAN 141:
Diepoxide surfactant for making water dispersible epoxies.

SYNTHRON, INC.: SYNTHRON Powder Coating Additives:

Powdered Catalysts for Epoxy Resins

ACTIRON NX-3 Powder:
Sixty-two percent active powder form of 2,4,6-tris (dimethylamino) methyl phenol. This catalyst can be used in lower temperature curing epoxy systems. ACTIRON NX-3 Powder also suitable for certain FDA regulated indirect food contact coatings.

ACTIRON NXJ-60 Powder:
Sixty percent active powder form of 2-propylimidazole. ACTIRON NXJ-60 Powder is a very easily dispersed powder. It lacks hard crystalline particles which may interfere with making a blemish free epoxy coating.

ACTIRON NX-91 Powder:
Sixty percent active powder of Benzyldimethylamine. ACTIRON NX-91 Powder is a good catalyst with excellent ease of dispersion for electrical epoxy molding powders.

Powdered Catalysts for Polyester or Urethane Resins

ACTIRON SNO-30:
Thirty percent active powder form of Stannous Octoate. ACTIRON SNO 30 is an encapsulated catalyst that improves storage shelf stability. This powder easily disperses in powder coatings and speeds production clean-up in contrast to the liquid form of Stannous Octoate.

ACTIRON DBT:
Sixty percent active powder form of Dibutyltindilaurate. ACTIRON DBT disperses easier and cleans up faster during production of polyester and urethane powder coatings than liquid Dibutyltindilaurate.

ACTIRON 303:
Encapsulated melamine catalyst in fine powder form.

TEGO CHEMIE SERVICE USA: De-Aerators:

TEGO AIREX 905:

De-aerater for emulsion paints and water thinnable systems.

Chemical composition: High molecular weight organic compound.

Physical data:

* Active matter content: 40%
* Density (20C): 0.97 g/cm3
* Flash point: 77F (25C)
* Non-volatile content: 40%+-1
* Solvent: ethyl glycol/crystal oil K 60 (3:1)
* Weight/gallon: 8.095 lbs.

Properties:

TEGO AIREX 905 accelerates the rise of the bubbles out of the film before the drying process is complete and thus prevents the formation of pinholes.

Areas of application:

Building protection
* high gloss emulsion paints based on pure acrylic resins

Wood/furniture varnishes
* water thinnable emulsions and resins

Anticorrosive coatings
* water soluble acrylic and polyester resins

Car finishes
* water soluble alkyd resins

Printing inks
* water thinnable gravure and flexographic inks

TEGO CHEMIE SERVICE USA: De-aeraters(Continued):

TEGO AIREX 900:

De-aerator for solvent containing and solvent-free systems based on epoxide and unsaturated polyester resins.

Chemical composition:
Polysiloxane modified with organic groups

Physical data:
* Active matter content: 100%
* Density (20C): 1.0 g/cm3
* Flash point: 145F (63C)
* Weight/gallon: 8.345 lbs

Properties:

TEGO AIREX 900 is soluble in alcohols, ketones, aromatic hydrocarbons and glycols.
TEGO AIREX 900 is also suitable for application in water based systems.

Areas of application:

Building protection
* floor coatings and primers based on unsaturated polyester and epoxy resins/epoxy-mortar

Wood/furniture varnishes
* top coats based on unsaturated polyester resins

Printing inks
* solvent and water based gravure and flexographic inks

Miscellaneous
* PVC plastisols

TEGO CHEMIE SERVICE USA: Mar & Slip Additives:

TEGO GLIDE 100:

Mar resistant and flow additive for water and solvent-based paints.

Chemical composition:
Polysiloxane-polyether-copolymer

Physical data:

* Active matter content: 100%
* Density (20C): approx 1.0 g/cm3
* Flash point: 106F (41C)
* Viscosity (25C): 800+-150mPa.s
* Colour index: <5 (Gardner)
* Surface tension (20C): approx. 21 mN.m-1
* Weight/gallon: 8.345 lbs.

Properties:

Because of its surface-active properties TEGO GLIDE 100 improves the flow of paints. Due to the special composition of the product a good compatibility with both water soluble and solvent soluble resins is achieved. TEGO GLIDE 100 is hydrolytically stable.

Areas of application:

TEGO GLIDE 100 has proved itself in systems based on the following resins:
* alkyd
* saturated polyester
* polyacrylate

Taking into account the properties of the product and the method of application, TEGO GLIDE 100 is recommended as an additive for the production of
* building protection coatings
* anticorrosive paints
* car finishes
* printing inks

CTFA adopted name: dimethicone copolyol

TEGO CHEMIE SERVICE USA: Mar & Slip Additives(Continued):

TEGO GLIDE ZG 400:
Mar resistant additive for water and solvent-based paints.

Chemical composition: Polysiloxane-polyether-copolymer

Physical data:

* Active matter content: 100%
* Density (20C): approx. 1.03 g/cm3
* Flash point: 181F (83C)
* Colour index: >3 (Gardner)
* Viscosity (25C): approx. 800 mPa.s
* Weight/Gallon: 8.595 lbs

Properties:

TEGO GLIDE ZG 400 reduces the slip resistance on the paint surface and therefore increases the mar resistance. Because of its surface-active properties TEGO GLIDE ZG 400 also improves the flow of paints.

Due to the special composition of the product a good compatibility with both water soluble and solvent soluble resins is achieved. TEGO GLIDE ZG 400 is hydrolytically stable.

Areas of application:

TEGO GLIDE ZG 400 has proved itself in systems based on the following resins:

* alkyd
* alkyd/melamine
* polyurethane
* vinyl

Taking into account the properties of the product and the method of application, TEGO GLIDE ZG 400 is recommended for the production of:

* building protection coatings
* anticorrosive paints
* wood/furniture varnishes
* printing inks

CTFA adopted name: dimethicone copolyol.

TEGO CHEMIE SERVICE USA: Mar & Slip Additives(Continued):

TEGO GLIDE A 115:

Mar resistant and flow additive for solvent-based systems.

Chemical Composition: Organically modified polysiloxane.

Physical data:

* Active matter content: 100%
* Density (20C): approx. 1.0 g/cm3
* Flash point: 140F (60C)
* Viscosity: 100+-20 mPa.s
* Surface tension (20C): 21.6+-0.2 mN.m-1
* Weight/gallon: 8.345 lbs.

Properties:

TEGO GLIDE A 115 reduces the slip resistance on the paint surface and therefore increases the mar resistance. Due to its ability to reduce the surface tension TEGO GLIDE A 115 promotes levelling and flow.

TEGO GLIDE A115 is soluble in aliphatic, aromatic and chlorinated hydrocarbons, alcohols, ethers and esters.

Areas of application:

TEGO GLIDE A 115 has proved itself in systems based on the following resins:

* polyurethane
* vinyl
* alkyd/melamine
* acrylate/isocyanate

Taking into account the properties of the product and the method of application TEGO GLIDE A 115 is recommended as an additive for the production of solvent-based.

* anticorrosive paints
* wood-/furniture varnishes, applied by the curtain coating process
* car finishes
* printing inks

TEGO CHEMIE SERVICE USA: Mar & Slip Additives(Continued):

TEGO GLIDE B 1484:

Mar resistant and flow additive for solvent-based systems.

Chemical composition: Polysiloxane-polyether-copolymer

Physical data:

* Active matter content: 100%
* Density (20C): approx. 1.0 g/cm3
* Flash point: 149F (65C)
* Colour index: <6 (Gardner)
* Surface tension (20C): 21.5+-1.0 mN.m-1
* Viscosity (25C): 450+-100 mPa.s
* Weight/gallon: 8.345 lbs.

Properties:

TEGO GLIDE B 1484 reduces the slip resistance on the paint surface and consequently increases the mar resistance. Due to its surface activity TEGO GLIDE B 1484 prevents the formation of Bernard cells on the surface of pigmented paint systems.

TEGO Glide B 1484 is soluble in most organic solvents but not in water.

In epoxy paints TEGO GLIDE B 1484 exhibits defoaming characteristics.

The overcoatability and the intercoat adhesion is not affected by TEGO GLIDE B 1484.

Areas of application:

TEGO GLIDE B 1484 has proved itself in paint systems based on the resins:

* epoxy
* polyurethane
* acrylate/isocyanate
* alkyd/melamine

Taking into account the properties of the product and the method of application TEGO GLIDE B 1484 is recommended as an additive for the production of solvent-based

* car finishes
* wood-/furniture varnishes, applied by spraying and roller coating
* floor coatings

TEGO CHEMIE SERVICE USA: Mar & Slip Additives(Continued):

TEGO GLIDE 410:

Mar resistant additive for water and solvent based paints.

Chemical composition: Polysiloxane polyether copolymer

Physical data:

* Active matter content: 100%
* Density at 25C: approx. 1.0 g/cm3
* Flash point: 178F (81C)
* Weight/gallon: 8.345 lbs.

Properties:

TEGO GLIDE 410 reduces slip resistance (coefficient of friction) at the paint surface, giving enhanced scratch resistance. The same coefficient of friction can be achieved with TEGO GLIDE 410 using only a tenth of the amount required with previously existing slip additives. If the same amount of TEGO GLIDE 410 is added, as is required with other additives, the friction coefficient is reduced by 1/3 to 1/2 of that previously obtained. This applies not only to water and solvent-based systems but also to air-dried paints.

Areas of Application:

Because of its very good compatibility with vehicles and solvents, TEGO GLIDE 410 can be used in all types of paints such as primers, extenders and coating varnishes.

With due regard for the product and the method of application, TEGO GLIDE 410 is recommended as an additive in the production of:

* Automobile and touchup paints
* Building protective paints
* Anti-corrosion paints
* Wood/Furniture varnishes
* Printing inks
* Industrial paints
* Package lacquers

CTFA adopted name: dimethicone copolyol

TEGO CHEMIE SERVICE USA: Mar & Slip Additives(Continued):

TEGO GLIDE 405:

Mar resistant additive for water and solvent based paints.

Chemical composition: Polysiloxane polyether copolymer.

Physical data:

* Active matter content: 50%
* Density at 25C: approx. 0.96 g/cm3
* Flash point: 149F
* Solvent: Butylglycol
* Colour index: <3 (Gardner)
* Weight/Gallon: 8.035 lbs.

Properties:

TEGO GLIDE 405 reduces the slip resistance on the surface and therefore increases the mar resistance. Because of its surface-active properties TEGO GLIDE 405 also improves the flow of paints.

Due to the special composition of the product a good compatibility with both water soluble and solvent soluble resins is achieved.

TEGO GLIDE 405 is hydrolytically stable.

Areas of Application:

TEGO GLIDE 405 has proved itself in paint systems on the following resins:

* alkyd resins
* combination of alkyd-melamine resins
* polyurethane resins
* vinyl polymers

Taking into account the properties of the product and the method of application TEGO GLIDE 405 is recommended as an additive for the production of:

* building protection coatings
* anticorrosive paints
* wood-/furniture varnishes
* printing inks

ULTRA ADHESIVES, INC.: ULASEIN 15:

ULASEIN 15:

ULASEIN 15 is a stabilized casein solution prepared under carefully balanced conditions of concentration, temperature and pH to assure constant quality, performance and package stability. ULASEIN 15 is recommended for use in the formulation of a variety of water based paints such as: paste casein, resin emulsion, styrene butadiene, acrylic emulsion and resin emulsion modifications of emulsion polymers used in compounding paint finishes, gloss and semi-gloss emulsion paints and paste type spackling compounds.

The unique properties of ULASEIN 15 simplify formulations and greatly improve the properties of latex and emulsion paints. ULASEIN 15 is a very effective thickener, viscosity control agent and protective colloid in most mildly alkaline emulsion and latex systems,

Specifications and Properties:
Color: Off White
pH: 6.5-7.5
Pounds Per Gallon: 8.6-8.7

Applications:

The excellent dispersing and wetting properties of ULASEIN 15 assure:

1) Faster pigment wetting and dispersion
2) Maximum color development
3) Improved flow and leveling
4) Improved film continuity
5) Better washability and resistance to stain penetration
6) Wet edge retention preventing lap marks and assuring uniformity of color
7) Controlled penetration helping equalize suction on surfaces of varying porosity

ULASEIN 15 is an additive in latex and resin emulsion gloss and semi-gloss finishes.

A) Aids flow and leveling
B) Helps eliminate brush and lap marks
C) Improves color uniformity
D) Adds adhesion and on drying reduces after tack

Paper Coatings:

ULASEIN 15 is an excellent binder for paper and paper board coatings.

Adhesives:

ULASEIN 15 by itself is an excellent adhesive for laminating wood, paper and cloth.

WESTVACO CHEMICALS: Paint Industry Products:

WESTVACO L-1 tall oil fatty acid:

A light colored tall oil fatty acid containing a low level of rosin acids. WESTVACO L-1 is an excellent raw material for the production of high quality, specialty industrial chemicals such as dimer acids, epoxidized tallate esters, and fatty acid esters for cosmetic, polymer, and textile applications.

Specifications
- Acid Number: 193
- Rosin Acids, %: 1.2
- Color, Gardner: 3

WESTVACO L-1A tall oil fatty acid:

A pale tall oil fatty acid containing less than 1% rosin acids. Used in applications where light color and low rosin acid content are critical, WESTVACO L-1A is an excellent raw material for dimers, alkyds, and epoxidized tallate esters. It replaces oleic acid in certain applications such as liquid soaps, fatty alcohols, amides, nitriles and formulations where a low titer, unsaturated fatty acid is required.

Specifications:
- Acid Number: 193.5
- Rosin Acids, %: 0.6
- Color, Gardner: 2+

WESTVACO L-5 tall oil fatty acid:

A liquid tall oil fatty acid containing 5% rosin acids, useful in applications not sensitive to light color: Industrial cleaners, anionic flotation reagents, alkyd resins, imidazoline corrosion inhibitors, metalworking fluids, general purpose tallate esters, and sulfated fatty acids are a few of the more important applications.

Specifications:
- Acid Number: 191
- Rosin Acids, %: 4.9
- Color: Gardner: 6

WESTVACO L-5A tall oil fatty acid:

An economical, pale tall oil fatty acid that combines light color with a moderate rosin acids content.

Specifications:
- Acid Number: 191
- Rosin Acids, %: 4.6
- Color, Gardner: 4 1/2

WESTVACO CHEMICALS: Paint Industry Products(Continued):

WESTVACO DIACID 1550 dicarboxylic acid:

* A unique cost saving hydrotrope
* Excellent hydrotrope in caustic systems
* Synergistic with nonionic surfactants
* Superior coupling agent on cost/performance basis

WESTVACO DIACID 1550 is a biodegradable and non-toxic C21 dicarboxylic acid. Salts of DIACID exhibit unusual solubility which allows one to make high solids soaps. These salts have excellent coupling ability and surface activity in concentrated alkali.

DIACID 1550 is a liquid monocylic C21 dicarboxylic acid derived from fatty acids via a patented Westvaco process, having applications in two primary areas: surfactants and chemical intermediates.

Specifications:
Acid Number: 272
Color, Gardner: 7+
C18FA, %: 12

WESTVACO M-28B distilled tall oil:

A medium colored tall oil distillate containing fatty acids and approximately 28% rosin. WESTVACO M-28B offers economy and improved quality in fatty acid applications where rosin content enhances product performance. It is used as an emulsifier for oil extended SBR latex and for metalworking oils, as a base for alkyd resins, as an anionic flotation reagent, and as an asphalt emulsion additive.

Specifications:
Acid Number: 185
Rosin Acids, %: 28.0
Color, Gardner: 9

Suppliers' Addresses

AO Polymer Corp.
44 Station Rd.
P.O. Box 793
Sparta, NJ 07871

Aceto Corp.
126-02 Northern Blvd.
Flushing, NY 11368

Agrashell, Inc.
5934 Keystone Drive
Bath, PA 18014

Alco Chemical Corp.
P.O. Box 5401
909 Mueller Drive
Chattanooga, TN 37406

Allied Colloids, Inc.
P.O. Box 820
2301 Wilroy Road
Suffolk, VA 23434

American Colloid Co.
1500 W. Shure Drive
Arlington Heights, IL 60004

Arakawa Chemical (USA) Inc.
625 North Michigan Ave.
Suite 1700
Chicago, IL 60611

Buckman Laboratories
1256 N. McLean Blvd.
Memphis, TN 38108

Burgess Pigment Co.
P.O. Box 349
Sandersville, GA 31082

Cabot Corp.
Cab-O-Sil Division
P.O. Box 188
Tuscola, IL 61953

Cargill, Inc.
Box 5630
Minneapolis, MN 55440

R.E. Carroll, Inc.
1570 North Olden Ave.
P.O. Box 139
Trenton, NJ 08601

Colloids, Inc.
394 Frelinghuysen Ave.
Newark, NJ 07114

Cook Paint & Varnish Co.
P.O. Box 389
Kansas City, MO 64141

Cosan Chemical Corp.
400 Fourteenth St.
P.O. Box 7
Carlstadt, NJ 07072

Cyprus Industrial Minerals Co.
Talc Division
Box 3419
Englewood, CO 80155

Daniel Products Co.
400 Claremont Ave.
Jersey City, NJ 07304

Davison Chemical Division
W.R. Grace & Co.
P.O. Box 2117
Baltimore, MD 21203-2117

Dayglo Color Corp.
4515 St. Clair Ave.
Cleveland, OH 44103

Degussa Corp.
Pigments Division
Route 46 at Hollister Road
P.O. Box 2004
Teterboro, NJ 07608

Dock Resins Corp.
1512 West Elizabeth Ave.
Linden, NJ 07036

Dover Chemical Corp.
P.O. Box 40
Dover, OH 44622

Dow Corning Corp.
Midland, MI 48686-0994

Drew Industrial Division
Ashland Chemical Co.
One Drew Plaza
Boonton, NJ 07005

DuPont Co.
Wilmington, DE 19898

Dynamit Nobel Chemicals
10 Link Drive
Rockleigh, NJ 07647

ECC America, Inc.
5775 Peachtree-Dunwoody Road,NE
Suite 200 G
Atlanta, GA 30342

Eastman Chemical Products, Inc.
P.O. Box 431
Kingsport, TN 37662

Engelhard Corp.
Performance Minerals Group
Menlo Park, CN 28
Edison, NJ 08818

Evans Clay Co.
P.O. Box 6
Summit, NJ 07901

Exxon Chemicals
Tomah Products
1012 Terra Drive
P.O. Box 388
Milton, WI 53563

FRP Co.
P.O. Box 349
Baxley, GA 31513

Floridin Co.
5380 Capital Circle N.W.
Tallahassee, FL 32303

Franklin Limestone Co.
612 Tenth Ave., North
Nashville, TN 37203

Franklin Mineral Products
635 Main St.
P.O. Box O
Wilmington, MA 01887

Freeman Chemical Corp.
217 Freeman Drive
P.O. Box 996
Port Washington, WI 53074

General Electric Co.
Silicone Products Division
Waterford, NY 12188

Genesee Polymers Corp.
Flint, MI 48507

Genstar Stone Products Co.
Executive Plaza IV
Hunt Valley, MD 21031

Georgia Kaolin Co., Inc.
2700 U.S. Highway 22 East
Union, NJ 07083

Georgia Marble Co.
2575 Cumberland Parkway N.W.
Atlanta, GA 30339

W.R. Grace & Co.
55 Hayden Ave.
Lexington, MA 02173

Great Lakes Chemical Corp.
P.O. Box 2200
Highway 52 N.W.
West Lafayette, IN 47906

Grefco, Inc.
3435 West Lomita Blvd.
Torrance, CA 90509

Harwick Chemical Corp.
60 S. Seiberling St.
P.O. Box 9360
Akron, OH 44305-0360

Henkel Corp.
Process Chemicals
350 Mt. Kemble Ave., CN 1931
Morristown, NJ 07960-1931

Hercules, Inc.
Wilmington, DE 19899

Hill Brothers Chemical Co.
1675 North Main St.
Orange, CA 92667

Hi-Tek Polymers, Inc.
Liquid Specialty Chemicals
1338 Coronet Drive
Dalton, GA 30722

J.M. Huber Corp.
P.O. Box 310
Havre de Grace, MD 21078

Humphrey Chemical Corp.
P.O. Box 2, Edgewood Area
Aberdeen Proving Ground,
MD 21010

ICI Americas, Inc.
Wilmington, DE 19897

Illinois Minerals Co.
2035 Washington
Cairo, IL 62914

O.G. Innes Corp.
21 East 40th St.
New York, NY 10016

Interez
10100 Linn Station Road
P.O. Box 37600
Louisville, KY 40233

International Sourcing, Inc.
555 Route 17 South
Ridgewood, NJ 07450

Interstab Chemicals
500 Jersey Ave.
P.O. Box 638
New Brunswick, NJ 08903

Johnson Wax
Specialty Chemicals Group
1525 Howe St.
Racine, WI 53403

Kelco
8355 Aero Drive
P.O. Box 23576
San Diego, CA 92123-1718

Kerr-McGee Chemical Corp.
Kerr-McGee Ctr.
Oklahoma City, OK 73125

King Industries
Science Rd.
Norwalk, CT 06852

Lonza, Inc.
22-10 Route 208
Fair Lawn, NJ 07410

M&T Chemicals, Inc.
Rahway, NJ 07065-0970

Malvern Minerals Co.
P.O. Box 1246
Hot Springs National Park,
AR 71902-1246

Manville
P.O. Box 5108
Denver, CO 80217

Milliken Chemicals
P.O. Box 1927
Spartanburg, SC 29304

The Milwhite Co., Inc.
P.O. Box 15038
Houston, TX 77020

Mooney Chemicals, Inc.
2301 Scranton Road
Cleveland, OH 44113

NL Chemicals, Inc.
P.O. Box 700
Hightstown, NJ 08520

National Mineral Products
P.O. Box 1938
Salinas, CA 93902

Nyco
Willsboro, NY 12996-0368

Ohio Lime Co.
Woodville, OH 43469

C.J. Osborn Chemicals Inc.
820 Sherman Ave.
Pennsauken, NJ 08110

C.J. Patterson Co.
3947 Broadway
Kansas City, MO 64111

Pfizer Minerals, Pigments & Metals Division
640 North 13th St.
Easton, PA 18042-1497

Pfizer Pigments Inc.
235 East 42nd St.
New York, NY 10017

Polyvinyl Chemicals
730 Main St.
Wilmington, MA 01887

Ranbar Technology Inc.
1114 Wm. Flinn Hwy.
Glenshaw, PA 15116-2638

Raybo Chemical Co.
P.O. Box 2155
Huntington, WV 25721

Reichhold Chemicals, Inc.
525 N. Broadway
White Plains, NY 10603

Resinall Corp.
3065 High Ridge Rd.
P.O. Box 8149
Stamford, CT 06905

Reynolds Metals Co.
P. O. Box 35030
Louisville, KY 40232

Rohm and Haas Co.
Independence Mall W
Philadelphia, PA 19105

Ross Chem, Inc.
303 Dale Drive
Fountain Inn, SC 29644

Sandoz Chemicals Corp.
Charlotte, NC 28205

Sanncor Industries, Inc.
300 Whitney St.
P.O. Box 703
Leominster, MA 01453-0703

Sartomer Co.
Marshall Bldg.
West Chester Pike
West Chester, PA 19382

Saturn Chemicals, Inc.
State Road & Cottman Ave.
Philadelphia, PA 19135

Shamrock Chemicals Corp.
Foot of Pacific St.
Newark, NJ 07114

The Sherwin-Williams Co.
P.O. Box 1028
Coffeyville, KS 67337

Silbrico Corp.
6300 River Road
Hodgkins, IL 60525

Smith Chemical & Color Co., Inc.
104-20 Dunkirk St.
Jamaica, L.I., NY 11412

Sonneborn Division
Witco Corp.
520 Madison Ave.
New York, NY 10022-4236

StanChem, Inc.
East Berlin, CT 06023

Synray Corp.
209 N. Michigan Ave.
Kenilworth, NJ 07033

Synthron, Inc.
P.O. Box 1111
Morganton, NC 28655

Tammsco, Inc.
P.O. Box J
Tamms, IL 62988

Tego Chemie Service USA
P.O. Box 1299
914 E. Randolph Rd.
Hopewell, VA 23860

Thibaut and Walker
P.O. Box 296
Newark, NJ 07101

Thiele Kaolin Co.
P.O. Box 1056
Sandersville, GA 31082

Thompson, Weinman & Co.
P.O. Box 130
Cartersville, GA 30120

Tri-Star Chemical Co.
P.O. Box 38627
Dallas, TX 75238

Troy Chemical Corp.
One Avenue L
Newark, NJ 07105

Ultra Adhesives, Inc.
Park Station P.O. Box 98
Paterson, NJ 07543-0098

Union Carbide Corp.
Old Ridgebury Rd.
Danbury, CT 06817-0001

Unocal Chemicals Division
Unocal Corp.
1345 North Meacham Road
Schaumburg, IL 60196

U.S. Silica
P.O. Box 187
Berkeley Springs, WV 25411

R.T. Vanderbilt Co., Inc.
30 Winfield St.
Norwalk, CT 06855

The Vinings Co.
3950 Cumberland Parkway
Atlanta, GA 30339-4501

Westvaco Chemicals
P.O. Box 70848
Charleston Heights, SC 29415-0848

Trade Name Index

ACID REFINED	NL Chemicals, Inc.
ACLARAT	Sandoz Chemicals Corp.
ACRAWAX C	Lonza Inc.
ACRYLOID	Rohm and Haas Co.
ACRYLRON	Synthron, Inc.
ACRYMUL	Synthron, Inc.
ACTIRON	Synthron, Inc.
ACTIV	R.T. Vanderbilt Co., Inc.
ADMEROL	NL Chemicals, Inc.
ADTAC	Hercules, Inc.
AEROSIL	Degussa Corp.
AGED	NL Chemicals, Inc.
ALCOGUM	Alco Chemical Corp.
ALGEL	Raybo Chemical Co.
ALSYNOL	Daniel Products Co.
ALTOWHITE	Georgia Kaolin Co., Inc.
AMA	Vinings Industries
AMBERLAC	Reichhold Chemicals, Inc.
AMERFLOC	Drew Industrial Division
AMERFLOC PLUS	Drew Industrial Division
AMORWAX	AO Polymer Corp.
ANTI-FOAMING AGENT	Interstab Chemicals
ANTI-LIVERING AGENT	Interstab Chemicals
ANTISAG	Raybo Chemical Co.
ANTISAG	Troy Chemical Corp.
ANTI-SAGGING AGENT	Interstab Chemicals
ANTISETTLE	Troy Chemical Corp.
ANTI-SETTLING AGENT	Interstab Chemicals
ANTISILK	Raybo Chemical Co.
ANTI-SKIN	Troy Chemical Corp.
ANTISTAT	Raybo Chemical Co.
AOCURE	AO Polymer Corp.
AOTHANE	AO Polymer Corp.
APEX	Evans Clay Co.
AQUA-DRY	Synray Corp.
AQUAMIX	Harwick Chemical Corp.
AQUA-STAIN	Synray Corp.
AQUAWET	Raybo Chemical Co.
ARAKYD	Arakawa Chemical (USA) Inc.
AROFLAT	NL Chemicals, Inc.
AROFLINT	NL Chemicals, Inc.
AROLON	NL Chemicals, Inc.
AROPLAZ	NL Chemicals, Inc.
AROTHIX	NL Chemicals, Inc.
A-R VARNISH AND GRINDING	NL Chemicals, Inc.
ATH NY COAT	NYCOAT
ASARCO	R.E. Carroll, Inc.
ASHSTAT	Synthron, Inc.

COSPERSE	Cosan Chemical Corp.
COTIN	Cosan Chemical Corp.
COURLOSE	Aceto Corp.
CYKELIN	NL Chemicals, Inc.
CYRACURE	Union Carbide Corp.
DAPER NOVACITE	Malvern Minerals Co.
DAPRO	Daniel Products Co.
DARVAN	R.T. Vanderbilt Co., Inc.
DAXAD	W.R. Grace & Co.
DEE FO	Ultra Adhesives, Inc.
DEGUMMED	NL Chemicals, Inc.
DIAMOND "K"	NL Chemicals, Inc.
DIANAL	Cargill/Mitsubishi Rayon Co., Ltd.
DIATOMITE	Grefco, Inc.
DICALITE	Grefco, Inc.
DICAPERL	Grefco, Inc.
DISLON	King Industries
DISPERSE	Raybo Chemical Co.
DISPERSE-AYD	Daniel Products Co.
DISPEX	Allied Colloids
DORESCO	Dock Resins Corp.
DOVER	R.E. Carroll, Inc.
DOW CORNING	Dow Corning Corp.
DREWFAX	Drew Industrial Division
DREWFLOC	Drew Industrial Division
DREWGARD	Drew Industrial Division
DREWPLUS	Drew Industrial Division
DRIKALITE	Thompson, Weinman & Co.
DRIKALITE/MICRO-WHITE	ECC America, Inc.
DRI-RX	Mooney Chemicals, Inc.
DRISOY	NL Chemicals, Inc.
DUPONT	Du Pont
DURAMITE	ECC America, Inc.
DURAMITE	Thompson, Weinman & Co.
DYMSOL	Henkel Corp.
DYNACOLL	Dynamit Nobel Chemicals
DYNAPOL	Dynamit Nobel Chemicals
EASTACEL	Eastman Chemical Products, Inc.
EASTMAN	Eastman Chemical Products, Inc.
EBONY NOVACITE	Malvern Minerals Co.
ELVACITE	Du Pont
EPI-CURE	Interez
EPI-REZ	Interez
EPI-TEX	Interez
EPOTUF	Reichhold Chemicals, Inc.
ERVOL	Sonneborn Division

INTERCAR	Interstab Chemicals
INTERCIDE	Interstab Chemicals
INTERDRY	Interstab Chemicals
INTERFOAM	Interstab Chemicals
INTERLITE	Interstab Chemicals
INTERSPERSE	Interstab Chemicals
INTERSTAB	Interstab Chemicals
INTERWAX	Interstab Chemicals
INTERWET	Interstab Chemicals
ISOESTER	FRP Co.
I.T.	R.T. Vanderbilt Co., Inc.
JONCRYL	Johnson Wax
KAOPAQUE	Georgia Kaolin Co., Inc.
KAOPLATE	Thiele Kaolin Co.
KATHON	Rohm and Haas
KAYDOL	Sonneborn Division
K-CRYL	King Industries
K-CURE	King Industries
KELECIN	NL Chemicals, Inc.
KELLIN	NL Chemicals, Inc.
KELLOX	NL Chemicals, Inc.
KELMER	NL Chemicals, Inc.
KELPOL	NL Chemicals, Inc.
KELPOXY	NL Chemicals, Inc.
KELSOL	NL Chemicals, Inc.
KELTROL	NL Chemicals, Inc.
KELZAN	Kelco Division
KETOMUL	AO Polymer Corp.
KEVLAR	Du Pont Co.
K-FLEX	King Industries
KLEAROL	Sonneborn Division
KOTAMITE	ECC America, Inc.
KOTAMITE	Thompson Weinman & Co.
KRISTALEX	Hercules, Inc.
K.V.O.	NL Chemicals, Inc.
LEUCOPHOR	Sandoz Chemicals Corp.
LIN-ALL	Mooney Chemicals, Inc.
LINAQUA	NL Chemicals, Inc.
LUMIFLON	ICI Americas, Inc.
LUSTRALITE	Reichhold Chemicals, Inc.
LUSTRASOL	Reichhold Chemicals, Inc.
MACROBASE	Sartomer Co.
MACROMER	Sartomer Co.
MACROMITE/MICRO-WHITE	ECC America, Inc.

NOURY-DRY	Interstab Chemicals
NOVACITE	Malvern Minerals Co.
NOVACURE	Interez
NOVAMER	Interez
NOVEPOX	Synthron, Inc.
NYAD	Nyco
NYCOAT	Nyco
NYCOR	Nyco
NYFLAKE	Nyco
NYGLAS	Nyco
NYSPHERES	Nyco
NYTAL	R.T. Vanderbilt Co., Inc.
OHSO	Ohio Lime Co.
OKO	NL Chemicals, Inc.
ONCOR	NL Chemicals, Inc.
OPACICOAT	ECC America, Inc.
OPACIMITE	ECC America, Inc.
OPTISPERSE	Raybo Chemical Co.
OPTIWHITE	Burgess Pigment Co.
ORZOL	Sonneborn Division
PARCO	Thibaut and Walker
PARCRYL	Thibaut and Walker
PAROIL	Dover Chemical Corp.
PATCO	C.J. Patterson Co.
PATCOTE	C.J. Patterson Co.
PEERLESS	Franklin Limestone Co.
PEERLESS	R.T. Vanderbilt Co., Inc.
PENTEX	Colloids, Inc.
PERLITE	Grefco, Inc.
PFERRITAN	Pfizer Pigments Inc.
PHOSFAC	Hi-Tek Polymers, Inc.
PICCO	Hercules, Inc.
PICCODIENE	Hercules, Inc.
PICCOFYN	Hercules, Inc.
PICCOLASTIC	Hercules, Inc.
PICCOLYTE	Hercules, Inc.
PICCOPALE	Hercules, Inc.
PICCOTAC	Hercules, Inc.
PICCOTEX	Hercules, Inc.
PICCOVAR	Hercules, Inc.
PLASTOFLEX	Interstab Chemicals
PLUS	Engelhard Corp.
POLY BD	Sartomer Co.
POLYBOND	Reichhold Chemicals, Inc.
POLYLITE	Reichhold Chemicals, Inc.
POLY/TEX	Interez

SNOBRITE	Evans Clay Co.
SNOWFLAKE	ECC America, Inc.
SNOWFLAKE P.E.	Thompson, Weinman & Co.
SNOWFLAKE WHITE	Thompson, Weinman & Co.
SNOWFLAKE WHITE/MICRO-WHITE	ECC America, Inc.
SNOW*TEX	Smith Chemical & Color Co., Inc.
SNOW*TEX	U.S. Silica
SNOW WHITE	Tammsco, Inc.
SOLKA-FLOC	Grefco, Inc.
SONNEBORN	Sonneborn Division
SOY SOLINEX	NL Chemicals, Inc.
SPANGLE	Raybo Chemical Co.
SPECIAL	Engelhard Corp.
SPECIAL T-BLOWN	NL Chemicals, Inc.
SPENKEL	NL Chemicals, Inc.
SPENLITE	NL Chemicals, Inc.
SPENSOL	NL Chemicals, Inc.
SPURSO	Mooney Chemicals, Inc.
STABILSAL	Interstab Chemicals
STANCHEM	StanChem Inc.
STONELITE	Ohio Lime Co.
STYRESOL	Reichhold Chemicals, Inc.
SUPER-ALKYD	Thibaut and Walker
SUPER-BECKAMINE	Reichhold Chemicals, Inc.
SUPERCOAT	ECC America, Inc.
SUPERCOAT	Thompson, Weinman & Co.
SUPER FINE	Manville
SUPERFINE BESTONE	Smith Chemical & Color Co., Inc.
SUPERFLO	NL Chemicals, Inc.
SUPER FLOSS	Manville
SUPERIOR	NL Chemicals, Inc.
SUPERMITE	Thompson, Weinman & Co.
SUPERMITE/MICRO-WHITE	ECC America, Inc.
SUPER POLYESTER	Thibaut and Walker
SUPER THANE	Thibaut and Walker
SUSPEND-AYD	Daniel Products Co.
SUSPERSE	Raybo Chemical Co.
SWANEE	R.E. Carroll, Inc.
SYLOID	Davison Chemical Division
SYN-AQUA	Synray Corp.
SYNCONE	Synray Corp.
SYN FAC	Milliken Chemicals
SYNOTEX	Daniel Products Co.
SYNPLAST	Synray Corp.
SYNPLEX	Synray Corp.
SYNPOXY	Synray Corp.
SYNRAY	Synray Corp.

VANCIDE	R.T. Vanderbilt Co., Inc.
VAN GEL	R.T. Vanderbilt Co., Inc.
VANSIL	R.T. Vanderbilt Co., Inc.
VANTALC	R.T. Vanderbilt Co., Inc.
VEECOTE	R.T. Vanderbilt Co., Inc.
VEEGUM	R.T. Vanderbilt Co., Inc.
VEPRIN	Saturn Chemicals, Inc.
VERFLAKE	White Pigment Corp.
VERSAFLEX	W.R. Grace & Co.
VERSAFLOW	Shamrock Chemicals Corp.
VERWHITE	White Pigment Corp.
VICRON	Pfizer Minerals, Pigments & Metals Division
VISCALEX	Allied Colloids
VISCASIL	General Electric Co.
VISC-AYD	Daniel Products Co.
VISCON	Raybo Chemical Co.
VISCOTROL	Mooney Chemicals, Inc.
VISTAC	Interstab Chemicals
VOLCLAY	American Colloid Co.
WALLKYD	Reichhold Chemicals, Inc.
WESTVACO	Westvaco Chemicals
WESTVACO DIACID	Westvaco Chemicals
WHITE MIST	Manville
WINGDALE WHITE	Georgia Marble Co.
WOLLASTOKUP	Nyco
XAMA	Sanncor Industries, Inc.
YARMOR	Hercules, Inc.
YORK WHITE	R.E. Carroll, Inc.
ZEO	J.M. Huber Corp.
ZEODENT	J.M. Huber Corp.
ZEOFREE	J.M. Huber Corp.
ZEOLEX	J.M. Huber Corp.
ZEOMATT	J.M. Huber Corp.
ZEOSYL	J.M. Huber Corp.
ZEOTHIX	J.M. Huber Corp.
ZINPLEX	Ultra Adhesives, Inc.
ZIRCO	Interstab Chemicals
ZONYL	Du Pont Co.

Other Noyes Publications

ENGINEERING RESINS
An Industrial Guide

by

Ernest W. Flick

This volume describes almost 2500 engineering resins which are currently available for industrial usage. The definition of engineering resins differs depending on whom one asks. The book contains those products which are termed "engineering plastics" as well as "upgraded commodity resins." The book will be of value to industrial technical and managerial personnel involved in the specification and use of these products. It has been compiled from information received from numerous industrial companies and other organizations.

The data included represent selections from manufacturers' descriptions made at no cost to, nor influence from, the makers or distributors of the materials.

Only the most recent information has been included. Products covered include blends, composites, alloys, and resins with superior impact resistance, heat or cold resistance, and/or stiffness. The products have been manufactured to meet high performance requirements in major end-use industries—automotive, packaging, electronics, computers and business equipment, aerospace and medical supplies. Many can be tailored to specific end-uses on request. It is believed that all of the products are currently available, which will be of interest to readers concerned with product discontinuances.

Products are presented by company, and the companies are listed alphabetically. Also included are a **Trade Name Index** and a list of **Suppliers' Addresses.**

The book lists the following product information, as available, in the manufacturer's own words: company name and product category, trade name and product numbers, and a description of the product as presented by the supplier. Companies represented in the book are:

Allied-Signal
Amoco Performance Products
Applied Composites
Arco Chemical Co.
Ashley Polymers, Inc.
Atochem, Inc.
BASF Corp.
Borg Warner Chemicals, Inc.
The Budd Co.
Celanese Engineering Resins
Colorite Plastics Co.
Continental Polymers, Inc.
Cosmic Plastics, Inc.
Custom Resins
Dartco Manufacturing, Inc.
Dexter Plastics
Dow Chemical U.S.A.
DuPont Co.
Eagle-Picher Plastics Division
Eastman Chemicals Division
El Paso Products Co.
Enron Chemical Co.
Eval Co. of America
Fiberite
Fina Oil and Chemical Co.
Freeman Chemical Corp.
General Electric Co.
Georgia Gulf Corp.
B.F. Goodrich
Haysite Reinforced Plastics
Hexcel
Huntsman Chemical Corp.
ICI Americas, Inc.
Industrial Dielectrics, Inc.
Koppers Co., Inc.
MKB Industries, Inc.
Mitsui Petrochemicals(America)
Mobay Corp.
Mobil Chemical Co.
Monmouth Plastics, Inc.
Monsanto Chemical Co.
Nuodex, Inc.
Orthane Division/Ohio Rubber
Owens-Corning Fiberglas Corp.
Pennwalt Corp.
Perstorp, Inc.
Phillips 66 Co.
Plaskon Electronic Materials
Plaslock Corp.
Plastics Engineering Co.
RTP Co.
Resinoid Engineering Corp.
Rhone-Poulenc, Inc.
Richardson Polymer Corp.
Rogers Corp.
Rohm and Haas
Rostone
A. Schulman, Inc.
Standard Oil
Thermofil, Inc.
U.S.I. Chemicals
Union Carbide Corp.
Valite Division/
Valentine Sugars, Inc.
Washington Penn Plastic Co.
Wellman, Inc.
Wilson-Fiberfil International

ISBN 0-8155-1172-8 (1988) 6"x9" **838 pages**

CONSTRUCTION AND STRUCTURAL ADHESIVES AND SEALANTS
An Industrial Guide

by

Ernest W. Flick

This book describes more than 1600 construction and structural adhesives and sealants currently available for industrial, commercial and consumer usage. It will be of value to industrial, technical, and managerial personnel involved in the specification and use of these products.

The adhesives and sealants described in the book can be used for major construction and structural applications as well as in commercial businesses, schools, and offices, and by the home-repair handyman. The products can be applied to such diverse materials as concrete, ceramics, vinyl, polyethylene, foamed plastics, wood, plywood, fiberboard, gypsum board, prefinished panels, leather, and cotton.

The data included represent selections from manufacturer's descriptions, made at no cost to, nor influence from, the makers or distributors of these materials. Only the most recent information has been included. It is believed that all of the products are currently available, which will be of interest to readers concerned with product discontinuances.

Product Information is presented by company, alphabetically, and there is also a section of **Sample Formulations.** Another useful section contains **Suppliers' Addresses,** and a **Trade Name Index** completes the book. The following product information, as available, is given in the manufacturer's own words: company name and product category, trade name and product numbers, and product description. Companies represented in the book are:

Air Products and Chemicals
Alkemi Plastics, Inc.
American Cyanamid Co.
American Resin & Chemical
Andek Chemical Corp
Anti-Hydro Co.
Apple Adhesives, Inc.
Aremco Products, Inc.
Ashland Chemical Co.
Atlas Minerals & Chemicals
Bostik Construction Products
3C
The Cambridge Tile Mfg. Co.
Ceilcote
Chemical Development Corp.
Chemseco
Chicago Adhesive Products Co.
Childers Products Co.
Columbus Adhesive & Chemical
Conap, Inc.
DAP, Inc.
Darworth Co.
Dexter Division/Hysol Corp.
Dow Corning Corp.
Durabond/USG Industries
Durez Resins & Molding Materials
E-Bond Epoxies, Inc.
Emerson & Cuming, Inc.
The Epoxylite Corp.
The Euclid Chemical Co.
Evans-St. Clair, Inc.
Fiber-Resin Corp.
Franklin International
H.B. Fuller Co.
Furane Products Co.
Gaco Western, Inc.
Garon Products, Inc.
Gates Engineering Co., Inc.
GC Electronics
General Electric Co.
General Sealants
Gibson Homans Co.
Gilsonite, Inc.
Goal Chemical Sealants Corp.
Hardman, Inc.
Henkel Adhesives Co.
Henkel Corp.
Hernon Manufacturing, Inc.
Highside Chemicals, Inc.
Industrial Polychemical Service
Karnak Chemical Corp.
Larsen Products Corp.
Leech Products, Inc.
Loctite Corp.
Lord Corp.
Macco Adhesives
Marston Bentley, Inc.
McGhan NuSil Corp.
Miracle Adhesives Corp.
Goodloe E. Moore, Inc.
Morton Thiokol, Inc.
National Starch and Chemical
NPC Sealants
Oneida Electronic Mfg. Inc.
Pacer Technology
Parr, Inc.
Pecora Corp.
Palmer Laboratories, Inc.
Permalite Plastics Corp.
Pierce & Stevens Corp.
Plastics & Resins, Inc.
Polymeric Systems, Inc.
Polymer Plastics Corp.
Preco Industries, Ltd.
Products Research & Chemical
Protective Coating Co.
Protective Treatments, Inc.
Reichhold Chemicals, Inc.
ResCon Technology Corp.
Rock-Tred
Rutland Products
Sandell Manufacturing Co.
Sherwin-Williams Co.
Sika Corp.
Smooth-On, Inc.
Sonneborn Building Products
Superior Sealants, Inc.
Swift Adhesives
3M
TACC International Corp.
Tanner Chemical Co., Inc.
Texaco Chemical Co.
Thermoset Plastics, Inc.
Trumbull Asphalt
Uniroyal Plastics Co., Inc.
Uniseal
United Gilsonite Laboratories
U.S. Industrial Chemicals Co.
Vimasco Corp.
Warren Paint and Color Co.
Wilhold Glues, Inc.

ISBN 0-8155-1149-3 (1988) **754 pages**

Other Noyes Publications

WATER-BASED TRADE PAINT FORMULATIONS

by

Ernest W. Flick

This collection of 562 up-to-date water-based trade paint formulations will be of value to technical and managerial personnel in paint manufacturing companies and firms which supply raw materials or services to these companies, and to those interested in less hazardous, environmentally safer formulations. The book will be useful to both those with extensive experience as well as those new to this ever expanding $10 billion/year industry.

The data in the book consist of selections of manufacturers' suggested formulations made at no cost to, nor influence from, the makers or distributors of these materials. The information given is presented as supplied; only the most recent data supplied have been included. Any solvent contained is minimal.

Included in the descriptive information for each formulation, where available, the following properties may be listed: viscosity, solids content, % nonvolatiles, pigment volume concentration, density, pH, spatter, leveling, sag resistance, scrub cycles to failure, contrast ratio, ease of dispersion, fineness of grind, heat stability, freeze-thaw stability, ease of application, gloss, foaming, cratering, brightness, opacity, water spotting, adhesion to chalk, brush clean-up, reflectance, and sheen.

The formulations described are divided into three major sections, which cover exterior, interior, and exterior and/or interior water-based paints, enamels, and coatings. Further subdivision into chapters is as indicated below. Parenthetic numbers indicate the number of formulations per section or chapter. Also included are a **Trade-Named Raw Materials** section and a list of **Suppliers' Addresses.**

ISBN 0-8155-1147-7 (1988) 697 pages

Other Noyes Publications

INDUSTRIAL WATER-BASED PAINT FORMULATIONS

by

Ernest W. Flick

This collection of 220 up-to-date industrial water-based paint formulations will be of value to technical and managerial personnel in paint manufacturing firms, and firms which supply raw materials or services to these companies, and those generally interested in less hazardous, environmentally safer formulations.

U.S. paint manufacturers ship over 950 million gallons of paints and coatings per year, for a total dollar volume approaching $10 billion. The industry seems to thrive on competition, and technical challenges are being met with new generation finishes with improved durability and drying time, and, of course, a water base.

The data in the book consist of selections of manufacturers' suggested formulations made at no cost to, nor influence from, the makers or distributors of these materials. The information given is presented as supplied. Only the most recent data supplied have been included. Any solvent contained is minimal.

The book is divided into the following Sections:

1. **Air Dry Coatings (14)**
2. **Air Dry or Force Dry Coatings (10)**
3. **Anti-Skid or Non-Slip Coatings (2)**
4. **Bake Dry Coatings (47)**
5. **Clear Coatings (10)**
6. **Coil Coatings (7)**
7. **Concrete Coatings (6)**
8. **Dipping Enamels (5)**
9. **Lacquers (27)**
10. **Primers (44)**
11. **Protective Coatings (3)**
12. **Spray Enamels (3)**
13. **Topcoats (18)**
14. **Traffic and Airfield Paints (7)**
15. **Miscellaneous (17)**

Parenthetic numbers indicate the number of formulations in each section. In addition to the above, there are two other Sections which will be helpful:

16. **A chemical trade name section where trade-named raw materials are listed with a chemical description and the supplier's name. The specifications which the raw materials meet are included, if applicable.**
17. **Main office addresses of the suppliers of trade-named raw materials.**

Included in the descriptive information for each formulation, where available, the following may be listed:

1. **Type of paint or coating**
2. **End use**
3. **Ingredients, by weight and/or volume**
4. **Mixing suggestions**
5. **Properties such as viscosity, total solids (by weight and/or volume), gloss, pencil hardness, pH, nonvolatiles, pigment/binder ratio, density, flash point, adhesion**
6. **Formulation source**

ISBN 0-8155-1146-9 (1988) 277 pages